HISTOCHEMISTRY IN FOCUS
A Sourcebook of Techniques and Research Needs

HISTOCHEMISTRY IN FOCUS
A Sourcebook of Techniques and Research Needs

K. Shyamasundari

Formerly, Professor of Zoology
Andhra University
Visakhapatnam

K. Hanumantha Rao

Formerly, Professor of Zoology, and Honorary Professor
Andhra University
Visakhapatnam

MJP PUBLISHERS
Chennai 600 005

ISBN 10: 81-8094-030-6
ISBN 13: 978-81-8094-030-9

Cataloguing-in-Publication   Data
Shyamasundari, K (1928 – 1967).
    Histochemistry in Focus : a sourcebook
of techniques and research needs / by
K. Shyamasundari and K. Hanumantha Rao. –
Chennai : MJP Publishers, 2007.
    xii, 848p. ; 23 cm.
    Includes References Author Index and
Subject Index.
    ISBN 81-8094-030-6 (pbk.)
    1. Histochemistry I. Hanumantha Rao, K II. Title.
572 SHA MJP 026

ISBN 81-8094-030-6   **MJP PUBLISHERS**
© Publishers, 2007   47, Nallathambi Street
All rights reserved   Triplicane
                         Chennai 600 005

         Publisher : J.C. Pillai
  Managing Editor : C. Sajeesh Kumar
    Project Editor : P. Parvath Radha
 Assistant Editors : B. Ramalakshmi, S. Revathi
     Composition : M. Uma, Lissy John, Yamuna Devi
  Cover Designer : M. Uma Maheswari
        CIP Data : Prof. K. Hariharan

ANDHRA UNIVERSITY
(NAAC "A" —85-90 Grade Points)

Fax   : 0891 - 2525611
Grams   : "UNIVERSITY"
Phone   : Off. : 0891-2755547 ;
          2844333 ; 2844222
Res.   : 0891-2755516
E-mail : l.venugopalreddy@gmail.com
          l.venugopalreddy@yahoo.com

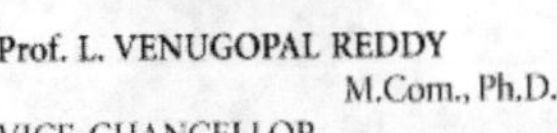

Prof. L. VENUGOPAL REDDY
          M.Com., Ph.D.
VICE-CHANCELLOR

VISAKHAPATNAM (A.P.), INDIA
Date : 14-03-2007

MESSAGE

It is indeed a pleasure and privilege to bear testimony to such a meticulously prepared compendium on histochemistry embracing most if not all techniques to localize chemical substances in tissues. This I believe is a much needed one, keeping in view the various constraints an Indian researcher of many branches of Life Sciences faces. To be able to say that a particular tissue has so and so chemical constituents from DNA to most complicated scleroproteins of mucopolysaccharides, etc. is most necessary for a biochemist biotechnologist, pathologist, physiologist and general biologist. The care taken to put in maximum information within the scope of a single book neither too small nor too large and supported by excellent original colour photographs drawn from investigations of the authors is commendable. Professors K. Shyamasundari and K. Hanumantha Rao have considerable experience in this field as evidenced by the manner and scope of presentation. I congratulate them and wish as wide an audience as possible for this much needed book "Histochemistry in Focus".

Foreword

Histochemistry remains one of the cornerstones of biological and medical training and research, even in this age when molecular studies are producing many exciting advances in our understanding of living systems. The many techniques now being employed by histochemists provide us with fundamental information about cell, tissue and organ structure that complements the work, for example, of biochemists, molecular biologists, and physiologists. This volume *Histochemistry in Focus* with subtitle *A Sourcebook of Techniques and Research Needs*, presents a broad array of histochemical techniques applicable to the study of the many cell and tissue types (e.g. neurosecretory cells and connective tissue) as well as to the constituents (e.g. nucleic acids and carbohydrates) that are present both in invertebrate and vertebrate organisms. In order to advance our knowledge of the structure of organisms down to the cellular and subcellular levels, there was a need for the development of tools to visualize cells and their organelles. This need led to the invention of the light microscope and subsequently to the electron microscope. Along with the appearance of these microscopes came the invention of several types of microtomes, including the ultramicrotome; and with this new "hardware" came the need to develop the many staining techniques described in this volume.

The authors of this volume, Drs. K. Shyamasundari and K. Hanumantha Rao, are recognized not only in India but also worldwide for having made important, fundamental contributions to the field of histochemistry. These two investigators are as highly qualified to write a book on this subject as any two other histochemists of whom I am aware.

The techniques described in this volume will find use not only in research laboratories where basic questions are being investigated, but also in medical and pathological laboratories where disease stages are being studied; and these techniques will help to provide a better understanding of the changes that bring about pathological conditions.

This volume will find a welcoming home wherever the techniques of histochemistry, as described herein, are being utilized. Certainly, this book will become a standard text among practitioners of histochemistry.

Milton Fingerman
Department of Ecology and Evolutionary Biology
Tulane University
New Orleans, Louisiana 70118, USA

Preface

The objective of this treatise is to provide a more or less simplified compendium to enable investigators to practice histochemistry in either of two directions. One direction is to learn techniques and some theory behind each one of them as a course in histochemistry. The other direction is to use it as a sort of vademecum although it cannot be claimed that A to Z have been touched. For teachers, researchers and students, a book of this kind as an Indian counterpart of more sophisticated expensive foreign volumes may be within easy reach costwise. With the present increased proliferation of credit courses in semester systems and maximum ascent on molecular, biochemical, genetic and biotechnological aspects general histological training not to speak of histochemistry has recorded into the background. Histology and histochemistry form the backbone of biochemistry and many aspects of life sciences subjects. Excellent treatises and reference books on histochemistry are available. Spicer *et al.* (1967), Pearse (1980, 1985), Bullock and Petrusz (1982, 1983, 1985), Chayen and Bitensky (1991), Pearse and Stoward (1991), Lyon (1991), Churikian (1993a, 1993b), Bancroft and Cook (1996) and Presnell and Schreibman (1997) are almost the bibles of histochemistry. Recent years have witnessed the publication of many excellent books and periodicals such as *Journal of Histochemistry and Cytochemistry* that continue to enlighten eager workers on many aspects of interest.

We have supported many techniques with results from our studies in the form of colour photographs. No apology is needed for our effort to provide colour photographs and this has perhaps to some extent escalated the cost of the book. Can there be a plum pudding without plums! Another feature of this book is the emphasis on invertebrate material. Many histochemical techniques are highly capricious even in the hands of an experienced researcher. The batch number of the dye, components of the dye, the company of its origin, etc., are matters of some importance. Then come the accurate schedule starting from the preparation of the reagent, through time constraints, etc., to mounting (placement of coverslip) in a suitable medium. Maintenance of adequate controls is necessary. In the first chapter in respect of some specific cases we have outlined how meaningful interpretations could be arrived at in the matter of histochemistry/ histophysiology. Histochemistry alone could enable us to arrive at proper solutions to some intractable problems. Biochemistry can only stand on the shoulders of histochemistry. Histochemistry can afford to divorce biochemistry but biochemistry perhaps cannot ignore histochemistry. My appeal to biochemists is to bring a slice of histochemistry into their studies and not to be satisfied with "Alice in Wonderland" position.

We are beholden to Prof. L. Venugopal Reddy, Vice-Chancellor, Andhra University for his encouragement and support. We have derived immense satisfaction from our association with our research students. We are highly grateful to them for helping us to increase our patience. Grateful acknowledgements are also due to

Professors M. Fingerman, Tulane University, New Orleans, USA, G.P. Verma, Department of Zoology, Berhampur University, Berhampur, India, P. Jayakar, Professor of Microbiology, Andhra Medical College, Visakhapatnam, India for scrutinizing several chapters and giving valuable and critical suggestions. We are grateful to heads of the Department of Zoology, Professors A.V. Raman, C. Vijayalakshmi, A. Joseph and G. Gnanamani who occupied the chair since 1997 for providing accommodation and encouragement throughout the completion of the book. We sincerely thank Prof. C. Kalavathi who graciously helped us in the preparation of scanned photographs. Sri S. Nageswara Rao has to be thanked immensely for preparing the hard copy and disc in time. We will be failing in our duty if we forget the technical skills of Dharmavarapu Ram Babu, Lab Technician who helped us in several ways.

K. Shyamasundari

K. Hanumantha Rao

Contents

INTRODUCTION

Microanatomy or histology has been an important part of training for every biology student. To know the constituents of the different kinds of tissues including blood has been an important follow-up to gross anatomy. With histology came the microscope, the microtome and a host of techniques like staining and mounting procedures. Some familiarity with all these aspects is expected by the time the student comes to an advanced level of study (M.Sc., M.D., M.S., Ph.D., etc.). Perhaps actual operation of the microtome and all that is necessary for section cutting, processing and mounting come up in the post-graduate life of a life sciences or medical student. For those who wish to busy themselves in a career of microanatomy/histology, it will be a lifetime pursuance of histological techniques. Without doubt this is the background mainstay for one who needs to be led on to histochemistry. Histological oversight methods may be enough for a gross distinction of aspects of a section of any tissue. The most familiar iron–haematoxylin–eosin procedure is more than a standard in all college laboratories and medical institutions. In addition Delafield's and Ehrlich's haematoxylins have proved their worth and value. Routine analyses of sections pertaining to biopsies and sections in college laboratories are done by the three variants of haematoxylin mentioned above. Most often, eosin is the choice counterstain. More complicated procedures are Azan and Mallory's tricolour techniques. A lot of information can be drawn out of these techniques and one can gain some insight into gross histochemical information also. For instance aniline blue stains mucoid material, collagen, etc. blue. Some structural proteins are stained red with azocarmine. Coupled with orange G, some more information can be envisaged. Likewise Mallory's stain has some advantages. It seems that these staining procedures (Azan and Mallory) are useful for histochemical interpretation but with caution.

The investigation should now proceed to more histochemical than empirical methods. Even in the matter of substances such as glycogen in sections, the time-honoured method has been Best's carmine procedure. This again is empirical but useful. Histochemical procedures require rigorous and precise methods of fixation of tissues, often no more than 3 to 5 micrometres thickness and most immediately after the death of the animal. Most of the methods have been carefully chosen by us and most importantly experimentally. The photographs such as the ones we selected are mostly invertebrate material coming as they do from our research studies. A book on histochemistry would be meaningless without being punctuated by appropriate, clear colour photographs.

There are three ways of looking at histochemistry:

1. To try to discern and locate chemical substances (viz. glycogen, various types of proteins, lipids in tissues, etc.).

2. To be able to add information to functional morphology of organ systems.

3. To modify or introduce new techniques or new combinations of staining procedures, and perhaps give a chemical basis for the capture reaction (between tissue chemical and dye of choice).

One need not be perturbed by the range and plethora of techniques presented. There may be no more than a dozen techniques that may be needed for routine analyses and work. But if one has to go down to such aspects like whether it is arginine or dihydroxyphenol or a particular enzyme like alkaline phosphatase, or Ca, Fe, phospholipids or glycolipids and one of so many kinds of mucopolysaccharides, then certainly all these come within the framework of this book particularly meant to be handy to the Indian student. The main reason is to initiate and inculcate the necessary stimulus for going deeper into the chemical analysis of tissues so that an *in situ* picture emerges. This is most essential as a prelude to biochemistry and as an adjunct to clear functional morphology and physiology problems.

In histochemistry, it is essential to delve into and apply a number of techniques for one substance, what is known as a battery of tests/reactions. This is because any one technique may bring out in colour more than one chemical substance. An elimination process ensues and with the use of appropriate digestion tests, confirmation about a single substance is possible, for example the famous Schiff's reagent. Two procedures of prior hydrolysis with acid are in vogue.

1. The famous Feulgen technique is perhaps the only test for DNA in the entire gamut of histochemistry that is specific for just one substance, namely DNA.

2. With PAS technique about 5 different substances can be brought out in magenta colour.

Substance	Abolition
Glycogen	α-amylase digestion
Glycoproteins	Deamination
Mucopolysaccharide (neutral)	Enzyme digestion
Phospholipid	Bouin's fixation and pyridine extraction
Collagen	Collagenase
Keratin	Keratolytic enzyme

	Abolition	Other tests
Glycogen	Amylase digestion	Best's carmine ++ (Empirical) Formol calcium fixation
Phospholipids	Bouin's fixation and pyridine extraction Sudan black B	Sudan Black B ++ Acid haematin ++ Nile blue sulphate ++ Methanol Fast blue + Neutral red
Neutral mucopolysaccharide	Enzyme	AB pH 2.5/PAS
Collagen		Van Gieson ++ Verhoeff's ++ Aniline blue ++ Aldehyde fuchsin negative –
Keratin		Rhodamine B ++ Aldehyde fuchsin ++
Glycoproteins		Congo red ++ Bromophenol blue ++

HISTOCHEMISTRY IN HISTOPHYSIOLOGICAL STUDIES

The following are some examples of histochemistry being used in histophysiological studies.

Studies in zoology have been traditionally devoted to systematics involving morphology and anatomy together with some natural history. This of course is the foundation of zoology and must continue as new species and new genera come into our kin from time to time. Taxonomic revision of old species is inevitable. It has been made clear that our achievements in biodiversity studies, although considerable, remain totally inadequate especially in India. Groups like turbellarians and nematodes remain unattended and several other groups remain only partially studied. There has been a gradual loss of interest and sufficient build-up of indifference because of overriding influences like molecular myopathy or electrophoretic elution and immunological immunity. Meanwhile the list of endangered species is increasing.

This is one side of the picture. On the other hand, physiological and biochemical studies originally were the monopoly of medical colleges. Both these disciplines have enormously improved in scope, methodology and interpretation during the last fifty years. The infiltration of these subjects into university departments of zoology, botany and later genetics, etc. has been a much needed and welcome gesture. It was sufficient in the beginning to start with elementary methods and a moderate quantum. But with all branches of science undergoing revolutionary improvements both in ideas and sophistication of instrumentation, today we are neck-deep in a plethora of instruments, techniques and interdisciplinary ideas cutting across several subjects. At University Departments of Zoology, some at least took to physiology so seriously but many times the investigators have been unable to give the correct name of the species that they may be working with.

In the current phase of activity which is more molecular and/or biochemical, there has been no dearth of papers on these aspects. Linking these studies to the studies in toxicology which were initiated in the later half of the last century and are continuing even today, we have all kinds of physiological and biochemical studies. Side by side there has been little but cognizable progress in histological aspects interwoven into many of the foregoing aspects of study. However, on the whole histological studies have not attracted many investigators and did not receive the attention they deserve. For instance in most histological studies the standard techniques have been the meagerly three, (1) Iron haematoxylin–eosin, (2) Delafield's haematoxylin and (3) Ehrlich haematoxylin. But even for routine studies, including pathological aspects, these techniques have been very useful and to the trained and experienced eye, still much information of importance could be retrieved.

A step further in histology is the Azan, and Mallory techniques. No modern histology laboratory can afford to be without routine use of at least the Azan technique. Even as a histological tool, some histochemical use of the Azan method is evident. For instance with and without aniline blue staining or only with aniline blue staining, much preliminary histochemical information can be obtained. And if the Mallory technique is adopted in conjunction with Azan, much more confirmatory information can be envisaged.

Now when we plunge into regular histochemistry and a battery of techniques thereof, the seemingly static histology can be transformed into more dynamic histophysiology. We will see in the following excursion into histochemistry how drab and incomplete mere biochemical studies remain. Cases will be presented to demonstrate the strength, relevance and inevitability of a prior histochemical approach for every biochemical investigation. It is now well over 50 years that in the laboratories of the Zoology Department of Andhra University, histochemistry and its applications have been involved in several kinds of studies from a mere locating of chemical substance in tissues/cells to solving some very difficult problem of physiology. It is the purpose of this chapter to highlight and enlighten the use of histochemistry in the solving of some aspects of the origin, constitution and stabilization of scleroproteins.

1. Eggshell and Mehlis' gland
2. Byssus chemistry in some species

Helminth Egg Shell Scleroproteins vis-à-vis Mehlis' Gland

It has been known for a long time that the egg shell in trematodes and pseudophyllidean cestodes is secreted and shaped in the ootype—at one time considered as a junction of oviduct, vitelline duct, uterus with or without a Laurer's canal as depicted in Figure 1.

The ootype is surrounded by shell gland cells. The oocyte upon entering the ootype is surrounded by a specific number of vitelline cells (specific for each species). The shell gland cells secrete the shell material around the ensemble of oocyte and vitelline cells and the eggshell hardens and the complete egg is pushed into the uterus. Another egg is now under completion. The process goes on. Thus it was largely believed that the so called shell gland secretes the entire shell.

In 1909 Goldsmidt discovered that the shell material is secreted by the vitelline cells themselves. What then is the function of the "shell gland"? Some thought that the oocyte and vitelline cells on arrival into ootype are encapsulated by shell gland secretion. This capsule wall is a thin membrane believed to keep together the oocyte and vitelline. The shell material from vitelline cells is extruded underneath the shell gland membrane to fortify it as a thick amber-coloured egg shell. It was in 1912 that Cort named the "shell gland" as Mehlis' gland after the discoverer of the gland (Mehlis) because it does not secrete the shell. Now the crucial issue was to pinpoint the chemical nature of the shell and the chemical nature of Mehlis' gland. Mehlis' gland is a collection of gland cells surrounding the ootype. They are not aggregated into a compact gland but as discrete cells interspersed with parenchymal cells, and neurosecretory cells. It is not quite possible to separate this gland without an admixture of tissue with it and most importantly the vitelline reservoir with vitelline cells which contain shell material. A major breakthrough has been the demonstration of the fact that the shell material is a quinine-tanned protein (scleroprotein). The precursors of this are brought by the vitelline cells and they are,

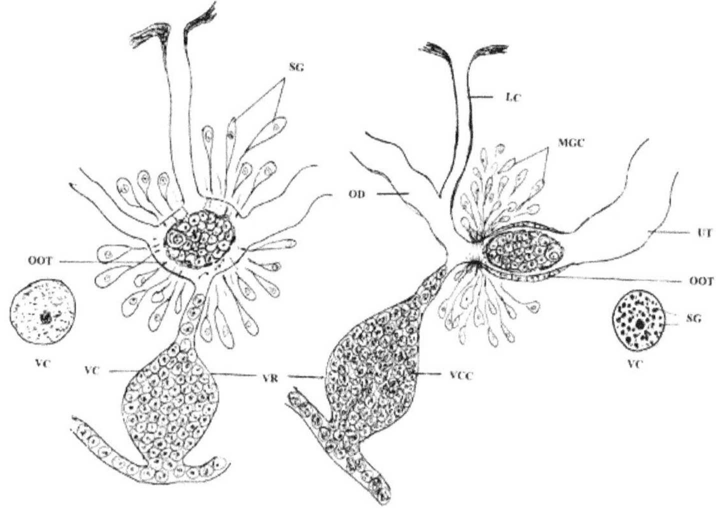

a. Before 1902 (Histological)

SG—Shell gland, OOT—Ootype,
VC—Vitelline cells, VR—Vitelline reservoir

b. 1906 (Histological)

LC—Laurer's canal, MGC—Mehlis' gland cells,
OD—Oviduct, Ut—Uterus, OOT—Ootype,
SG—Shell gland, VC—Vitelline cell,
VCC—Vitellocalycal cells, VR—Vitelline reservoir

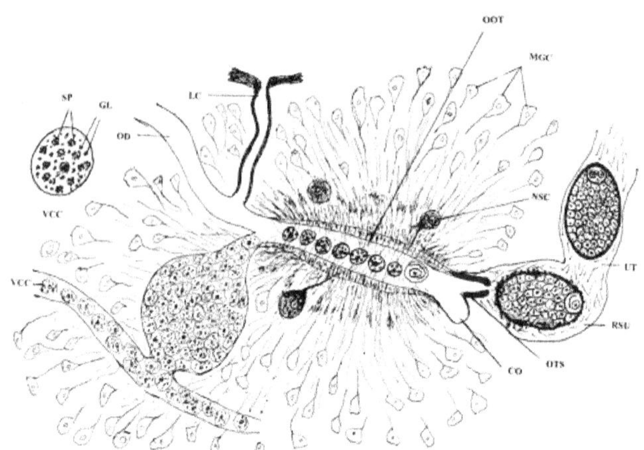

c. 1959 (Histochemical)

LC—Laurer's canal, OD—Oviduct, OOT—Ootype,
MGC—Mehlis' gland cells, NSC—Neurosecretory cells
Ut—Uterus, RSU—Receptaculum seminis uterinum,
OTS—Ootype sphincter, CO—Caecum of ootype,
VCC—Vitello calycal cells, Sp—Shell precursors,
GL—Glycogen

Figure 1   Mehlis' gland complex in *Fasciola hepatica*

(1) basic protein, (2) dihydroxyphenol, and (3) phenolase enzyme. It was histochemistry/cytochemistry that confirmed the presence of these precursors in the vitelline cells. In brief, 1) protein was stained by bromophenol blue, 2) dihydroxyphenol by Fast Red Salt B and 3) phenolase by incubation in catechol at 37°C. Necessary controls were maintained.

The same three reactions have been used to establish the presence of these substances in the just formed eggshell. As stabilization progresses, the egg shell becomes more and more refractory to the histochemical procedures mentioned above. A complete quinine-tanned protein is amber-coloured and has no radicals to react with the techniques mentioned above. What then is the chemical nature of Mehlis' gland secretion and what may be its role? That the secretion enters the ootype is clear even in histological preparation. Various conjectures have been made in respect of the role and chemistry of Mehlis' gland. It was in 1959 that Hanumantha Rao clearly demonstrated, purely from histochemistry, that Mehlis' gland cells are elaborating and secreting mainly phospholipids. The following reactions are pertinent in this connection.

Histochemical work on Mehlis' gland of a pseudophyllidean cestode *Penetrocephalus ganapatii* by Hanumantha Rao (1960) also confirmed the phospholipid nature of Mehlis' gland. Meanwhile Clegg (1965) reported that electrophoretic studies revealed that Mehlis' gland secretion is of two types, phospholipid as well as lipoprotein. As mentioned earlier, it is virtually impossible to eliminate adulteration by vitelline cells when Mehlis' gland is separated and hydrolysed. Then came the discovery from our laboratory that in the Paramphistomatidae (Trematodes) the eggshell is a glassy transparent instead of being amber-coloured. We then established that the egg shell in amphistomes is a keratin protein stabilized by —S —S disulphide bonds, what Lord Astbury of the Department of Biomolecular Studies of Leeds University called vulcanized protein.

In Paramphistomatidae, Mehlis' gland cells secrete phospholipids. In cyclophyllidean cestodes also the thin capsule (eggshell) surrounding the embryophore is a keratin. Only histochemistry could bring out this. Our interpretation is that the transmembrane expulsion of shell precursors, notably protein, out of vitelline cells is effected by phospholipids which are well known in carrier-mediated transport. The extrusion of shell precursor occurs only after the vitelline cells are influenced by Mehlis' secretion and not before. This histophysiological aspect of eggshell formation of helminths could be elucidated only with the techniques of histochemistry.

Byssus Chemistry

In pelecypods which lie attached to a substratum (rock, hull of a ship, etc.) the byssus threads are kneaded in the groove of the foot from secretion of glands in the foot (see recent review Hanumantha Rao and Shyamasundari, 2003). The chemistry of the complicated scleroprotein of the byssus could be unravelled only through histochemistry while biochemical analyses only gave a general information on its composition. The architecture of the foot itself in genera like *Perna* or *Mytilus* is a matter to wonder. The precise separate locations of the glands in the foot so as to be secreted in a coordinated way culminating in the formation of both the thread and adhesive disc could only be studied through histochemistry. Four mussels found in India can be considered.

1.   *Perna viridis* (green mussel)
2.   *Perna indica* (brown mussel)
3.   *Mytilopsis sallei*
4.   *Barbatia obliquata*

In *Perna viridis,* the byssus thread is compacted by elastin and keratin, whereas the attachment disc is purely keratin. In *Perna indica,* the byssus is composed of collagen and keratin but the attachment disc is keratin only. In both *P. viridis* and *P. indica,* additionally a quinone-tanned protein component occurs. We have called this formation of byssus as "scleroprotein alloy" (Hanumantha Rao and Shyamasundari, 2003). Such precise chemical characterization of glands and stabilization of scleroproteins could be possible only through histochemistry. In *Mytilopsis sallei* or *Barbatia obliquata,* the byssus thread as well as adhesive disc are of keratin only.

Another example of the great role histochemistry plays comes from a study of the curious myzostome parasites that are harboured by the crinoids of the intertidal environment. The crinoids (Echinodermata), *Tropiometra palmata* and *Tropiometra encrinus* harbour these small myzostomes as ectoparasites on their arms. The myzostomes (with annelidan affinities) are soft-bodied and prone to mutilation in the turbulent intertidal environment with pounding wave action.

In a histochemical study of these myzostomes we were struck with a dramatic feature of their dorsal body wall. The muscle layer beneath the epidermis is fortified with collagen. A similar situation is completely absent in the ventral musculature, the reason being these ectoparasites lie exposed on the dorsal side while the ventral side lies applied to the crinoid arm. It is the dorsal side that bears the brunt of tidal action and hence the musculature has been fortified with a collagen skeletal framework. The discovery was made possible through our routine histochemical work on myzostomes.

Collagen of myzostomes

PAS +++

Aniline blue +++

AF – – –

Van Gieson ++

Verhoeff ++

The staining results mentioned above suffice to say that these examples—Mehlis' gland chemistry vis-à-vis eggshell chemistry, varying chemistry of "scleroprotein alloys" of byssus in different species of mussels, and dorsal collagen in myzostomes—demonstrate the strength, utility, efficacy and preciseness of histochemical studies. Many more examples may be considered wherein histochemical studies must precede biochemical studies and in many cases histophysiological studies where biochemical studies may not be meaningful. Moreover, without a prior histochemical background or specific tissues in specific issues, biochemical results may be far from comprehensive and may even lead to blurred interpretation. Histochemistry as a discipline and tool in life sciences programmes of teaching and research is mandatory.

FIXATIVES AND METHODS OF FIXATION

Histological or histochemical investigations, except for certain activities like enzyme activities, cannot be conducted on living cells because the integrity of the tissue or cells may be damaged. It is therefore necessary to "fix" the tissue and subsequently determine the microanatomy and/or localize the chemical constituents. Tissues should be fixed immediately after death. Normally the fixative should be 20 times the bulk of the tissue. The aim of fixation is to preserve the tissue in as life-like a manner as possible.

After death, tissues are prone to degenerative changes. When a tissue is removed from a living condition, various changes occur in the cells. Autolysis sets in and there is immediate bacterial action resulting in putrefaction. Autolysis is dissolution of cells by enzymes due to rupture of lysosomes, and it is most common in the gut especially intestine where bacterial action is heavy. To avoid this, it is necessary to fix the tissues immediately after biopsy. In autolysis, enzymes play a reverse action by degrading proteins into amino acids, and amino acids diffuse out of the cell. Fixation overcomes most of these changes, especially distortion of cell and shrinkage, apart from improving the staining potential. All events in this procedure adopted are referred to as "Fixation" of the tissue. Selection of a fixative is important since it is not possible to conduct all studies with a single fixative or fixing reagent or a group of reagents. We call this a choice fixative and it should fulfil the study of gross anatomy (microscopical) of the tissue or its chemical constitution.

Fixing the tissue immediately after death or removal from the body is the most important point to be considered and remembered. If the fixative is not ready in hand, the tissue is kept in the freezing chamber of the refrigerator. It is always advisable to keep a jar ready with fixative (appropriate) at least 20 times the bulk of the tissue to be fixed. It is better to select thin pieces of tissue for fixing. The larger the tissue block, the more is the fixative required for complete immersion of the tissue. In addition to this, the thickness of the tissue determines the rate of penetration of the fixative. A moderately thin tissue is always advisable. If a whole organ is to be fixed, the fixative has to be injected into the organ followed by immersion. Experience will enable one to judge the volume of the tissue.

Most of the fixatives comprise more than one chemical. A single chemical may not fulfil all the requirements necessary for a fixative. No fixing solution is ideal. In general, any fixative is bound to cause changes in the protein structure of the cell. The type of investigation to be carried out dictates the type of fixative to be used. It may be advisable to prepare the fixative before use. Solutions may be kept ready separately to be mixed before use.

The hardening effect of a fixative is of great significance. It should preserve the tissue in such a way that all the details or forms should be maintained. The tissue should be sufficiently hard to enable cutting of thin slices. Appropriately fixed tissue withstands the various techniques that are to be employed later.

Fixation may interfere with refractive indices of the various cells and tissue components. Added to this, most of the fixatives also affect histochemical reactions. They may also act as mordants and improve staining qualities but at times they may act adversely.

Fixatives differ in their effects on tissue shrinkage. Barh *et al.* (1957) in their work on 600 different varieties of tissues could recognize certain differences in tissues fixed in formaldehyde and in osmium tetroxide. If the tissue is fixed for a longer time, there may be greater swelling. Subsequently dehydration causes shrinkage. Between methanol and ethanol, methanol was found to be superior. Infiltration causes further shrinkage. From this it is evident that fixation and paraffin processing resulted in total shrinkage in the order of 30–40 per cent. According to Hopwood (1967), glutaraldehyde alone engenders much shrinkage. Although the mechanism behind this phenomenon is far from clear, it may be a result of various factors like inhibition of respiration, membrane permeability and change in sodium transport activity.

The rate of penetration of a fixative is also important, as it is not the same with different fixatives. An independent or single fixative acts in a different way from the compound fixative. In a compound fixative, different ingredients penetrate at different timings. A variety of fixatives on a number of tissues have been used by Dempster (1960) and the depth of penetration was measured at different intervals. As soon as the tissue is fixed in a compound fixative, the ingredients begin to separate during penetration. Consider a tissue fixed in Zenker's fluid, the tissue would have been penetrated to a depth of 0.85 mm by the dichromate, 2.6 mm by the mercuric chloride and 5.8 mm by the acetic acid. The most rapidly penetrating component would have fixed the tissue quicker. Thus other components act as post-fixatives. The speed of the reaction between the fixative and tissue determines the speed of penetration in addition to the rate of diffusion (Flitney, 1966).

Fixation procedures are best at low temperatures but that may not be possible for routine work in most of the laboratories. Most fixatives have an acid base, therefore for histochemical studies the pH should be adjusted to neutrality.

The following chemicals are used as single fixatives or are used in combination with other ingredients: formaldehyde, mercuric chloride, chromic acid, osmium tetroxide. Formaldehyde (40 per cent) takes the pride of being the most popular and routinely available fixative for whole mounts or sections or in the preservation of museum specimens. Commercially it is available as 40 per cent formaldehyde, which may be considered as 100 per cent. To prepare 10 per cent formaldehyde, 90 parts of water is added to 10 parts of formaldehyde. Formaldehyde reacts with various constituents of animal tissue. It reacts with proteins (French and Edsail, 1945; Bowes and Cator, 1966; Hopwood, 1969) and insoluble products may result. Lysine, arginine, histidine, glutamine, asparagine, cystein, tyrosine and tryptophan are some of the amino acids which are likely to react with aldehydes. Some lipids like sphingomyelins, cerebrosides, cholesterol and sulphatides remain unaffected by formalin. Phospholipids can remain unaffected provided calcium is added to formalin (Baker, 1944; Lillie, 1954a).

The main constituent of 40 per cent commercial formaldehyde is formic acid. It occurs as an impurity or due to oxidation of formaldehyde. So it is advisable to neutralize it with phosphate buffer at pH 7.2. Baker (1944) modified the procedure by adding calcium chloride to formalin. It can be used for fixing tissue to demonstrate phospholipids in sections. Subsequently Lillie (1954a) recommended the addition of 2 per cent calcium acetate to formalin. This fixative preserves phospholipids. The following are some fixatives having formalin as a component.

- 10 per cent formol saline
- 10 per cent formalin
- 10 per cent neutral formalin
- 10 per cent formol calcium
- 10 per cent formalin in alcohol
- 10 per cent formol sucrose
- Formaldehyde vapour

The other chemicals which are ingredients of compound fixatives include the following.

Acetic Acid (CH_3COOH)

Glacial acetic acid solidifies at temperatures below 17°C. Acetic acid causes lot of swelling to the tissue and as such it cannot be used as a single fixative but it is a very common reagent used in compound fixatives which counteracts shrinkage. Perhaps this quality provides a place for it as one of the ingredients in most of the fixatives. It has some deleterious effects. It distorts mitochondria, Golgi, and nucleoproteins and metals like calcium. Lipids are soluble in it. It rapidly penetrates into the tissues.

Acetone (CH_3COCH_3)

Cold acetone or acetone are excellent preservatives for the demonstration of enzymes. It very rarely finds a place in a compound fixative. The problem with cold acetone is, it does not penetrate rapidly. At 4°C it fixes cryostat sections in an hour. Acetone also can cause considerable amount of shrinkage. This is the best fixative for the demonstration of phosphatases and lipases but glycogen is not well preserved.

Chromium Trioxide

This is one of the ingredients in Orths' and Zenker's fluid, and causes moderate hardening but penetration is very slow. Shrinkage is quite moderate. When chromium trioxide is dissolved in water it becomes chromic acid, a good oxidizing agent, and preserves carbohydrates. To such fixatives, formalin should be added before use. Polysaccharides are oxidized and carbohydrates and DNA are converted to aldehydes. This is the reason why DNA and carbohydrate take Schiff stain. The fixative needs to be washed in running water until the tissue is almost colourless.

Ethyl Alcohol (C_2H_5OH)

This slow fixative is rarely used alone. Shrinkage and hardening of tissue are common, but when it is combined with other reagents (Carnoy), fixation is very rapid. However, it is a choice fixative for enzyme histochemistry (4°C). It precipitates (denatures) proteins and glycogen, and dissolves lipids. In enzyme histochemistry ethyl alcohol retains them in the original state, and lipids are dissolved out.

Some of the fixatives which contain alcohol are:

1. Formol alcohol
2. Carnoy fixative
3. Clarkes' fixative
4. Gendre's fixative
5. Acetic Alcohol Formalin (AAF)
6. Walman's fixative
7. Apathy's fixative
8. Henning's fixative
9. Ohlmacher's fixative
10. Alcoholic lead nitrate

Mercuric Chloride (Hg_2Cl_2)

With mercuric chloride, shrinkage and hardening are mild. Thus as an individual fixative, it is not recommended. But it is one of the constituents of a number of fixatives, which are routinely used in histological and histochemical investigations. It can be combined with acetic acid, formalin and potassium dichromate. Some scientists avoid fixatives containing mercuric chloride. It precipitates proteins but this can be remedied by treating the tissue with iodine alcohol (few crystals of iodine in 70 per cent alcohol) and this is followed by immersion in 3 per cent sodium thiosulphate. According to Pearse (1968), it combines with acid groups of proteins and phosphoric acid groups of nucleoproteins and S–H groups. Its use for investigation of sulphydryl groups or nucleoproteins or for glycogen is minimal. A plus point is, mercuric-chloride-fixed tissue stains brilliantly with cytoplasmic stains.

Picric Acid [$C_6H_2(NO_2)_3OH$]

An advantage with this chemical is that it causes minimum shrinkage. It is a good fixative with sufficient hardening power, but penetration is very slow. It is mostly used as one of the ingredients in alcoholic Bouin's, Bouin's fluid, Gendre's fluid and so on. Tissues fixed in picric-acid-containing fixatives give excellent results with trichrome stain. It is an excellent fixative for the demonstration of glycogen. Picric acid precipitates proteins to form picrates some of which are soluble in water. It does not dissolve lipids and does not fix carbohydrates. Thorough washing in running water is necessary for the tissues fixed in either picric acid or picric-acid-containing fixatives till the yellow colour disappears. Usually Scot's tap water is used as a post-treatment fluid. Picric acid removes ferric iron and makes RNA resistant to ribonuclease digestion.

Potassium Dichromate ($K_2Cr_2O_7$)

This is not used alone as a fixative. A non-coagulant of proteins, its affinity for carboxyl and hydroxyl groups of proteins is great. It is suitable for chromaffin test only. Chromosomes are fixed well in Champy's fluid consisting of potassium dichromate, chromic acid and osmium tetroxide (pH 2.5). For lipids, tissues are put in potassium dichromate to ensure that lipids are well preserved. Subsequently, tissues must be made colourless by repeated washing before proceeding for dehydration. Potassium dichromate is sometimes added to mercuric chloride, picric acid or osmic acid but never to formalin.

Osmium Tetroxide (OsO_4)

It is an expensive chemical used extensively in electron microscopy as a 1 per cent solution. It gives excellent details of single cells or minute pieces but is not used in routine histology or histochemistry. Penetration is poor and tissue remains soft and as such sectioning becomes difficult. It is a good fixative for carbohydrates. Osmium tetroxide vapours are irritating and cause conjunctivitis, so it is stored in a dark cool place. Osmium tetroxide is known for its uneven fixation, i.e., the periphery is fixed and it takes a longer time for the central region to be fixed. As it oxidizes unsaturated bonds, unsaturated lipids reduce osmium tetroxide to show a black compound. Bahr (1954) opines that osmium tetroxide's reaction with proteins depends on histidine, cystein, and tryptophan content. It is used as a vapour fixative with freeze-dried material (Tock and Pearse, 1965).

Glutaraldehyde

Introduced first by Sabatini *et al.* (1963), it is mostly used in electron microscopy. Commercial glutaraldehyde contains a number of impurities. After treatment with activated charcoal at 4°C for several days and when amber colour disappears, it should be stored in the refrigerator. Among aldehyde fixatives used in electron microscopy, glutaraldehyde is superior. It reacts with amino groups (Bowes and Cator, 1967) and reacts with amino acids like tyrosine, tryptophan and phenylalanine (Hopwood, 1968). Chambers (1968) recommended the use of 4 per cent glutaraldehyde in phosphate buffer at pH 7.4 as a routine fixative in histopathology.

COMPOUND FIXATIVES

The choice of the fixative largely depends on the type of investigation to be carried out. A battery of fixatives and a battery of reactions (each with proper fixative) are necessary for a complete study. Anatomical fixatives, cytological fixatives or histochemical fixatives specifically differ.

ANATOMICAL FIXATIVES

For a long time formol saline was popular as a routine fixative. Buffered formalin (Lillie, 1954a) or formol calcium acetate (Lillie, 1965) are in vogue now. Baker (1944) used calcium chloride in 10 per cent formalin to preserve phospholipids.

Thus buffered formalin, formol saline and formol calcium are frequently used. Tissue can be left in these fixatives for longer periods (sometimes one year). Thick blocks of tissue (5 mm) can be kept for 6–12 hours at room temperature and 3-mm pieces for 1½ –2 hours at 55°C.

CYTOLOGICAL FIXATIVES

When preservation of intracellular structures or inclusions is important. Particular fixatives should be used.

HISTOCHEMICAL FIXATIVES

Care should be taken that minimal changes are undergone by the elements of the tissues during fixation. Though freeze-drying technique is most reliable, it is cumbersome for routine use and time-consuming. Some well-known fixatives for anatomical, cytological, cytoplasmic and histochemical investigations are given. Apart from these, fixatives to be used for a particular component are mentioned in the relevant chapters.

ANATOMICAL FORMALIN FIXATIVES

AZF Fixative

Zinc chloride	25.03
Formaldehyde (37–40 per cent)	150.0 ml
Glacial acetic acid	9.5 ml
Distilled water	Fill to 100 ml

Fix for 4 hours. Dissolve zinc chloride in 500 ml water and later add acetic acid and remaining water to make it to 1000 ml.

Formol Calcium (Lillie, 1965)

Formalin	10.0 ml
Calcium acetate	2.0 g
Distilled water	100.0 ml

Formol Calcium (Baker, 1944)

Formalin	10.0 ml
Calcium chloride	2.0 g
Distilled water	100.0 ml

This solution should be prepared with buffered formalin.

Buffered Formalin

Formalin	10.0 ml
Acid sodium phosphate monohydrate	400 mg
Anhydrous disodium phosphate	650 mg
Distilled water	100.0 ml

Buffered Formol Sucrose (Holt and Hicks, 1961)

Formalin	10.0 ml
Sucrose	7.5 g
M/15 phosphate buffer (pH 7.4)	100.0 ml

This fixative is recommended for cytochemistry and electron microscopy (combined). It has to be stored in the refrigerator. Mitochondria and endoplasmic reticulum are preserved. It is good for phospholipids and some enzymes.

Alcoholic Formalin

Formalin	10.0 ml
70 per cent alcohol	90.0 ml

0.5 of calcium acetate may be added to make it neutral.

Acetic Alcoholic Formalin (AAF)

Formalin	5.0 ml
Glacial acetic acid	5.0 ml
70 per cent alcohol	90.0 ml

This is a fixative for glycogen. It penetrates rapidly and fixation time is 4 hours.

The following are the fixatives commonly used in a laboratory. For general histological purposes 40 per cent formalin is preferred. It can be diluted to 10 per cent with 90 parts of water to 10 parts of formaldehyde. Lillie's (1954a) neutral buffered formalin is recommended for some histopathological studies.

Commercial formalin 40 per cent	100.0 ml
Distilled water	900.0 ml
Sodium acid phosphate monohydrate ($NaH_2PO_4H_2O$)	4.0 g
Disodium phosphate Anyhydrous (Na_2HPO_4)	6.5 g

Tissue especially kept at 4°C is fixed in this solution for 72 hours but overnight fixation is sufficient.

Zenker's Stock Solution

Potassium dichromate ($K_2Cr_2O_7$)	2.5 g
Mercuric chloride	5.0 g
Distilled water	100.0 ml

Dissolve potassium dichromate and mercuric chloride in water by heating. This is the stock solution.

Zenker's stock	100.0 ml
5 per cent glacial acetic acid	5.0 ml

Helly's fluid can also be prepared from Zenker's stock solution. Instead of acetic acid, formaldehyde is added.

Zenker's stock	100.0 ml
5 per cent formaldehyde	5.0 ml

Material fixed in these requires post-treatment with potassium dichromate (2 per cent) for 24 hours. Tissue requires washing for 10–15 hours. This is followed by the dehydration procedure.

Bouin's Fluid

This is well known as a routine fixative recommended especially for Masson's staining technique.

Picric acid 4 per cent aqueous	150.0 ml
40 per cent formaldehyde	50.0 ml
Glacial acetic acid	10.0 ml

It lasts long and is ready for use. Tissues can be kept in this solution for 12–16 hours depending on the thickness. The material has to be washed for several hours till the yellow colour of picric acid disappears. The tissue is now ready for carrying out the dehydration procedure.

Alcoholic Bouin's (Pantin, 1946)

95 per cent alcohol	150.0 ml
40 per cent formaldehyde	60.0 ml
Glacial acetic acid	15.0 ml
Picric acid crystals	1.0 g

This fixative has to be prepared just before use. It is preferred over Bouin's fluid. Fixation time is 16–18 hours.

Baker's (1944) formalin remains indefinitely.

Calcium chloride	1.0 g
40 per cent formaldehyde	10.0
Distilled water	100.0 ml

Scott's tap water

Sodium bicarbonate	2.0 g
Magnesium sulphate	20.0 g
Distilled water	1000 ml

Add a pinch of thymol to prevent moulds.

Baker's (1958) Formalin

40 per cent formaldehyde	10.0 ml
Calcium chloride (10 g in 100 ml water)	10.0 ml
Distilled water	80.0 ml

The tissue can remain in this fixative for a considerable time.

Gomori's 1-2.3 Fixative

40 per cent formaldehyde	1 part
Saturated mercuric chloride	2 parts
Distilled water	3 parts

This is good for general cell morphology. Wash for 6–8 hours. Post-treatment is necessary to remove mercury. Fixation time is 2–3 weeks.

Hollande Bouin's Fixative (Romeis, 1948)

Copper acetate	2.5 g
Picric acid crystals	4.0 g
40 per cent formaldehyde	10.0 ml
Distilled water	100.0 ml
Glacial acetic acid	1.5 ml

Dissolve copper acetate in water and to this add picric acid while stirring. Then filter and add formaldehyde and picric acid. This is good for calcified tissue (Hartz, 1947). Several hours of washing in water is necessary.

Fixation time ranges from 8 hours up to 3 days.

Orth's Fixative (Gatenby and Beams, 1950; Galigher, 1934)

40 per cent formaldehyde	10.0
Potassium dichromate	2.5 g
Sodium sulphate	1.0 g
Distilled water	100.0 ml

Mix before use. A routine fixative. Wash the tissue in running water for 6–8 hours. The fixation time is 12 hours at room temperature and 3 hours at 37°C.

Stieve's Fixative (Romeis, 1948)

Mercuric chloride saturated (Aqueous)	76.0 ml
40 per cent formaldehyde	20.0 ml
Glacial acetic acid	4.0 ml

Post-treatment for mercuric chloride is necessary. This fixative is most similar to Susa fixative. The fixation time is 18–24 hours.

Susa Fixative (Romeis, 1948)

Saturated mercuric chloride	50.0 ml
Trichloroacetic acid (TCA)	2.0 g
Glacial acetic acid	4.0 ml
40 per cent formaldehyde	20.0 ml
Distilled water	20.0 ml

It is a choice fixative, substitute for Zenker's post-treatment in iodine alcohol (few crystals of iodine in 70 per cent alcohol) is necessary to remove mercury. The fixation time is 18–24 hours.

Altman's Fixative (Gatenby and Beams, 1950)

5 per cent potassium dichromate	10.0 ml
2 per cent osmic acid	10.0 ml

Wash in running water overnight. Altman's fixative also acts as a stain. It fixes and stains mitochondria. Fixation time is 12 hours.

Dafano's Fixative (Romeis, 1948)

Cobalt nitrate	1.0 g
40 per cent formaldehyde	15.0 ml
Distilled water	100.0 ml

Wash in running water overnight. It gives good results for Golgi apparatus. Fixation time is 12–24 hours.

Flemming Fixative

1 per cent chromic acid (1 g in 100 ml water)	15.0 ml
2 per cent osmic acid (2 in 100 cc water)	4.0 ml
Glacial acetic acid	1.0 ml

Mix just before use. Wash in running water for 24 hours. Fixation time is 12–24 hours.

Formol Alcohol (Lillie, 1954a)

70 per cent ethyl alcohol	100.0 ml
40 per cent formaldehyde	5.0 ml
Glacial acetic acid	5.0 ml

It is best for insects and crustaceans. Transfer directly to 80 per cent alcohol. Fixation time is 1–24 hours.

Gendre's Fluid (Lillie, 1954a)

95 per cent alcohol saturated with picric acid	80.0–85.0 ml
40 per cent formaldehyde	15.0 or 10.0 ml
Glacial acetic acid	5.0 ml

Wash in several changes of 80 per cent alcohol. It gives good results for glycogen. Fixation time is 1–4 hours at 25°C.

Gilson's Fixative

Concentrated nitric acid	15.0 ml
Glacial acetic acid	4.0 ml
Mercuric chloride	20.0 g
60 per cent ethyl alcohol	100.0 ml
Distilled water	880.0 ml

Post-treatment is necessary for mercuric chloride. It is a good fixative for invertebrate material. Wash in 50 per cent alcohol. Fixation time is 24 hours.

Johnson's Fixative (Gatenby and Beams, 1950)

2.5 per cent potassium dichromate (2.5 g in 100 ml water)	70.0 ml
2 per cent osmic acid (2 g in 100 ml water)	10.0 ml
1 per cent platinum chloride (1 g in 100 ml water)	15.0 ml
Glacial acetic acid	5.0 ml

Wash in running water for 5 hours. Without acetic acid, it is good for mitochondria. Fixation time is 12 hours.

Kolmer's Fixative

5 per cent aqueous potassium dichromate (5 g in 100 ml water)	20.0 ml
10 per cent formaldehyde	20.0 ml
Glacial acetic acid	5.0 ml
50 per cent trichloroacetic acid (50 in 100 ml water)	5.0 ml
10 per cent uranyl acetate (10 in 100 cc water)	5.0 ml

It gives good results for nerve tissue. Wash in running water. Fixation time is 24 hours.

Lavdowsky Fixative (Gray, 1954)

Distilled water	80.0 ml
95 per cent ethyl alcohol	10.0 ml
2 per cent chromic acid (2 g in 100 ml water)	10.0 ml
Glacial acetic acid	0.5 ml

Transfer to 80 per cent alcohol. It is good for glycogen. Fixation time is 12–24 hours.

Navashin's Fixative (Randolph, 1935)

Solution 1

Chromic acid	1.0 g
Glacial acetic acid	10.0 ml
Distilled water	90.0 ml

Solution 2

    40 per cent formaldehyde   40.0 ml
    Distilled water   60.0 ml

Mix equal parts of solution 1 and 2 just before use. After 6 hours, change to a new solution for another 18 hours. Transfer to 75 per cent alcohol. Fixation time is 24 hours.

Sanfelice's Fixative (Baker, 1958)

    1 per cent chromic acid (1/100 ml water)   80.0 ml
    40 per cent formaldehyde   40.0 ml
    Glacial acetic acid   5.0 ml

Mix before use. It is good for chromosomes and mitochondria. Wash in running water for 6–12 hours. Fixation time is 4–6 hours.

Smith's Fixative (Galigher, 1934)

    Potassium dichromate   5.0 g
    40 per cent formaldehyde   10.0 ml
    Distilled water   87.5 ml
    Glacial acetic acid   2.5 ml

Mix before use. It is good for yolk material.

Formol Sucrose (Holt and Hicks, 1961)

    40 per cent formaldehyde   10.0 ml
    Sucrose   7.5 g
    0.2 M phosphate buffer (pH 7.4)   90.0 ml

Formol Sublimate

    Saturated (aqueous) mercuric chloride   90.0 ml
    40 per cent formaldehyde   10.0 ml

6 Per cent Glutaraldehyde

    Glutaraldehyde (25 per cent)   24.0 ml
    0.1 M Phosphate buffer (pH 7.4)   76.0 ml
    (pH should be adjusted to 7–7.2)

1 Per cent Acrolein

    Acrolein   1.0 ml
    0.1 M phosphate buffer (pH 7.4)   99.0 ml

The final pH should be 7.2–7.6.

Rossman's Fluid (Lillie, 1954a)

    Absolute alcohol saturated with picric acid   9 parts
    Formalin   1 part

Keep in a refrigerator and fix at refrigerator temperature for 24 hours.

Vapour fixatives	These are for cryostat sections of fresh tissue or frozen tissue.
Formaldehyde	By heating paraformaldehyde at 50°–80°C. Vapour is obtained.
Acetaldehyde	Heat "meta" aldehyde at 80°C for 1–4 hours.
Glutaraldehyde	Rost and Ewen (1971) used 50 per cent solution at 80°C for 2 minutes to 4 hours. Pearse (1968) suggests fixation at 60°C for 7 hours.
Acrolein chromyl chloride	This reagent is used at 37°C for 1–2 hours.

CYTOLOGICAL FIXATIVES

Biltat (Modified Carnoy)

Isopropyl alcohol	60.0 ml
Chloroform	30.0 ml
Formic acid	10.0 ml

Fixation time is 1–8 hours. It is good for worms and hard cuticled organisms like insects.

Carnoy's Fluid

Absolute alcohol	60.0 ml
Chloroform	30.0 ml
Glacial acetic acid	10.0 ml

Carnoy's fluid penetrates rapidly, and nuclear preservation is excellent. This fluid has been recommended for carbohydrates. It causes shrinkage and most of the cytoplasmic elements are destroyed.

Methacarn (Puchtler *et al.*, 1970)

Methyl alcohol	60.0 ml
Chloroform	30.0 ml
Glacial acetic acid	100 ml

Washing the tissue before fixation is not necessary, this will avoid shrinkage and structural alterations. Overexposure to alcohol will result in shrinkage and hardening. It is better than Carnoy. Before embedding transfer the tissue to methyl alcohol (2 changes in 2–4 hours) methyl benzoate (2 changes in 3 hours and 1 hour), methyl benzoate–xylene mixture (1 hour) and xylene. Later infiltrate and embed.

Clarke's Fluid

Absolute alcohol	75.0 ml
Glacial acetic acid	25.0 ml

It penetrates rapidly. Nuclear fixation is good. It proves good for smears and or chromosomes.

Newcomer's Fluid

Isopropanol	60.0 ml
Propionic acid	30.0 ml
Petroleum ether	10.0 ml
Acetone	10.0 ml
Dioxane	10.0 ml

It penetrates rapidly and preserves chromosomes better than Carnoy. This fixative was recommended by Saunder's (1964) for mucopolysaccharides. Fixation time is 12–18 hours.

Flemming's Fluid

1 per cent aqueous chromic acid	15.0 ml
2 per cent aqueous osmium tetroxide	4.0 ml
Glacial acetic acid	10.0 ml

It exhibits poor penetration. Fixation time is 12 hours.

Champy's Fluid

3 per cent potassium dichromate	7.0 ml
1 per cent chromic acid	7.0 ml
2 per cent osmium tetroxide	4.0 ml

Prepare just before use. It is a poor penetrant. It preserves mitochondria, fat, yolk and lipids. Tissue after fixation should be washed overnight. Fixation time is 12 hours.

Parenyi Fixative (Galigher, 1934)

1 per cent chromic acid	15.0 ml
10 per cent aqueous nitric acid	40.0 ml
95 per cent alcohol	30.0 ml
Distilled water	15.0 ml

Fixation time is 12–24 hours. Recommended for eyes. Instead of immersing the eye in the fixative inject a little into the chambers. Haematoxylin is the best stain. Washing is in 50 per cent or 70 per cent alcohol.

Regaud's Fluid

3 per cent potassium dichromate	80.0 ml
40 per cent formaldehyde	20.0 ml

Prepare just before use. It is a deep penetrant but it over-hardens tissue. It is good for mitochondria. Postchromation in 3 per cent potassium dichromate. Fixation time is 24 hours.

Muller's Fluid

Potassium dichromate	2.5 g
Sodium sulphate	1.0 g
Distilled water	100.0 ml

It is rarely used. This can also be used for postchromation instead of 3 per cent potassium dichromate.

Zenker's formol can also be used for cytoplasmic details.

Schaudinn's Fluid

Saturated mercuric chloride　　2 parts
Absolute alcohol　　　　　　1 part

It was popular as a cytoplasmic fixative for wet smears. It is not recommended for tissues since it causes a lot of shrinkage. It is excellent for protozoans.

B-5 Fixative

Stock solution
　　Mercuric chloride　12.0
　　Sodium acetate　　2.5 g
　　Distilled water　　200.0 ml

Working solution
　　Stock solution　20.0 ml
　　Formaldehyde　2.0 ml (Add just before use)

Zinc Formalin

Zinc formalin gained popularity in recent times. L'Hoste and Torres (1995) used this as a routine fixative for haematoxylin/eosin and it finds a place in immunostaining.

Formaldehyde　　10.0 ml
Distilled water　　90.0 ml
Zinc sulphate　　1.0 g

Final pH should be 4.2. This often forms formalin pigment in the tissue which can be easily removed with 10 per cent ammonium hydroxide. Taylor (1993) substituted zinc sulphate with zinc chloride. With increasing pH (6.2) formalin pigmentation can be prevented. Zinc formalin is recommended for histochemical and immunohistochemical procedures (Churikian, 1993).

ADDITIONAL FIXATIVES

1. Aoyama's Fixative

Cadmium chloride 1 per cent　　200.0 ml
40 per cent formalin　　　　　80.0 ml
Distilled water　　　　　　　220.0 ml

2. Apathy's Fixative

5 per cent mercuric chloride　　200.0 ml
10 per cent acetic acid　　　　25.0 ml
Distilled water　　　　　　　75.0 ml
Absolute alcohol　　　　　　　200.0 ml

3. Cetyl Pyridinium Chloride–Formaldehyde (Williams and Jackson, 1956) (For acid mucopolysaccharides)

Cetylpyridinium　　　　　　2.5 g
40 per cent formaldehyde　50.0 ml
Distilled water　　　　　　450.0 ml

Fixation for 48 hours.

4. Del Rio Hortega's Fixative

5 per cent uranium nitrate	75.0 ml
40 per cent formaldehyde	25.0 ml
Distilled water	15.0 ml

5. Gilson's Fixative

Absolute alcohol	100.0 ml
Glacial acetic acid	100.0 ml
Chloroform	100.0 ml

6. Hemming's Fixative

Saturated picric acid	60.0 ml
Mercuric chloride	16.0 g
Distilled water	112.0 ml
Absolute alcohol	250.0 ml
Chromium trioxide	400 mg
Concentrated nitric acid	72.0 ml

7. Hollande's Fixative

Distilled water	250.0 ml
Copper acetate	6.25 g
40 per cent formaldehyde	25.0 ml
Glacial acetic acid	2.5 ml

8. Jackson's Fixative

90 per cent alcohol	100.0 ml
40 per cent formaldehyde	12.5 ml
Glacial acetic acid	12.5 ml

9. Kaiser's Fixative

5 per cent mercuric chloride	166.0 ml
Glacial acetic acid	7.5 ml
Distilled water	84.0 ml

10. Kinyoun Fixative

Potassium dichromate	9.5 g
Copper sulphate	5.0 g
Distilled water	200.0 ml
40 per cent formaldehyde	50.0 ml

11. Kohn's Fixative

5 per cent mercuric chloride	64.0 ml
5 per cent potassium dichromate	116.0 ml
Glacial acetic acid	12.5 ml
Distilled water	70.0 ml

12. Kollmann's Fixative

5 per cent chromium trioxide	100.0 ml
10 per cent potassium dichromate	125.0 ml
Distilled water	25.0 ml
Nitric acid	5.0 ml

13. Mann's Fixative

1 per cent osmic acid	100.0 ml
Sodium chloride	0.75 mg

14. Muller's Fixative

Potassium dichromate	2.5 g
Sodium sulphate	1.0 g
Distilled water	100.0 ml

15. Novak's Fixative

5 per cent mercuric chloride	106.0 ml
1 per cent chromium trioxide	75.0 ml
Glacial acetic acid	7.5 ml
Distilled water	44.0 ml
40 per cent formaldehyde	25.0 ml

16. Ohlmacher's Fixative

Absolute alcohol	64.0 ml
Chloroform	12.0 ml
Glacial acetic acid	4.0 ml

Immediately before use add mercuric chloride.

17. Romeis Fixative

95 per cent alcohol	56.0 ml
Distilled water	54.0 ml
40 per cent formaldehyde	12.5 ml
Glacial acetic acid	2.5 ml

18. Sansom's Fluid

Absolute alcohol	65.0 ml
Glacial acetic acid	5.0 ml
Chloroform	30.0 ml

Mercuric chloride crystals to saturation.

19. Swank and Davenport's Fixative

Solution 1

40 per cent formaldehyde, 3 times the bulk of the tissue

Solution 2

1 per cent osmic acid	82.0 ml
1 per cent potassium chlorate	62.5 ml

10 per cent acetic acid	25.0 ml
Distilled water	80.5 ml

Fix fresh tissues in solution 1 for 24 hours. Transfer without washing to solution 2 and fix for one week.

20. Turchini's Fixative

Saturated picric acid	200.0 ml
Ammonium molybdate	17.0
Distilled water	50.0 ml
40 per cent formaldehyde	50.0 ml

21. Worcester Fixative

Mercuric chloride	14.0
Distilled water	200.0 ml
40 per cent formaldehyde	22.5 ml
Acetic acid	25.0 ml

FIXATIVES FOR SPECIAL PURPOSES

Glycogen

No formalin-containing fixatives are advisable to demonstrate glycogen (except Bowie's fixative). For histochemical purposes, Lison's Gendre fluid is superior to many (*see* page 11) (Fixation time 18 hours.). Before fixing, the solution is cooled to –73°C. These results are better than in the material which is freeze-dried.

Nucleic Acids

For routine work on nucleic acids in paraffin sections, Carnoy's fixative gives satisfactory results. For freeze-dried material and smears, absolute methanol is quite suitable. An alternative for Carnoy is AAF (Lillie, 1949). It gives excellent results.

40 per cent formaldehyde	10.0 ml
Acetic acid	5.0 ml
Ethanol	85.0 ml

Fix for 24 hours at 4°C.

Chromosomes

For the preservation of chromosomes, Newcomer (1953) recommends his experiment with isopropanol. All histochemical reactions can be carried out. Hydrolysis varies from plant to animal tissues. Fixation time is 12–24 hours.

Isopropanol	6 parts
Propionic acid	3 parts
Light petroleum	1 part
Acetone	1 part
Dioxane	1 part

Mucopolysaccharides and Mucoproteins

Lead acetate combinations are good for acid mucopolysaccharides.

Lillie's (1954a) Alcoholic Lead Nitrate

Lead nitrate	8.0 g
40 per cent formaldehyde	10.0 ml
Water	10.0 ml
Ethanol	80.0 ml

Fix for 24 hours at room temperature or for 2–3 days at 40°C.

Another alternative for this is Mota *et al.*, (1956) fluid.

Lead subacetate	1.0 g
Ethanol	50.0 ml
Water	50.0 ml
Acetic acid	0.5 ml

Fix for 24 hours at room temperature.

Williams and Jackson (1956) suggested two fixatives for mucopolysaccharides.

40 per cent formaldehyde	10.0 ml
Cetyl pyridinium chloride	500 mg
Water	90.0 ml

Fix for 48 hours.

Ethanol	50.0 ml
5-aminoacridine hydrochloride	400 mg
Water	50.0 ml

Fix for 48 hours.

Enzymes Special fixatives are used.

For cholinesterases, formalin–sucrose–ammonia (Pearson, 1968) is suggested.

40 per cent formaldehyde	10.0 ml
Sucrose	15.0 g
0.880 ammonia	1.0 ml
Distilled water	100.0 ml

Final pH is 6.7. Fix for 18–24 hours.

Lipids Generally Baker's (1944) formol calcium is used.

Another alternative is Elfman's (1957) fixative, provided enzymes are not required.

Mercuric chloride	5.0 g
Potassium dichromate	2.5 g
Water	100.0 ml

Fixation time is 36 hours at room temperature.

VAPOUR FIXATIVES

Formaldehyde vapour Paraformaldehyde is heated to 50–80°C. Tissues are fixed for 3–5 hours at 56–60°C for hour.

Osmium tetroxide Fix for 1 hour at 37°C in crystalline osmium tetroxide.

Glutaraldehyde Add 10 ml of 25 per cent glutaraldehyde to 100 ml liquid paraffin. Shake for several hours. Allow to stand and then remove the upper paraffin layer with separation funnel. Centrifuge remaining water. Liquid paraffin contains 10 per cent glutaraldehyde.

Acetaldehyde Melt-aldehyde on heating at 80°C yields acetaldehyde. Fixation time is 1–4 hours. Rost and Ewen (1971) did not show satisfactory results.

Acrolein or chromyl chloride can be used at 37°C for 12 hours.

FIXATIVES FOR VARIOUS GROUPS OF INVERTEBRATES

Protozoa

- Schaudinn's fluid
- Bouin's fluid
- Champy's fluid
- Flemming's fixative
- Worcester fixative

Porifera

- Osmic mercuric chloride

    | Osmic acid | 2.5 g |
    | Mercuric chloride | 9.5 g |
    | Distilled water | 250.0 ml |

- Gilson's fluid
- Carnoy

Coelenterata

Hydrozoa

- Mercuric chloride–acetic acid mixture

    | Saturated mercuric chloride | 95.0 ml |
    | Glacial acetic acid | 5.0 ml |

- Susa
- Bouin's
- Mercuric chloride combination
- 10 per cent formalin

Jelly Fishes

- 10 per cent formalin

Medusae

- 10 per cent formalin
- 5 per cent acetic acid

Corals

- Hot saturated mercuric chloride
- 5 per cent acetic acid

Platyhelminthes

Planarians

- Gilson's fluid
- Saturated mercuric chloride

Trematodes

- Gilson's fluid
- Saturated mercuric chloride

Cestodes

- Gilson's fluid
- Saturated mercuric chloride

Nemathelminthes

Nematoda

- Hot glycerol alcohol

  (70–80 per cent alcohol + 10 per cent glycerol)

Small Nematodes

- Formalin acetic alcohol

95 per cent alcohol	12–20 ml
Glacial acetic acid	1.0 ml
Formalin	6.0 ml
Distilled water	40.0 ml

Microfilaria

Smears to be fixed in any mercuric chloride fixative.

Nemertinea

Saturated mercuric chloride + acetic acid (95/5).

Bryozoa

Marine Bryozoa are fixed in the following solution.

10 per cent chromic acid	7–10 ml
(10 g + 100 ml water)	
10 per cent acetic acid	10.0 ml
Distilled water	80.0 ml

Freshwater Bryozoa are fixed in 10 per cent alcohol.

Annelida

Earthworms

- Bouin's fluid
- Saturated mercuric chloride in 80 per cent alcohol + acetic acid

Leeches

- Bouin's fluid
- Susa fixative

Arthropoda

Crustaceans

- Helly's fluid
- Methanol + formalin + acetic acid (MFA)

Insects

- Carnoy
- Bouin's fluid

Echinodermata

- Mercuric chloride + acetic acid is good.

Preparation of these fixatives has been given in the beginning of this chapter.

FIXATIVES FOR VERTEBRATE TISSUES

Alimentary Canal

- Susa fixative
- Formol saline
- Zenker's fluid

Ear

- 5 per cent aqueous trichloroacetic acid for 3 days
- Zenker's fluid

Embryos

- Susa fixative
- 4 per cent formaldehyde
- Zenker's fluid

Eye

- 4 per cent formaldehyde
- Zenker's fluid

- Mann's fluid
- Kolmer's fixative

Hair

Equal volumes of ether and absolute alcohol (for whole mounts).

For sections:

- Fix as mentioned above.
- Transfer to 2 per cent alcohol (6 vol.) + 5 per cent ammonia (4 vol.) for 5 minutes.
- Transfer to 10 per cent potassium hydroxide (50°C) for 2 minutes.
- Wash in 5 per cent sulphuric acid and absolute alcohol.
- Clear in benzene for 15 minutes.
- Dry with filter paper.
- Embed by impregnation in wax for 15 minutes.

Kidney

Susa fixative

Mercuric chloride + formaldehyde

5 per cent formaldehyde in 0.9 per cent sodium chloride

Zenker's fluid

Ligaments

Susa fixative. Softening of tissues is carried out in Lendrum's phenol.

Liver

- Susa fixative
- Formol saline
- Zenker's fluid

Lung

- Susa fixative
- Formol saline
- Bouin's fluid
- Zenker's fluid

Lymph glands

- Mercuric chloride + formaldehyde
- Zenker's fluid
- 4 per cent formaldehyde in 0.9 per cent sodium chloride

Muscle

- Susa fixative
- Zenker's fluid
- Flemmings' fluid

Ovary

- 5 per cent formaldehyde
- Zenker's fluid
- Bouin's fluid
- Sansom's fluid

Pancreas

- Zenker's fluid
- Formol saline
- Flemming's fluid
- Helly's fluid

Pituitary

- Formol saline
- Zenker's fluid

Plasma cells

- Absolute alcohol

Skin

- Susa fixative
- Zenker's fluid
- Sanfelices fluid
- Formol saline

Teeth

- Sansom's fluid

Testis

- Equal volumes of ether + absolute alcohol

Thyroid

- Zenker's fluid
- 5 per cent formaldehyde

Uterus

- Bouin's fluid

TISSUE PROCESSING

For microscopic investigations, tissues have to be sectioned. For this it is necessary to impregnate the tissue with a medium which after solidification facilitates sections of desired thickness to be cut. Sections can also be cut with frozen tissue on cryostats or on freezing microtomes. The advantage here in impregnating and keeping the tissue in a solid medium is that tissues could be stored this way readily cut to be used at a later stage.

It is very important that fixation is complete, before further processing. If the tissue is in small fragments, dehydrating alcohols may also aid in fixation. Tissues fixed in certain fixatives require special post-treatment. This should be done before proceeding further.

1. Tissues fixed in formol calcium have to be washed thoroughly in running tap water and then treated with 2.5 per cent potassium dichromate first at room temperature for 24 hours and at 60°C for 24 hours.

2. Tissues fixed in potassium dichromate require thorough washing in running water till the tissue is colourless. This is essential since an insoluble green-brown pigment can be produced due to its reaction with dehydrating alcohols.

3. Fixatives like alcoholic Bouin's fluid contain saturated picric acid. Prior to dehydration, tissues should be washed thoroughly in running tap water and later treated with Scott's tap water for 24 hours (*see* page 19) prior to dehydration.

4. When tissues are fixed in Carnoy's fluid, they can be directly transferred to 95 per cent alcohol since Carnoy's fixative contains absolute alcohol as one of the ingredients.

Most of the fixatives are in aqueous media, thus a large amount of water has to be removed before the tissue is put to further processing. This dehydration is necessary because the tissue as it is lacks ideal consistency for sectioning. If the tissue is soft or a lumen is present, it may get deformed while sectioning, because due to loss of fluid content, the epithelial wall collapses. To avoid all these problems, a medium is necessary and the aim in tissue processing is to embed the tissue in a solid medium which encloses the tissue extracellularly and intracellularly, and gives sufficient rigidity to enable thin sections to be cut and yet soft enough to enable the knife to cut the sections with little damage to the tissue. For histology, the most routinely used medium for embedding with satisfactory results is paraffin wax. Gelatin and water-soluble wax are also used. This embedding medium thoroughly permeates the tissue in fluid form so that it solidifies with little damage to the tissue. Before cutting the sections, the tissue must pass through the following procedures.

1. complete fixation
2. removal of water by dehydration
3. clearing with a clearing agent which is completely miscible with both dehydrating agent and embedding agent
4. embedding in paraffin wax

Usually when tissue is immersed in fluid, interchange occurs between the tissue fluid and the surrounding medium. The tissue is always kept in a fluid and allowed to sink to the base of the container. The tissue is agitated in the liquid so that when the fluid in contact with the upper surface of the tissue is contaminated, it is replaced by fresh immersing fluid. Agitation is done mechanically, also by shakers and mixers or processing machines which engender rotation or vertical oscillation for agitation. Agitation should neither be too low (ineffective) nor too high (tissue damage occurs). Agitation reduces 25 to 35 per cent of the processing time.

Another way of hastening the process is by heat, which increases the rate of penetration. Heat hastens paraffin wax processing but care must be taken not to overheat the tissue which may cause damage to tissue by cooking up. Overheating is also hazardous if inflammable fluids are being used.

Viscosity of the fluid is also important, which determines the rate at which fluids penetrate the tissues. The larger the molecule the higher is the viscosity and the slower the rate of penetration. Viscosity is not very important (except cedar wood) in the dehydrating and clearing stages, but is important in the impregnating stage. For celloidin, and nitrocellulose and resins, heat will cause premature hardening of the impregnating medium.

The use of vacuum at dehydrating and clearing stages is not much important but at the same time it removes air bubbles trapped within the tissue. Some of the automated processing machines incorporate the use of reduced pressure at all stages of processing.

Some hard tissues must be given special treatment before processing. For example, fibrous tissues, nails and tendons require this special treatment for good sections. There are a number of solutions which are used to soften the hard tissues, and the most recommended is a mixture of 4 per cent phenol in 70 per cent alcohol. Some commercial agents such as Mollifex and a mixture of glycerol alcohol and aniline oil are also used.

DEHYDRATION

The first stage in the processing of tissue is the removal of aqueous and lipid tissue fluids by a variety of agents of which alcohol is the best. Since these agents are hydrophilic, they attract water from the tissue. Alcohols are of varying types. This dehydration stage is common to all embedding techniques except when a water-miscible substance is used as the embedding medium.

Dehydration is achieved by passing the tissue in graded increasing series of alcohol: 50 per cent $\rightarrow$ 70 per cent $\rightarrow$ 90 per cent $\rightarrow$ 95 per cent $\rightarrow$ Absolute alcohol. The time required for dehydration depends upon the type and size of tissue. Delicate tissue requires slow dehydration. The tissue is passed gradually. After aqueous fixation, tissue should be taken through graded series of alcohol (starting from 30%). This will not cause much shrinkage whereas tissues fixed in Carnoy's fluid (containing alcohol) can be placed in higher grades of alcohols (70%, 90%). To remove the acid content of the fixative, several changes of alcohols may be required. For delicate tissues, particularly embryonic tissues like hepatopancreas, it is recommended to start with 30 per cent.

Bigger chunks of tissue have to be started from 50 per cent through 100 per cent for 24 hours in each grade depending on the bulk, using 3 changes in each strength. Smaller tissue pieces can remain for 24 hours in each grade. For cytological purposes, the same graded series could be used limiting the time to 24 hours in each grade.

Some of the dehydration fluids include the following.

Ethanol (C_2H_5OH) It is an inflammable colourless clear liquid, miscible with water as well as many organic solvents and a very good dehydrating agent but it is not preferred because of its heavy excise duty and high cost.

Methylated spirit (denatured spirit) As far as its physical characters are concerned, it is similar to ethanol with a pungent odour. It contains ethanol to which methanol is added. It is obtained as methylated spirit.

Acetone (CH_3COCH_3) It is a highly inflammable fluid, colourless, clear with a pungent odour, easily mixes with water, ethanol and other solvents. It is much more volatile than the other dehydrates and causes brittleness of the tissue if dehydration is prolonged, but it works more rapidly than ethanol. It is not preferred owing to its volatility, inflammability and hardening nature. Only when tissue has to be processed rapidly, acetone is preferred as a dehydrant.

Methanol (CH_3COH) It is a clear, colourless, inflammable, poisonous fluid with an unpleasant odour. It easily mixes with water, ethanol and other solvents. But as a dehydrant, it is not as popular as ethanol.

Propan-2-al-isopropyl alcohol (CH$_3$CHOH CH$_3$)   This easily mixes with water, ethanol and other organic solvents. Although it does not harden the tissues, it is not preferred in many laboratories. In recent years, Muller and Jacks (1975) used 2,2-dimethoxypropane. Tissues may be rapidly dehydrated in a mixture of 100 ml of 2,2-dimethoxypropane and 0.5 ml of concentrated hydrochloric acid.

CLEARING

Between dehydration and embedding, there is one more step. This is the process of clearing. A medium is chosen to remove the dehydrating agent and it should mix with alcohol as well as the embedding media. Many of the fluids have a similar refractive index. There are several reagents suitable for this. The most common and frequently used is xylene. When xylene is used as a dealcoholizing agent, the translucency indicates whether small pieces of tissue are cleared. There are a number of clearing agents which fulfil this purpose but only a few are in regular use.

Most of these clearing agents are inflammable and one should be cautious while handling them. Clearing agents used in paraffin wax processing have an effect on the quality of the sections. Most of the clearing agents give satisfactory results.

Some of the clearing agents are listed below.

Toluene   It is less damaging even if tissues are immersed for longer periods. It is useful for automated tissue processing. As it is inflammable, it is dangerous.

Chloroform   Though chloroform is used as a clearing agent, its action is much slower than xylene or toluene. Even a 1-cm thick block may be processed. But even traces of chloroform in the tissues after embedding comes in the way of sectioning. It is not inflammable. When heat is applied, it releases the toxic gas phosgene.

Benzene   As far as the properties are concerned, it is similar to xylene but it is not preferred as a clearing agent in the laboratory because of its carcinogenic characteristics.

Carbon tetrachloride   It is almost similar to chloroform but it is more toxic than chloroform and releases phosgene on oxidation.

Carbon disulphide   It has an unpleasant odour but is rapid in action. As it is highly inflammable, it is not recommended.

Petrol   It has properties similar to xylene. It is not recommended because of the presence of various additives.

Amylacetate   Its odour is unpleasant and objectionable. It removes alcohol fairly rapidly but it is costlier than other clearing agents.

Methyl benzoate and methyl salicylate   The speed of action is moderate and they do not cause much damage to the tissue. They are more preferred for specialized museum techniques. Because of their odour, they are not recommended in the histophysiology laboratories.

Cedar wood oil   Though it causes less shrinkage and hardening, it is very slow in action because of its low volatility. Because of its low viscosity, cedar wood oil is preferred for tissue processing.

Clove oil   It is rarely used in the laboratory because of its cost. Both clove oil and cedar wood oil will tolerate small percentage of water in tissue.

1, 1, 1 trichloroethane (methyl chloroform)   It was actually proposed by Maxwell (1978) as a clearing agent. It is non-inflammable with fairly low viscosity. It causes little brittleness of tissue and is moderately a fair remover of dehydrant.

Citrus fruit oils   These oils are extracted from orange and lemon rinds and available as clearing agents. They are non-toxic and fairly good as clearing agents, but the problem is that they may leach out dyes if used to clear sections prior to mounting.

For satisfactory results, biodegradable clearing agents are available commercially (Wynnchuk, 1993; Andre *et al.*, 1994; Langman, 1995).

EMBEDDING

Embedding is the process whereby dehydrated and cleared material, after infiltration by molten paraffin wax at 58°C has to be entomed in solid paraffin at room temperature. The tissue is put in a rectangular cup containing molten wax which solidifies.

Different researchers adopt different methods of embedding. A number of receptacles are suggested to be used as moulds. The most commonly used is 'L' piece usually used for routine purposes. Paper boats are also used. They are time-consuming, and excess wax has to be trimmed away before sectioning. Embryo cups, petri dishes, watch glasses, shallow dishes having sloping sides are all suitable as moulds.

Orientation of the tissue is very important. Tissues of tubular nature are cut in a transverse plane, skin in a plane right angles to the surface of the tissue, muscle either transversely or longitudinally. It is advisable to mark the tissue before commencing processing.

When the wax has solidified, remove the mould. If several specimens or tissues are embedded in one large mould they are separated by breaking.

The surface of the tissue towards the mould base is the one from which sections are to be cut. This surface is nicely trimmed with a scalpel. The opposite face is attached to a wooden or metal block. This face is evenly trimmed so that it is parallel to the first. Even the side face of the block should be parallel to the knife edge and they are also trimmed leaving at least 2 or 3 mm of wax between the edge and the tissue.

Attachment of wax to the wooden or metal block is carried out in the following manner. Wooden or metal blocks should have their upper surface previously impregnated with wax. While attaching the wax block to the wooden block, both are heated simultaneously by means of the spatula and pressed with molten wax and the edge is sealed.

Paraffin Wax

Although several media are available causing least damage to the tissue, paraffin finds a popular place in the laboratory due to the ease with which large number of tissue blocks are processed in a short time. Sectioning can be done with ease and later staining presents few difficulties. Above all, paraffin wax is cheaper than other media.

It is a mixture of hydrocarbons obtained as a by-product in the cracking of mineral oil. The melting point varies (40–70°C) and it influences the histological properties of the tissue. There is hard as well as soft wax and a suitable wax is chosen to produce ribbons consistently. In India, wax with a melting point of 56–58°C is most suitable and most commonly used. If wax is melted for prolonged periods there is likelihood of the tissue being cooked. In addition to this, when wax is heated for prolonged periods much above the melting point, some changes take place due to oxidation resulting in yellowish soapy wax. The crystallinity is lost. To produce easier sections, ribboning is achieved by using paraffin wax additives.

Paraffin Wax Additives

To modify the consistency of wax, a number of substances can be added to paraffin wax. These additives primarily increase the hardness so that thin sections can be produced, and secondly they alter the crystalline structure of wax, to give additional support in sectioning hard tissues. Of all these changes, the most significant is the reduction in crystalline size, perhaps one per cent. If it exceeds this it will result in undesirable side effects such as excessively high melting point or greatly increasing viscosity. If melting point of embedding medium has to be altered, large amounts of additives have to be used. Additives which have been commonly used are beeswax, ceresin, rubber, bay berry wax, stearic acid and diethylene glycol distearate.

Microcrystalline wax is obtained from petroleum distillation and it has very fine microcrystalline structure with a very high melting point (62–88°C) unlike paraffin wax. Up to 5 per cent of this could be added to paraffin wax which gives a fine-structured hard medium. Anything above this leads to high melting point which in turn hardens processed tissue.

A number of paraffin wax embedding media are commercially available. These contain resins. The following are the few formulae.

Paraffin wax (melting point 58°C)	70 parts
Rubber	5 parts
Beeswax	5 parts
Spermaceli	5 parts
Nevillite '5'	15 parts

The melting point of this medium is 50°C. If greater hardness is required, the resin concentration may be increased. There are two varieties of waxes, soft and hard. Melting point of soft paraffin is 50–52°C or 53–55°C and that of paraffin hard is 56–58°C or 60–62°C. Soft paraffin is used for soft tissues and vice versa. Generally the tissue may not lose its structure. Sometimes room temperatures play a role. Summer temperatures prefer hard wax. To obtain good results, paraffin with a melting point of 56–58°C is recommended.

Metal jugs containing paraffin wax are stored in the thermostatically controlled ovens. Nowadays paraffin wax of high standard is commercially available, there is little need to filter the wax before use. If wax clippings are to be reused, they should be filtered through a funnel with No. 1 filter paper which is permanently kept in the wax oven.

Precautions

1. Wax to be used must be free from dust and there should not be any trace of clearing agent.
2. Immediately after tissue embedding, the wax is suddenly plunged into water to reduce the wax crystal size.

PROCESSING (DIOXANE METHOD)

Tissues after fixation are transferred to dioxane. In cases where tissues are fixed in chromate solutions, washing is required. Fused calcium chloride or oxide is kept at the base of the container, and to avoid direct contact with the tissue, a metal gauze is used. Small pieces of tissue are dehydrated for few a hours and larger pieces overnight. Three changes of fluid are required. Dehydration is slower than with ethanol. After dehydration, the tissue is placed in an intermediate bath of equal parts of dioxane and paraffin wax. This is followed by a change of paraffin wax. One problem in using dioxane as dehydrant is that the tissue tends to fall away from the wax ribbon. If dehydration is complete, this should not happen. Cellosolve is less toxic and less volatile than dioxane. It is slow in action and it is better suitable with ester wax processing (Steedman, 1960) than paraffin wax processing.

Mixtures of oil of wintergreen (methyl salycilate), aniline, creosote or toluene also candidates as clearing agents. The volume of the clearing agent should be 50–100 times that of bulk of the tissue. Tissues from alcohols are blotted briefly on filter paper to remove alcohol and then transferred to clearing agent.

Using a clearing agent is very important and necessary because it clears the opacity from dehydrated tissues making them transparent. If tissue is opaque, it is difficult to obtain clear-cut sections. Therefore, the process of clearing has to be repeated.

VACUUM IMPREGNATION

In this technique when once the tissue is completely cleared, it is transferred to a container with molten wax and suction is applied to the container. While applying suction the tissue should not be damaged, so the degree of vacuum should not exceed 400–500 mm of mercury. Vacuum impregnation removes air bubbles in the tissue if any, especially in lung tissue. It also removes the clearing agent rapidly. Vacuum impregnation hastens the procedure to one-third. Vacuum should not be applied immediately after the tissue is transferred to molten wax. Few minutes are allowed. Vacuum impregnation is recommended for lung tissue, muscle, spleen, skin, and decalcified bone. The overall processing of tissue is schematically represented in Figure 2.1.

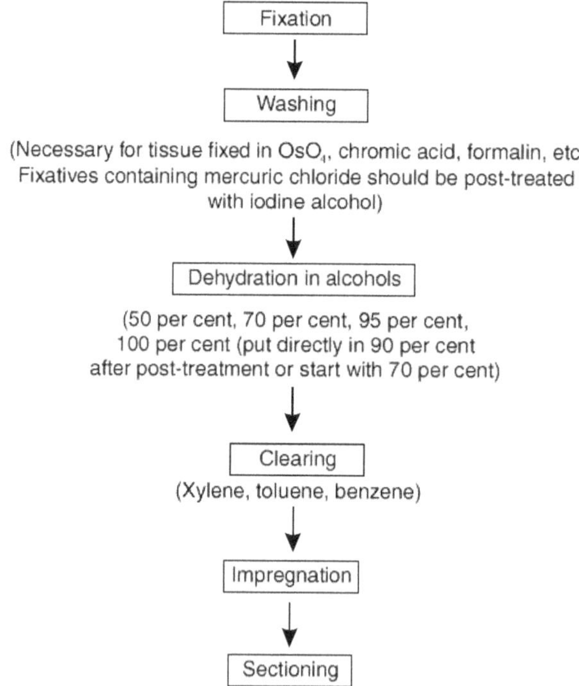

Figure 2.1   Schedule for dehydration, clearing and embedding

AUTOMATED TISSUE PROCESSING

Of late, most of the histopathology laboratories have started using machines to process tissue. The results obtained are far superior to manual tissue processing. The advantage with these machines is that they have the facility to use vacuum and heating at all stages and this will speed up the processing. But if high temperatures are maintained, the delicate tissues like spleen, muscle, skin and decalcified tissues become hardened. Temperature during fixation, dehydration and clearing should be at 37°C.

If the processing speed is to be increased, the following points have to be adopted.

1. The fixative should be warm (40–50°C) which completes fixation thoroughly.

2. Start dehydration from 95 per cent alcohol.
3. Always, a fast clearing agent like xylene is to be used.
4. Vacuum infiltration is to be used at all wax stages.
5. Agitation is to be done at all steps including fixation.
6. Maintain minimum of time at all stages.

Fluid replacement depends upon the size of the tissue processed. Any odour of clearing agent indicates that a change is necessary.

Fluid and wax containers must be filled to appropriate level and located in the machine. Fluid spillage, wax from beaker covers, lids and surrounding areas should be wiped off. Thermostats should be at satisfactory level gradually 3 degrees above the melting point of paraffin wax. Paraffin wax baths are checked now and then to see that wax is molten.

MANUAL TISSUE PROCESSING

The advantage of manual processing is its flexibility so that in each fluid, the tissues can be treated for the optimum time. For any tissue, the optimum processing schedule is one in which complete impregnation of the tissue is effected in minimum time without damage or distortion to the tissues. This will minimize the sectioning difficulties, and final staining. We cannot draw a rigid schedule for processing a tissue block. Experience will teach us to use minimum time. Too many variables are present, so experience alone will set right all problems of processing. Some of the fluids (for example acetone is inflammable, volatile and in some cases the reagent is expensive) are not recommended for automated techniques.

While processing a single specimen or block, the tissue is kept in a small container (screw-capped). It solves the space problem and is economical with fluids. If a number of specimens are to be processed, a permanent arrangement for space is made with big containers.

In each container, the volume of the fluid should be 3-fold. Sometimes fixation time may be reduced if formalin is used by warming it but there is every chance of the tissue being distorted. If fixatives other than formalin are used, appropriate post-treatment methods are followed.

The duration of time in each grade of alcohol depends not only on the size of the tissue but also on the type of tissue, for example, soft tissues like liver, pancreas, hepatopancreas, lungs, brain, ovary, testes, kidneys do not require longer time in each grade, whereas stomach, intestine, rectum, heart muscles, skin, cartilage and so on, require longer duration in each grade. Big blocks of tissue say whole brain of frog or bird (though brain is soft) require longer duration.

While transferring material from one grade to another, forceps are used for bigger tissues, and a camel hair brush for small tissues. Blot the tissue on filter paper to ensure that all alcohol of previous grade is absorbed and then transfer it to the next grade. Keep for longer time in 100 per cent alcohol say for 24 hours or for two or three days if it is a bigger block. Slowly transfer the block to xylene. First put it on a filter paper, blot it and later transfer it to xylene. Usually as soon as tissue from 100 per cent alcohol is directly transferred to xylene, it becomes turbid first and turns to amber colour. To avoid this (from our experience) have a dish with 50 ml of alcohol and 50 ml of xylene and transfer the block first to this solution and later to xylene. Keep in xylene till the material is transparent. Time schedule depends on the size and type of tissue. If the material has not cleared properly leave for some more time. Now the material is ready for impregnation. Transfer to molten wax. Time schedule again varies from size to size and tissue to tissue. One example is the skin of fish or any vertebrates requires longer periods say 24 hours or even 48 hours, whereas hepatopancreas of a crab requires only 20 minutes impregnation. Most soft parts as mentioned above require short duration for impregnation. If soft parts are impregnated for longer period, the tissue gets cooked up.

Alternate schedule for isopropyl alcohol   Previous steps up to 95 per cent are similar.

1.   Transfer to isopropyl alcohol for 1 hour (2 changes of 1 hour each).
2.   Transfer to toluene for 1 hour (2 changes of 1 hour each).
3.   Transfer to molten wax for 1 hour (2 changes of 1 hour each).
4.   While passing the tissue from one grade to another gently blot the tissue on filter paper.

If the tissue is well-hardened, it is not necessary to dehydrate through a series of graded alcohols (50 per cent and 70 per cent can be avoided). You can directly put in isopropyl alcohol.

Paraffin Processing at 4°C

1.   Select a tissue block of the size of 1.0 × 1.5 × 0.5 cm.
2.   Place in pre-cold acetone at 4°C (overnight).
3.   Transfer to petroleum ether at 4°C for 1 hour (2 changes of 1 hour each).
4.   Transfer to petroleum ether at 20°C for 1 hour.
5.   Place in 40–45°C paraffin wax at 42°C for 15 minutes.
6.   Embed in 40–45°C wax.

This method causes shrinkage but some of the hydrolytic enzymes could be demonstrated.

Paraffin Processing Double Embedding at 4°C (Gomori, 1952a)

1.   Select tissue and trim block to 1.0 × 0.5 cm.
2.   Place tissue blocks in pre-cooled acetone at 4°C for 18 hours.
3.   Transfer to fresh acetone at 4°C for 6 hours.
4.   Dehydrate in acetone at 4°C for 2 hours.
5.   Place tissue blocks in 2 per cent celloidin (in alcohol for ether, 1:1).

6. Place tissue blocks in 4°C for 12 hours.

7. Blot blocks to remove excess celloidin.

8. Transfer blocks to chloroform (1) at 20°C for 1 hour.

9. Transfer blocks to chloroform (2) at 20°C for 2 hours.

10. Embed in paraffin wax.

Some hydrolytic enzymes could be demonstrated. This method causes less shrinkage.

SECTIONING

Now the embedded blocks are ready for sectioning. The blocks can be trimmed into squares. Trim the side edges. Use metal block holders or wooden block holders.

Metal and wooden block holders are first coated with layers of paraffin. Keep the block containing the tissue on the block holder with a spatula. Melt the wax and pour along the sides of the block along the base. Ensure that the block firmly sits on the block holders without tilting. It is always better to pour excess so that the block does not give way while sectioning. After this plunge the block holder into a bowl containing water.

Sometimes blocks can be kept in ice water at least for 15–20 minutes. Palmer and McDonald (1963) designed a method where both microtome knife and blocks are kept cold throughout sectioning.

Adjust the microtome for section thickness (5–6 mm). Raise the tissue carrier and place the block holder with its mounted tissue block, tighten the clamp of the tissue carrier onto the stem of the block holder.

Insert the microtome knife and tighten its clamps. The knife should be at a proper angle so that the sections adhere to each other serially. This requires some practice and trial and error are editable. The knife should be tilted just enough so that the cutting facet next to the block clears the surface of the block. Tilt must be sufficient so that the surface of the block is not pressed down with the wedging effect of the facet. If such a thing happens for the next stroke of knife, a thick section is cut. If the tilt is great, the edge of the knife acts like a chisel so that a thick section is cut. Now adjust the knife in such a way that the face of the tissue block barely touches the cutting edge of the knife. Start trimming the block until the desired area is reached. Fragments of paraffin can be flicked off with camel hair brush. Scratches appearing in sections can be remedied by rubbing the knife edge with finger tip upwards or clean the tissue by dipping in xylene. When the block has been completely sectioned, use xylene to clean the knife.

Sectioning can be done in such a way that the sections move down in the form a ribbon (serial sections), each section adhering to the preceding as well as following sections. As the ribbon forms, hold it with a wet camel hairbrush away from the knife. Now place the ribbon on a sheet of black paper. Arrange the ribbons of sections from left to right and top to bottom (like a printed page). Cut the ribbon into convenient pieces and place on albuminized slides with few drops of water. Serially number each slide with glass marking pencil. This system will enable to sort out sections.

Stretching the sections in an art. Stretch as much as possible. Drain off excess water and dry the slide in an oven adjusted to 37°C. Most of the processes involved lie more or less in the realm of art although everyone must obtain sufficient practice. Some of difficulties encountered while sectioning and the remedial measures are listed in Table 2.1.

Table 2.1   Difficulties encountered while sectioning and remedies

Error	Remedy
Ribbon curves either upwards instead of straight	Upper or lower surface of the block to be trimmed so as to be parallel.
Sections are cut but do not form ribbon	Paraffin is hard so keep a layer of soft wax on the upper surface and cut sections rapidly.
Crumpled sections	Paraffin is soft. Keep the block in a freezing chamber for 30 minutes.
Scratch in sections	Clean knife edge to remove waste wax.
Thin and thick sections alternate	Knife tilts; tighten knife and object holder.
Section of unequal thickness	Again tilt of knife is great. Reduce tilt.

Properly mounted sections will have a smooth clear appearance. Sometimes there is a tendency for sections to loosen from slides. Certain types of tissue require special treatment (yolky tissue, brittle tissues). After drying, the tissue is ready for staining. Deparaffinize either in xylene or toluene. Now place the slide for one or two minutes in ethyl alcohol followed by 1–2 minutes in a dilute solution of (0.5 per cent) nitrocellulose in ether absolute alcohol (50:50). Drain off extra nitrocellulose, wave the slide in air and place in 70 per cent or 80 per cent alcohol and proceed for staining.

Paraffin sections may be used for the demonstration of proteins, carbohydrates, nucleoproteins, pigments and phospholipids.

Proteins   Though freeze-drying is ideal, it is possible to demonstrate structural proteins in routine paraffin slide style. Tissues are fixed either in formalin or acetic ethanol. Dehydration and clearing procedures are kept to a minimum.

Nucleoproteins   Carnoy's fluid is used as a fixative. Fix for 4 hours and after washing, transfer the tissue to 70 per cent alcohol. This is the best fixative. Nucleoproteins can be demonstrated by routine paraffin method.

Carbohydrates   Best results can be achieved by the freeze-drying technique. But at the same time, routine paraffin sections also yield satisfactory results and carbohydrate could be easily demonstrated. Bouin's or alcoholic Bouin's could be used as fixatives.

Pigments   Routinely processed paraffin material gives good results for pigments.

Lipids   Paraffin sections are suitable only for phospholipids and not for other lipids. It is always preferable to use frozen sections.

Although paraffin wax is the choice embedding medium and has been in use since a long time in many laboratories and suited to automation, under certain circumstances it is not the technique of choice. Due to the following factors alternative media are used.

1. To some tissues the paraffin wax medium does not give proper support to the tissue and is not sufficient.
2. Tissues at 58°C may get affected.
3. Some dehydrating and clearing agents may distort the complete morphology of the tissue.
4. Due to inadequate adhesion between the wax and tissue, it may fall off at the time of sectioning.
5. Sometimes amorphous medium is necessary for certain tissues instead of paraffin wax with crystalline structure.
6. When thin sections are required.

Even with additives, paraffin wax is not suitable in many instances, so alternatives are suggested. Some of the alternatives are listed below.

Celloidin   It is a slightly yellowish amorphous substance which is insoluble in water. It is also known as celloidion and parlodion. It can be dissolved in 50 : 50 diethyl ether or methyl benzoate. It is mostly recommended for the central nervous system.

Agar   Agar gel works very well when it is used in combination with other bases. When employed alone, it does not give sufficient support to the tissue. So its role in double embedding technique with ester wax and paraffin wax is satisfactory. It is good on the freezing microtome.

Gelatin   As its melting point is lower than agar, it does not find a place in double embedding. For whole organ sections in frozen sectioning, gelatin is satisfactory.

Resins   As an embedding medium, it finds a place in electron microscopy, for bone sections (undecalcified).

SPECIALIZED EMBEDDING TECHNIQUES

Celloidin/Paraffin Wax Double Embedding (Peterfi technique)

Celloidin embedding is advantageous in giving great support and adhesion to very hard tissue, fibrous tissues and decalcified tissue especially bone.

1. After fixation place in 70 per cent alcohol for 12 hours.
2. Transfer the tissue to 90 per cent alcohol for 4 hours.
3. Place in absolute alcohol for 4 hours.

4. Transfer to a solution containing 1 per cent celloidin in methyl benzoate for 24 hours giving 3 changes.
5. Transfer to toluene for 8 hours giving 3 changes.
6. Finally embed in paraffin wax.

An adhesive is necessary so that sections do not get detached while staining.

Agar/Ester Wax Double Embedding

For hard fibrous tissue, agar/ester wax is preferred over celloidin/paraffin wax. It is less time-consuming:

1. Fix the tissue.
2. Wash thoroughly in running water.
3. Place the tissue in 5 per cent agar solution at 60°C (agar powder is dissolved in water by autoclaving at 105°C for 30 minutes).
4. Embed tissue in 5 per cent agar in normal moults (as in paraffin wax embedding).
5. Trim off excess agar.
6. Dehydrate in ethylene–glycol–monoethyl ether.
7. Immerse in equal parts of alcohol–ethylene–glycol–monoethyl ether.
8. Place in methylene glycol–monoethyl ether–ester wax (equal parts).
9. Embedding of the agar blocks in ester wax is done in the usual manner.
10. Trim excess wax and keep the block ready for sectioning.

As in the previous case a section adhesive is necessary.

Agar/Paraffin Wax Double Impregnation

In this method instead of 5 per cent, 2 per cent agar is used. This is suggested for many whole animals or pieces of tissues in a block. After preparing a block in agar, the same procedure has to be adopted for dehydration and embedding in paraffin wax. For this method, automated processing is preferred.

Paraffin Wax/β-pinene Polymer Wax Double Embedding

To get 1–2 mm sections, a double embedding method is preferred. First impregnate in paraffin wax followed by blocking out in paraffin wax containing β-pinene polymers (Hammond and Beckman, 1978).

Ester Wax Embedding

Ester wax is harder than paraffin wax and has certain advantages over paraffin and celloidin (melting point is 46–48°C). Steedman (1947) first described this method. Cutting is similar to that of tissue in paraffin wax. 1–2 μ sections could be cut easily.

Tissue after clearing should not be transferred directly to ester wax. It should be kept in a mixture of clearing agent and ester wax for at least 3–6 hours.

Impregnation time varies with the size of the block and at least 3 changes of wax should be used before embedding and in each change, the tissue should be kept for at least 3–6 hours. Wax is heated to 68–70°C. Wax is poured into a mould as in the case of paraffin wax and the tissue block is placed and oriented in a correct plane. The mould is cooled in water rapidly. When the mould solidifies, the block is cut from the mould and trimmed. As the block is very hard, sections have to be cut with a sharp knife on a sledge-type microtome.

Sections are flattened as for paraffin embedding and should be dried at 37°C to avoid crumbling. Staining is the same as for paraffin embedding. Sometimes sections can be stained with the wax still present (Kiernan, 1996) but the staining time should be increased. Float the sections on staining reagents and later they can be fixed to albuminized slides.

Water-soluble Wax

Water-soluble waxes are polyethylene glycols. An advantage with this is that the tissue does not require dehydration and clearing before infiltration and added to this, the amount of shrinkage is much less than that caused by paraffin wax.

Water-soluble wax is more suitable for the demonstration of lipids and enzymes after freeze-drying.

Ester Wax

The technique for ester wax is as follows.

Ester wax is different from water wax.

1. Wash the tissue after fixation thoroughly to remove all traces of fixatives.
2. Leave in a mixture of polyethylene glycol 900 in distilled water for about 15 minutes.
3. Transfer to molten polyethylene glycol 900 giving 3 changes of at least 45 minutes in each bath at 28–30°C.
4. Place the tissue in a mixture of polyethylene glycol 900 and Nonex 6313 at 39°C for 40 minutes.
5. The next step is to keep the tissue in a mixture of three parts of Nonex 6313 and one part of polyethylene glycol 900 for 15 minutes.
6. Next transfer it to Nonex 6313 (39°C) giving 3 changes of 45 minutes in each change.
7. Embed in 'L' pieces (of convenient size).
8. Store the blocks in a desiccator.

Sections are cut in the usual manner.

Sections are floated on water to which a trace of soap or teepol or 10–20 per cent polyethylene glycol 900 is added. Wax-free sections thus floated may be floated on slides in the usual manner and stained.

Gelatin Embedding

1.   Fix the tissue in formol saline.
2.   Wash the tissue for 6–12 hours to remove the formalin.
3.   Transfer the tissue to a mixture of 10 per cent gelatin in 1 per cent phenol for 24 hours at 37°C.
4.   Place tissue in 20 per cent phenol gelatin for 12 hours at 37°C.
5.   Embed now the tissue in 20 per cent gelatin using a mould as for paraffin wax embedding.
6.   Trim excess gelatin as much as possible leaving a margin of 3 mm around the tissue and as little as possible on the surface.
7.   Immerse the trimmed block in 10 per cent formalin for 12–24 hours.
8.   Cut the frozen sections as thin as possible to avoid background staining by the gelatin.

Some workers prefer to avoid step 6. After fixing sections to the slide remove gelatin by using warm water. Float the sections on cold water and transfer sections immediately to albuminized slides. Slides are dried on the table, and gently heated to coagulate the albumin.

Celloidin Embedding

Celloidin embedding is required for exceptionally hard tissues. For this, a purified form of celloidin or nitrocellulose is used. Celloidin does not require heat at any stage of processing and its consistency is rubbery as a result of which it gives support to the tissue. Because of this consistency, even sections could be cut. For example when sections of eye are required, only celloidin embedding gives good results.

Celloidin embedding was very much preferred for sections of large pieces of tissue but now paraffin sections are thin and easier to process.

Some disadvantages of celloidin embedding are as follows.

1.   It is difficult to cut thin sections. Only sections that are 10 or 15 microns thick can be cut.
2.   It is difficult to prepare serial sections. This can be overcome by double embedding.
3.   It is a laborious process requiring several weeks.
4.   Blocks and sections should be stored in 70 per cent alcohol.

Preparation of solution   Celloidin supplied is celloidin wool fully damped in absolute alcohol. This wool is dissolved in a mixture of equal parts of ether and absolute alcohol.

Thick solution	8 per cent
Medium solution	4 per cent
Thin solution	2 per cent

Thick solution takes some time to dissolve, so it is better to prepare well before it is required. Solutions of celloidin should be kept in well-stoppered preferably ground-glass-stoppered bottles to prevent evaporation of ether alcohol and it should also be free from contamination. As ether is highly inflammable, care should be taken to keep it away from naked light.

Necol or Necoloidin is almost similar to celloidin and is cheaper. It is supplied as strong pyroxylin solution.

Thick solution   16 per cent
Medium solution   8 per cent
Thin solution   4 per cent

The solution usually supplied is 8 per cent. Dissolve in equal parts of ether and alcohol. It is evaporated to dryness in a fume cupboard and 16 per cent solution is obtained. Medium and thin solutions are prepared from stock (16 per cent).

Necol blocks are easy to cut.

Technique   Dehydration procedure is the same as for paraffin wax impregnation. After passing through successive grades of alcohol, the tissue is placed in a mixture of equal parts of absolute alcohol and ether for 24 hours.

The time schedules for 10-mm thick tissue are as follows:

Alcohol + ether	24 hours
2 per cent celloidin or 4 per cent necol	5–7 days
4 per cent celloidin or 8 per cent necol	5–7 days
8 per cent celloidin or 10 per cent necol	2–3 days

Impregnation requires quite a long time.

Tissue for impregnation is placed in a mould containing 8 per cent celloidin. This mould should be 1–1¼ inches deep so as to prevent the tissue being exposed when celloidin contracts. The moulds may be of the same model as used for paraffin wax. Sometimes 2″ depth petri dishes can be used for celloidin embedding. Celloidin hardens on evaporation. Prevent air bubbles. Hardening of celloidin may be hastened by the use of chloroform in a separate container under a bell jar. To avoid air bubbles on the cutting surface of the block, the block is placed in thick celloidin for 12 hours in an air-tight container. Hardening of celloidin takes a long time and it should be continued until the celloidin is of rubber consistency and when it is pressed with the thumb, there should not be any impression. When the celloidin is sufficiently hard trim the block leaving a margin of ¼ inch all round. The block is then fixed to a hard wooden block and the surface is roughened. This block is thin, hardened in 70 per cent alcohol for ½ an hour. Celloidin and necoloidin blocks should be stored in 70 per cent alcohol.

Nitrocellulose

In some laboratories, celloidin is being replaced by nitrocellulose due to its low viscosity and greater speed of impregnation. It has great water tolerance, and 95 per cent alcohol or ether serves as a solvent. A great disadvantage of this as an embedding medium is that there is a tendency for sections to crack while staining. This can be rectified or minimized by adding 1 per cent tricresyl phosphate or 1 per cent celloidin or 0.5% oleum ricin (castor oil) (Moore, 1951). Due to these reasons nitrocellulose did not gain popularity in laboratories. Another drawback with nitrocellulose is its highly explosive nature, nitrocellulose solutions used are 5, 10 and 20 per cent in a mixture of 5 parts of absolute alcohol to six parts of ether, with a little tricresyl phosphate, celloidin or castor oil as a plasticizer.

Time schedule is as follows.

5 per cent   3–5 days
10 per cent   2–4 days
20 per cent   1 day

This is for small pieces (7 mm). Larger pieces require increase in time of impregnation.

Dry Celloidin Method

Celloidin blocks are prepared as described previously but instead of fixing it to vulcanite block, place in Gilson's mixture, i.e., equal parts of cedar wood oil and chloroform. Cedar wood oil is added twice daily for a period of 10 days until the mixture is composed of 90 per cent cedar wood oil. Celloidin now becomes transparent and the block should be exposed to air. Now the block is fixed to a vulcanite block as described previously.

Double Embedding

Very fragile and very hard objects crumble when they are processed by the paraffin method. Double embedding technique remedies these troubles:

1. Fix and wash as small pieces.
2. Dehydrate in graded series of alcohols (2 hours each).
3. Keep in absolute alcohol for 18 hours.
4. Immerse in methyl benzoate–celloidin solution for 24 hours and pour off. Replace solution and keep for 48 hours. If tissue is not clear, repeat and keep for 72 hours.
5. Place in pure benzene giving 3 changes for 94 hours or in HISS for 12 hours.
6. Place in a mixture of equal parts of paraffin and benzene in an embedding oven for 1 hour.

7. Place in paraffin giving 2 changes for ½ to 6 hours, depending upon thickness and nature of tissues (3 hours for 5 mm thickness).

8. Embed as in paraffin method.

Methyl Benzoate–Celloidin Solution

Celloidin flakes (air-dried)   1 g
Methyl benzoate   100 ml

Allow to stand for 1 hour. Invert for an hour. Lay bottle aside for an hour and turn it upright. Repeat this several times.

Brigg's double embedding (1958)

1. Fix, wash and dehydrate, giving each process 2 hours duration.
2. Keep in absolute alcohol for 24 hours.
3. Transfer to 4 per cent celloidin (or nitrocellulose) for 48 hours.
4. Immerse in alcohol ether to remove excess celloidin. Harden in chloroform vapour for few minutes. Drop in chloroform and leave until tissue sinks for 12–24 hours.
5. Infiltrate in paraffin giving 2 changes of 12–24 hours each.
6. Embed.

Banny and Clark method (1949)

This is almost similar to Brigg's method. In step 3 place tissue in 2 changes of nitrocellulose 3 per cent and 6 per cent. After step 4, place in benzene overnight.

Dioxane in double embedding

1. Fix, decalcify and wash.
2. Place in dioxane for 2 hours.
3. Keep in dioxane–nitrocellulose for 3 days.

   Dioxane   70 ml
   2 per cent nitrocellulose   30 ml

4. Place in dioxane–paraffin (50 : 50) for 2 hours.
5. Add fresh paraffin 3 times (20 minutes, 30 minutes, 1 hour).
6. Embed and section immediately.

Crabb (1949) combined resin with celloidin and then double embed with paraffin. Salthouse (1958) double embeds with tetrahydrofuran for insects.

MOUNTING

If albumen fails as an adhesive agent Masson's gelatin fixative (*see* Chapter 5) can be used. Spread on slides, allow them to cool, blot excess gelatin and place in formalin vapour (overnight) when slides are ready for staining place the slides in chloroform, before you put in 95 per cent alcohol chloroform removes paraffin and hardens celloidin.

THEORY OF STAINING

3

The theory of staining is not properly understood. This phenomenon of staining arouses certain questions.

- When stains are applied why should a tissue component stain?
- Why should this stained tissue remain stained?
- Why should only certain parts of the tissues stain whereas other regions remain unstained?

While staining a section or a smear two independent variables are involved, i.e., solid tissues and cells and solutions of staining reagent. The main aim or idea of staining a tissue is to impart a colour so that the tissue constituents are well in contrast to their original colour. Staining methods mostly depend on selective uptake of reagents into the tissues and selective losses of products.

Most often the tissue component is not stained (stain does not penetrate into the tissue) but it is surrounded with stain. For this examples range from visualising individual macromolecules for electron microscopy to the demonstration of the canaliculi of bone matrix.

Sometimes staining reagents can be taken into the cells in various ways which in turn depends upon the physiological activities of the organism. This condition is called vital staining or supravital staining which may be either within the body (*in vivo*) or in preparation (*in vitro*). For treating neuronal pathways, this method was applied by Aschoff *et al.* (1982).

Uptake of stain by tissue is mainly due to the affinity between the dye and tissue or reagent and tissue. There is much to be said about the term affinity. When we say that a particular tissue has a high affinity for certain dyes, it means that under specifications of use, the tissue component gets intensely stained. The term affinity is also used to describe the binding of the dye to the tissue. So affinity is considered as a measure of the tendency of dyestuff to transfer from a dye bath onto a section and its magnitude of the affinity depends on every factor, i.e., solvent–solvent, dye–solvent, dye–dye interactions and finally the interaction between dye and tissue.

Though the theory of staining is not completely understood, several theories have been put forward.

PHYSICAL THEORIES OF STAINING

These theories depend upon (i) simple solubility, and (ii) absorption.

Examples of solubility method are fat stains, because these stains are more soluble in fat than in 70 per cent alcohol or other solvents in which it may be dissolved. Baylis (1906) developed this theory and called it electrical theory. The principle behind is that a large body attracts to itself minute particles from the surrounding medium.

CHEMICAL THEORIES OF STAINING

According to this, a radical of the dye binds to the pertinent chemical substance of the tissue. Acid dyes stain basic elements (cytoplasm) and basic dyes stain acidophil material (nucleus). For instance haematoxylin is an acid dye and it does not stain the cytoplasm but stains nuclei. It is the most widely used nuclear stain. If the staining is a purely chemical reaction, it would be expected to continue until one of the reagents were exhausted. Yet no matter how dilute a stain is used the solution is never completely decolorized by the tissue (Conn, 1946).

Dyes are classified as acid and basic dyes and they combine or react with acid or basic tissue components. This reaction can be modified by altering the pH of the solution in which they are employed. Their ability to bind each stain is due to the fact that they have the capacity to ionize as acids or bases depending upon the medium immediately surrounding them. So an acid dye will stain with acid dye if the dye solvent is acid and vice versa.

pH of the dye bath is an important factor especially in the differentiation of COOH and sulphate groups. When the pH of Alcian blue is below 2.5, the ionization of the COOH groups is suppressed and at pH 1.0 only sulphate groups are stained and carboxyl groups are not ionized and do not stain, whereas sulphate groups are demonstrated. At pH 2.5, COOH groups stain while sulphated groups take a very light shade. This indicates that staining is functionally chemical in nature.

TYPES OF STAINING

In 1714 Leeuwenhoek first used the dye as a biological stain. He applied safranin to muscles. Later in 1849 Goppert and Cohn used carmine. In 1863, Waldeyes used haematoxylin.

Subsequently in 1856 a number of aniline dyes were introduced in textile industry. In 1891 Heidenhain introduced haematoxylin technique and it is a popular stain till date. Biological stains and commercial dyes are more or less the same. But the techniques of staining are many.

An element can be considered to be stained when it acquires colour when treated with one reagent or more. The stained element is transparent.

Vital Staining

It is the staining of structures in living condition. This staining may be either in the body (*in vivo*) or in preparation (*in vitro*). Janus green stains mitochondria in living cells.

Phagocytic cells from colloidal suspension take up particles of coloured matter. These particles do not demonstrate specific structures. This does not come under true staining.

Histochemical Reactions

With the help of these reactions the chemical composition of elements could be confirmed, the reaction sometimes as in Feulgen technique may take place between colourless fluid and tissue.

Fat Stains

Sudan-III is a stain for neutral fats. Here the colouring agent is more soluble in the element to be demonstrated than in the solvent in which it is applied.

Histological Stains

These stains appear to be specific for a particular tissue element or group of elements. They stain either killed or non-living tissue elements. The mechanism of staining is not clearly understood.

DYE CHEMISTRY

In earlier days dyes employed in biology were mostly known as aniline dyes though they do not have any connection with aniline. These days a bulk of dyes are called aniline dyes. Some of them are derivatives of coal tar though they are not termed as coal tar dyes.

These coal tar dyes are derivatives of benzene. Let us see the structure of benzene to enable a clearer understanding of the dye chemistry.

Benzene molecule has 6 carbon atoms and 6 hydrogen atoms. The arrangement is such that carbons form a ring, and each carbon atom has a hydrogen atom attached to a free valency. As a carbon atom is considered to have four valency bonds, in the hexagon arrangement the carbon atom has a double bond on one side and a single bond on the other (alternate single and double bonds). When two O_2 bonds replace the hydrogen bonds, a readjustment of the double bonds takes place resulting in the formation of a new compound, quinone. This change from benzene to quinone is very important because benzene derivatives absorb light only in the ultraviolet band and are colourless whereas quinone compounds have absorption bands in the visible spectrum and are coloured.

Chromophores

The characteristic feature of these is to absorb certain wavelengths of light and confer colour. Chromphoric group is a linear series of atoms along which there is an alternation of double and single bonds. This linear sequence may be single or branched or a closed ring. This type of sequence is called a resonance system. Quinoid arrangement is a chromophore, the nitro group is also a chromophore and in this when three nitro groups replace three hydrogen bonds in a benzene molecule a yellow trinitrobenzene is formed. It is a chromogen containing chromophoric group. These chromogens are not dyes since they have no affinity to cloth fibres or tissue components.

Auxochromes

A dye compound in addition to containing chromophoric group must possess another group which has a salt-forming property. These are auxochrome. If to the trinitrobenzene a hydroxyl group is added (OH), which is an auxochrome, the chromogen is converted to a dye picric acid. This is capable of forming a salt and has affinity for fibres and certain tissue components. Some auxochromic groups such as hydroxyl group are acidic and some are basic such as aminogroups (NH_2) and some chromophoric groups.

Sulphonic Group (SO_3H)

This is a very important group in dye chemistry because it forms acid salts. With its help, an insoluble dye is made soluble in water or a basic dye is converted into acidic dye. Basic fuchsin is converted to acid fuchsin by the introduction of a sulphonic group.

Leuco Compounds

Leuco compound is a colourless compound and this is very important.

Certain dyes become colourless after removing acid radical for example triphenyl methane (basic fuchsin) and the xanthenes (phenolphthalein) and this reaction is reversible.

CLASSIFICATION OF DYES

Classification of dyes is based not on colours but on their chromophoric groups. [For this, consult Conn (1969).]

Natural Dyes

Majority of the dyes used are synthetic dyes but there are a few natural dyes namely carmine (cochineal), orcein, litmus and haematoxylin.

Carmine   It is a favourite dye in the laboratory and is produced from cochineal after treatment with alum. It is actually extracted from the female cochineal insect (*Coccus cacti*). As an individual dye it is not of great use. It should always be employed with a salt or a metal or a mordant. It is slightly soluble in water (neutral pH) and is employed in acid or alkaline solution.

Orcein and litmus are extracted from lichens. Of these two, litmus is an indicator and orcein is an elastin stain.

Haematoxylin   It is the most popular stain which is used in many laboratories as a routine stain, a very good histological stain showing nuclei prominently. It is a constituent of the heart wood of the logwood tree *Haematoxylin compectuanum* which grows in Mexico. Haematoxylin should be ripened either by exposure to air for 6–8 weeks or by adding oxidizing agent such as sodium iodate, potassium permanganate or mercuric oxide. While ripening, Haematoxylin loses ($C_{16}H_{14}O_6$) two hydrogen atoms to become haematein ($C_{16}H_{12}O_6$) which has a quinoid arrangement in one of the rings.

When oxidation is continued for a longer period, haematein will form a colourless compound. Even ripened haematoxylin is rarely used alone since it has no affinity for tissue components. Prior to staining with haematoxylin tissue should be mordanted. These mordants are salts of metals. Some of the mordants used are aluminium, iron, tungsten, chromium, copper, iron, molybdenum and lead.

Acid, Basic and Neutral Dyes

All ordinary stains contain an acid and a base.

Acid dyes   In these, the acid component is coloured, and the basic is colourless. One is sodium which is colourless and the other is rose aniline trisulphuric acid which is coloured. These stain basic components like cytoplasm, e.g. acid fuchsin.

Basic dyes   The two components are the colourless acid component and coloured basic component. For example basic fuchsin is composed of coloured base rose aniline and a colourless acid radical. Basic dye usually stains acid component such as nuclei.

Neutral dyes   Both acid and basic radicals are coloured, it is a neutral stain and certain tissue components have no affinity for neutral stain. So the structures are termed neutrophilic.

The first neutral dye was obtained by mixing methylene blue and eosin for blood films. When both acid and basic coloured radicals are mixed a precipitate results. The precipitate is not soluble in water and gives best results when mixed with methyl alcohol. In modern laboratories only modified methylene blue and eosin will be combined and they are called azure eosin and azure methylene blue.

MORDANTS

There are certain dyes which will not stain tissue in a simple solution unless a third element is involved. The third element is called the mordant. This is a link between the stain and the tissue. So the combination of the dye and the mordant forms a compound which is termed as a lake which firmly gets attached to the tissue. Some lakes are unstable and some are stable.

Mordant is applicable to salts and hydroxides of divalent and trivalent metals. Most often the mordant used in histological stains are salts usually sulphates of aluminium, chromium and iron. Generally double sulphate or alums are used.

Accentuators

Accentuators are groups of substances which do not act as mordants forming lakes with the dyes or do they take past in chemical union or increase the staining power of the dyes which are capable of staining without the accentuators. Examples are potassium hydroxide, in Loeffler's methylene blue and phenol in carbol thionin and carbol fuchsin. Effect of staining is based on change in pH of the staining solution.

The term mordant is strictly applicable to salts and hydroxides of divalent and trivalent metals and they should not be used to indicate any substance that improves staining.

Metachromasia

Most of the metachromatic dyes stain tissues in different intensities of the same colour (original colour of the dye). For example acid fuchsin or basic fuchsin stain the tissue in various shades of red and light green in different shades of green only. Some tissue components on the other hand in the presence of certain basic dyes stain in a colour other than that of the dye. In this phenomenon, absorption of the dye at different wavelengths depends on its concentration and surroundings. For example toluidine blue when in low concentration transmits in the blue (orthochromatic) whereas in high concentration when it is absorbed into cartilage it transmits in the purple (metachromatic). These variations are due to dye–dye interactions, that is, when two or more dye molecules come into close proximity, there is interaction between their electronic energy levels resulting in a new set of spacing and new wavelength of light absorption. Orthochromatic colour results from unperturbed levels and metachromatic colour results from new energy levels produced by dye–dye interaction with toluidine blue mucins stain in a red shade and rest of the tissue stains in shades of blue. This is metachromasia and the stain is metachromatic stain. Mucin, cartilage and most cell granules exhibit metachromasia. Paddy (1970) gave the definition of metachromasia as follows "A characteristic reversible colour change that any dye may undergo by a virtue of a change in its environment not involving chemical reaction of the dye". According to Pearse (1968) in metachromasia staining of tissue component, the absorption spectrum of the resulting tissue dye complex differs sufficiently from that of the original dye and gives a marked contrast in colour from its ordinary tissue complex.

Thionin stain exists in two forms 1) In normal blue colour, and 2) metachromatic red colour. The red is a polymerized form of blue and the colour change is due to interaction between the chromophores present. With increased temperature, low pH, low concentration of the dye or addition of alcohol, acetone or salts, the blue colour is prominent, whereas the prominence of red is with decrease in temperature, increased pH, high concentration of the dye (Bergeron and Singer, 1958). Thionin stains chromatin blue, and the granules of mast cells, ground substance of cartilage and mucus red. There are three types of metachromasia, alpha with blue colour, beta with violet and gamma with red colour. Kramer and Windrum (1954) have shown that metachromasia could be induced in polysaccharides by sulphation, i.e., treatment with sulphuric acid for 30–60 seconds. The majority of dyes do not stain metachromatically but are orthochromatic. Orthochromatic dyes may give r00 to metachromatic impurities by allochromacy. For example Nile blue is allochromatic. It has cations of the dye (blue) anions of the dye (sulphate), Nile red (red or rose) and amino base of Nile blue (orange yellow).

Metachromatic Dyes

Some of the metachromatic dyes are azure A, B, C, methyl violet, toluidine blue, brilliant cresyl blue, methylene blue, crystal violet, safranin, bismarck brown, basic fuchsin.

A chromatrope is a substance that can alter the colour of a metachromatic dye. Examples of chromatropic tissue substances are the matrix of cartilage granules of mast cells and sections of some mucous glands.

Use of control sections in the identification of chemical structure or tissue elements By using chemical action or enzyme action, it is possible to specifically remove or modify chemical groups and/or tissue components. For example diastase removes or abolishes glycogen, ribonuclease removes ribonucleic acid, methylation esterifies carboxyl groups and removes sulphate groups, acetylation renders 1:2 glycols non-reactive to the PAS reaction.

If a section treated in this manner is stained by a routine stain in parallel with a similar but untreated control section, the absence of colour in the treated section will serve to identify it in the control section. However, it must be remembered that both sections should be treated as near identically as possible.

3

FURTHER READINGS

Baker (1958, 1966), Carson (1990), Chayen and Bitensky (1991), Clark and Allard (1983), Gabe (1976), Green (1990), Gurr (1969), Horobin (1982, 1988, 1990, 1996), Kiernan (1990), Lillie (1977), Lillie and Fullmer (1976), Lyon (1991) Pearse (1972, 1980, 1985), Pearse and Stoward (1991), Summer (1988) and Thomson (1966).

DECALCIFICATION

4

Some tissues especially bones in vertebrates, and molluscan shells, crustacean cuticle or exoskeleton, tissues with spicules, and small setae among invertebrates, all have excess calcium. In these, sectioning in good condition is difficult. Such tissues require decalcification. The size of the tissue should be kept to a minimum say 4–5 mm to ensure adequate fixation. Sometimes calcium deposits are sparse. For such tissues, overnight soaking in water usually softens the tissue. Heavy calcification definitely requires decalcification. After decalcification and washing, the cut surface should be trimmed nicely to remove damaged surfaces.

Before decalcifying the tissue, it should undergo routine fixation. Generally formalin is preferred but Zenker's formol is also recommended. There are a number of decalcifying agents with or without acid interference. Cook and Erzacolin (1962) are of the opinion that acid decalcification causes a lot of damage to the tissue. So it is better to avoid acid content. If necessary, very dilute acids may be used.

There are several ways of decalcifying the tissue. It may be effected by dilute mineral acids or chelating agents or by electrophoretic removal of calcium. Decalcification takes place at a rapid rate when it is heated (56–60°C) but heating has its own effect. It may cause unusual swelling.

While using decalcifying agents, it should be ensured that calcium is completely removed without causing damage to tissue, cells or fibres and it should not interfere with staining. While decalcifying the tissues, precautionary measures are to be taken not to expose for longer periods to decalcifying agents. The complete removal of calcium can be tested by one method. For a long time, radiography of the tissue was in use but this method cannot be followed with tissues fixed in fluids containing mercuric chloride. The following is the best way to test whether calcium is completely removed. 5–6 ml of decalcifying fluid is neutralized with N_2NaOH and then 1 ml of 5 per cent sodium or ammonium oxalate is added. If the fluid is still turbid, it means that calcium is not completely removed and decalcification is not complete. Absence of turbidity after five minutes indicates that the tissue is free from calcium.

Most often hydrochloric acid, acetic acid and nitric acid are used as decalcifying agents resulting in unsatisfactory histochemical results. Enzymes, nucleic acids and polysaccharides give unsatisfactory results. Various opinions are expressed regarding decalcification of tissue and decalcifying fluids.

ACID REAGENTS

When acids are used as decalcifying agents transfer the tissue directly to 70 per cent alcohol to prevent swelling.

Formic Acid A

Formic acid	5–25 ml
Formalin	5.0 ml
Distilled water	100.0 ml

If formic acid content is 5.0 ml, the tissue should be kept in the solution for 5 days. If it is 25 ml, less time is required.

Formic Acid B

50 per cent formic acid (50 ml/50 ml water)	50.0 ml
15 per cent sodium citrate	50.0 ml
Concentrated formalin	10.0 ml
Nitric acid (68.70 per cent)	5.0 ml
Distilled water	85.0 ml
60–70 per cent nitric acid	5.0 ml
Distilled water	950.0 ml

Kristensen Fluid (1948)

8N formic acid	50.0 ml
1N sodium formate	50.0 ml

Tissue is treated for 24 hours and later washed in running water.

Fixation time is 12–24 hours and it varies with the size of the tissue. It is washed in running water and the usual procedure followed.

If acid interferes with staining, treat with 2 per cent lithium carbonate or 5 per cent sodium sulphate for 6–12 hours and then transfer to 70 per cent alcohol.

Strong acids like nitric acid are found to alter histochemical behaviour. So care must be taken while using them (Schajowicz and Cabrini, 1955). According to Morris and Benton (1956) 1–2 M hydrochloric acid is a rapid decalcifier (3 hours). Staining becomes perfect provided the slides are pretreated with 5 per cent ammonium chloride for 3 hours before staining (i.e., mordanting).

Sodium citrate can be replaced by a cation-exchange resin which accelerates decalcification (Dotti *et al.*, 1951). The liberated calcium could be removed rapidly from the solution. For this, the author recommends 40 per cent resin in formic acid or 10 per cent resin.

CHELATING AGENTS FOR DECALCIFICATION

Chelating agents are compounds (organic) which have the power of binding certain metals. EDTA (ethylene diamine tetra acetic acid), sequestrene and versene (disodium salt) have the power of binding calcium. These maintain good fixation and sharp staining and these organic compounds have the power of binding certain metals like calcium iron. In this case decalcification is low.

Hilleman and Lee (1953) used 200 ml of 5.5 solution of either versene and sequestrene in 10 per cent formalin. The time required for complete decalcification is 3 weeks. The solution should be changed every week and after some time it is changed every day. The tissue can be directly transferred to 70 per cent alcohol.

Vacek and Plackova (1959) recommended 0.5 M solution of EDTA at pH 8.2–8.5 which yielded better results. Of versene and sequestrene, versene seems to be better at pH 7.0 with NaOH and HCl.

DECALCIFYING FLUIDS

Gooding and Stewart's Fluid

Formic acid	5–25 ml
Formalin	5.0 ml
Distilled water	100.0 ml

It is a good decalcifying fluid causing minimum damage to the tissue. Decalcification is complete in 2–4 days, depending on the thickness of the tissue.

Citrate Citric Acid Buffer (pH 4.5)

7 per cent citric acid (monohydrate)	5.0 ml
7.54 per cent ammonium citrate (anhydrous)	95.0 ml
1 percent zinc sulphate	0.2 ml
Chloroform	A few drops

Calcium ions are soluble at pH 4.5, so buffer solutions are used to decalcify tissues.

Lillie's Fluid (1954a)

2 per cent picric acid (2 g/100 ml water)	85.0 ml
Concentrated formaldehyde	10.0 ml
Formic acid (90–95 per cent) (95 ml/5 ml water)	5.0 ml

The tissue is kept for one or two days in fluid and for 2–3 days in 70 per cent alcohol till yellow colour disappears.

Alternate Method (Lillie, 1954a)

Add 5 per cent of 90 per cent formic acid to Zenker's fixative.

McNamara *et al.* Fluid (1940)

Solution 1

Mercuric chloride	10.0 g
Distilled water	300.0 ml

Heat, cool and add.

Solution 2

Trichloroacetic acid	30.0 g
Distilled water	300.0 ml

Dissolve and then add.

Concentrated nitric acid	5.0 ml
95 per cent ethyl alcohol	50.0 ml
Concentrated formalin	40.0 ml

Mix solution 1 and 2 and keep for a week. Every day the solution should be changed. The tissue is washed overnight.

Perenyi's Fluid (see Chapter 1)

It is used as a fixative. It does not have hardening effect and is good for calcified tissues especially arteries and glands. The fixation time is 12–14 hours. The tissue is washed in 50–70 per cent alcohol.

Schmidt's Fluid (1956)

4 per cent formalin (4 ml/96 ml water)	100.0 ml
+ 1 g sodium acetate	
Disodium versenate (EDTA)	10.0 g

Fix for 24–48 hours. Transfer directly to 70 per cent alcohol. No washing is required.

Jenkin's Fluid (Culling, 1957)

Absolute alcohol	73.0 ml
Distilled water	10.0 ml
Chloroform	10.0 ml
Glacial acetic acid	3.0 ml
Hydrochloric acid	4.0 ml

The volume of the fluid should be 40–50 times the bulk of the tissue. After decalcification, transfer to absolute alcohol, fixation 16–48 hours. Then post-fix in neutral formalin for 48 hours. Avoid prolonged treatment with decalcifiers. Jenkin's fluid not only decalcifies but also dehydrates. Large amounts of this fluid should be used.

DECALCIFICATION FOR ENZYMES

This method (Lorch, 1946) is particularly good for alkaline phosphatase.

1. Fix in absolute acetone at 4°C and give 2 changes.
2. Decalcify in Lorch's fluid for 3–10 days.

Citric acid	14.7 g
0.2N NaOH	700.0 ml
0.1N hydrochloric acid	300.0 ml
1 per cent zinc sulphate	2.0 ml
Chloroform	0.1 ml

3. Wash in distilled water.
4. Reactivate the enzyme by treating with 0.1 M vernol acetate buffer (pH 9.0) for 6–8 hours.
5. Dehydrate and clear in chloroform.
6. Embed *in vacuo* in 48° wax.

DECALCIFICATION OF MOLLUSCAN SHELL

The shell is first cleared of foreign materials adhering to it and thoroughly washed with distilled water. Then the shell is fixed in 5 per cent formaldehyde overnight and decalcified with 2 per cent acetic acid. If the shell is fragile, only 1 per cent acetic acid is used but for heavy shells, higher concentrations are used. After the decalcification process, the material is again treated with 5 per cent formaldehyde for 5 hours and washed in running water overnight. Then it is transferred to 70 per cent alcohol and dehydrated.

From practice, it has been found that gastropod shells with or without animal softened when put in Susa for longer periods (more than 20 days), and no further decalcification was necessary for sectioning. No cellular or structural or functional damage was noticed though the animals were in Susa for one month.

DECALCIFICATION WITH ENZYME PRESERVATION

This method was given by Greep *et al.*, 1948, and includes the following steps.

1. Fix small pieces in absolute alcohol at 4°C for 24 hours.
2. Decalcify in a solution containing 2 per cent formic acid and 20 per cent sodium citrate for 20 days. Change solution every three days.
3. Rinse in tap water.
4. Reactivate by immersion in 1 per cent sodium diethyl barbiturate for 24–48 hours.
5. Rinse in tap water.
6. Dehydrate, clear in benzene, embed *in vacuo* at 48°C.

Some important works on preservation of acid and alkaline phosphatases are those of Greep *et al.* (1948), Zorzoli (1948), Lillie *et al.* (1951), Majno and Rouiller (1951) and Schajowicz and Cabrini (1954, 1958).

Takada *et al.* (1960), Mori *et al.* (1962), Yoshiki (1962) and Balogh (1965) made observations on acid phosphatase.

Strong acids such as nitric acid are likely to alter the histochemical behaviour of the tissue. Formic acid does not have this disadvantage. Morris and Benton (1956) found that 1–2N hydrochloric acid is a rapid decalcifier (approximately 3 hours). But better results could be obtained by mordanting in 5 per cent aqueous ammonium chloride for 30 minutes before staining. Formic acid penetrates with a fairly good speed causing minimum damage to tissue.

Culling (1949) has shown that paraffin sections may be stained before removal of the wax. Paraffin blocks must be placed with face downwards in 5 per cent hydrochloric acid but not for long. After sufficient exposure, the block may be cut.

ELECTROPHORETIC DECALCIFICATION

The principle of this method is based on the theory that there is attraction of calcium ions to a negative electrode. It is believed that there is considerable decrease in the length of time, and decalcification may be achieved by this method. Clayden (1952) demonstrated that similar decrease in the length of time may be achieved by increasing the temperature of the electrolyte and it is evident that rise in temperature was the principal reason for speed of decalcification. In their experiments with two pieces of bone, one was immersed in a bath containing equal parts of 8 per cent hydrochloric acid and 10 per cent formic acid at 40°C and the other was decalcified by a normal electrolyte method. Decalcification was completed almost at the same time in both the cases.

POST-FIXATION TREATMENTS

Chromatization

For tissues fixed in formol calcium, chromatization is necessary. The tissue was kept after washing in 2 per cent potassium dichromate for 24 hours at room temperature and another for 24 hours at 60°C. It is washed thoroughly in distilled water or running water (overnight for large specimens) for few hours.

Deformalization

Formalin should be completely removed especially in silver impregnation methods (Cajal and del Rio-Hortega).

Lhotka and Ferriera Method (1950)

1. Wash the tissue in distilled water for 15 minutes.
2. Keep in 20 per cent aqueous chlorolhydrate, giving two changes (24 hours each).

3. Wash in distilled water for 15 minutes.
4. Proceed to silver stains.

Krajian and Gradwhol Method (1952)

1. Keep in ammonia water (40 drops in 100 ml water).
2. Wash in running water.
3. Fix in special fixative for 1 hour.
4. Wash in running water.
5. Proceed for staining.

PREPARATION OF STAINS

5

SOME PRECAUTIONS TO BE TAKEN BEFORE STAINING

1. Stain preparation should be handled by a senior technician.
2. Inaccurate weighing, absence of cleanliness and using tap water should be totally avoided.
3. Glassware should be perfectly clean, rinsed in distilled water and dried.
4. Only distilled water should be used, unless otherwise specified.
5. Flasks and pipettes used for silver techniques should be marked.
6. Silver and osmic acid should be kept in amber-coloured bottles.
7. Ammonia should be kept in a refrigerator.
8. The stain should be prepared in the order given in the formula.
9. Alcoholic stains should be kept in tight-stoppered bottles.

SOME IMPORTANT DEFINITIONS

Molecular Solution

A molecular solution is the molecular weight in grams plus 1000 ml water, i.e., M oxalic acid (molecular weight) 126 g in 1 litre of water. Molecular weight is expressed in grams called the gram molecular weight or mole. A millimole is 1/1000 of a mole.

Normal Solution

This solution contains 1.0 g molecular weight of substance dissolved in 1 litre of solution.

Percentage Solution

Percentage is determined by weight or volume, i.e., w/v means weight in grams in a 100 volume solution v/v means volume in millilitres in 100 ml total volume of solution.

Preparation of any solution is done by dilution, taking for granted that the concentration of a reagent is 100 per cent, i.e., 1 per cent solution of acetic acid is 1 ml of glacial acetic acid in 99 ml of distilled water.

To make a "normal solution" add to the ml of the right-hand side column, required distilled water to make up to 1000 ml, i.e., NHCl will be 85.9 ml of HCl + 914.1 ml distilled water.

DILUTION OF 1N SOLUTIONS

	Molecular weight	Percentages assay	Specific gravity	Grams per litre	Millilitre per litre
Acetic acid CH_3COOH	60.05	99.7–100.0	1.0498	1050	57.2
		99.0	1.0524	1042	57.6
		98.0	1.0549	1034	58.0
Hydrochloric acid (HCl)	36.46	36	1.1789	424.4	85.9
			1.188–1.192	451.6	80.4
			1.1980	479.2	76.0
Nitric acid (HNO_3)	63.02	69	1.4091	972.3	64.8
		70	1.4134	989.4	63.6
		71	1.4176	100.6	62.6
		72	1.4218	102.4	61.5
Sulphuric acid (H_2SO_4)	98.075	95	1.8337	1742	28.1
		96	1.8355	1762	27.8
		97	1.8364	1781	27.5
Ammonium hydroxide (NH_4OH)	17.03	26	0.94040	235	72.4
		28	0.8980	251.4	67.7
		30	0.8920	267.6	63.63
Formic acid (HCOOH)	46.33	96	1.217	1180	39.3
		98	1.2183	1194	38.4
		99	1.2202	1208	38.6
		100	1.2212	1221	37.6

SOME PREPARATIONS

Acid Alcohol

70 per cent ethyl alcohol	100.0 ml
Conc. HCl	1.0 ml

Alkaline Alcohol

70 per cent ethyl alcohol   100.0 ml
Ammonia (concentrated)   1.0 ml

Carbol–xylol

Phenol (carbolic acid)   1 part
Xylol   3 parts

Gold Chloride Solution

Gold chloride   1.0 mg
Distilled water   100.0 ml

Lugol's Iodine

Concentrated Solution
Iodine crystals   1.0 mg
Potassium iodide   2.0 mg
Distilled water   12.0 ml

Weigert's Variation
Iodine   1.0 g
Potassium iodide   2.0 g
Distilled water   100.0 ml

Gram's Variation
Iodine   1.0 mg
Potassium iodide   2.0 mg
Distilled water   300.0 ml

In all these first dissolve potassium iodide and then add iodine crystals.

Physiological Saline–Sodium Chloride in Distilled Water

For mammals   0.9 per cent (9 g in 1000 ml water)
For birds   0.75 per cent (7.5 g in 1000 ml water)
For salamanders   0.80 per cent (8 g in 1000 ml water)
For frogs   0.64 per cent (6.4 g in 1000 ml water)

Earle's Solution

Sodium chloride   680 mg
Calcium chloride   20 mg
Magnesium sulphate   10 mg
Potassium chloride   40 mg
Sodium bicarbonate   14 mg
Sodium phosphate (monobasic)   220 mg
Glucose   100 mg
Distilled water   100.0 ml

Hank's Solution

Sodium chloride   800 mg
Calcium   20 mg

Magnesium sulphate	20 mg
Potassium chloride	40 mg
Potassium phosphate (monobasic)	10 mg
Sodium bicarbonate	127 mg
Sodium phosphate (dibasic)	10 mg
Glucose	200 mg
Distilled water	100.0 ml

Locke's Solution

Sodium chloride (for poikilotherms)	9.0 mg
Potassium chloride	42 mg
Sodium bicarbonate	30 mg
Calcium chloride	24 mg
Glucose	100 mg
Distilled water	100.0 ml

Ringer's Solution

Sodium chloride (for poikilotherms)	9.0 mg
Potassium chloride	42 mg
Calcium chloride	0.025 mg
Distilled water	100.0 ml

Artificial Sea Water (Hale, 1958)

$NaCl$	23.991 g
KCl	742 mg
$CaCl_2$	1.135 g
$MgCl_2$	5.102 g
Na_2SO_4	4.012 g
$NaHCO_3$	197 mg
$NaBr$	85 mg
$SrCl_2$	11 mg
H_3BO_3	27 mg

Dissolve in distilled water and make up to 1 litre. This solution is not advisable for aquaria.

Synthetic Spring Water (Kirby, 1947)

	Formula 1	**Formula 2**
Na_2SO_3	15.0 mg	100.0 mg
$NaCl$	12.0 mg	12.0 mg
Na_2SO_4	6.0 mg	6.0 mg
$CaCl_2$	6.5 mg	6.5 mg
$MgCl_2$	3.5 mg	3.5 mg
$FeCl_3$	4.0 mg	4.0 mg
Distilled water	1000 ml	1000 ml

Adjust with HCl to pH 6.8–7.0.

Dose formula 1 for first transfer from nature and for succeeding transfer use 2.

Chatton's Agar for Blocking Small Organisms (Gray, 1954)

Agar   1.3 g
Add to boiling water   100.0 ml

Stir until dissolved, then add,

40 per cent formaldehyde   2.5 ml
Store in refrigerator.

Sodium Thiosulphate (hypo)

Sodium thiosulphate   5.0 g
Distilled water   100.0 ml

Cleaning Solution (glassware)

Potassium dichromate 20.0 g
Water   200.0 ml

When dissolved add
    Concentrated sulphuric acid   20.0 ml
    Potassium dichromate   9 parts
    Sulphuric acid   1 part

For weak solution
    2 per cent $K_2Cr_2O_7$ 9 parts
    H_2SO_4 1 part

Subbed Slides (Boyd, 1955)

Dissolve 1 g of gelatin in 1 litre of hot distilled water, cool, then add 0.1 mg of chromium potassium sulphate. Store in refrigerator. Dip slides in this solution. Drain and dry.

Land's Adhesive (Gray, 1954)

Solution 1
    Gum Arabic   500 mg
    Distilled water 50.0 ml

Solution 2
    Potassium dichromate   500 mg
    Distilled water   100.0 ml

Smear a small amount of solution 1 on a slide. Flood slide with solution 2, place and flatten over slide heater. Drain excess water.

Egg Albumen

Beat white of egg with eggbeater and pour into a graduated jar. Allow to settle until suspended material settles on top. Pour off this and add equal volume of glycerine. A piece of thymol or sodium salicylate prevents growth of moulds.

Masson's Gelatin Fixative

Gelatin   50.0 mg
Distilled water 25.0 ml

Float sections on slide and place on warm plate. When the sections spread, remove excess gelatin, blot-dry and fix in formalin vapour.

Hupt's Gelatin Fixative (1930)

Gelatin   1.0 g
Distilled water   100.0 ml

Dissolve at 30°C in water bath and then add the following.

Phenol crystals   2.0 g
Glycerol   15.0 ml

Stir well and filter.

Weaver's Gelatin Fixative (1955)

Solution 1
    Gelatin   1.0 g
    Calcium propionate   1.0 g
    Roccal (1 per cent benzalkonium chloride)   1.0 ml
    Distilled water   100.0 ml

Solution 2
    Chrome alum   1.0 g
    Distilled water   90.0 ml
    Formalin   10.0 ml

Mix 1 part of solution 1 and 9 parts of solution 2. Flood slide with adhesive fixative and paraffin ribbons and allow to stretch. Drain off excess adhesive and blot.

Blood Serum Fixative (Preman, 1954)

Fresh human blood serum   15.0 ml
Distilled water   10.0 ml
5 per cent formalin   6.0 ml

Filter, and use as albumen fixative.

PREPARATION OF ONE LITRE OF A MOLAR SOLUTION

1. Calculate the gram molecular weight (using automatic weights) of the solute.
2. Weigh out one gram molecule of the solute.
3. Measure out 1000 ml of distilled water.
4. Dissolve the solute in a small quantity of the water.
5. Add the remaining distilled water to make the solution up to one litre.

For solutions of different molarity, e.g. 2 M and 0.1 M

2 M solution = 2 × gram molecular weight in 1000 ml water

0.1 M solution = 0.1 × gram molecular weight in 1000 ml water

For example,

Sodium β-glycerophosphate
Molecular weight of sodium β-glycerophosphate = 315.13

Molar solution (M) = 315.13 g in 1000 ml distilled water

0.1 M = 31.513 g in 1000 ml distilled water

The chemical is dissolved in a small quantity of the water and the volume is made up to 1 litre.

PREPARATION OF MOLAR SOLUTIONS OF LIQUIDS

1. Calculate the molar weight of the liquid.
2. Find the liquid's (a) valency, (b) specific gravity, (c) concentration.
3. Using the formula:

   Number of ml of liquid per 1000 ml of distilled water =

$$\frac{\text{Molecular weight}}{\text{Valency} \times \text{Specific gravity} \times \text{Concentration}}$$

   Substitute the values found in (2) to the formula to find out the volume of liquid required to make up a molar solution.
4. Measure out the required volume of liquid and make up to 1000 ml by adding distilled water.

   For example

   For a 2 M solution, the volume required for 1 M solution has to be doubled then made up to 1000 ml by adding distilled water.

   For a 0.1 M solution the volume required for a 1 M solution has to be divided by 10 and then made up to a litre with distilled water.

PREPARATION OF NORMAL SOLUTIONS

A normal solution contains one gram molecular weight of the substance, divided by the hydrogen equivalent of the substance.

For example:

1. A Sodium hydroxide (NaOH)

   Molar weight of NaOH = 40.0; valency = 1

   Normal solution = 40 g of NaOH in 1000 ml distilled water

   Divided by 10, = 4.0 g in 1000 ml distilled water

   Normality 2:
2. For liquids the following formula may be used. 1 ml of fluid per 1000 ml of distilled water.

   Molecular weight = Valency × Specific gravity × Concentration
3. Hydrochloric acid (HCl)

   Valency = 1, Sp. Gravity = 1.18; Conc. = 36 per cent, Mol. weight 36.4

$$\frac{36.4}{1 \times 1.8 \times 0.36} = 85.6\,\text{ml}$$

NHCl = 85.6 ml + 914.4 distilled water

4. Hydrogen sulphate (H_2SO_4)

Valency = 3, Sp. gravity = 1.835, Conc. = 98 per cent, Molecular weight = 98.08

$$\frac{98.08}{3 \times 1.835 \times 0.98} = 29\%\ \text{ml}$$

NH_2SO_4 = 29.1 ml H_2SO_4 + 970.9 ml distilled water

0.1 NH_2SO_4 = 2.91 ml H_2SO_4 + 997.09 ml distilled water

BUFFER TABLES

1. 0.2 M Acetate Buffer (Gomori) (pH 3.8–5.6)

Stock solution

Acetic acid   12.0 ml made up to 1000 ml with distilled water

Sodium acetate   27.2 g made up to 1000 ml with distilled water

Add few crystals of camphor to both solutions.

For the desired pH, mix correct amounts as indicated below.

pH	Acetic acid	Sodium acetate
3.8	87.0	13.0
4.0	80.0	20.0
4.2	73.0	27.0
4.4	62.0	38.0
4.6	51.0	49.0
4.8	40.0	60.0
5.0	30.0	70.0
5.2	21.0	79.0
5.4	14.5	85.5
5.6	11.0	89.0

2. Acetate–Acetic Acid Buffer (Walpole) (pH 2.7–6.5)

Stock solution

M/5 acetic acid   12.0 ml (99 per cent assay) made up to 1000 ml with distilled water

M/5 sodium acetate   27 g made up to 1000 ml with distilled water

For desired pH, mix correct amounts as indicated below.

pH	M/5 Acetic acid (ml)	M/5 Sodium acetate (ml)
2.70	200.0	—
2.80	199.0	1.0
2.91	198.0	2.0
3.08	196.0	4.0
3.14	195.0	5.0
3.20	194.0	6.0
3.31	192.0	8.0
3.41	190.0	10.0
3.59	185.0	15.0
3.72	180.0	20.0
3.90	170.0	30.0
4.04	160.0	40.0
4.16	150.0	50.0
4.27	140.0	60.0
4.36	130.0	70.0
4.45	120.0	80.0
4.53	110.0	90.0
4.62	100.0	100.0
4.71	90.0	110.0
4.80	80.0	120.0
4.90	70.0	130.0
4.99	60.0	140.0
5.11	50.0	150.0
5.22	40.0	160.0
5.38	30.0	170.0
5.57	20.0	180.0
5.89	16.0	190.0
6.21	5.0	195.0
6.50	0.0	200.0

3. 0.05M Barbital Buffer (Gomori) (pH 8.7–6.9)

Stock solution

Sodium barbital   1.03 g in 50 ml distilled water

Add 0.1N HCl according to the table below. Dilute to a total of 100 ml.

pH	0.1 NHCl (ml)
8.7	5.0
8.5	7.5
8.3	11.0
8.1	15.0
7.9	19.0
7.65	26.0
7.45	31.0
7.30	36.0
7.15	41.0
6.90	43.5

4. Boric Acid–Borax Buffer (Holmes) (pH 7.4–9.0)

Stock solution

M/15 boric acid   12.36 g made up to 1000 ml with distilled water

M/20 Borax   19.071 mg made up to 1000 ml with distilled water

For desired pH, mix correct amounts as indicated below.

pH	M/5 Boric acid	M/20 Borax
7.4	90.0	10.0
7.6	85.0	15.0
7.8	80.0	20.0
8.0	70.0	30.0
8.2	65.0	35.0
8.4	55.0	45.0
8.7	44.0	60.0
9.0	20.0	80.0

5. Bicarbonate Buffer (pH 9.1–10.6)

Stock solution

0.1 M Na$_2$CO$_3$ 10.6 g made up to 1000 ml with distilled water

0.2 M Na$_2$HCO$_3$ 8.4 g made up to 1000 ml with distilled water

For desired pH add correct amounts as indicated below.

pH	0.1 M Na_2CO_3 (ml)	0.1 M Na_2HCO_3 (ml)
9.1	11.3	98.7
9.2	14.0	86.0
9.3	18.0	82.0
9.4	22.0	78.0
9.5	27.0	73.0
9.6	32.5	67.5
9.7	38.5	61.5
9.8	45.0	55.0
9.9	51.5	48.5
10.0	58.0	42.0
10.1	64.0	36.0
10.2	69.5	30.5
10.3	74.5	25.5
10.4	79.0	21.0
10.5	83.5	16.5
10.6	80.0	12.0

6. 0.2 M phosphate Buffer (Gomori) (pH 5.9–7.7)

Stock solution

Monobasic sodium phosphate   27.6 g made up to 100 ml with distilled water

Dibasic sodium phosphate   53.6 g made up to 1000 ml with distilled water

For desired pH mix correct amounts as indicated below.

pH	Monobasic sodium phosphate (ml)	Dibasic sodium phosphate (ml)
5.9	90.0	10.0
6.1	85.0	15.0
6.3	77.0	23.0
6.5	68.0	32.0
6.7	53.0	47.0
6.9	45.0	55.0
7.1	33.0	67.0
7.3	23.0	77.0
7.4	19.0	81.0
7.5	16.0	84.0
7.7	10.0	90.0

7. Phosphate Buffer (Sorensen) (pH 5.29–8.04)

Stock solution

M/15 Dibasic sodium phosphate   9.465 g made up to 1000 ml with distilled water

M/15 potassium acid phosphate   9.07 g made up to 1000 ml with distilled water

For desired pH mix correct amounts as indicated below (some technique will require dilution up to 1000 ml).

pH	M/15 dibasic sodium phosphate (ml)	M/15 potassium acid phosphate
5.29	2.5	97.5
5.59	5.0	95.0
5.91	10.0	90.0
6.24	20.0	80.0
6.47	30.0	70.0
6.64	40.0	60.0
6.81	50.0	50.0
6.98	60.0	40.0
7.17	70.0	30.0
7.38	80.0	20.0
7.73	90.0	10.0
8.04	95.0	5.0

8. Standard Buffer (McIlvaine) (pH 2.2–8.0)

Stock solution

0.1 M citric acid (anhydrous)   19.212 g made up to 1000 ml distilled water

0.2 M disodium phosphate (anhydrous)   28.39 per cent (7 H_2O, 53.628 g) made up to 1000 ml with distilled water

For desired pH, mix correct amounts as indicated below.

pH	Citric acid (ml)	Disodium phosphate (ml)
2.2	19.6	0.40
2.4	17.76	1.24
2.6	17.82	2.18
2.8	16.83	3.70
3.0	15.89	4.11

(Contd.)

Table (Continued)

pH	Citric acid (ml)	Disodium phosphate (ml)
3.2	15.06	4.94
3.4	14.30	5.70
3.6	13.56	6.64
3.8	12.90	7.10
4.0	12.29	7.71
4.2	11.72	8.28
4.4	11.18	8.82
4.6	10.65	9.35
4.8	10.14	9.86
5.0	9.70	10.30
5.2	9.28	10.72
5.4	8.35	11.15
5.6	8.40	11.60
5.8	7.91	12.09
6.0	7.37	12.63
6.2	6.78	13.22
6.4	6.15	13.85
6.6	5.45	14.55
6.8	4.55	15.45
7.0	3.53	16.47
7.2	2.61	17.39
7.4	1.83	18.17
7.6	1.27	18.73
7.8	0.85	19.15
8.0	0.55	19.45

9. 0.2 M Tris Buffer (Hale) (pH 7.19–9.1)

Stock solution

0.3 M tris (hydroxymethyl) aminomethane 24.228 g made up to 1000 ml with distilled water

0.1N HCl (38 per cent assay) 8.08 ml made up to 1000 ml with distilled water

To 25 ml 0.2 M tris acid, add 0.1 N HCl as indicated in the following table below and dilute to 100 ml.

pH	0.1 N HCl (ml)
7.19	45.0
7.36	42.5
7.54	40.0
7.66	37.5
7.77	35.0
7.87	32.5
7.96	30.0
8.05	27.5
8.14	25.0
8.23	22.5
8.32	20.2
8.41	17.5
8.51	15.0
8.02	12.5
8.74	10.0
8.92	7.5
9.10	5.0

10. Tris Maleate Buffer (Gomori) (pH 5.8–8.2)

Stock solution

Maleic acid	29.0 g
Tris (hydroxymethyl) aminomethane	30.3 g
Distilled water	500 ml

Add 2 mg charcoal, shake, let stand for 10 minutes and filter.

To 40 ml of stock solution add 0.1 N NaOH (4 per cent) as indicated below and dilute to 100 ml.

pH	Sodium hydroxide (ml)
5.8	9.0
6.0	10.5
6.2	13.0
6.4	15.0
6.6	16.5
6.8	18.0
7.0	19.0

(Contd.)

Table (Continued)

pH	Sodium hydroxide (ml)
7.2	20.0
7.6	22.5
7.8	24.2
8.0	26.0
8.2	29.0

pH 0.65 to 5.20 Walpole (1914)

50 ml N sodium acetate + x ml NHCl made up to 25 ml

pH	x ml HCl
0.65	100.0
0.75	90.0
0.91	80.0
1.09	70.0
1.24	65.0
1.42	60.0
1.71	55.0
1.85	53.5
1.99	52.5
2.32	51.0
2.64	50.0
2.72	49.75
3.09	48.5
3.29	47.5
3.49	46.25
3.61	45.0
3.79	42.5
3.95	40.0
4.19	35.0
4.39	30.0
4.58	25.0
4.76	20.0
4.92	15.0
5.20	10.0

pH 3.6 to 5.6 Walpole (1914)

200 ml mixtures of 0.1N acetic acid and 0.1N sodium acetate

pH	0.1N acetic acid	0.1N sodium nitrate
3.6	185	15
3.8	176	24
4.0	164	36
4.6	102	98
4.8	80	120
5.0	59	141
5.2	42	158
5.4	29	171
5.6	19	181

pH 1.5 to 3.5 Lewis (1962) Acid phosphate buffer stock solution

0.4 M acid potassium phosphate, 27.22 g potassium dihydrogen phosphate to 1 litre

Molar proportion of pH values at:

HCl	0.02M	0.1M
0.05	3.55	3.35
0.10	3.20	3.00
0.15	3.00	2.75
0.20	2.85	2.55
0.25	2.70	2.40
0.30	2.60	2.25
0.35	2.50	2.10
0.40	2.40	2.00
0.45	2.30	1.85
0.50	2.20	1.75
0.60	2.10	1.55
0.75	2.00	

MOUNTANTS

6

Sections, whether they are histological or histochemical, should be preserved for a long time and for this they are to be stored. Sections stored on slides should be covered under a cover glass. For this, some media are necessary. The media in which the sections are generally mounted fall under two categories—aqueous media and resinous media. Aqueous media are generally used for unstained or metachromatically stained or fat-stained sections. Resinous media are used routinely in laboratories.

AQUEOUS MOUNTING MEDIA

Aqueous media usually have a low refractive index (1.40–1.42). Glycerine is usually used. Aqueous media are of three types,

1. syrups
2. gelatin media
3. gum arabic

For syrups and gelatin, glycerine is incorporated to prevent cracking after drying. A ringing medium is used round the coverslip.

There are some metachromatic stains which diffuse into the mounting medium. For such type of stains, fructose syrup or potassium acetate gum syrup could be used. The setting qualities of the syrups could be improved by the addition of gelatin (12 per cent). All aqueous mounting media are prone to bacterial action. To overcome this, thymol crystals or 25 per cent phenol or sodium merthiolate should be added. Sections or whole mounts are removed from water and placed on a slide. Then the mounting medium is added (adsorptive) and the cover glass is placed on it. The slide should always be kept horizontal. If the slides are kept vertical, the specimen or sections are likely to shift to a corner.

SOME AQUEOUS MOUNTING MEDIA

Glycerine Jelly (R.I. 1.47)

Gelatin	10.0 g
Distilled water	60.0 ml
Glycerine	70.0 ml
Phenol	250 mg

Dissolve gelatin in water and heat it in a water bath. Once gelatin is melted, add glycerine and phenol. Then stir well and store in screw-capped bottles. Sometimes, instead of phenol, 0.025 per cent sodium merthiolate may be added which acts as preservative. If heated too much, gelatin gets transformed into metagelatin which will not harden at room temperature.

Apathy's Gum Syrup (R.I. 1.52)

Gum arabic	50.0 g
Sucrose	50.0 g
Distilled water	50.0 ml
Formalin	10.0 ml

It is a wonderful medium for fluorescent microscopy.

Kaiser's Glycerine Jelly

Water	52.0 ml
Gelatin	8.0 g
Glycerol	50.0 ml

Allow gelatin to soak in water for 2–3 hours. Add glycerol and preservative. Warm (do not exceed 75°C) and stir. Add phenol 0.1 g or merthiolate 0.01 g or zephiran 1.0 ml.

Highman's Modification of Apathy's Medium

Gum arabic	20.0 g
Cane sugar	20.0 g
Potassium acetate	20.0 g
Sodium merthiolate	10.0 ml
Distilled water	40.0 ml

Dissolve the ingredients using mild heat.

Farrant's Medium

Gum arabic	40.0 g
Glycerol	20.0 ml
Distilled water	40.0 ml
Phenol	100 mg

Dissolve gum arabic in distilled water with gentle heat, add glycerine and arsenic trioxide.

Fructose (Levulose) Syrup (R.I. 1.47)

This syrup is useful as a temporary or special mountant.

Fructose   30.0 g
Distilled water   20.0 ml

Gelatin may be added.

Gray and Wess Medium (PVA)

Polyvinyl alcohol (PVA) 17–24[13]   21.0 g
70 per cent acetone   7.0 ml
Glycerol   5.0 ml
Lactic acid   5.0 ml
Distilled water   10.0 ml

Make a paste of dry alcohol and acetone. Mix half of water with glycerol and lactic acid. Add remaining water drop by drop while stirring. To start with, the solution is cloudy but later becomes transparent when it is warmed.

Kirk Patric and Lendrum's DPX (R.I. 1.52)

Distrene 80   10.0 g
Dibutyl phthalate   5.0 ml
Xylene   35.0 ml

This is the most favoured mountant. The advantage of this over Canada balsam is that excess mountant can be easily cleared by stripping it off after cutting around the edges of the coverslip.

OTHER MOUNTING MEDIA

Euparol (R.I. 1.48)

Euparol is semi-synthetic composed of sandare resin dissolved in a mixture of eucalyptus, paraldehyde and camsal (a liquid composed of camphor and phenyl salicylate).

Neutral Mounting Medium (R.I. 1.52)

This medium which is a mixture of coumarone and other resins dissolved in eucalyptus oil is specially prepared for mounting blood films and sections stained by Ramanovsky methods. Blood films are mounted dry.

Caro Corn Syrup (Patrick, 1936)

Corn syrup   1 part
Distilled water   1 part

Add crystals of thymol and store at 4°C.

Chrome Glycerine Jelly (Zwemer, 1933)

Distilled water   80.0 ml
Glycerine   20.0 ml
Gelatin (powder)   3.0 g
Chrome alum   200 mg

Dissolve the chrome alum in 30 ml of water and the gelatin in 50 ml of water. Mix and add the glycerine, test pH before use and adjust to pH 7.4 with $NaHCO_3$. This mountant shrinks and sets very hard.

Apathy's Medium (Modified by Lillie and Ashburn, 1943)

Gum arabic	50.0 g
Cane sugar	50.0 g
Distilled water	100.0 ml
Thymol	100.0 mg

Gum arabic and cane sugar are dissolved in distilled water by heating to 60°C.

Hoyer's Medium (Beek, 1955)

Gum arabic	30.0 g
Glycerol	16.0 ml
Chlorol hydrate	200.0 g
Distilled water	50.0 ml

First dissolve gum arabic in water by little warming and then add chlorol hydrate and glycerine.

Polyvinyl Alcohol Mounting Medium

PVA	15.0 g
Water	100.0 ml

Add PVA to water. Slowly warm it. The solution is viscous. Filter the solution. Remove undissolved lumps and allow it to stand for a few hours.

Berlese Mounting Medium (Gray, 1952)

Dextrose syrup	5.0 ml
Glacial acetic acid	3.0 ml
Water	10.0 ml
Gum arabic	8.0 g
Chlorol hydrate	75.0 g

Mix syrup, glacial acetic acid and water and then add gum arabic. When the solution is clear (it will take a week) add chlorol hydrate.

Monk's Karo Medium (1938)

White karo syrup	5.0 ml
Certe (fruit pectin)	5.0 ml
Water	3.0 ml

This is good for small parts.

Lactophenol Mounting Medium

Phenol (solution)	30.0 ml
Lactic acid	1.0 ml
Glycerol	2.0 ml
Distilled water	1.0 ml

Yetwins Mounting Medium for Nematodes and Ova (1944)

10 per cent bacti-gelatin	15.0 ml
Glycerol	50.0 ml
1 per cent chrome alum	100.0 ml
Phenol	1.0 ml

Fluorescent Mounting Medium (Rodriquez and Deenhardt, 1960)

Elvanol 51-05 (Du Pont)	20.0 g
0.14 Sodium chloride buffered with	
$\quad$ 0.01 M KH_2PH_4–Na_2HPO_4·12 H_2O (pH 7.2)	80.0 ml
Glycerine	40.0 ml

Agitate for 16 hours and centrifuge at 12000 rpm and remove undissolved particles of elvanol (Final pH should be between 6 and 7).

There are certain standard principles to be applied to process a tissue on slides.

For example in paraffin sections, wax must be removed (deparaffinization) because stains do not penetrate through wax. Usually xylene is used to remove paraffin. Since it is not possible to stain the tissue in xylene medium, xylene has to be removed with absolute alcohol. Then it is customary to transfer the slides to a medium. Comparable to the solvent of the dye, i.e., if the dye is in water solution, the slides have to be hydrated through a graded series of alcohol (100 per cent, 95 per cent, 90 per cent, 70 per cent) and finally to water. Suppose the dye is dissolved in 50 per cent alcohol solution, then the slides are carried up to 50 per cent alcohol and then placed in the staining solution.

Counterstaining must be done in proper sequence allowing each dye or chemical to maintain its specific effect on the tissue elements. Improper sequence may result in poor staining, haematoxylin/eosin staining being a simple example. If eosin is applied before haematoxylin stain, the eosin stain will be completely removed during the action of haematoxylin.

Several types of slide holders are available, Coplin's jar can hold 10 slides, two in each groove. To make permanent preparations, all water and alcohol have to be extracted and the slides have to be passed on to a medium which makes the tissue clear and transparent without altering the colour or intensity of stain and holds the cover glass in place. Water is removed by passing through graded series (50 per cent, 70 per cent, 90 per cent, 95 per cent, 100 per cent) of alcohol. The final reagent is xylene (or a similar solvent). Xylene removes all traces of alcohol, and slides will be crystal clear. Now a mounting medium is placed on the tissue and then a cover glass is placed completely covering it. While placing the cover glass, care must be taken to allow air to be displaced from under the cover glass. For this the cover glass is pressed gently so that the medium will be distributed evenly. If by any reason dull black splits appear on the mounted slide, it means that the clearing solution has evaporated out of the sections before the cover glass was in place. Such slides have to be returned to xylene. The slides are kept in xylene so that the mounting medium is dissolved.

The air is allowed to leave the section and the slide is remounted. If the slides appear milky, that means water is present. First dissolve the mounting medium in xylene, transfer the slides to absolute alcohol preferably a fresh solution, clear and mount.

RESINOUS MOUNTING MEDIA

There are various types of mounting media. The most common mounting medium used for routine work in the laboratories is Canada balsam. Now a number of natural and synthetic resins are being used. These media are composed of a resin either in its natural solvent or dissolved in a solvent such as xylol. The viscosity of these media is the choice of the technician. The choice of the mounting medium should not fade the stain and it should have the correct refractive index.

Natural Resins

In olden days natural resins were used as mounting media. Canada balsam, Gumdamar and gum arabic were used. Canada balsam is from a Canadian fir tree (*Abies balsamea*). It is composed of terpenes, carboxylic acid and their esters. Gum damar is composed of unsaturated resin acids and their esters and gum sandarac of an unsaturated acid resin. They dry slowly. They develop acidity, fade stains, turn yellow and crack after few years. Euparol (R.I. 1.483) a mixture combination of eucalyptus, gum sandarac, solol, paraldehyde, menthol and camphor is used for blood smears. Dry smears can directly be mounted with euparol. Canada balsam is a thick resin which usually dissolves in xylol. Haematoxylin/eosin-stained slides are well preserved and aniline stains fade out. Prussian blue slowly bleaches. Usually ready-made balsam is purchased but some laboratories still prefer to prepare Canada balsam.

Damar balsam is almost similar to Canada balsam but contains impurities and dirt.

Colophonium resin is rarely used as a mountant. It is dissolved either in xylene or in turpentine.

Synthetic Resins

Synthetic resins have proved to be more superior over natural resins. They can be either purchased or prepared in the laboratory. They are stable, inert, dissolve in xylene or toluene. They do not require long period of drying and adhere tightly to glass. They are pale and do not turn yellow with age.

The commonly used synthetic resins are β-pineni polymers terpene resins such as permount, picolyte with a R.I. of 1.51 to H.R.S. with R.I. 1.5202, bioloid syrup resin R.I. 1.5396, Klur mount H and so on. These resins should have a R.I. of 1.53–1.54. They are quite effective and soluble in toluidene, xylene, aromatic hydrocarbon solvents and chlorinated hydrocarbons such as chloroform.

Most of the synthetic resins are allowed to air-dry. If quick drying is required, the slides are placed in metal trays and kept in an oven at 160°–170°C for 2 minutes. They are removed and chilled in a freezer for another 2 minutes. Now the slides can be marked and stored.

New unimount (R.I. 1.50) contains neither oxidizing nor bleaching contaminants. It is used as a routine mountant. It is mostly used for fluorescent antibody preparation.

HAEMATOXYLIN STAINING

Haematoxylin is a powerful dye, staining purple, blue and blue-black. If mordanted with iron, it is exceedingly good for mitotic study and chromatin takes a black or blue-black colour.

Various methods are available for oxidizing haematoxylin. Sodium iodate oxidation has been recommended by Baker (1958). It is better to use less chemicals than is necessary for complete oxidation. The solution will then combine with atmospheric oxygen and thus maintain its strength. Brilliant results can be expected from solutions allowed to ripen for six weeks or so (Mayer's haematoxylin). Oxidants will range from 0.1 per cent sodium iodate ($NaIO_3$), to sodium metaperiodate ($NaIO_4$) and potassium permanganate ($KMnO_4$). Thymol and salicylic acid are used as preservatives. Alcohol also prevents moulds and glycerine prevents over-oxidation. Usually the colour of the stock solution is an indicator of its efficiency. The colour may range from white through purple to brown. At the purple stage, it is the best giving excellent results; red stage, indicates the weakening of the strength of the solution and brown stage indicates that the solution has to be discarded.

Usually the type of mordant used influences the type of tissue that takes the stain. Some haematoxylin staining procedures require the tissues to be pretreated with a mordant (e.g. Hedenhain's iron haematoxylin). Most of these mordants get incorporated into the haematoxylin solution. Based on these factors, haematoxylin solutions can be categorized into the following types.

1. Alum haematoxylin
2. Iron haematoxylin
3. Tungsten haematoxylin
4. Molybdenum haematoxylin

5.   Lead haematoxylin

6.   Haematoxylin without mordants

ALUM HAEMATOXYLIN

This includes Ehrlisch, Delafield, Mayer's, Harris, Cole, Gills and Carazzis methods.

With alum haematoxylin, ageing is a big problem. Nuclear staining deteriorates with age of the solution. Added to these, a precipitate is formed on the surface of the solution which has to be filtered and the staining time should be increased.

In this procedure, the mordant is either aluminium ammonium sulphate (ammonia alum) or aluminium potassium sulphate (potash alum). First the nuclei are stained in a red shade but turn blue-black when sections are treated with a weak alkali such as lithium carbonate or 0.05 per cent ammonia in distilled water or Scot's tap water.

Ehrlisch Haematoxylin (Ehrlisch, 1886; Gurr, 1956)

Ehrlisch haematoxylin is not only an excellent stain for nuclei, but also stains mucins including the mucopolysaccharides of cartilage.

Haematoxylin	2.0 g
Ammonia alum	3.0 g
Ethyl alcohol	100.0 ml
Glycerol	100.0 ml
Distilled water	100.0 ml

Dissolve haematoxylin in distilled water using gentle heat. Then add alum, use slight heat. When the alum is dissolved, add the sodium iodate followed by citric acid and chlorol hydrate. This solution requires 6–8 weeks for ripening. It will stain nuclei selectively. It is good for demonstration of amoebae if it is diluted with equal quantity of water.

Harris Haematoxylin (Harris, 1900; Mallory, 1944)

This haematoxylin is ripened with mercuric oxide. Nuclear staining is very clear and is especially useful in exfoliative cytology.

Haematoxylin	2.5 g
Absolute alcohol	50.0 ml
Ammonia or potash alum	50.0 ml
Distilled water	500.0 ml
Mercuric oxide	1.5 g
Glacial acetic acid	20.0 ml

Dissolve the haematoxylin in absolute alcohol and the alum in the water using heat if necessary. Now mix the two solutions together. Heat the mixture to boiling and then add mercuric oxide. This solution is ready for use. It is better to add glacial acetic acid to give precise results.

The advantage of Harris haematoxylin is that it can be readily used as soon as it is prepared. The staining capacity decreases after 2–3 months. When a precipitate forms at the bottom of the bottle, this indicates that the solution is deteriorating.

Delafield's Haematoxylin (Carleton and Leach, 1947)

The longevity of this haematoxylin is the same as that of Ehrlisch.

Solution 1
Haematoxylin 6.0 g
Ethyl alcohol 50.0 ml

Solution 2
Ammonia alum 55.0 g
Distilled water 100.0 ml

Solution 3
Glycerol 150.0 ml
Ethyl alcohol 150.0 ml

First mix haematoxylin in ethyl alcohol and then ammonia alum in water separately. Now mix solutions 1 and 2 for 2–3 days. Filter and then add solution 3.

Mayer's Haematoxylin

Add 1.0 g of haematoxylin to 1000 ml of distilled water, heat gently, then add 0.2 g of sodium iodate and 50 g of potash alum. Heat until dissolved and then add 1.0 g of citric acid and 50 g of chlorol hydrate. Allow it to ripen for 8 weeks.

This haematoxylin is chemically ripened with sodium iodate. It is used as a nuclear counterstain in the demonstration of glycogen, and also in enzyme histochemical technique. When it is used as a counterstain, staining time is shortened for 5–10 minutes.

Cole's Haematoxylin (Cole, 1943)

This haematoxylin is an alum haematoxylin and it an be ripened artificially with an alcoholic iodine solution.

Preparation of Solution

Haematoxylin   1.5 g
Saturated aqueous potash alum   700 ml
1 per cent iodine in 95 per cent alcohol   50 ml
Distilled water   200 ml

The haematoxylin is dissolved in warm distilled water and mixed with iodine solution. Then alum solution is added and boil the mixture, cool and filter. The solution is ready for immediate use. Filter before use.

Carazzi's Haematoxylin (Carazzi, 1911)

This haematoxylin is ripened chemically with potassium iodate.

Haematoxylin   5.0 g
Glycerol   100.0 ml
Potash alum   25.0 g
Distilled water   400.0 ml
Potassium iodate 100 mg

First, dissolve haematoxylin in glycerol. Dissolve alum in most part of water and slowly add haematoxylin glycerine solution. Stir well. Add potassium iodate to the rest of the water with gentle heat and then add it to the haematoxylin–glycerol–alum mixture. Now it is ready for use.

Carazzi's haematoxylin is good for frozen sections.

Gill's Haematoxylin (Gill *et al.*, 1974)

Haematoxylin	2.0 g
Sodium iodate	200 mg
Aluminium sulphate	17.6 g
Distilled water	750.0 ml
Ethylene glycol	250.0 ml
Glacial acetic acid	20.0 ml

First mix ethylene glycol and distilled water, then add haematoxylin and aluminium sulphate. Finally add glacial acetic acid.

IRON HAEMATOXYLIN

In these haematoxylins, iron solutions are used both as oxidizing agents and as mordants. Ferric chloride and ferric ammonium sulphate are the iron salts. Because of the strong oxidizing ability of the solutions, it is also used as a differentiating solution after haematoxylin staining.

Weigert's Haematoxylin (Weigert, 1904; Lillie and Henderson, 1960)

Solution 1

Haematoxylin	1.0 g
Absolute alcohol	100.0 ml

Solution 2

30 per cent ferric chloride	4.0 ml
(30 g + 100 ml water)	
Distilled water	95.0 ml
Hydrochloric acid	1.0 ml

Mix equal parts of 1 and 2 immediately before use. The staining solution should be violet-black in colour. Solution 1 must be allowed to ripen and must be prepared about a week before it is required. The time required is 4–5 weeks. Nuclei stain black.

Heidenhain's Haematoxylin (Heidenhain, 1890)

In this solution, ferric ammonium sulphate is used as oxidant and also as mordant and the same solution can be used as a differentiator. First the section is mordanted in iron solution, stained with haematoxylin (overstain it) and then differentiated in iron solution.

Solution 1

Iron alum	4.0 g
Distilled water	100.0 ml

Solution 2
    Haematoxylin   10.0 g
    Absolute alcohol 100.0 ml

Allow it ripen for 6 months

Working Solution
    Solution 2   5.0 ml
    Distilled water   95.0 ml

Procedure

1. Bring slides to 95 per cent alcohol.
2. Place in solution 1 for 2 hours.
3. Wash.
4. Transfer to working solution for 45 minutes.
5. Rinse rapidly.
6. Differentiate in 2 per cent iron alum.
7. Wash in running water.
8. Dehydrate, clear and mount.

Results

Muscle striations, myelin and chromatin stain grey-black.

Slidder's Iron Haematoxylin (1969)

Solution 1
    Haematoxylin   1.0 g
    95 per cent alcohol   100.0 ml
    Aluminium chloride hydrate 10.0 g

Solution 2
    Ferrous sulphate hydrate   10.0 g
    Distilled water   100.0 ml

Solution 3 Sodium iodate (9 per cent aqueous)
    Combine solution 1 and 2 and add 2.0 ml concentrated hydrochloric acid.

Solution is ready for use after 2 days. Staining time is 5–10 minutes. Slides can be differentiated in 0.5 per cent hydrochloric acid in 70 per cent alcohol and neutralized with 2 per cent potassium acetate.

Krutsay Iron Haematoxylin (1962)

Haematoxylin   1.0 g
Potassium sulphate 50.0 g
Potassium iodate   0.2 g
Hydrochloric acid   5.0 ml
Distilled water   100.0 ml

This solution can either be used as Mayer's or converted to iron haematoxylin as follows:

Add 100 volumes of the above to 8 volumes of

Ferric alum	2.0 g
Hydrochloric acid	0.5 ml
Distilled water	100.0 ml

A brownish solution is formed. Staining time is five minutes. When solution turns orange, it has to be discarded.

Groat's Variation of Weigert's Haematoxylin (1949)

Distilled water	50.0 ml
Sulphuric acid (sp.gr. 1.84)	0.8 ml
Ferric alum	1.0 g
95 per cent alcohol	50.0 ml
Haematoxylin	500 mg

Mix in the order given. Filter. Staining time is 9–10 minutes.

Loyez Haematoxylin (Loyez, 1910)

In this haematoxylin, ferric ammonium sulphate is used as a mordant. In this the differentiator is by Weigert's differentiator (Borax, potassium ferricyanide). This technique can be applied to paraffin as well as frozen and nitrocellulose sections.

Verhoeff's Haematoxylin (Verhoeff, 1908)

This is the most important stain for elastic fibres. Ferric chloride is included in the haematoxylin stain with Lugol's iodine and 2 per cent aqueous ferric chloride as differentiator (*see* Chapter 19).

TUNGSTEN HAEMATOXYLIN

This is Mallory PTAH technique. Mallory (1897, 1900) used phosphotungstic acid as mordant. Haematein can also be used instead of haematoxylin without intervention of oxidation processes. Haematoxylin can be oxidized by $KMnO_4$ solution.

PTAH is usually used to demonstrate fibrin, muscle striations, glial fibres, cilia, myelin, etc. CNS stains with PTAH (*see* chapter 19).

PTAH Solution using Haematein (Shum and Hon, 1969)

Solution 1

Haematein	800 mg
Distilled water	1.0 ml

Grind haematein to a paste with distilled water. The colour should be chocolate brown.

Solution 2

Phosphotungstic acid	900 mg
Distilled water	99.0 ml

Mix solution 1 and solution 2 and boil, cool and filter.

Procedure

1. Deparaffinize and bring slides to water.
2. Oxidize in permanganate (0.5 per cent) for 5 minutes.
3. Rinse in tap water.
4. Bleach with oxalic acid (5 per cent).
5. Wash.
6. Treat with staining solution for 12–24 hours.
7. Wash.
8. Dehydrate rapidly, clear and mount.

PTAH Solution Chemically Oxidized with Potassium Permanganate

Haematoxylin	500 mg
Phosphotungstic acid (PTA)	10.0 g
Distilled water	500.0 ml
Aqueous $KMnO_4$ (0.25 per cent)	25.0 ml

Dissolve haematoxylin in 100 ml of distilled water, and PTA in 400 ml water. Mix both solutions. The stain is ready after 24 hours, but preferably after 1 week.

PTAH Solution Naturally Oxidized

Haematoxylin	500 mg
PTA	5.0 g
Distilled water	500.0 ml

Dissolve haematoxylin in 100 ml water. In another dish dissolve PTA in 400 ml water. Mix these two solutions. Allow it to ripen for few months (6 months).

Procedure

1. Deparaffinize and hydrate slides to water.
2. Post-chromate the section in 3 per cent dichromate in 10 per cent HCl (dichromate 36 + 12 ml HCl).
3. Wash in tap water.
4. Oxidize in permanganate (0.5 per cent) for 1 minute.
5. Bleach in 1 per cent oxalic acid for 1 minute.
6. Rinse in tap water.
7. Treat with PTAH, overnight.
8. Dehydrate rapidly, clear and mount.

Result

Fibrin stains dark blue, nuclei stain blue and myelin stains light blue.

MOLYBDENUM HAEMATOXYLIN
(Thomas, 1941; McManus and Mowry, 1964)

This is rarely used but is preferred for the demonstration of collagen (*see* Chapter 19) and also for argentaffin cell granules.

Phosphomolybdic Acid Haematoxylin (Thomas, 1941)

Solution 1 (Haematoxylin solution)
Haematoxylin   2.5 g
Dioxane   49.0 ml
Hydrogen peroxide   1.0 ml

Solution 2 (PMA solution)
Phosphomolybdic acid (PMA)   16.5 g
Distilled water   44.0 ml
Diethylene glycol   11.0 ml

Filter and take 50.0 ml of solution 2 and add to solution 1. Allow it to stand for 24 hours.

Procedure

1. Deparaffinize and hydrate slides to water.
2. Stain with staining solution for 2 minutes.
3. Wash.
4. Immerse in picroacetic alcohol (500 mg picric acid + 0.5 ml acetic acid + 100 ml 70 per cent alcohol).
5. Rinse in tap water.
6. Dehydrate, clear and mount.

Results

Collagen stains violet to black, argentaffin cells stain black and nuclei stain blue.

LEAD HAEMATOXYLIN

These solutions are best used to demonstrate endocrine cells of the alimentary tract (*see* Chapter 21). It is also used in localizing gastrin-secreting cells in the stomach (Beltrami *et al.*, 1975).

HAEMATOXYLIN WITHOUT A MORDANT

The basis of the Mallory methods is the ability of ripen haematoxylin method to form blue-black lakes with these metals (Lillie and Fullmer, 1976).

Delafield Haematoxylin

Fixation

Any general fixative.

Procedure

1. Deparaffinize and hydrate slides to water.
2. Transfer to Delafield's haematoxylin for 20 minutes.
3. Wash in running water.
4. Transfer to 70 per cent alcohol.
5. Differentiate in acid alcohol (2 to 3 drops of hydrochloric acid in 60 ml of 70 per cent alcohol).

6. If nuclei are still deep blue, repeat the process.
7. Place in 70 per cent alcohol.
8. Counterstain if desired.
9. Wash, dehydrate, clear and mount.

Results

Nuclei stain deep blue whereas cytoplasm takes up the colour of the counterstain.

Mayer's Haematoxylin

Fixation

Any general fixative.

Procedure

1. Deparaffinize and hydrate slides to water.
2. Place in staining solution, Mayer's haematoxylin, for 10 minutes.
3. Wash in running water.
4. Counterstain with eosin for 1 minute.
5. Wash quickly, dehydrate, clear and mount.

Result

Nuclei stain deep blue whereas cytoplasm stains deep pink.

Heidenhain's Iron Haematoxylin

Fixation

Avoid mercuric-chloride-containing fixatives.

Procedure

1. Deparaffinize and hydrate slides to water.
2. Mordant in 4 per cent iron alum for 30 minutes.
3. Wash in running water.
4. Stain in haematoxylin for 30 minutes.
5. Wash in running water.
6. Destain in 2 per cent iron alum until nuclei are sharply coloured.
7. Wash in running water.
8. Counterstain in eosin for 20 seconds.
9. Rapidly dehydrate, clear and mount.

Result

Nuclei stain deep black whereas cytoplasm stains eosin colour.

Weigert's Haematoxylin

Fixation

Any general fixative.

Procedure

1. Deparaffinize and hydrate slides to water.
2. Stain in haematoxylin for 3–5 minutes.
3. Wash in running water.
4. Counterstain with eosin for 30 seconds.
5. Wash, dehydrate, clear and mount.

Result

Nuclei stain black whereas other elements take up the eosin colour.

Phosphotungstic Acid Haematoxylin (PTAH) (Puchtler *et al.*, 1963)

Fixation

Any general fixative.

Procedure

1. Deparaffinize and hydrate slides to water.
2. Wash in running water.
3. Oxidize in 0.5 per cent potassium permanganate for 5 minutes.
4. Bleach in 2 per cent oxalic acid.
5. Mordant in 4 per cent iron alum for 1 hour.
6. Treat with staining solution for 2–24 hours.
7. Transfer directly to 95 per cent alcohol.
8. Dehydrate, clear and mount.

Result

Nuclei stain blue whereas other elements stain yellowish to brown.

Lillies Variation of Weigert's Haematoxylin

Solution 1
  Ferric alum   20.0 g
  Distilled water 200.0 ml

Solution 2
  Haematoxylin   2.0 g
  Absolute methyl alcohol   60.0 ml

Mallory's Iron Chloride

Solution 1
  Iron chloride   5.0 g
  Distilled water 100.0 ml

Solution 2
  Haematoxylin   50.0 g
  Distilled water 100.0 ml

Ripening can be done by rapidly bubbling through haematoxylin (Rawlin and Takahashi, 1947). Ripening can be done rapidly (Hance and Green, 1961).

HAEMATOXYLIN SUBSTITUTE PROCEDURES

A good substitute for haematoxylin is gallocyanin. The preparation of solutions using gallocyanin is not cumbersome and it is ready for use. No ripening is required. At times, it is better than haematoxylin especially for the central nervous system. Negri bodies. The fixatives recommended are Zenker or formalin.

Gallocyanin (Berube *et al.*, 1966; Einarsson, 1951)

Gallocyanin	150 mg
Chrome alum	100.0 ml

Boil for 2 minutes, cool, filter, wash the precipitate and make up to 100 ml. It lasts for a week.

Procedure

1. Dewax and hydrate slides to water.
2. Immerse in stain for 3 hours at 56°C or overnight at room temperature.
3. Wash and counterstain.
4. Dehydrate, clear and mount.

Result

Nuclei stain blue.

Haematein (Kornhauser, 1930)

Hematein	500 mg
95 per cent alcohol	10.0 ml
Aluminium potassium	500.0 ml
sulphate (5 per cent aqueous)	

Grind haematein with alcohol and then alum. It is ready for use. Any counterstain can be used.

Procedure

1. Dewax and hydrate slides to water.
2. Treat with haematein for 5 minutes.
3. Wash in running water.
4. Counterstain.
5. Dehydrate, clear and mount.

Result

Nuclei are stained blue.

Eriochrome Cyanin RC (Chapman, 1977)

Fixation

Any general fixative.

Reagents

Eriochrome cyanin R	200 mg
Ferric chloride	900 mg
Hydrochloric acid	5.0 ml
Distilled water	100.0 ml

Procedure

1. Dewax and hydrate slides to water.
2. Rinse in 0.5N HCl for 5 seconds.
3. Treat with staining solution.
4. Repeat step 2 for 2 seconds.
5. Wash in running water.
6. Treat with 0.1 per cent sodium bicarbonate for 1 minute.
7. Wash in running water.
8. Counterstain if desired.
9. Dehydrate, clear and mount.

Result

Nuclei stain blue.

RED NUCLEAR STAINING

Darrow Red (Powers and Clark, 1963; Powers *et al.*, 1960)

Fixation

Any general fixative.

Reagents Preparation

Darrow red	50 mg
0.2 M glacial acetic acid (pH 2.7)	200.0 ml

Boil, cool and filter. It lasts for a month.

Procedure

1. Dewax and hydrate slides to water or frozen sections.
2. Treat with staining solution for 20–30 minutes.
3. Wash in distilled water.
4. Remove excess stain in 70 per cent and 95 per cent alcohol.
5. Dehydrate, clear and mount.

Result

Nuclei are stained red.
This is a good nuclear stain especially for Nissl substance at pH 2.7.

Scarba Red (Slidders *et al.*, 1958)

Fixation

Any general fixation.

Reagents Preparation

Scarba red	2.0 g
Phenol (melted)	2.0 g
Neutral red	1.0 g

Mix and dissolve in 15 ml of 95 per cent alcohol and then add 85 ml of 2 per cent aniline in water. Then add glacial acetic acid (1.0 to 3.0 ml). Mix and filter. It lasts for 6 months.

Differentiator

70 per cent alcohol	85.0 ml
Formalin	15.0 ml
Glacial acetic acid	15 drops

Procedure

1. Deparaffinize and hydrate slides to water.
2. Treat with staining solution for 5 minutes.
3. Rinse in distilled water.
4. Immerse in 70 per cent alcohol for 3 minutes.
5. Keep in differentiator till nuclei are clear.
6. Dehydrate, clear and mount.

Result

Chromatin and calcium stain red.

Eosin (Putt, 1948)

Eosin Y	1.0 g
Potassium dichromate	500 mg
Saturated aqueous picric acid	10.0 ml
Absolute alcohol	10.0 ml
Distilled water	80.0 ml
Acetic acid (if necessary)	1 drop

Eosin orange G

1 per cent eosin Y in 95 per cent alcohol	10.0 ml
Orange G saturate solution in 95 per cent alcohol (0.5 g + 100 ml)	5.0 ml
95 per cent alcohol	45.0 ml

Orange G

Orange G	1.0 g
PTA	5.0 g
95 per cent alcohol	100.0 ml

OTHER COUNTERSTAINS

1. 5 per cent aqueous acid fuchsin
2. Orange G saturated in 95 per cent alcohol
3. Van Gieson
4. Bordeaux red
5. Biebrich scarlet (1 per cent aqueous)
6. Eosin Y, eosin B, erythrosin B
7. Phloxine (0.5 per cent aqueous)
8. Congo red (0.5 per cent aqueous)
9. Light green (0.2–0.3 per cent in 95 per cent alcohol)
10. Aniline blue
11. Fast green (same as light green)
12. Wood green (0.5 aqueous)
13. Methylene blue (1 per cent aqueous)

FROZEN METHODS

8

The first few chapters of the book dealt with the methods involving standard routine procedures of dehydration, embedding the tissue in paraffin wax, mounting and so on. This chapter is solely devoted to methods to produce sections without the involvement of dehydration solutions, clearing agents, or embedding media. Previously, frozen sections were used for rapid histological standard processing methods which had artifacts of their own. This resulted in deleterious effects on tissue constituent. With the expansion of immunocytochemistry, frozen sectioning techniques came into vogue. Frozen techniques surpass all other techniques. These techniques are mostly used for demonstrating enzymes, fats and some radioisotopes.

FROZEN SECTIONS

Of late both fixed and unfixed frozen sections are much in vogue and are by far the most common and important techniques. With the help of freezing microtome, a cryostat or a standard microtome modified by the addition of thermomodules frozen sections can be produced. Based on the choice of the technique, the tissues are fixed or unfixed.

FREEZING MICROTOME

This method is most common when enzymes and lipids are to be demonstrated histochemically. Usually tissues fixed in cold (4°C) formol calcium can be used. When structures in the nervous system are to be demonstrated, formol ammonium bromide is used as a suitable fixative. Fixatives containing mercury or dichromate or alcohol as one of the ingredients are usually not recommended for frozen sectioning since there is a tendency for the tissue to become brittle. Moreover the freezing point of tissue is changed by the alcohol.

For frozen sections, a number of microtomes are available, specially designed for this purpose. The paraffin and frozen microtomes are basically similar. But the knife usually passes over the surface of the block, whereas for paraffin sections, it is the reverse. The freezing microtome is provided with a thermomodule unit by using direct electrical current or it is attached to the liquid nitrogen or carbon dioxide cylinder where carbon dioxide is fed to the microtome block stage, which keeps the tissue in a frozen state. The thermomodule method is to keep the tissue in a frozen condition.

Knives

Wedge-shaped knives are used for frozen sectioning. These knives give the best results.

Process of Sectioning on Freezing Microtome

On freezing microtomes, thin sections cannot be expected. The thermomodule unit incorporation can give sections from 5 µ onwards, but using carbon dioxide, sections from only 15 µ upwards can be cut. Attempts to cut thin sections results in alternate thick and thin sections. For easy sectioning, gelatine is the best supporting medium.

Gelatin–Glycerine Solution

Gelatine	16.0 g
Glycerine	15.0 ml
Distilled water	70.0 ml
Thymol	one crystal

Store at 4°C, and before use, warm it in water bath till it becomes liquid.

Procedure

1. Fix tissue in formol calcium (10 per cent).
2. Wash overnight.
3. Place the tissue in gelatine–glycerol solution at 37°C for 6 hours.
4. After 6 hours, transfer to fresh solution.
5. Embed and store in refrigerator.
6. Trim excess gelatine from the block.
7. Harden the block by immersing in 10 per cent formol calcium overnight.

The advantage of frozen sections is that the procedure time is minimized. All diffusible substances could be demonstrated and there is no danger of shrinkage. But there are certain disadvantages which includes thick sections without serial order and damage to tissue at times due to freezing as well as cold microtome.

Sectioning

An experienced technician can get excellent sections with freezing microtome and it is relatively easy. The tissue block is attached onto the microtome stage with a drop of water and a blast of carbon dioxide from the cylinder. The tissue gets frozen and the knife passes across the face of the block. The tissue block should always be maintained at an optimum temperature. If there is a slight rise in the temperature, the block will not be sectioned well and the sections will tend to disintegrate on cutting. Further blasts of carbon dioxide are required. The sections adhere to the knife and they are removed with a wet camel-hair brush and the sections are floated onto distilled water.

Section Handling

The sections may be picked up onto a slide for staining or manoeuvered free-floating throughout the various staining solutions. These floating sections are usually used for demonstrating components of nervous system and lipids.

CRYOSTAT

Cryostats are laboratory equipment used in routine work to produce good quality of frozen sections at a faster rate.

It is a refrigerated cabinet in which a modified microtome is set. This is specially designed. A number of microtomes can suit a cryostat, and its controls are operated from outside.

Ever since the introduction of cryostat in 1954 with an adjustable temperature of –5 to –30°C there have been improvements in design. After that many cryostats were developed with current electronics with which section thickness and cabinet temperature can be easily regulated and are digitally demonstrated.

For cryostat sectioning, the tissue must be prepared to suit it. It must be fresh and rapidly frozen and the conditions in the cryostat must be correct.

Tissues for freezing should be fresh and freezing should be as rapidly as possible because if freezing is done slowly, distortion of the tissue results. The most common freezing techniques are

Liquid nitrogen	–90°C
Isopentane cooled by liquid nitrogen	–150°C
Carbon dioxide gas	–70°C
Aerosol sprays	–50°C

Of all these, liquid nitrogen is commonly used in many laboratories since it is the most rapid of all freezing agents but the main disadvantage is that soft tissue is liable to crack due to sudden expansion of the ice within the tissue and there will be damage both to the block and knife especially if it is cut above –75°C.

Process of Sectioning on Cryostat

Preparation of Tissue for Cryostat Cutting

The major advantage of cryostat sections is that substances lost or affected can be demonstrated. The effect of freezing unfixed tissue is to cause diffusion of substances and this is enhanced when the tissue is cut in the cryostat. To get accurate localization of some hydrolytic enzymes and antigens, it is necessary to fix the tissue before sectioning in the cryostat. The tissue must be fresh, and placed in formol calcium at 4°C. The block is fixed at this temperature for 18 hours.

Procedure

1. Fix tissue block in formol calcium at 4°C for 18 hours.
2. Rinse in running tap water.
3. Blot and dry.
4. Place tissue in gum sucrose solution at 4°C for 18 hours. The following agents are required.

Gum acacia	2.0 g
Sucrose	60.0 g stored at 4°C
Distilled water	200.0 ml

5. Blot and dry.
6. Freeze tissue on the block holder moderately and slowly to cryostat temperature.

Cryostat Sectioning

Before sectioning, the temperatures of the microtome and cryostat are checked. Most unfixed material section well between –15°C and –23°C. Tissues containing large amounts of water section best at warmer temperature and harder tissue at the cooler end of the range. Unfixed tissues containing moderate to large amounts of fat require as cold a temperature as possible, whereas most fixed tissues give good sections at a range of –7° to –12°C depending upon the hardness of the tissue.

Cutting sections of variable thickness (think or thin) may not be a problem. Then it is better to check with customary sectioning on the bench.

The knife is so designed that it suits the cryostat microtome. Sometimes inaccurate sharpening of the knife can lead to problems. The finish required on the knife edge seems to be less than for paraffin wax sectioning, certainly a very fine edge is rapidly removed by a hard frozen block of tissue and for this reason, many workers feel that stropping is unnecessary.

Disposable blades are routinely used in many laboratories. Stainless steel and Teflon-coated blades are used. One advantage is that they are rapidly cooled as a result of their size.

Speed of cutting is an important factor. It is better to cut soft tissues at a slow rate while the harder tissues will often section better at a slightly faster speed. The cut section rests on the knife face after cutting and may be picked upon to a slide or a coverslip. Sections of fixed blocks have a tendency to float off the slide or coverslip during incubation or staining. This can easily be avoided by coating the slide or coverslip with gelatine–formaldehyde mixture before picking up the sections.

Gelatine–formaldehyde mixture

1 per cent gelatine	5.0 ml
2 per cent formaldehyde	5.0 ml

Coat the slides and coverslips with this mixture and dry them at 37°C for 1 hour.

Staining Procedure

1. Fix sections in formol acetic formaldehyde for 20 seconds.
2. Rinse in water.
3. Place in Harris haematoxylin for 2 minutes.
4. Rinse in saturated lithium carbonate.
5. Rinse in water.
6. Stain in eosin for 10 seconds.
7. Rinse in tap water.
8. Dehydrate, clear and mount in DPX.

FREEZE-DRYING

Generally it is not used routinely in laboratories though it has applications in immunocytochemistry. Due to its cost and the time involved with the method, its use is restricted to histochemical and research laboratories. Freeze-dried tissue has to be fixed either before embedding or as individual sections. There are some disadvantages with this method such as loss of soluble substances, displacement of cell constituents, destruction of enzyme, denaturation of proteins and chemical alterations of reactive groups.

This was actually introduced by Altman in 1890, and Gersh modified it in 1932. Pearse (1963) changed the technique of freeze-drying with a thermoelectric dryer.

This procedure is actually quenching of fresh tissue at $-160°C$ and subsequent removal of water molecules by sublimation in vacuum at a higher temperature ($-40°C$). The freeze-dried blocks are raised to room temperature. Freeze-drying involves "quenching", "drying", "fixation" "embedding" subsequent treatment. Quenching stops the chemical reactions within the tissue and brings the tissue to a solid state, turning the water in the tissue to ice. This way the diffusion of chemical constituents within the tissues are stopped. Freezing may produce ice crystals which may lead to damage of the tissue, and if large crystals are formed the morphological structure is changed. The more rapid the freezing, the smaller are the ice crystals formed. Liquid nitrogen at $-190°C$ has a low temperature but its conductivity is low due to the formation of vaporized air around the tissue, slowing down the transfer of heat from the tissue to the freezing solution. Freon 22 has a high rate of thermal conductivity and if this is cooled to $-160°C$ by placing it in a bath of liquid nitrogen, then rapid freezing will be obtained.

Drying is the most time-consuming process as some tissues contain as much as 70–80 per cent water by weight and this has to be removed without causing damage to the tissue. Drying can be done in 3 stages.

1. Introduction of heat to the tissue to cause sublimation of ice.
2. Transfer of water vapour from the ice crystals through the dry part of the tissue.
3. Removal of water vapour from surface of the specimens (Merryman, 1960).

Drying of tissue takes place when heat is supplied to the frozen tissue in a vacuum of 133 mpa. The heat vaporizes the water molecules and they pass through the tissue to the surface.

Vapour Fixation

A number of fixatives can be used in their vapour form including formaldehyde, glutaraldehyde and osmium tetroxide. The most important is formaldehyde. It fixes excellently and tissues fixed in formaldehyde vapour can be used for all histochemistry except enzymes.

Embedding follows fixation. A number of waxes are used but paraffin wax usually gives good results. Celloidin embedding is recommended for enzyme (Burstone, 1960).

Embedding

When the tissue is dry, it is allowed to come up to room temperature at atmospheric pressure. It will not absorb moisture unless drying is complete when tissue will rapidly reabsorb water. The dried tissue is friable and even slight pressure will cause the tissue to disintegrate into a fine powder. The delicate tissue is ready for re-embedding and sectioning.

Freeze-drying was suggested for

- Demonstration of hydrolytic enzymes
- Fluorescent antibody studies
- Autoradiography
- Microspectrofluorometry of autofluorescent substances
- Formaldehyde-induced fluorescence
- Mucosubstances
- Proteins
- Scanning electron microscopy

FREEZE SUBSTITUTION

This technique was originated by Simpson (1941). It involves rapid freezing of small pieces of tissues in a similar manner for freeze-drying and the substitution of ice in such tissue by placing them in dehydrating agents at sub-zero temperatures. Freeze substitution is basically of two types:

1. Quenched tissue substituted in a dehydrating fluid.
2. Tissue dehydrated in a fluid which also contains a fixative.

In both the cases it is important to remember the action of fixatives and dehydrating fluids at these sub-zero temperatures. Absolute alcohol will not fix the tissue at −70°C. Following substitution at −70°C for 5 days, the tissue is placed in a fresh substituting fluid at 4°C before embedding in wax with a low melting point wax. The preservation obtained by freeze-substitution is excellent especially for mucosubstances, glycogen, proteins, etc., but the drawback is the amount of shrinkage caused.

Freeze-substitution Technique (Feder and Sidmon, 1958)

1. Quench small tissue blocks in isopentane cooled to −170°C with liquid nitrogen.
2. Transfer blocks to bottles containing 1 per cent mercuric chloride in ethanol at −70°C. Maintain at −70°C for 5 days with frequent agitation.
3. Rinse specimen with fresh ethanol at −70°C to remove fixative.
4. Transfer tissues and ethanol to 4°C until this temperature is attained.
5. Transfer tissue to absolute alcohol giving 3 changes of 8 hours each.
6. Place tissue in chloroform at 4°C for 12 hours.
7. Transfer to fresh chloroform at 4°C and then allow chloroform to reach room temperature.
8. Transfer tissue to paraffin wax (52°C MP) at 54°C giving 2 changes of 15 minutes each.
9. Embed tissue in fresh wax.

The demonstration of enzymes, mucosubstances and proteins have been shown to be good. The preservation of lipids however is poor. The freeze-substitution technique involves the rapid freezing of the tissue to −160°C in isopentane super cooled by liquid nitrogen. Cryostat sections are cut at 8–10 μm and sections are placed in a cold container maintained at cryostat temperature. Then the sections are transferred to water-free acetone and cooled to −70°C for 12 hours. The sections are floated onto coverslips or slides and allowed to dry. The histochemical methods are then applied.

CARBOHYDRATES

9

Histochemical differentiation of polysaccharide–protein complexes is of importance for studying both living and fixed tissues. Carbohydrates were defined initially as compounds containing only carbon, hydrogen and oxygen. However, more recently other compounds, which are similar both structurally and functionally to classical carbohydrates, such as their ester sulphates, are also considered to be carbohydrates. Some histochemists and histopathologists, while discussing carbohydrates, treated them under mucins and mucopolysaccharides. Spicer *et al.* (1965) first coined the term "mucins" as substances that are rich in carbohydrate and can be characterized biochemically. Mucins can be histochemically analysed. Mucosubstances could not be characterized biochemically, and this led to controversies as far as their histochemical and biochemical classifications are concerned. Some investigators (Meyer, 1953, Jeanloz, 1960; Spicer *et al.*, 1965; Pearse, 1968, Zugibe, 1970) suggested that the prefix "mucin" should not be used to describe or classify carbohydrates. The classification scheme for carbohydrates formulated by Meyer (1953, 1966), Spicer *et al.* (1965), Pearse (1968) and Cook (1972, 1974), has been followed for a long time.

Carbohydrate-containing mucosubstances consist of hexosamine sugars, proteins and lipids in differing amounts. Neutral mucopolysaccharides can be demonstrated by the use of periodic acid/Schiff (PAS) reaction, and acid mucosubstances by the use of acid radicals. These mucopolysaccharides can be classified as connective tissue mucosubstances, epithelial mucosubstances and those that are sensitive to enzyme digestion by using, for example, testicular hyaluronidase.

POLYSACCHARIDES

Glycogen is a glucose polymer that is stored in the hepatocytes of the liver. It is easily and clearly demonstrable histochemically. After death, tissue glycogen breaks down to glucose. Unfixed cryostat sections are excellent for demonstration of glycogen. They are far superior over routine paraffin sections (Laske and Mayersbach, 1969; Lake, 1970). Mercury fixatives should be avoided. The best results are obtained when the tissue is fixed either in formol saline with prolonged washing (Lillie, 1947) or Gendres fixative at 4°C (Bancroft, 1975). Glycogen can be demonstrated histochemically by the periodic acid/Schiff (PAS) or the empirical Best's carmine techniques. Though the PAS technique is the best one for glycogen, there are other substances which are also PAS-positive. The amylase digestion test, which abolishes the PAS reaction, is confirmative. Silver techniques and iodine have also been used, but they never received much attention and hence are little used today. The enzyme diastase digests glycogen in a microscopic section whereas other PAS-positive carbohydrates are resistant. Amylase is also used in a 0.5–1.0 per cent solution in phosphate buffer, pH 6, or distilled water to digest glycogen. Sodium chloride as a coenzyme could be added. This solution should be freshly prepared. Human saliva which contains amylase can also be used. The sections should be kept in amylase for 1 hour at 37°C. Diastase could be used in conjunction with Best's carmine before or after the staining procedure. But it should be used only prior to oxidation with the periodic acid in the PAS technique.

HISTOCHEMICAL METHODS FOR GLYCOGEN

Fixation of Glycogen

Fixatives containing large amounts of picric acid, like Gendre's, are considered best. Formol saline is also a very good fixative as is alcohol-picric acid. However, freeze-drying is far superior to using such fixatives. Best (1906) suggested celloidin embedding immediately following alcohol fixation. But Lillie (1947) and Vallance-Owen (1948) stated that the tissues require a 24-hour wash after fixation.

Bauer–Feulgen Method (Bauer, 1933)

Reagents Required

    Chromium trioxide
    Schiff's reagent
    Potassium or sodium metabisulphite
    Hydrochloric acid
    Distilled water

Preparation of Reagents

    Solution 1 (Chromium trioxide solution)
        Chromium trioxide   4.0 g
        Distilled water   100.0 ml

Solution 2 (Schiff's reagent)

Boil 1 g of basic fuchsin in 100 ml of distilled water. Shake the flask, allow it to cool, filter. To the filtrate add 20 ml of 1N hydrochloric acid and 1 g of sodium or potassium metabisulphite. Store the solution in dark for 24 hours. After 24 hours add 200–300 mg of activated charcoal. Filter and store in an amber-coloured bottle at 4°C.

Solution 3 (Sulphurous acid rinse)

10 per cent sodium metabisulphite (10 g/100 ml water)	5.0 ml
1N Hydrochloric acid	5.0 ml
Distilled water	90.0 ml

Procedure

1. Deparaffinize and hydrate slides to water.
2. Place sections in solution 1 for 30 minutes.
3. Wash in tap water for 5 minutes.
4. Transfer sections to solution 2 (Schiff's reagent) for 15 minutes.
5. Shift to solution 3 giving 3 changes of 3 minutes each.
6. Wash in tap water.
7. Counterstain with 0.1 per cent light green.
8. Wash in distilled water.
9. Dehydrate, clear in xylol and mount in Canada balsam or DPX.

Result

Glycogen	Red
Nuclei	Green

This gives superior results with frozen sections.

Bauer (1933) introduced this method for the demonstration of glycogen. Glycogen is first oxidized by 4 per cent chromic acid and this is followed by treating with Schiff's reagent. All carbohydrates except 1-2 glycols are over-oxidized and do not take up the stain. This method is excellent with cryostat or frozen sections.

Carbohydrates other than 1 : 2 glycol groups are over-oxidized and will not affect staining.

Best's Carmine Method (Best, 1906) (Plate 1, Figures 1–3)

Fixatives

Paraffin, freeze-dried and frozen sections can be used. Paraffin sections should be fixed in Bouin's.

Reagents Required

Carmine
Potassium carbonate
Potassium chloride
Ammonia (–880)

Methyl alcohol
Absolute alcohol
Distilled water

Preparation of Reagents

Solution 1 (Best's carmine stock)

Carmine	2.0 g
Potassium carbonate	1.0 g
Potassium chloride	5.0 g
Distilled water	60.0 ml

Boil the solution for 5 minutes (gently), then cool and filter. Add 20 ml of ammonia to the filtrate.

Solution 2 (Best's carmine working solution)

Stock solution	12.0 ml
Ammonia	18.0 ml
Methyl alcohol	18.0 ml

Solution 3 (Best's differentiator)

Absolute alcohol	8.0 ml
Methyl alcohol	4.0 ml
Distilled water	10.0 ml

Procedure

1. Deparaffinize and bring the sections to 70 per cent alcohol.
2. Keep the sections in 1 per cent celloidin for 5 minutes.
3. Keep the sections in tap water.
4. Transfer to alum haematoxylin.
5. Wash in tap water.
6. Place sections in solution 2 for 30 minutes.
7. Rinse briefly in solution 2 giving 3 changes of 20 seconds each.
8. Wash in 90 per cent alcohol.
9. Place in absolute alcohol.
10. Clear in xylene and mount in DPX.

Result

Glycogen	Red
Nuclei	Blue

Remarks

This method, first introduced by Best (1906), has been modified by several investigators. Mucin also stains but not as intensively as glycogen. Any tissue section known to contain glycogen should be used as a control. It is better to use diastase in the incubating medium as in the previous case.

Carbohydrate–PAS reaction
(Plate 1, Figures 3 and 4; Plate 2, Figures 1 and 2; Plate 3, Figures 1 and 2)

Fixatives

All types

Reagents Required

Periodic acid
Basic fuchsin
Hydrochloric acid
Potassium metabisulphite

Preparation of Reagents

Solution 1 (Periodic acid)
Periodic acid   1.0 g
Distilled water   100.0 ml

Solution 2 (Schiff's Reagent) (*see* page 119)

Procedure

1. Deparaffinize and bring sections down to water.
2. Place in solution 1 (1 per cent periodic acid) for 10 minutes.
3. Wash in tap water.
4. Treat with Schiff's reagent for 30 minutes.
5. Wash in tap water.
6. Counterstain in 0.5 per cent light green for 5 minutes.
7. Wash in distilled water.
8. Dehydrate, clear and mount in Canada balsam.

Result

PAS-positive material   Magenta
Nuclei and proteins   Green

Rationale

Periodic acid attacks 1, 2-glycols and 1, 2-amino alcohols and oxidizes adjacent aldehydes by breaking the carbon chain. When Schiff's reagent is prepared, sulphurous acid breaks the quinoid structure of the basic fuchsin, and as a result, the solution appears clear. When sections containing aldehyde are treated with Schiff's reagent, the quinoid structure is restored and the resulting product colours the sites of aldehydes.

Diastase Digestion for Glycogen (Lillie, 1949)

Reagents Required

Malt diastase
Sodium chloride

Preparation of Reagents

Solution 1 (Malt diastase)

0.02 M phosphate buffer (pH 6.0)	50.0 ml
Malt diastase	500 mg
Sodium chloride	400 mg

Procedure

1. Bring sections down to water (2 sections).
2. Treat one section with solution 1(diastase solution) for one hour at 37°C and keep another section in distilled water at 37°C for one hour.
3. Rinse in water.
4. Treat both sections with Schiff's reagent.
5. Wash, dehydrate, clear and mount.

Result

Glycogen in sections treated with diastase is abolished. Glycogen in sections immersed in distilled water is intact.

Rationale

Diastase digests glycogen, hence the negative result in sections treated with this enzyme. In the section not subjected to enzyme digestion, the glycogen-containing regions remain PAS-positive.

Lead Tetra-acetate Schiff Method (Shimazu and Kumamoto, 1952)

Fixation

Any general fixative can be used.

Reagents Required

Lead tetra-acetate
Acetic acid
Sodium acetate
Schiff's reagent

Preparation of Reagents

Solution 1

Lead tetra acetate	1.0 g
Glacial acetic acid	30.0 ml
Saturated sodium acetate	70.0 ml

Solution 2

1.0 M Sodium acetate solution (*see* Chapter 5)

Solution 3

Schiff's reagent (*see* page 119)

Procedure

1. Deparaffinize and hydrate the slides to water.
2. Place in solution 2 for 3 minutes.
3. Transfer to solution 1 for 10 minutes.
4. Place again in solution 2 for 5 minutes.
5. Wash in distilled water.
6. Treat with Schiff's reagent for 15 minutes.
7. Differentiate in sulphate water (sulphurous acid).
8. Wash in running water.
9. Dehydrate, clear and mount.

Result

Glycogen and other mucosubstances stain reddish purple.

Acrylamine–Aldehyde Condensation (Lillie, 1962)

Fixation

Formalin

Reagents Required

Periodic acid
Glacial acetic acid
m-Aminophenol
Fast black salt K
Hydrochloric acid
Veronol

Preparation

Solution 1 (Periodic acid solution)
Periodic acid   1.0 g
Distilled water   100.0 ml

Solution 2 (Aminophenol solution)
m-Aminophenol   11.0 g
Glacial acetic acid   100.0 ml

Solution 3 (Fast red salt K solution)
Fast black salt K   300 mg
0.1 M Veronol–HCl buffer   100.0 ml
(pH 8.0)

Solution 4
0.1N Hydrochloric acid

Procedure

1. Deparaffinize and hydrate slides to water.
2. Place in solution 1 for 10 minutes.

3. Wash in running water for 10 minutes.
4. Place slides in glacial acetic acid for 1 minute.
5. Transfer to solution 2 and keep for 1 hour.
6. Wash in running water for 10 minutes.
7. Azo couple for 2 minutes at 3°C in solution 3.
8. Differentiate in solution 4 for 5 minutes going 3 changes.
9. Wash in running water.
10. Dehydrate, clear and mount.

Result

PAS-reactive sites	Black
Other structures	Grey

Mucicarmine Method for Mucosubstances (Mayer, 1896) (Modified by Southgate, 1927)

Fixation

Any general fixative can be used.

Reagents Required

Carmine
Aluminium hydroxide
Aluminium chloride
Absolute alcohol
Distilled water

Preparation of the Reagent

Solution 1

Carmine	1.0 g
Aluminium hydroxide	1.0 g
Distilled water	50.0 ml
Absolute alcohol	50.0 ml

Shake well and then add 500 mg aluminium chloride. Boil the solution for just 3 minutes, then cool and restore to original volume by adding 50 per cent alcohol. Filter. This solution is stable for 12 months. From the stock solution, prepare a working solution by mixing 1 part of the stock with 4 parts of distilled water. Store at 4°C.

Procedure

1. Deparaffinize and hydrate slides to water.
2. Stain in haematoxylin for 10 minutes.
3. Wash in tap water.
4. Differentiate in 1 per cent alcohol.
5. Wash in running water for 5 minutes.

6. Transfer to solution 1 (staining solution) for 30 minutes.
7. Wash, dehydrate, clear and mount in Canada balsam.

Results

Mucosubstances	Red
Nuclei	Blue

Remarks

Although this is a good method for acid mucosubstances as originally described, the modification of adding aluminium hydroxide enhances the contrast between acid mucosubstances and the background.

Fluorescent Method (Hicks and Matthaei, 1950)

Fixation

10 per cent neutral buffered formalin

Reagents Required

Ferric ammonium sulphate
Acridine orange

Preparation of Reagents

Solution 1 (Iron alum)
Ferric ammonium sulphate   4.0 g
Distilled water   100.0 ml

Solution 2 (Acridine orange)
Acridine orange   100 mg
Distilled water   100.0 ml

Procedure

1. Dewax and hydrate slides to water or cut frozen sections and place them in water.
2. Transfer to solution 1 for 5–10 minutes.
3. Wash in distilled water.
4. Treat with solution 2 for 1–5 minutes.
5. Rinse in distilled water, blot and mount in fluorescent mountants.

Result

Mucin fluoresces with a reddish orange shade.

Modification of Mayer's Mucihematein
(Laskey, 1950; Lillie, 1965)

Fixation

Absolute alcohol   8 hours
Sublimate solution   5 hours

Reagents Required

Solution 1

Haematoxylin	1.0 g
70 per cent alcohol	100.0 ml
Aluminium chloride	500 mg
Sodium iodate	100 mg

Make up to 500 ml with 70 per cent alcohol.

Procedure

1. Deparaffinize and hydrate slides to water.
2. Flood slides with solution 1 for 10 minutes.
3. Drain off excess stain.
4. Wash in distilled water.
5. Dehydrate, clear and mount.

Results

Mucin	Deep violet
Nuclei	Grey
Connective tissue	Pale grey

Mucicarmine with Haematoxylin and Metanil Yellow (Meyer, 1896; Modified by Masson 1923)

Fixation

10 per cent formalin

Reagents Required

Carmine
Aluminium chloride

Preparation of Reagents

Solution 1

Carmine	1.0 g
Aluminium chloride	0.5 ml
Distilled water	2.0 ml

Dissolve carmine and aluminium chloride in 2 ml water. Heat gently until the mixture becomes dark. Then add 100 ml of 50 per cent alcohol. Stir continuously. Allow the solution to stand for 24 hours, then filter. This forms the stock solution. Then dilute it with 10 vol. of distilled water.

Solution 2

Metanil yellow	250 mg
0.25 per cent acetic acid	1000.0 ml

Procedure

1. Deparaffinize and hydrate slides to water.
2. Transfer to Weigert's haematoxylin (*see* Chapter 7) for 1 minute.

3. Wash in distilled water.
4. Stain sections in solution 2 for 30 seconds.
5. Wash in distilled water.
6. Transfer to solution 1 for 45 minutes.
7. Wash quickly, dehydrate, clear and mount.

Results

Connective tissues	Yellow
Mucin	Red
Nuclei	Black

MUCOPOLYSACCHARIDES

One of the most important carbohydrate-containing substances are the "mucins". Mucin, a slimy substance, is a secretion of epithelial cells in many tissues, especially respiratory, alimentary and even epidermal. The term "mucin" is frequently used; however, in recent years it has been increasingly referred to as "mucosubstance" or "glycosaminoglycol" by various investigators.

Glycosamines and galactosamines, which are the hexosamine units, are constituents of mucins. These are monosaccharides containing nitrogen and are called amino sugars. These monosaccharides combine with sialic acid or glucuronic acid. In addition they may also contain sulphate esters.

Mucins according to Peterson and Leblond (1964), are mostly secreted by cells of connective tissue and some epithelia. But they are also found in columnar cells, cuboidal cells of epithelia, osteoblasts, mast cells (Curran and Collins, 1957) and fibroblasts (Curran and Kennedy, 1955).

Mucins can be broadly categorized into (a) mucopolysaccharides, (b) mucoproteins and (c) mucolipids.

1. *Mucopolysaccharides*   These are polysaccharide–protein complexes. Carbohydrates are the main components of mucins.

2. *Mucoproteins*   As the name suggests, they are predominantly protein in nature. They are complex combinations of polysaccharides and proteins (e.g. glycoproteins), occurring mostly in beta cells, the thyroid, the pituitary and basement membranes, and are reactive to protein tests and are PAS-positive. Their polysaccharide reactivity is masked to some extent by the protein component.

3. *Mucolipids*   Mucolipid is synonymous with glycolipid. A mucolipid is a combination of polysaccharide and fatty acids resulting in the formation of a complex substance. These fatty acids are predominantly cerebrosides and gangliosides. Mucolipids are PAS-positive as well as being positive to all lipid techniques.

Histochemically, mucopolysaccharides have been further subdivided into:

a)   Neutral mucopolysaccharides
b)   Acid mucopolysaccharides

Neutral Mucins

Neutral mucins do not have free acidic groups, but possess hexosamine and hexose units, and have immunologically reactive neutral glycoproteins. Fucomucins are mannose-rich mucosubstances found in connective tissue and epithelia. Neutral mucins are common in mammals in the epithelial lining of the stomach and Bruner's glands of the duodenum, and in the accessory and salivary glands of most invertebrates. Neutral mucins are PAS-positive and periodate paradiamine-reactive, but alcian-blue-negative.

Acid Mucins

These contain hexosamine units like neutral mucins but are linked with glucuronic acid, iduronic acid or sialic acid in addition to sulphate radicals. Acid mucins are subdivided into two groups, (a) sulphated and (b) non-sulphated groups and/or carboxylated group.

1. Strongly sulphated connective tissue mucins   These are connective tissue mucins, such as chondroitin sulphate, heparin sulphate and keratin sulphate, which are alcian-blue-positive at pH 0.5 and below and are PAS-negative or unreactive and metachromatic at pH 1.0 and below with Azure A.

Connective tissue mucins   These are PAS-negative mucins which react at low pH values with suitable cationic dyes. They occur in mast cells, chondrocytes, osteocytes, endothelial cells and fibroblasts. The subtypes are:

Chondroitin sulphate A   It contains D-galactosamine and D-glucuronic acid.

Chondroitin sulphate B   It is similar to 'A' but in addition contains ioduronic acid.

Chondroitin sulphate C   It has a chemical composition similar to 'A' but is sulphate-esterified.

Hyparin/heparin sulphate (Synonym of hepartin)   Both contain N-acetyl-D-glucosamine, N-sulphate-D-glucosamine and D-glucuronic acid. Heparin in addition contains a sulpho-amino group and iduronic acid.

Keratin sulphate   It contains sulphate-esterified N-acetyl glucosamine and galactose but no uronic acid.

Hyalurono sulphate   It contains N-acetyl-D-galactosamine and D-glucuronic acid.

2. Strongly sulphated epithelial mucins   They are epithelial in origin. They react at low pH levels with cationic dyes just like connective tissue mucins, but unlike them these are PAS-positive.

Sulphated (weakly acidic)   In contrast to the strongly acidic sulphated mucins, these are periodate-reactive and have a weak basophilia (Spicer *et al.*, 1965). The reactive acid of these mucins is the sulphate radical. They are positive to alcian blue at pH 0.5–1.5.

Non-sulphated (carboxylated) sialic-acid-containing mucins   The acid radical is sialic acid, an acetylated derivative of neuraminic acid, and can be digested by the enzyme neuraminidase. Alcian blue reactive groups are the carboxyls which take alcian blue at pH 2.5 and above. They are metachromatic at pH 3.0 and above with Azure A. According to Spicer (1963) sialomucins contain no uronic acid or sulphate esters. Their frequent occurrence in goblet cells of lungs, intestine and submandibular salivary gland mucous cells has enabled extensive investigation.

Non-sulphated uronic acid-containing mucins   The acid radical is hyaluronic acid. It reacts as an acid mucin by virtue of the carboxyl group on D-glucuronic acid and is degraded by the enzyme hyaluronidase. Like other sulphated mucins (as far as alcian blue pH levels are concerned), they are metachromatic with Azure A and with toluidine blue but are negative to PAS.

Fixation of Mucins

Aqueous fixatives such as formalin are preferred (Lamb, 1969; Allison, 1973). Hyaluronic acid can be demonstrated better when the tissue is fixed in 10 per cent formol alcohol, but at the same time there is a loss of sialomucins from alcohol-fixed tissue (Sorvari and Lauren, 1973). Other fixing solutions suitable for mucin histochemistry are 1 per cent cetyl pyridinium chloride in 10 per cent formalin and 3 per cent calcium acetate in 10 per cent formalin. However, freeze-drying gives the best results. Fresh frozen tissues clearly give better results than paraffin sections (Lake, 1970).

Newcomer's and dioxane are not inferior to the above-mentioned fixatives. Susa and Zenker–acetic acid solutions are not suitable here according to Bancroft and Cook (1994). However, this view is not shared by the present authors. In our extensive studies of invertebrate tissues ranging from annelids to echinoderms, we have found that Susa and Carnoy fixatives almost always gave good results. We mostly used paraffin sections.

ALCIAN DYES

Alcian dyes are Alcian blue 8GX, Alcian yellow, Alcian green 2GX and Alcian green 3BX.

These are the most commonly used histochemical dyes. Earliest among these is alcian blue, introduced by Haddock (1948), which is a copper phthalocyanin. This dye was extensively used in cotton dyeing. Steedman (1950) introduced it to the histological world. This stain has several important qualities: 1) It stains acid mucins specifically and intensely. 2) Once the tissue is stained with alcian blue, the binding of stain to the section is so great that it is not affected by subsequent water or alcohol treatment (personal observation). When the foot of a pelecypod or gastropod, where mucins are abundant, is stained and processed in the usual manner, the sections appear colourless but when observed under a microscope the mucin sites are stained brilliantly. 3) Sections retain the stain even after several years of storage (We have a collection of sections of the foot of molluscs and salivary and accessory glands of crustaceans that were stained 25 years ago and the stain has still not faded).

Any tissue component will be intensely stained if the dye is used at the pH at which the reacting groups are fully ionized. Sulphated mucins react at a lower pH than do carboxylated mucins. Alcian blue allows pH-dependent identification of mucins. For example, strongly sulphated mucins react more consistently at low pH levels (below 1.0). Weakly sulphated mucins will stain well at pH 1.0–2.5. Hyaluronic acid and *N*-acetyl sialomucin will stain well between pH 1.7 and 3.2. *N*-acetyl O-acetyl sialomucin stains well at a pH below 1.5.

Preparation of alcian blue at various pH levels

pH of AB	Reagents used	Quantity
AB at pH 0.2	Alcian blue	1.0 g
	10 per cent sulphuric acid	100.0 ml
pH 0.5	Alcian blue	1.0 g
	0.2N hydrochloric acid	100.0 ml
pH 1.0	Alcian blue	1.0 g
	0.1N hydrochloric acid	100.0 ml
pH 2.5	Alcian blue	1.0 g
	3 per cent acetic acid	100.0 ml
pH 3.2	Alcian blue	1.0 g
	0.5 per cent acetic acid	100.0 ml

HISTOCHEMICAL TECHNIQUES FOR MUCOPOLYSACCHARIDES

Haddock (1948) introduced alcian blue to cotton dying. Later Steedman (1950) introduced alcian blue to histochemical techniques. By varying the pH levels of alcian blue solutions, it is possible to show the presence of acid mucins [70 per cent sulphuric acid (pH 0.2), N-5 hydrochloric acid (pH 0.5)], N-10 hydrochloric acid (pH 1.0) because alcian blue at pH 2.5 is specific for acid mucins.

Alcian Blue (pH 2.5) (Plate 4, Figures 2 and 3)

Reagents Required

Alcian blue 8 GX

Preparation of Reagents

Solution 1 (Alcian blue solution)
Alcian blue 8 GX   1.0 g
3 per cent acetic acid   100.0 ml

Procedure

1. Deparaffinize and hydrate slides to water.
2. Stain in solution 1 (alcian blue) for 30 minutes. With ageing, more time is required for staining.
3. Wash in running tap water.
4. Dehydrate, clear and mount in Canada balsam.

Result

Weakly acidic sulphated mucopolysaccharides, hyaluronic acid and sialomucins — Dark blue

Strongly acidic chromotropes (those metachromatic at pH 5.0) — Light blue

Alcian Blue (pH 2.8) (Putt, 1971)

Fixation

Any general fixative, but Rossman's fluid or formalin/99 per cent alcohol (1:9) is preferable.

Reagents Required

Calcium chloride
Alcian blue 8GX
Kernechtrot
Aluminium sulphate

Preparation of Reagents

Solution 1

Calcium chloride	500 mg
Alcian blue	1.0 g
Distilled water	100 ml

Dissolve calcium chloride and filter. Add a crystal of thymol.

Solution 2 (Kernechtrot)

Kernechtrot (nuclear stain)	100 mg
5 per cent aluminium sulphate	100.0 ml
(5 g + 100 ml water)	

Procedure

1. Deparaffinize and hydrate slides to water.
2. Place in 3 per cent acetic acid for 3 minutes.
3. Transfer to solution 1 for 30 minutes.
4. Rinse in water and place in 0.3 per cent sodium bicarbonate for 30 minutes.
5. Wash in running water.
6. Counterstain with solution 2 for 5 minutes.
7. Rinse in distilled water.
8. Dehydrate, clear and mount.

Result

Acid mucopolysaccharides	Blue-green
Nuclei	Red

Alcian Blue (pH 1.0) (Luna, 1968) (Plate 4, Figure 1; Plate 5, Figure 1)

Reagents Required

Alcian blue
Hydrochloric acid

Preparation of the Reagent

Alcian blue solution

Alcian blue 8GX	1.0 g
0.1N hydrochloric acid	100.0 cc

(1 ml HCl in 100 ml distilled water)

Procedure

1. Hydrate slides.
2. Stain in alcian blue solution (pH 1.0) for 30 minutes.
3. Blot and dry with filter paper (No rinsing).
4. Dehydrate, clear and mount.

Result

Sulphated mucosubstances	Greenish blue
Non-sulphated mucosubstances	Remain unstained

Remarks

Steedman (1950) first introduced alcian blue for acid mucosubstances. In these techniques, alcian blue stains with salt linkage to the acidic group in acid mucosubstances. By altering the pH of the alcian blue solution, different types of acid mucosubstances can be distinguished. At pH 0.2, only strongly sulphated mucosubstances stain. By slightly increasing the pH to 1.0, both weakly and strongly sulphated mucosubstances take the stain. At pH 2.5, only acid mucosubstances take the blue colour. According to Cook (1974), even these are not satisfactory techniques. It is better to use the critical electrolyte concentration method (CEC) to obtain precise results. This method was first introduced by Scott and Dorling (1965).

Combined Alcian Blue/Alcian Yellow Method
(Ravetto, 1964) (Plate 11, Figures 1 and 3)

Reagents Required

Alcian blue
Alcian yellow
Hydrochloric acid
Neutral red

Preparation of Reagents

Solution 1 (Alcian blue solution)
Alcian blue 8GX   1.0 g
0.2N Hydrochloric acid 0.5 per cent   100.0 ml

Solution 2 (Alcian yellow solution)
Alcian yellow   3.0 g
3 per cent acetic acid   100.0 ml

Procedure

1. Hydrate slides.
2. Rinse in N/5 hydrochloric acid.
3. Transfer to solution 1 (alcian blue solution) for 30 minutes.
4. Rinse in N/5 hydrochloric acid and then in water.
5. Stain in solution 2 (alcian yellow solution) for 30 minutes.
6. Wash and stain with neutral red.
7. Wash in distilled water.
8. Dehydrate, clear and mount.

Result

Sulphated mucins   Blue
Carboxylated mucins   Yellow
Mixture of the above   Green

Remarks

This method is meant to differentiate between sulphated (blue) and acid (yellow) mucopolysaccharides and glycoproteins. Though Carlo (1964) and Staple (1967) recommended this method, its specificity has been questioned (Sorvari and Sorvari, 1969). In fact, we too have found this method to be not of much use.

Rationale for Alcian Dye

This is the most specific dye for acid mucosubstances. It is actually a copper phthalocyanin dye, giving a brilliant green-blue colour. Alcian blue stains by forming a salt linkage with acidic groups. If the staining time is short, only acid mucosubstances will stain. Different acid mucosubstances can be differentiated by altering the pH, i.e., at pH 2.5, acid mucosubstances stain, at pH 1.0, both weakly and strongly sulphated mucosubstances stain. At a still lower pH of 0.2, only strongly sulphated mucosubstances will take the stain.

Alcian Blue with Varying Electrolyte Concentrations (CEC) (Scott and Dorling, 1965) (Plate 9, Figures 1–4)

Reagents Required

Alcian blue
Magnesium chloride
Acetate buffer

Preparation of Reagents

Alcian blue	50 mg
Acetate buffer pH 5.8 (*see* Chapter 5)	100.0 ml

This forms the stock solution. Suggested concentrations of magnesium chloride for different substances (see result, below) are 0.06 M, 0.2 M, 0.3 M, 0.5 M, 0.7 M and 0.9 M. To obtain these molarities, the amount of magnesium chloride to be added to 100 ml alcian blue stock is as follows:

0.06 M	1.2 g magnesium chloride
0.3 M	6.1 g magnesium chloride
0.5 M	10.15 g magnesium chloride
0.7 M	14.2 g magnesium chloride
0.9 M	18.3 g magnesium chloride

Procedure

1. Deparaffinize and hydrate slides to water.
2. Stain in alcian blue containing magnesium chloride for 4 hours.
3. Rinse in distilled water.
4. Stain with 0.5 per cent neutral red.
5. Wash in tap water.
6. Dehydrate, clear and mount in Canada balsam.

Result

The results showed positive at the following magnesium chloride concentrations:

0.06 M	Carboxyl and sulphated mucins
0.2 M	Most sulphated mucosubstances including those positive which are metachromatic with Azure A at pH 0.5
0.3 M	Weakly and strongly sulphated mucosubstances
0.5 M	Strongly sulphated
0.7 M	Highly sulphated connective tissue mucins
0.9 M	Keratin sulphate

Remarks

Various sulphated mucosubstances lose alcianophilia at different levels of pH with increasing quantity of magnesium chloride.

Rationale

In this technique magnesium chloride when incorporated into the alcian blue solution will combine with alcian blue molecules for binding to the reactive constituents of acid mucopolysaccharides. This technique was utilized, for example, by Goldstein and Horobin (1974).

Alcian Blue at pH 2.5 and pH 1.0/PAS (Mowry, 1956)

(Plate 4, Figures 4–6; Plate 5, Figures 2–4)

Reagents Required

Alcian blue
Periodic acid

Schiff's reagent
Sodium metabisulphite

Preparation of Reagents

Solution 1
Alcian blue pH 2.5 (*see* page 127)
Alcian blue pH 1.0

Solution 2
Periodic acid   1.0 g
Distilled water   100.0 ml

Solution 3
Schiff's reagent (*see* page 115)

Procedure

1.   Deparaffinize and hydrate slides to water.
2.   Place in solution 1 (alcian blue) for 30 minutes.
3.   Wash in distilled water.
4.   Treat with solution 2 (periodic acid) for 10 minutes.
5.   Wash in water.
6.   Stain in solution 3 (Schiff's reagent) for 10 minutes.
7.   Rinse in sodium metabisulphite.
8.   Wash in distilled water.
9.   Dehydrate, clear and mount in Canada balsam.

Result

Hyaluronic acid and sialomucins (both the strongly and weakly acidic)	Blue (at pH 2.5)
Sulphated mucosubstances	Blue (at pH 1.0)
Neutral mucosubstances	Red (at pH 2.5)
Sialomucins	Red (at pH 1.0)

Remarks

It is a very good technique to distinguish acid from neutral mucins.

Combined Deamination–Alcian Blue/PAS Technique (Lillie, 1954a; Cook, 1974)

Preparation of Reagents

Sodium nitrite	6 g
Distilled water	35 ml
Mix and add glacial acetic acid	5 ml

Procedure

1.   Deparaffinize and hydrate slides to water.
2.   Treat both control and test sections with freshly prepared reagent for 15 hours at room temperature.

3. Wash all sections for 15 minutes.
4. Apply alcian blue/PAS technique to treated as well as to two slides which have not been treated with the deaminating fluid.
5. Rinse in absolute alcohol.
6. Clear in xylene and mount.

Result

Treated sections show negative results because acid mucins are masked.
Untreated sections exhibit positive alcian blue/PAS reactivity.

Rationale

When sections are treated with sodium nitrite and acetic acid, nitrous acid is formed, and as a result, amino groups are displaced by nitrogenation. After the alcian blue/PAS treatment, the tissue sections will show a change from PAS to alcian blue reactivity. Deamination will also destroy tissue acidophilia by blocking protein amino groups.

Aldehyde Fuchsin Method (Plate 3, Figure 3; Plate 6, Figures 1–2)

Reagents Required

Basic fuchsin
Paraldehyde
Hydrochloric acid

Preparation of Reagents

Solution 1

60 per cent alcohol	100.0 ml
Basic fuchsin	500 mg
Paraldehyde	1.0 ml
Conc. hydrochloric acid	1.0 ml

Allow the solution to ripen for 48 hours at room temperature. Keep in refrigerator. As the solution ages, staining time has to be increased.

Procedure

1. Deparaffinize and bring slides to water.
2. Rinse in 70 per cent alcohol.
3. Transfer to solution 1 for 30 minutes.
4. Rinse in 70 per cent alcohol.
5. Dehydrate, rapidly clear and mount.

Result

Sulphated mucosubstances	Deep purple
Non-sulphated mucosubstances	Light blue

Aldehyde Fuchsin/Alcian Blue Method for Sulphated and Carboxylated Mucosubstances (Spicer and Meyer, 1960)
(Plate 6, Figures 3 and 4)

Reagents Required

Basic fuchsin,
Conc. hydrochloric acid
Paraldehyde
Alcian blue
3 per cent acetic acid

Preparation of Reagents

Solution 1
Aldehyde fuchsin (*see* page 136)

Solution 2
Alcian blue (*see* page 130)

Procedure

1. Deparaffinize and hydrate slides to water.
2. Rinse in 70 per cent alcohol.
3. Stain in solution 1 (Halmi's aldehyde fuchsin) for 30–60 minutes.
4. Rinse in 70 per cent alcohol.
5. Transfer to solution 2 [alcian blue (1 per cent alcian blue in 3 per cent acetic acid)] for 30 minutes.
6. Rinse in water.
7. Dehydrate, clear and mount.

Result

Strongly acidic sulphated mucosubstances	Deep purple
Weakly acidic sulphated mucosubstances and carboxylic mucosubstances	Bluish purple
Non-sulphated mucosubstances	Blue

Rationale

The aldehyde–dye complex techniques were first introduced by Gomori (1950a) as an elastic stain. It has great affinity for sulphated mucosubstances. In the usual method, oxidation in a potassium permanganate solution precedes application of the aldehyde fuchsin technique. The rationale is not clear, but when aldehyde combines with basic fuchsin, organically bound sulphur, i.e., sulphated mucins and insulin (b cells of pancreas) (Rhinehart, 1952–1953) are demonstrated.

Aluminium Sulphate Method (Heath, 1961)

This is a confirmatory technique.

Reagents Required

Alcian blue
Aluminium sulphate
Neutral red

Preparation of Reagents

Solution 1 (Alcian blue/aluminium sulphate)
Alcian blue 8GX   100 mg

Dissolve this in a 5 per cent of solution aluminium sulphate (5 g in 100 ml of water), boil, cool and filter.

Solution 2 (Neutral red solution)
0.5 per cent neutral red (500 mg in 100 ml water)

Procedure

1. Deparaffinize and bring slides to water.
2. Stain in solution 1 (alcian blue/aluminium sulphate solution) for 30 minutes.
3. Wash in distilled water.
4. Counterstain in neutral red.
5. Wash in distilled water.
6. Dehydrate, clear and mount.

Result

Sulphated mucins	Blue
Non-sulphated acid mucins	Red
Mixtures of sulphated and non-sulphated acid mucins	Purple

Rationale

According to Heath (1961), mordant-type chelates are formed between aluminium and dyes such as thionin, toluidine blue, methylene blue, nuclear fast red and alcian blue. When alcian blue is dissolved in an aluminium sulphate solution, the dye will bind with mucins containing sulphates. As with the CEC technique, the staining is due to electrolyte competition with the dye ions.

Metachromatic Staining Azure A (Hughesdon, 1949) (Plate 11, Figure 2)

Reagents Required

Oxalic acid
Potassium permanganate
Azure A
Uranyl acetate

Preparation of Reagents

Solution 1 (Oxalic acid)
Oxalic acid   5.0 g
Distilled water   100.0 ml

Solution 2 (Permanganate solution)

    Potassium permanganate   1.0 g

    Distilled water   100.0 ml

Solution 3

    0.2 per cent Azure A aqueous (200 mg Azure A in 100 ml distilled water)

Solution 4

    0.2 per cent uranyl acetate aqueous (200 mg uranyl acetate in 100 ml distilled water)

Procedure

1. Deparaffinize and hydrate slides to water.
2. Treat with solution 2 (1 per cent potassium permanganate) for 5 minutes.
3. Wash in tap water.
4. Bleach in solution 1 (5 per cent oxalic acid) for 5 minutes.
5. Wash in water.
6. Transfer to solution 3 (0.2 per cent Azure A) for 5 minutes.
7. Rinse briefly in water.
8. Differentiate in solution 4 (0.2 uranyl acetate) for 10–20 seconds.
9. Wash briefly in water, blot and dry.
10. Dehydrate, clear and mount.

Result

Acid mucosubstances	Purple-red
Nuclei and background	Blue

Rationale

There are a variety of dyes which are popularly known as metachromatic dyes, such as Azure A, toluidine blue, thionin, etc. Most of the acid mucosubstances are capable of exhibiting metachromasia. These metachromatic dyes when used in dilute solution react with acid mucosubstances, which will result in the production of a colour that is different from the original colour of the dye. The remaining tissues take the colour of the dye. Thionin, toluidine blue and Azure A are all blue, but acid mucosubstances stain red. This red staining is called "metachromatic" and the blue staining is "orthochromatic". As the metachromatic dyes dissolve in alcohol, the sections instead of being passed through a graded series of alcohol should be mounted in glycerine jelly. It is best to examine the slide before mounting.

1. Sulphate mucosubstances alone are metachromatic at pH 2 or below.
2. At higher pH levels, i.e., between 3.5 and 4.5, sulphated mucosaccharides with masked alcianophilia are stained.
3. Sialomucins stain metachromatically at pH 3.0 and above.
4. In some sites, hyaluronic acid stains metachromatically at pH 4.0 or above. Sulphated or sialic-acid-rich epithelial secretions with masked azurophilia stain metachromatically at pH 4.0–4.5.

Technique for Frozen (fresh) Sections
((Dorling, 1980) (Modified by Crown *et al.*, 1983))

Preparation of Reagents

Saturated Azure A in absolute alcohol
Saturated Azure A in 10 per cent alcohol

Procedure

1. Cut 15-μm thick sections (frozen).
2. Treat them with Azure A in absolute alcohol for 10 minutes.
3. Again treat with Azure A in 70 per cent alcohol for 10 minutes.
4. Dehydrate (rapid) in absolute alcohol, clear and mount in synthetic medium.

Results

Glucosaminoglycans   Red-purple
Background   Blue

Toluidine Blue Method (Kramer and Windrum, 1955)
(Plate 7, Figures 1–6; Plate 8, Figure 1)

Reagents Required

Toluidine blue
Ethyl alcohol

Preparation of Reagent

Solution 1
Toluidine blue   100 mg
Absolute alcohol   30.0 ml
Distilled water   70.0 ml

Procedure

1. Deparaffinize and hydrate slides to water.
2. Transfer to solution 1 (toluidine blue) for 15 minutes.
3. Wash in distilled water.
4. Mount in glycerine jelly.

Result

Acid mucopolysaccharides   Pink (metachromatic)
Nuclei   Blue
Toluidine blue at
    pH 3.0   Sulphated mucopolysaccharides are positive
    pH 4.0   Phosphated and carboxylated mucosubstances are positive
    pH 5.0   Carboxylated mucosubstances are positive
    pH 7.0   Carboxylated mucosubstances are positive

Toluidine Blue for Permanent Preparations (Hess and Hollander, 1947)

Fixation

Zenker's solution

Reagents

Toluidine blue
Potassium dichromate
Mercuric chloride
Borax

Preparation of Reagents

Solution 1
Toluidine blue   250 mg
Borax   250 mg
Distilled water 100.0 ml

To a 0.25 per cent borax solution, add 250 mg of toluidine blue.

Solution 2
Prepare a saturated mercuric chloride solution and to this add 0.5 per cent potassium dichromate.

Procedure

1.   Hydrate slides to water.
2.   Cover the slides with solution 1 for 30 seconds.
3.   Apply solution 2 for 30 seconds.
4.   Add fresh dichromate sublimate for 2 minutes.
5.   Blot dry.
6.   Immerse in absolute alcohol for 10 seconds.
7.   Clear in xylene.
8.   Clear in two changes of benzene.
9.   Mount in Canada balsam.

Result

The red (Y) metachromasia is preserved.

Alcian Blue Ruthenium Red Method

Reagents Required

Alcian blue
Hydrochloric acid
Ruthenium red
Acetic acid

Preparation of Reagents

Solution 1 (Alcian blue solution)
    Alcian blue 8GX   500 mg
    0.1N HCl   100.0 ml

Solution 2
    Ruthenium red   500 mg
    3 per cent acetic acid   100.0 ml

Procedure

1.   Deparaffinize and bring sections to water.
2.   Rinse twice in 0.1N hydrochloric acid (pH 1.0).
3.   Stain in solution 1 (alcian blue) for 30 minutes.
4.   Rinse in 0.1N hydrochloric acid.
5.   Rinse in 3 per cent acetic acid.
6.   Transfer to solution 2 (Ruthenium red) for 5–20 minutes.
7.   Rinse twice in 3 per cent acetic acid.
8.   Dehydrate, clear and mount.

Result

Sulphated groups   Blue
Carboxyl groups   Red

Alcian Blue/Safranin Method (Spicer *et al.*, 1967)
(Plate 10, Figures 1 and 2)

Reagents Required

Alcian blue
Acetic acid
Safranin
Hydrochloric acid

Preparation of Reagents

Solution 1 (Alcian blue solution)
    Alcian blue   500 mg
    3 per cent acetic acid   100.0 ml

Solution 2 (Safranin solution)
    Safranin   250 mg
    0.125N hydrochloric acid   100.0 ml

Procedure

1.   Deparaffinize and bring slides to water.
2.   Place in solution 1 (alcian blue) for 30 minutes.
3.   Wash in water.
4.   Stain in solution 2 (safranin) for 30 seconds.

5. Wash rapidly.
6. Dehydrate, clear and mount.

Result

Strongly acid mucosubstances   Red
Other mucosubstances   Blue

Colloidal Iron Method (Mowry, 1958)

Reagents Required

Ferric chloride
Glacial acetic acid
Potassium ferrocyanide
Hydrochloric acid

Preparation of Reagents

Solution 1 (Colloidal iron stock)
    29 per cent ferric chloride 4.4 ml
    Distilled water   250 ml

Boil water and add ferric chloride solution. Solution is dark red. Remove from heat and cool it. Solution must be clear dark red.

Solution 2 (Working solution)
    Glacial acetic acid 5 ml
    Distilled water   15 ml
    Solution 1   20 ml

pH should be 1.1 to 1.3. It lasts for a day.

Solution 3 (Hydrochloric acid–ferrocyanide)
    2 per cent HCl (2 ml/98 ml water)   50 ml
    2 per cent potassium ferrocyanide   50 ml
    (2 g + 100 ml water)

It should be prepared just before use.

Procedure

1. Dewax and hydrate slides to water.
2. Rinse in 12 per cent acetic acid (12 ml acetic acid + 88 ml water) for 30 seconds.
3. Treat with solution 2 for 1 hour.
4. Rinse in 12 per cent acetic acid giving 4 changes of 3 minutes each.
5. Transfer to solution 3 for 20 minutes.
6. Wash in running water.
7. Counterstain with Van Gieson.
8. Dehydrate, clear and mount.

Result

Acid mucopolysaccharides stain prussian blue which include mucins of connective tissue, epithelium and mast cells.

Hale's Colloidal Iron Method
(Hale, 1946; Muller, 1955; Mowry, 1958)

Reagents Required

Ferric chloride
Acetic acid
Hydrochloric acid
Potassium ferrocyanide
Distilled water

Preparation of Reagents

Solution 1 (Colloidal iron)

29 per cent ferric chloride	2.2 ml
Distilled water	25.0 ml

Boil water and then add ferric chloride. This solution turns red. Allow it to cool.

Solution 2 Working solution of colloidal iron

Glacial acetic acid	5.0 ml
Distilled water	15.0 ml
Stock solution (Solution 1)	20.0 ml

Solution 3 Acid ferrocyanide mixture

Potassium ferrocyanide	2.0 g
Conc. hydrochloric acid	2.0 ml
Distilled water	98.0 ml

Potassium ferrocyanide is dissolved in distilled water and hydrochloric acid is added.

Procedure

1. Deparaffinize and hydrate slides to water.
2. Rinse in 12 per cent acetic acid.
3. Stain in solution 2 for 1 hour.
4. Rinse in 12 per cent acetic acid giving 4 changes of 3 minutes each.
5. Treat section with solution 3 (acid ferrocyanide) for 20 minutes.
6. Wash in distilled water.
7. Counterstain in Mayer's haemalum.
8. Wash, dehydrate, clear and mount in DPX.

Results

Acid mucosubstances	Blue
Nuclei	Red

Staining is same as that of alcian blue (pH 2.5). Strongly acidic mucosubstances do not stain. The less acidic mucopolysaccharides stain more intensely.

Colloidal Iron/PAS Method

Proceed as above for the colloidal iron step and then follow steps 4–9 in the alcian blue/PAS procedure.

Fixation and Processing

1. Fix small pieces of tissue in New Comer's fluid for 12–24 hours.
2. Transfer to equal parts of New Comer's fluid and n-butanol for 30 minutes.
3. Transfer to n-butanol for 30 minutes giving 3 changes.
4. Immerse in equal parts of n-butanol and wax for 30 minutes.
5. Impregnate in wax giving 3 changes of 30 minutes each.
6. Embed.

Result

Staining resembles the AB (pH 2.5)/PAS method.

Combined Dialysed Iron and PAS Stain for Polysaccharides (Ritter and Oleson, 1950)

Fixation

Helly, Formol alcohol (fresh tissue in 10 per cent formalin in 90 per cent alcohol.)

Reagents Required

Periodic acid
Schiff's reagent
Meyer's haemalum
Acetic acid
Iron

Preparation of Reagents

Solution 1
Hale's dialysed iron   1 vol
2 M acetic acid (*see* Chapter 5)   1 vol
    or
Colloidal iron (*see* page 143)
Solution 2 (Schiff's reagent (*see* page 119))
Solution 3 (Periodic acid solution (*see* page 121))
Solution 4 (Meyer's haemalum (*see* Chapter 7))

Procedure

1. Deparaffinize and hydrate slides to water.
2. Apply Hale's method.
3. Wash in water.
4. Transfer to solution 3 for 10 minutes.
5. Wash in running water.
6. Transfer to solution 2 for 30 minutes.
7. Wash in distilled water for 15 minutes.
8. Counterstain with solution 4, if desired, for 5 minutes.
9. Wash in water.
10. Dehydrate, clear and mount.

Result

Acid mucosubstances	Blue (Bright)
Proteins	Pale blue
Nuclei	Pale blue
Neutral mucosubstances	Red

Colloidal Iron–Haematoxylin for Acid Mucosubstances (After Muller 1955, Mowry, 1958, 1963)

Fixation

Any general fixative like Bouin's, Susa, etc.

Reagents Required

Colloidal iron
Acetic acid
Hydrochloric acid
Potassium ferrocyanide
Harris haematoxylin
Picric acid

Preparation of Reagents

Solution 1 (Stock colloidal iron (*see* page 143))

Solution 2 (30 per cent acetic acid)
    Acetic acid   30.0 ml
    Distilled water   100.0 ml

Solution 3
    Potassium ferrocyanide   20.0 g
    Distilled water   100.0 ml

Add 50.0 ml of 20 per cent HCl to 50.0 ml of the above solution. This should be prepared fresh.

Solution 4 (Harris Haematoxylin) (*see* Chapter 7)

Solution 5 (0.5 per cent picric acid solution)
    Picric acid   500 mg
    Distilled water   100.0 mlz

Procedure

1. Deparaffinize and hydrate slides to water.
2. Transfer to (colloidal iron) solution 1 for 2 hours.
3. Rinse in solution 2 giving 3 changes of 10 minutes each.
4. Wash in distilled water.
5. Transfer to solution 3 for 20 minutes.
6. Wash in running water.
7. Counterstain if desired in solution 4 for 5 minutes.
8. Differentiate in acid alcohol for 30 seconds.

9. Wash in distilled water.
10. Stain in solution 5 for 60 seconds.
11. Wash.
12. Dehydrate, clear and mount.

Result

Acidic mucosubstances are stained prussian blue. These include sialomucins, goblet cell mucins, connective tissue mucins, mast cell granules.

Tissue stains in the same manner as it does at AB pH 2.5.

Strongly acidic mucosubstances remain unstained.

Acridine Orange Method

Procedure

1. Deparaffinize and bring slides to water.
2. Treat three parallel slides with one per cent cetyl trimethyl ammonium chloride for 3 minutes.
3. Wash in tap water for 10 minutes.
4. Treat with ribonuclease (1 mg per 1 ml distilled water) at 45°C for 2 hours.
5. Treat slide 1 again for 3 minutes with 1 per cent trimethyl ammonium chloride. Wash in tap water, stain in 1 per cent acridine orange in distilled water.
6. Stain slide 2 in 0.1 per cent acridine orange in 0.01 per cent acetic acid (pH 3.2), wash in running water, differentiate in 0.3 M NaCl in 0.01 M acetic acid for 10 minutes. Wash in running water, dry and mount.
7. Treat slide 3 in the same way as slide 2 but replace 0.6 M NaCl by 0.3 M NaCl.

Result

In slide 1, red fluorescence is due to hyaluronic acid.

In slide 2, red fluorescence is due to chondroitin sulphate and heparin.

In slide 3, red fluorescence is due to heparin.

Low Iron Diamine Method for Acid Mucosubstances
(Spicer, 1965; Leppi and Spicer, 1967)

Reagents Required

N, N-dimethyl *m*-phenylenediamine dihydrochloride
N, N-dimethyl *p*-phenylenediamine dihydrochloride
40 per cent ferric chloride
3 per cent acetic acid
Alcian blue

Preparation of Reagents

Solution 1 Diamine Solution
    N, N-dimethyl *m*-phenylenediamine dihydrochloride   30 mg
    N, N-dimethyl *p*-phenylenediamine dihydrochloride   5 mg

Dissolve the salts in 50 ml of distilled water and add 0.5 ml of a 40 per cent solution of ferric chloride. This solution should be freshly prepared before use.

Solution 2 [Alcian Blue (pH 2.5)] *(see page 130)*

Procedure

1. Deparaffinize and hydrate through a graded series of alcohol and bring to water.
2. Treat with solution 1 (diamine solution) for 18 hours.
3. Rinse rapidly in distilled water.
4. Stain in solution 2 (alcian blue) for 5 minutes.
5. Briefly rinse in tap water.
6. Dehydrate, clear and mount in DPX.

Result

Sites of sulphated and non-sulphated mucosubstances	Black
Sites of some acid non-sulphated mucosubstances	Blue

Neutral mucosubstances can also be demonstrated on the same slides. An additional section is oxidized in 1 per cent periodic acid followed by a wash in tap water before staining in the diamine solution. Neutral mucopolysaccharides in the oxidized section will stain purple grey.

Rationale

First introduced by Spicer (1961), the method involves oxidation of a mixture of *meta* and *para* forms of the salt mentioned above by ferric chloride to form a black substance which is bound to some mucosubstances. When a low concentration of diamine salt is used, most sialomucins and sulphated mucosubstances stain.

High Iron Diamine/Alcian Blue Method for Sulphated Mucosubstances (Spicer, 1965; Leppi and Spicer, 1967)

Reagents Required

N, N-dimethyl *m*-phenylenediamine dihydrochloride
N, N-dimethyl *p*-phenylenediamine dihydrochloride
Ferric chloride 40 per cent

Preparation of Reagents

Solution 1

N, N-dimethyl *m*-phenylenediamine dihydrochloride	120 mg
N, N-dimethyl *p*-phenylenediamine dihydrochloride	20 mg

First dissolve the salts in 50 ml of distilled water, then add 1.4 ml of 40 per cent ferric chloride 1 : 4. The pH should be 1.5. This solution must be freshly prepared just before use.

Procedure

1. Hydrate slides to water.
2. Transfer to solution 1 (diamine solution) for 18 hours.

3. Wash in water.
4. Place in alcian blue solution for 5 minutes.
5. Wash in water.
6. Dehydrate, clear and mount.

Result

Sulphated mucosubstances   Black
Non-sulphated mucosubstances   Blue

Rationale

Higher concentrations of the diamine salts stain only sulphated mucosubstances. Both methods are followed by staining with alcian blue at pH 2.5 to show non-sulphated acid mucosubstances.

Periodic Acid/Schiff Method

Reagents Required

Periodic acid
Schiff's reagent (*see* page 115)
Sodium bisulphate

Procedure

1. Deparaffinize and hydrate slides to water.
2. Oxidize in one per cent periodic acid for 10 minutes.
3. Wash in water.
4. Stain in Schiff's reagent for 30 minutes.
5. Rinse in 0.5 per cent sodium bisulphate ($Na_2S_2O_3$) giving 3 changes of 2 minutes each.
6. Wash in running water.
7. Dehydrate, clear and mount.

Result

Connective tissue acid mucopolysaccharides are not reactive. Neutral connective tissue mucosubstances and a number of acidic epithelial mucins and glycogen stain red to magenta.

Mixed Diamine Method (Spicer, 1965; Leppi and Spicer, 1967)

This is a very good test for acid mucosubstances.

Reagents Required

N, N-dimethyl *m*-phenylenediamine dihydrochloride
N, N-dimethyl *p*-phenylenediamine dihydrochloride

Preparation of Reagents

Mix 30 mg of (1) and 5 mg of (2) in 50 ml of distilled water, and adjust pH to 3.4–4.0 with 0.2 M disodium hydrogen phosphate (0.15–0.65 ml).

Procedure

1. Hydrate 2 slides to water.
2. Hydrolysis is carried out at 60°C for 10 minutes.
3. Wash in distilled water.
4. Oxidize one section in one per cent periodic acid for 10 minutes.
5. Rinse in running water.
6. Stain both sections in mixed diamine for 24 hours.
7. Wash, dehydrate, clear and mount.

Result

Periodic-unreactive acid mucosubstances	Purple
Periodic-reactive neutral and acid mucosubstances	Grey to grey-brown

Both types of acidic mucosubstances are purple in unoxidized sections.

Rationale

Spicer (1961a, 1965) introduced the low iron, high iron and mixed diamine methods. In these techniques, the diamines react with periodate-engendered aldehyde groups to form yellow to brown Schiff bases. High iron diamine is specific for ester sulphate (Goad and Sylven, 1969, Reid *et al.*, 1972).

Periodic Acid–Diamine Procedure

Preparation of Reagents

Solution 1

Add 50 mg of N, N-dimethyl *p*-phenylenediamine hydrochloride just before use to 50 ml of citric phosphate buffer (M/10 citric acid 4.8 cc, M/5 sodium phosphate 7.2 cc, distilled water 38 cc). Alternatively, dissolve 100 mg of the *p*-diamine in 50 cc of distilled water and adjust pH to 5.0 with 0.5 M Na_2HPO_4.

Procedure

1. Deparaffinize and hydrate slides to water.
2. Oxidize in 1 per cent periodic acid for 10 minutes.
3. Rinse in running water for 10 minutes.
4. Transfer to solution 1 (*p*-diamine solution) for 24–48 hours.
5. Differentiate in 1 per cent HCl in 70 per cent alcohol for 8 seconds.
6. Wash in water.
7. Dehydrate, clear and mount.

Result

Neutral mucosubstances with large quantities of fructose, galactose or mannose take a brown stain.

Acid periodiate-reactive polymers are stained purple or grey to brown.

Acid periodate-reactive mucosubstances are stained black. Glycogen does not take the stain.

Periodic Acid–Phenylhydrazine–Schiff Procedure

Procedure

1. Deparaffinize and hydrate slides to water.
2. Oxidize in 1 per cent periodic acid for 10 minutes.
3. Wash in water for 10 minutes.
4. Transfer to 5 per cent aqueous phenylhydrazine for 30–60 minutes at 25°C.
5. Wash in water for 10 minutes.
6. Keep slides in Schiff's reagent for 10 minutes.
7. Rinse in 0.5 per cent sulphuric acid giving 3 changes of 2 minutes each.
8. Wash in running water.
9. Dehydrate, clear and mount.

Result

Phenylhydrazine blocks Schiff's reactivity of periodate-engendered aldehydes more effectively in neutral mucosaccharides. Sulphomucins and sialomucins stain selectively.

Ferric Alum–Coriphosphine for Sulphomucins (Stoward, 1967)

Fixation

Formalin

Reagents Required

Ferric alum
Coriphosphine
Isopropyl alcohol

Preparation of Reagents

Solution 1 (4 per cent aqueous ferric alum)
Ferric alum   4.0 g
Distilled water   100.0 ml

Solution 2 (0.01 per cent coriphosphine)
Coriphosphine   10 mg
Distilled water   100.0 ml

Procedure

1. Deparaffinize and hydrate slides to water.
2. Transfer to solution 1 for 10 minutes.
3. Wash in distilled water.
4. Transfer to solution 2 for 20 minutes.
5. Rinse in distilled water.
6. Dehydrate in isopropyl alcohol.
7. Clear in xylene, mount.

Result

Sulphomucins fluoresce green or dull red.

Coriphosphine–Thiazol Yellow for Sulphomucins (Stoward, 1967)

Fixation

Formalin

Reagents Required

Coriphosphine
Thiazol yellow

Preparation

Solution 1 (0.01 per cent coriphosphine)
    Coriphosphine 10 mg
Solution 2
    Thiazol yellow 10 mg
    Distilled water 100.0 ml

Procedure

1. Deparaffinize and hydrate slides to water.
2. Place in solution 1 for 20 minutes.
3. Wash in distilled water.
4. Transfer to solution 2 for 1 minute.
5. Rinse in distilled water.
6. Dehydrate in 3 changes of isopropyl alcohol.
7. Clear in xylene and mount.

Result

Sulphomucins	Fluoresce red or orange-yellow
Nuclei	Green

Para-anisidine Reaction for Sialic Acids (Cerbilis and Zittle, 1961)

Fixation

Formalin
Cryostat

Reagents Required

p-Anisidine
Phosphoric acid
Methanol

Preparation of Reagent

Solution 1 (0.5 per cent *p*-anisidine)

p-Anisidine	500 mg
H_3PO_4	3 ml
80 per cent methanol	100.0 ml

Procedure

1. Deparaffinize and hydrate slides to water.
2. Rinse in 30 per cent H_3PO_4.
3. Transfer to 60 per cent H_3PO_4 for 1½ hours.
4. Transfer to solution 1 at 95°C for 30 minutes.
5. Bring them back to 60 per cent H_3PO_4.
6. Wash and then mount in water.

Result

Sites of sialic acid activity stain brown.

Tetrazolium Method for Sulphate Groups (Geyer, 1962)

Fixation

The tissues may be fresh and unfixed or fixed in paraffin.

Procedure

1. Hydrate slides to water.
2. Immerse in acetic fast blue B solution for 10–30 minutes (50 mg fast blue B salt in 10 ml 5 per cent acetic acid).
3. Rinse in pre-cooled (0–5°C) distilled water for 60 seconds.
4. Treat for 2–5 minutes in pre-cooled (0–5°C) saturated solution of 1–naphthol in borax buffer at pH 9.4.
5. Wash in distilled water and mount in glycerine jelly.

Result

Sulphate esters and sulphonic acid groups induced by oxidation are stained reddish violet.

BLOCKING TECHNIQUES

Blocking of Aldehyde Groups by Acetylation (Lillie, 1965; McManus and Cason, 1950)

Reagents Required

Acetic anhydride
Pyridine
Potassium hydroxide

Preparation of Reagents

Solution 1

Acetic anhydride	16 ml
Dry pyridine	24 ml

Solution 2

Potassium hydroxide	1 g
Absolute alcohol	70 ml
Distilled water	30 ml

Procedure

1. Deparaffinize and hydrate slides to water and label sections 1, 2, 3.
2. Place sections 1 and 2 in solution 1 (acetic anhydride solution) for 1–24 hours. Leave section 3 in distilled water.
3. Rinse sections 1 and 2 in distilled water.
4. Treat section 2 in solution 2 (potassium hydroxide) for 30 minutes.
5. Rinse section 2 in distilled water.
6. Apply Schiff's stain to all 3 sections.
7. Wash, dehydrate, clear and mount.

Result

Section 1 and 3 give positive results and section 2 gives negative result indicating that the reaction is due to 1:2 glycol groups.

Mild Methylation Blocking of Carboxylated Mucins (Spicer, 1960)

Reagents Required

Methanol
Hydrochloric acid
Alcian blue
Acetic acid

Preparation of Reagents

Solution 1

Methanol	50 ml
Conc. hydrochloric acid	0.4 ml

Solution 2

Alcian blue pH 2.5 (*see* page 130)

Procedure

1. Deparaffinize and bring sections to absolute alcohol.
2. Methylate the slide in solution 1 for 4 hours at 37°C and leave another slide in distilled water at 37°C.
3. Wash all sections in tap water.
4. Stain both sections in solution 2 (alcian blue).
5. Wash, dehydrate, clear and mount.

Result

Four-hour methylation at 37°C blocks alcianophilia of non-sulphated mucosubstances. Methylation for 1 hour at 60°C blocks alcianophilia of sulphated mucosubstances.

Sulphated mucosubstances at 37°C stain blue.

Rationale

This method blocks the reactivity of acid mucosubstances. Only carboxyl groups are blocked at 37°C by forming methyl esters. At 60°C, all carboxyl and sulphate esters are blocked. Carboxyl groups are methylated and the sulphate groups are hydrolysed.

Methylation/Saponification/Alcian Blue for Carboxylated Mucosubstances (Spicer and Lillie, 1959)

Reagents Required

Potassium hydroxide
Methanol
Alcian blue

Preparation of Solutions

Solution 1
Methylation solution as in previous method

Solution 2
Potassium hydroxide   1.0 g
Absolute alcohol   70.0 ml
Distilled water   30.0 ml

Solution 3
Alcian blue solution pH 2.5 (*see* page 130)

Procedure

1. Deparaffinize 3 slides and hydrate to water.
2. Place sections 1 and 2 in methyl alcohol at 60°C for 1 hour. Place section 3 in distilled water at 60°C.
3. Wash all sections in distilled water.
4. Treat sections in solution 2 for 30 minutes Leave sections 2 and 3 in 70 per cent alcohol for 30 minutes.
5. Wash.
6. Place in 70 per cent alcohol.
7. Wash.
8. Stain section with alcian blue for 5 minutes.
9. Wash, dehydrate, clear and mount.

Result

In section 1, carboxylated mucins stain blue.

In section 2, carboxylated and sulphated mucosubstances are blocked and there is no blue staining.

In section 3, all acid mucosubstances stain blue.

Rationale

With saponification (with strong alkali like potassium hydroxide), mucosubstances blockage is reversible. Methylation restores the carboxyl groups. The sulphate groups which are hydrolysed are irreversibly blocked.

Sulphation (Lewis and Grille, 1959)

Procedure

1. Immerse section in a solution containing 10 cc of concentrated sulphuric acid and 30 cc of glacial acetic acid at 25°C for 30 minutes.
2. Wash in water.
3. Stain with any basic dye.

Result

Most sulphated components, except collagen, lack affinity for pH 2.5 alcian blue.

Rationale

This technique introduces sulphate esters into mucosubstances to engender a subsequent reaction to a metachromatic dye.

Enzyme Digestion Method

Reagents Required

Testicular hyaluronidase
Alcian blue
Phosphate buffer (pH 6.7)

Preparation of Reagent

Testicular hyaluronidase (600 units/mg)   0.001 g
Phosphate buffer (pH 6.7)   10.0 ml
Alcian blue (pH 2.5 or 0.5)

Procedure

1. Deparaffinize 2 test and 2 control sections to water.
2. Treat one test and one control section with hyaluronidase solution at 37°C. The other two sections are placed in buffer for 3 hours at 37°C.
3. Wash all sections.
4. Treat all with alcian blue solution for 10 minutes.
5. Wash in water.
6. Counterstain with Mayer's haemalum for 5 minutes.
7. Wash, dehydrate, clear and mount.

Result

Hyaluronic acid and chondroitin sulphate   Negative
Other acid mucosubstances   Blue

Rationale

Hyaluronidase digests hyaluronic acid and chrondroitin sulphates A and C.

Sialidase Digestion/Alcian Blue (Spicer *et al.*, 1962)

Reagents Required

Sialidase
Acetate buffer (pH 5.5)
Calcium chloride
Alcian blue

Preparation of Reagents

Solution 1

Sialidase or neuraminidase	1.0 ml
Acetate buffer (pH 5.5)	4.0 ml
Calcium chloride	50 mg

Solution 2
Alcian blue (pH 2.5) *(see page 130)*

Procedure

1. Hydrate slides to water (2 control and 2 test).
2. Treat section 1 test and one control in sialidase solution at 37°C for 18 hours. Incubate the other slide in buffer solution only at 37°C for 18 hours.
3. Wash in tap water.
4. Treat with solution 2 [alcian blue (pH 2.5)] for 5 minutes.
5. Wash in tap water.
6. Counterstain if desired in Mayer's haemalum.
7. Wash, dehydrate, clear and mount.

Result

Sialidase-labile sialomucins remain unstained.

Other acid mucins	Blue
Nuclei	Red

Remarks

Sialidase-resistant sialomucins may sometimes display loss of alcian blue staining.

Deacetylation–Sialidase Digestion Method (Ravetto 1968)

Deacetylating Solution

Concentrated ammonia	20 ml
Ethanol	70 ml
Distilled water	10 ml

Procedure

1. Deparaffinize 3 sections and bring them to distilled water, label them as 1a, 2a, 3a.
2. Treat 1a with deacetylation solution for 24 hours at 37°C. The other two sections, 2a and 3a, to be placed in distilled water for the same length of time.
3. Wash 1a in distilled water and treat all three with 0.2 M acetate buffer at pH 5.5.
4. Treat 1a and 2a with sialidase for 16 hours at 37°C. The section 3a is placed in buffer at room temperature.
5. Wash in running water and follow the routine alcian blue procedure.
6. Dehydrate, clear and mount.

Result

In section 1a, both sialidase-labile and sialidase-resistant mucins are not stained with alcian blue.

In section 2a, only sialidase-labile sialomucins remain negative to alcian blue.

In section 3a, all acid mucins stain normally.

Borohydride Blockade–Saponification–PAS to Demonstrate Sialidase-resistant Sialomucins

This technique was originally developed by Culling *et al.*, (1975) to demonstrate acetylated sialomucins.

Reagents Required

Sodium borohydride
Disodium hydrogen phosphate
Potassium hydroxide
Ethanol

Preparation of Reagents

Solution 1

Sodium borohydride	1000 mg
Disodium hydrogen phosphate	1.0 g
Distilled water	100.0 ml

First dissolve disodium hydrogen phosphate in distilled water and then add potassium hydroxide. Final pH should be 9.4. Prepare the solution afresh.

Solution 2

Potassium hydroxide	500 mg
Ethanol	70.0 ml
Distilled water	100.0 ml

Procedure

1. Deparaffinize 3 sections and bring them down to distilled water, label them as 1a, 2a and 3a.
2. Treat all sections with 1 per cent periodic acid for 30 minutes.
3. Wash in distilled water. Transfer section 1a and 2a to solution 1 for 30 minutes.

4. Wash in running water and then immerse section 1a in 70 per cent alcohol followed by immersion in solution 2 for 30 minutes Again wash in 70 per cent alcohol and then treat with 1 per cent periodic acid for 5 minutes.
5. Treat all sections with Schiff's reagent (*see* page 119).
6. Wash in running water.
7. Dehydrate, clear and mount.

Result

In section 1a, O acetylated sialidase-resistant siolomucins, and glycogen take a magenta colour.

In section 2a, there is no PAS reactivity.

In section 3a, all PAS-positive carbohydrates stain in magenta shade.

Blocking Aldehyde Groups (Lillie and Glenner, 1957)

After oxidizing the sections in periodic acid and washing, they are treated with a solution of 10 ml aniline and 90 ml acetic acid. The sections are washed before applying PAS technique.

Result

Only non-aldehyde PAS-positive material will be stained red.

There are a number of techniques for identifying sialic acid. Identification depends upon staining with alcian blue/PAS, both before and after sialidase digestion, hydrolysis and methylation. These remove sialic acid from tissues (Quintarelli *et al.*, 1964; Schmitz-Moorman, 1969). The best way to demonstrate the presence of sialic acid is the thiobarbituric acid assay technique of Warren (1959). This has also been used to assay the enzyme solution (Spicer and Warren, 1960; Mohos and Skoza, 1970). Sialomucins differ in their reactivity to periodic acid, phenyl hydrazine/Schiff (PAS) and in their lability to sialidase treatment. Some sialomucins which are stable to sialidase, become labile after saponification.

FURTHER READINGS

West *et al.* (1982), Latham and Atkins (1983) and Latham *et al.* (1983).

Table 9.1 Histochemical techniques applied for the demonstration of mucopolysaccharides

Technique	Fixative	Mucopolysaccharides				
		Acids				Neutral
		Sulphated		Non-sulphated		
		Weakly acidic	Strongly acidic	Hyalomucins	Sialomucins	
Alcian blue (pH 2.5)	1% cetylpyridinium chloride in 10% formalin	Dark blue		Dark blue	Dark blue	
Alcian blue (pH 2.8)	Rossman's fluid or formalin 199% alcohol 1% cetylpyridinium chloride in 10% formalin	Blue green				
Alcian blue (pH 1.0)	1% cetylpyridinium chloride in 10% formalin	Greenish blue		Unstained		
Alcian bluc/alcian yellow	3% calcium acetate in 10% formalin	Blue		Yellow		
Alcian blue with varyng electrolyte concentration (CEC)						
0.06 M	3% calcium acetate in 10% formalin	Sulphated and carboxyl mucins positive				
0.3 M	3% calcium acetate in 10% formalin	Weakly and strongly sulphated positive				

0.5 M	3% calcium acetate in 10% formalin	Blue strongly sulphated positive				
0.7 M	3% calcium acetate in 10% formalin	Highly sulphated connective tissue mucins positive				
0.9 M	3% calcium acetate in 10% formalin	Keratin sulphate positive				
Alcian blue (pH 2.5)/PAS	1% cetylpiridinium chloride in 10% formalin			Blue	Blue	Red
Alcian blue (pH 1.0)/PAS	1% cetylpiridinium chloride in 10% formalin		Blue		Red	
Aldehyde fuchsin (AF)	1% cetylpiridinium chloride in 10% formalin	Deep purple	Deep purple	Light blue	Light blue	
Aldehyde fuchsin/alcian blue	1% cetylpiridinium chloride in 10% formalin	Blue-purple	Deep purple	Blue	Blue	Blue purple
Aluminium sulphate	1% cetylpiridinium chloride in 10% formalin	Blue	Blue	Red	Red	
Azure A	1% cetylpiridinium chloride in 10% formalin	Mixtures of sulphated and non-sulphated–Purple Acid mucosubstances purple-red				
Toluidine blue	Zenker	Acid mucopolysaccharides – Pink (Metachromatic)				
Alcian blue/Ruthenium Red	1% cetylpyridinium chloride in 10% formalin		Blue		Red	
Alcian blue/safranin	1% cetylpyridinium chloride in 10% formalin		Red	Other mucosubstances blue		Blue

(Contd.)

Table 9.1   (Continued)

Technique	Fixative	Mucopolysaccharides				Neutral
		Acids				
		Sulphated		Non-sulphated		
		Weakly acidic	Strongly acidic	Hyalomucins	Sialomucins	
Colloidal iron	1% cetylpyridinium chloride in 10% formalin	Acid mucosubstance – Prussian Blue				
Collidal iron/PAS	1% cetylpyridinium chloride in 10% formalin	Acid mucosubstances – Blue				Red
Colloidal iron/haematoxyl in	Any general fixative	Sulphated acidic Unstained – others Prussian blue				
Low iron diamine	1% cetylpyridinium chloride in 10% formalin	Sulphated and non-sulphated – Black; Some acid non-sulphated – Blue				
High iron diamine	1% cetylpyridinium chloride in 10% formalin	Sulphated mucosubstances – Black		Non-sulphated mucosubstances – Blue		
Mixed diamine method	3% calcium acetate in 10% formalin	Periodic unreactive acid mucosubstances – Purple Periodic reactive neutral and acid mucosubstances – Grey to grey-brown				
Periodic acid-diamine	3% calcium acetate in 10% formalin	Black				Brown
Periodic acid/phenyl hydrazine schiff	3% calcium acetate in 10% formalin	Stain selectively		Stain selectively		Blocked

Ferric alum-Coriphosphine	Formalin	Sulphomucins fluoresce	Green or dull red
Coriphosphine-thiazol yellow	Formalin	Sulphomucins fluoresce	Red or orange-yellow
Para-anisidine	Formalin or cryostat		Sialic acid – Brown
Tetrazolium method	Fresh, unfixed or fixed with parrafin	Sulphate esters and sulphonic acid groups – reddish violet	
Acriflavine method	Formalin – frozen	Yellow fluorescence	
Acridine orange	New comers	Orange-red	
Lead tetraa-cetate Schiff	Any general fixative	Mucosubtances – Reddish purple	
Toluidine blue at different pH levels			
pH 2.0		Violet	
pH 3.0		Violet	
pH 4.0		Phosphated mucosubstances – Violet	
pH 5.0		Violet	

PROTEIN

10

Proteins are ubiquitous components of every living tissue. Because of their complexity, histochemical determination is not easy. In any living cell, the cytoplasm comprises important components like proteins, carbohydrates and lipids. Among these, proteins are complex organic nitrogenous compounds and are widely distributed in all plant and animal tissues. The word protein means "first importance" and this has been derived from the Greek word *proteios*. A protein is comparable to a string of beads, each bead being an amino acid. Amino acids in a protein are held together by peptide bonds. Proteins are formed by sending out water (H_2O) from the α-amino group (NH_2) of one amino acid and terminal carboxyl (COOH) of the adjacent amino acid. This resulting linkage (CO–NH) after expulsing H–OH is called "the peptide bond". Identification of proteins in general is not at all a problem, but difficulty arises with the identification of a particular protein and this leads to many problems, because of their intermingling or conjugation with carbohydrates and lipids.

Proteins could broadly be categorized into simple proteins, and conjugated proteins.

SIMPLE PROTEINS

Simple proteins on hydrolysis yield amino acids. These in turn are classified as (a) fibrous proteins, or scleroproteins (collagen, elastin, fibrin and keratin), (b) globular proteins (albumins, globulins). Fibrous proteins are insoluble in water whereas globular proteins are soluble. Globular proteins also include histones and protamines. Fibrous proteins have a supporting function or protective function in the body of an organism, e.g. keratin of hair, elastin, collagen of skin. These are eosinophilic. They take a red stain when treated with haematoxylin–eosin. They are easily characterized with a good number of histochemical procedures.

Albumins are water-soluble and could be precipitated in an aqueous medium by saturating with an acid salt such as ammonium sulphate or by a neutral salt like sodium sulphate.

Globulins are insoluble in water but soluble in dilute salt solutions such as 5 per cent sodium chloride solution. They occur in serum and tissue.

Histones are soluble proteins restricted to the nuclei of the cell. They could be precipitated with ammonium hydroxide.

Protamines are also soluble proteins, basic in nature. They form a sort of crystalline salt with mineral acids and form insoluble salts with more acidic proteins. An example of a protein salt is the drug protamine-insulin.

CONJUGATED PROTEINS

As the name implies, these are complex compounds. They may occur in conjugation with non-protein materials, e.g. nucleoproteins in combination with nucleic acids, glycoproteins in combination with carbohydrates (e.g. mucins), lipoproteins in conjunction with lipids (i.e., serum), chromoproteins with pigments (e.g. haemoglobin) and phosphoproteins when combined with phosphoric acid, (e.g. casein of milk). There are several histochemical stains to demonstrate these proteins.

Among fibrous proteins, collagen could be demonstrated with Van Geison, aniline blue and other trichrome stains. Similarly reticulin with silver reduction method and fibrin with Mallory PTAH method where it takes a dark blue stain, or MSB technique where it takes a bright red colour. Fibrin contains tryptophan and is demonstrated with *p*-dimethylaminobenzaldehyde/nitrite which is specific for tryptophan. Since keratin contains sulphur, it could be demonstrated with dihydroxydinaphthyl disulphide or mercury orange method.

Within the protein molecule, there are groups or linkages and these linkages reside within the amino acids making up the proteins. There are a number of histochemical techniques to demonstrate protein-bound amino groups (e.g. lysine): Ninhydrin/Schiff or dinitrofluorobenzene (DNFB), Hydroxy.naphthaldehyde.

The most simple and at the same time reliable methods are ninhydrin/Schiff (Yasuma and Ichikawa, 1953), dinitrofluorobenzene (Sanger, 1945, 1950) and hydroxy naphthaldehyde (Weiss *et al.*, 1954) techniques.

Guanidyl Groups

The only amino acid which has guanidyl group is arginine. A specific test for arginine is Sakaguchi reaction. Arginine reacts with α-naphthyl and hypochlorite in alkaline medium, and a red colour develops. Slides must be examined immediately.

Indole groups

Amino acids like tryptophan and tryptamine contain indole groups. The most specific method for the demonstration of tryptophan is the *p*-diphenylaminobenzaldehyde/nitrite technique (Adams, 1957). An intense blue colour develops with DMAB-nitrite. Freeze-dried sections are preferable.

Phenyl Groups

Amino acid tyrosine contains the hydroxyphenyl group and can be demonstrated by the specific Millon's reaction. This technique was modified by Baker (1956). A fairly specific method is the diazonium coupling method of Glenner and Lillie (1959).

Disulphide and Sulphydryls

Lillie (1965) stated that it is somewhat difficult to distinguish the two groups. There are precise methods to demonstrate these. Dihydroxy dinaphthyl disulphide (DDD) method of Barrnett and Seligman (1952), mercury orange method of Bennett and Watts (1958), performic acid/alcian blue method of Adams and Sloper (1955, 1956) are used to demonstrate disulphides. These sulphur-containing groups are present in the amino acids cystine, cysteine and methionine.

FIXATION OF PROTEINS

Freeze-dried sections which yield good results may be preferred. The choice of the fixative depends upon the reacting group involved in the method. Fixation with aldehydes is likely to interfere. Fixatives containing mercuric chloride are to be avoided for the demonstration of sulphydryl groups. A non-formalin fixative is preferable for hydroxynaphthaldehyde method. Barnett and Roth (1958) who experimented with a number of fixatives, suggested that osmium tetroxide completely inhibited any of these reactions. A recommended method of fixation is appropriate for each technique.

HISTOCHEMICAL TECHNIQUES FOR PROTEINS

Mercury Bromophenol Blue Method for General Proteins
(Plate 12, Figures 1 and 2)

Fixation

Any general fixative may be used, but Susa is preferable. Formalin is not recommended as it interferes with amino groups.

Reagents Required

Mercuric chloride
Bromophenol blue

Preparation of Reagents

Solution 1 (Staining solution)

Mercuric chloride	10.0 g
Bromophenol blue	1.0 g
95 per cent alcohol (or distilled water)	100.0 ml

Procedure

1. Hydrate slides to water.
2. Transfer the slides to solution 1 (bromophenol blue solution) for 30 minutes.
3. Wash in tap water for 10 minutes.
4. Dehydrate, clear and mount.

Result

Basic proteins stain deep blue.

Mercuric Bromophenol Blue Method (Chapman, 1975)

(Though it is histochemically not a valid technique, bromophenol blue is still used).

Fixation

Any general fixative may be used.

Reagents Required

Mercuric chloride
Sodium bromophenol blue
Acetic acid

Preparation of Reagents

Solution 1

Mercuric chloride	1.0 g
Sodium bromophenol blue	50 mg
2 per cent acetic acid (aqueous)	100.0 ml

Procedure

1. Deparaffinize and hydrate slides to water.
2. Place in solution 1 for 30 minutes.
3. Rinse in 0.5 per cent acetic acid giving 2 changes of 20 minute each.
4. Dehydrate and clear in xylene.
5. Treat with a mixture of xylene and *n*-butylamine (0.5 ml per 100 ml).
6. Clear again in xylene and mount.

Result

Proteins stain blue.

Acrolein–Schiff Technique

Fixation

Susa is a suitable fixative.

Reagents Required

Acrolein
Ethanol
Basic fuchsin
Potassium metabisulphite

Preparation of Reagents

Solution 1 (Acrolein solution)
   5 per cent acrolein in 95 per cent ethanol in screw-capped coplin jars. Acrolein must be fresh.
Solution 2 (Schiff's Reagent) (*see* Chapter 9)

Procedure

1. Bring sections to 95 per cent alcohol.
2. Transfer to solution 1 (5 per cent acrolein) for 15–60 minutes.
3. Wash in three changes of 95 per cent alcohol.
4. Wash in distilled water.
5. Transfer to solution 2 (Schiff's reagent) for 10–30 minutes.
6. Wash in running water.
7. Dehydrate, clear and mount.

Result

Sites of protein are stained red.

Rationale

In this technique, the double bonds of acrolein will react with SH, NH_2, NH and imidazoles leaving the free aldehydes to react with Schiff's reagent.

Biebrich Scarlet Technique for Basic Proteins
(Spicer and Lillie, 1961; Lillie, 1965)

Fixation

Carnoy, buffered mercuric chloride, alcohol.

Reagents Required

Biebrich scarlet
Glycine
Sodium hydroxide

Preparation of Reagents

Spicer and Lillie (1961) prefer 0.01 per cent Biebrich scarlet in glycine, NaOH-buffered at pH 8.0, 9.5 and 10.5. But Lillie prefers using 1 ml of 1 per cent Biebrich scarlet in 49 ml of each buffer solution (0.02 per cent) for a shorter period.

Procedure

1. Hydrate slides to water.
2. Stain for 30–90 minutes in each of the 0.01 per cent staining solution or stain for 20 minutes in Lillie's 20 per cent solution.
3. Dehydrate, clear and mount in synthetic resin.

Result

At pH 9.5, basic proteins stain strongly.

Acid Solochrome Cyanine Method for Basic Proteins

Fixation

Frozen or fresh tissue is fixed in Susa or Bouin's fixative.
Paraffin sections are fixed in Carnoy's fluid.

Reagents Required

Solochrome cyanine R
Orthophosphoric acid
Citric acid

Preparation of Reagents

Solution 1 (Solochrome cyanine R solution)
    Orthophosphoric acid   1.0 g
    Distilled water   100.0 ml

To this, add 1.0 g solochrome cyanine R.

Procedure

1. Bring paraffin sections to water.
2. Stain in solution 1 (staining solution) for 5–20 minutes at room temperature.
3. Wash in running water until sections change from orange through red and turn blue.
4. Pass rapidly through 70 per cent alcohol to absolute alcohol.
5. Clear in xylene and mount in Canada balsam.

Result

Nuclear chromatin is stained dark steel blue.

Ninhydrin/Schiff Method (Yasuma and Itchikawa, 1953)

Fixation

Paraffin, Susa freeze-dried, cryostat

Reagents Required

Ninhydrin
Schiff's reagent (*see* Chapter 9)
Hematoxylin

Preparation of Reagents

Solution 1 (Ninhydrin 0.5 per cent)
    Ninhydrin   500 mg
    Absolute alcohol 100 ml

Solution 2 (Schiff's reagent) (*see* Chapter 9)

Procedure

1. Deparaffinize sections and bring them down to absolute alcohol.
2. Treat with solution 1 (ninhydrin) at 37°C for 18 hours.
3. Wash in running water.
4. Immerse the slides in solution 2 (Schiff's reagent) for 45 minutes.
5. Wash in running tap water.
6. Counterstain if required with haematoxylin.
7. Wash in tap water.
8. Dehydrate, clear and mount.

Result

α-amino groups stain pink to red.
(As alternative, 1.0 g alloxan in absolute alcohol instead of ninhydrin can be used.)

Hydroxynaphthaldehyde Method for NH$_2$ Groups (Weiss *et al.*, 1954)

Fixation

Non-formalin fixed, Susa

Reagents Required

3-hydroxy 2-naphthaldehyde
Acetone
Veronol acetate
Fast blue B

Preparation of Reagents

Solution 1 (Incubating medium)
3-hydroxy 2-naphthaldehyde	20 mg
Acetone	20.0 ml
0.1 M veronol acetate buffer (pH 3.5) (*see* Chapter 5)	30.0 ml

The 3-hydroxy 2-naphthaldehyde is dissolved in acetone and the buffer is added.

Solution 2
Diazonium salt is prepared fresh using fast blue B and veronol acetate buffer (pH 7.4).

Procedure

1. Deparaffinize and hydrate slides.
2. Place sections in solution 1 (incubating medium) for 1 hour at room temperature.
3. Wash in 3 changes of distilled water.
4. Transfer sections to solution 2 [0.1 M veronol acetate buffer, (pH 7.4)]. To this solution add 25 mg of Fast blue B and shake well for 5 minutes).
5. Wash well in tap water.
6. Dehydrate, clear and mount.

Result

NH_2 groups are stained blue.

Rationale

Weiss *et al.* (1954) first introduced this method of using 3-hydroxy 2-naphthaldehyde to demonstrate protein-bound amino groups at pH 8.5. First the resulting product is coupled with a diazonium salt at pH 7.4 to produce a blue colour. It is advisable to avoid formalin fixation.

Chloramine T-Schiff for Protein-bound NH_2 (Chu *et al.*, 1953; Burstone, 1955)

Fixation

Carnoy or cold acetone

Reagents Required

Chloramine T
Sodium thiosulphate
Schiff's reagent (*see* Chapter 9)
Sodium bisulphite

Preparation of Reagents

Solution 1 (Chloromine T solution)
    Chloramine T   1.0 g
    Distilled water   100.0 ml

Solution 2 (Sodium thiosulphate)
    Sodium thiosulphate   5.0 g
    Distilled water   100.0 ml

Solution 3 (Sodium bisulphite)
    Sodium bisulphite   10.0 g
    Distilled water   100.0 ml

Procedure

1. Hydrate paraffin slides to water.
2. Place in solution 1 for 6 hours at 37°C.
3. Wash briefly.
4. Treat with solution 2 for 20–30 minutes.

5. Rinse in distilled water.
6. Place in Schiff's reagent.
7. Rinse in solution 3 (sodium bisulphite).
8. Wash in tap water.
9. Dehydrate, clear and mount.

Result

Protein-bound NH_2 groups stain pink or magenta.

Diacetylbenzene Reaction for Protein-bound NH_2 (Voss, 1940; Wartenberg, 1956)

Fixation

Susa

Reagents Required

Veronol acetate
O-diacetyl benzene

Preparation of Reagents

Solution 1
0.1 M acetate buffer (*see* Chapter 5)
o-diacetylbenzene

Procedure

1. Hydrate slides to water.
2. Immerse sections in 0.1 M veronol acetate buffer (pH 8.2).
3. Immerse in solution 1 (2 per cent o-diacetylbenzene in 70 per cent ethanol with an equal amount of 0.1 M veronol acetate buffer) (pH 8.2) for 30–60 minutes.
4. Wash briefly in buffer.
5. Dehydrate, clear and mount.

Result

Protein-bound amino groups stain reddish.

Rationale

o-diacetylbenzene reacts with free NH_2 groups and α-amino groups dissociate from NH_3 to NH_2. This reaction is more specific than Ninhydrin–Schiff, couple tetrazolium, acrolein–Schiff and so on.

Alkaline Salicylaldehyde Method for NH_2 (Stoward, 1963)

Fixation

Freeze-dried, Carnoy, alcohol, paraffin, frozen.

Reagents Required

Salicylaldehyde
Potassium hydroxide

Preparation of Reagents

Solution 1 (Potassium hydroxide)
    Potassium hydroxide   1.0 g
    Alcohol   100.0 ml

Solution 2 (Salicylaldehyde)
    Salicylaldehyde   1.0 g
    Solution 1   100.0 ml

Procedure

1. Bring sections to 95 per cent alcohol.
2. Place in solution 2 for 20 minutes.
3. Wash in 95 per cent alcohol.
4. Dehydrate, clear and mount in fluoromount.

Result

NH_2 groups give green fluorescence.

Acid Azide Reaction for NH_2 Groups (Geyer, 1965)

Fixation

Bouin's, Carnoy, alcohol-paraffin

Reagents Required

α-nitrobenzene acid azide
Veronol acetate buffer, pH 8.2 (*see* Chapter 5)
Trichloroacetic acid (TCA)
Sodium citrate buffer 0.5 M, pH 4.5
Sodium nitrate
Sulphuric acid
H-acid (8-amino 1-naphthol 3,6-disulphonic acid)
Borax

Preparation of Reagents

Solution 1 (Acid azide reagent)
    α-nitrobenzene acid azide   50 mg
    Absolute alcohol   10.0 ml
    Veronol acetate buffer   10.0 ml

Solution 2 (Nitrous acid)
    Sodium nitrite   5.0 g
    Distilled water   100.0 ml

Take 9 ml of the above solution and add 1 ml of $N\text{-}H_2SO_4$.

Solution 3 (Reducing reagent)

    15 per cent trichloroacetic acid   2 ml
    ($TiCl_3$)
    Sodium citrate buffer   8.0 ml
    (0.5 M, pH 4.5) (*see* Chapter 5)

Procedure

1. Deparaffinize and hydrate slides to water.
2. Incubate in solution 1 at 22°C for 2 hours.
3. Wash twice in absolute alcohol.
4. Bring sections to 70 per cent, 50 per cent alcohol and then to water.
5. Transfer to reducing agent solution 3 at 37°C for 30 minutes.
6. Wash in citrate buffer.
7. Diazotize at 4°C in a mixture of 8 ml of solution 2 and 1 ml of $N\text{-}H_2SO_4$ for 5 minutes.
8. Wash in distilled water.
9. Couple with H acid at 4°C for 5 minutes, pH 9 (saturated solution in borax or veronol acetate).
10. Wash, dehydrate, clear and mount.

Result

NH_2 groups stain bright red.

Diazotization Coupling Method for Tyrosine (Glenner and Lillie, 1959)

Fixation

Carnoy, etc.

Reagents Required

Sodium nitrite
Acetic acid
8-amino 1-naphthol 5-sulphonic acid (S-acid)
Ammonium sulphamate
Hydrochloric acid
Potassium hydroxide

Preparation of Reagents

Solution 1 (Diazotizing agent)

    Sodium nitrite   6.9 g
    Acetic acid   5.8 ml
    Distilled water   94.0 ml

Dissolve the nitrite in water and then add acetic acid.

Solution 2 (Alkaline coupling agent)

    8-amino 1-naphthol 5-sulphonic acid   1.0 g

Potassium hydroxide	1.0 g
Ammonium sulphamate	1.0 g
70 per cent alcohol	100.0 ml

Procedure

1. Deparaffinize and hydrate slides.
2. Place sections in Solution 1 (incubating medium) at 4°C overnight in the dark.
3. Rinse in distilled water at 4°C.
4. Transfer sections to solution 2 at 4°C for 1 hour (also in the dark).
5. Rinse in 3 changes of 0.1N HCl of duration 5 minutes each.
6. Wash in tap water for 10 minutes.
7. Dehydrate, clear in xylene, mount in Canada balsam.

Result

Tyrosine-containing proteins stain purple-red.

Rationale

This method was first introduced by Lillie (1957a). In this method diazonium nitrates are formed after prolonged nitrosation of tyrosine. Diazonium nitrate when in an alkaline medium couples with amines and the end products are coloured. S-acid (8-amino-1-naphthol-5-sulphonic acid) is the coupling amine.

Millons reaction is the most characteristic method for aromatic group found in the amino acid tyrosine. This is supposed to be the most specific for tyrosine. Millon's variations are as follows:

Millon's Reaction (Gomori, 1952a)

Fixation

Any general fixative.

Reagents Required

Mercuric acetate
Trichloroacetic acid
Sodium nitrite

Solution 1

Mercuric acetate	5.0 g
Distilled water	100.0 ml
Trichloroacetic acid	15.0 g

Solution 2

| Sodium nitrite | 1.0 g |
| Distilled water | 100.0 ml |

Procedure

1. Deparaffinize and hydrate slides to water.
2. Place slides in solution 1 for 5–10 minutes at 37°C.
3. Add 10 ml of solution 2 to solution 1 and keep for 30 minutes.

4.   Rinse in 70 per cent alcohol.

5.   Dehydrate, clear and mount.

Result

Tyrosine stain pink to brick red.

Millon's Reaction for Tyrosine (Baker, 1956)

Fixation

Formalin, alcohol or any other fixative.

Reagents Required

Sodium nitrite
Sulphuric acid
Mercuric sulphate

Preparation of Reagents

Solution 1 (Mercuric sulphate solution)
Concentrated sulphuric acid   10.0 ml
Distilled water   100.0 ml

To this solution add 10 g of mercuric sulphate and heat until dissolved. Cool to room temperature and add 100 ml distilled water.

Solution 2 (0.25 per cent sodium nitrite solution)
Sodium nitrite   250 mg
Distilled water   10.0 ml

Solution 3 (Staining solution)
Solution 1   10.0 ml
Solution 2   1.0 ml

Procedure

1.   Deparaffinize and bring down slides to water.

2.   Place sections in a beaker, add solution 3 and boil gently.

3.   Allow it to cool to room temperature for 2 minutes.

4.   Wash sections in distilled water (3 changes each) for 2 minutes.

5.   Dehydrate, clear and mount.

Result

Tyrosine stains red, pink or yellowish red.

Rationale

Biochemists have adopted this method for quite some time. Bensley and Gersch (1933) and Baker (1956) modified it for histochemical purposes. Baker's modification is mostly in vogue. Meyer (1864) was the first to experiment and has shown that mercuric chloride when applied to tyrosine in an acid medium in the presence of potassium nitrite, develops a red colour. Baker (1956) found that Folin's reagent gave a strongest reaction (Folin and

Ciocalteu, 1927; Foli and Marenzi, 1929). Baker (1956) also stated that formalin-fixed material gave excellent results. He further stated that strong colour was produced by heating mercuric sulphate solution. At 70°C a reddish pink colour develops at the sites of tyrosine. Other important methods to detect tyrosine are diazotization coupling method of Glenner and Lillie (1959) and DNFB method of Danielli (1950).

Dinitrofluorobenzene for Tyrosine, Sulphydryls and NH$_2$ Groups (Danielli, 1950; Burstone, 1955)

Fixation

Carnoy and other fixatives, alcohol

Reagents Required

2-4 Dinitrofluorobenzene
Sodium hydrogen carbonate
Sodium nitrite
Hydrochloric acid
H-acid (8-amino 1-naphthol 3,6-disulphonic acid)
Sodium hydrosulphate
Veronol acetate buffer

Preparation of Reagents

Solution 1

2-4 dinitrofluorobenzene saturated in 90 per cent alcohol saturated with disodium hydrogen carbonate

Solution 2
Nitrous acid
5 per cent sodium nitrite   10.0 ml
(5 g in 100 ml water)
2N Hydrochloric acid   40.0 ml

Solution 3
'H' acid saturated in 0.1 M veronol acetate buffer pH 9.4 (8-amino 1-naphthol 3,6 disulphonic acid)

Procedure

1. Bring down sections to absolute alcohol, dry in air.
2. Stain in solution 1 (DNFB) (2–16 hours) overnight.
3. Rinse in 90 per cent alcohol (3 changes), then in tap water.
4. Treat with 5 per cent sodium hydrosulphite at 45°C for 30 minutes.
5. Wash in cold distilled water.
6. Transfer to solution 3 ('H' acid solution) at 4°C for 30 minutes.
7. Wash in cold distilled water.
8. Transfer to solution 3 ('H' acid solution) at 4°C for 15 minutes.
9. Wash in tap water.
10. Dehydrate, clear and mount.

Result

The sites of DNFB attachment to the tissues appear reddish purple.

Rationale

This technique (DNFB) is commonly used as a reaction for protein end groups. It reacts with free amino groups in proteins, the amino groups of lysine and hydroxylysine, the hydroxylphenyl groups of sulphydryl groups of cysteine. The resulting product of DNFB is yellow coloured. On the other hand the end product of DNFB tyrosine reaction is colourless. This colourless product when diazotized and simultaneously coupled with phenol or aromatic amine, becomes coloured. The most appropriate coupling agent is H-acid (8-amino 1-naphthol-3,6-disulphonic acid). This gives a reddish purple colour to the end product. According to Burstone (1955), DNFB and H-acid combination is the best for lysine (primary amino group), tyrosine (hydroxy group). Most satisfactory results could be obtained by using a specific blocking agent before demonstrating DNFB technique. For further details consult Pearse (1968).

DDD Reaction for Disulphides
(Barrnett and Seligman, 1952, 1954; Pearse, 1968)

Fixation

Carnoy, formalin, etc.

Reagents Required

N-ethyl maleimide
Phosphate buffer
Potassium cyanide
Veronol acetate
Fast blue B
DDD reagent

Preparation of Reagent

Solution 1 Blocking solution

N-ethyl maleimide	1.25 g
0.1 M phosphate buffer (pH 7.4)	100.0 ml

Solution 2 Reducing agent

Potassium cyanide	10.0 ml
Distilled water	90.0 ml

Solution 3 Incubating medium

DDD reagent	250 mg
Absolute alcohol	15.0 ml
0.1 M veronol acetate buffer	35.0 ml

Solution 4 Fast blue B Solution

Fast blue B	50 mg
0.1 M phosphate buffer pH 7.4	50.0 ml

This solution has to be prepared just before use.

Procedure

1. Deparaffinize and bring slides to water.
2. Coat sections with 0.5 per cent celloidin.
3. Rinse in 70 per cent alcohol.
4. Wash in tap water.
5. Transfer sections to solution 1 at 37°C for 4 hours.
6. Wash in 1 per cent acetic acid for 2 minutes.
7. Rinse in tap water.
8. Treat sections with reducing agent (potassium cyanide) solution 2 for 2 hours at 60°C.
9. Wash in tap water.
10. Incubate sections in solution 3 (DDD reagent) at 50°C for 1 hour.
11. Allow to cool at room temperature.
12. Wash in distilled water.
13. Rinse sections in distilled water acidified with acetic acid (pH 4.5) for 5 minutes.
14. Rinse again in acidified water. Repeat the previous step.
15. Transfer sections to 70 per cent, 90 per cent and absolute alcohol (3 changes) and then 90, 70 per cent and distilled water for 2 minutes each.
16. Transfer sections to staining solution 4 Fast blue B for 2 minutes.
17. Wash section in tap water.
18. Dehydrate, clear and mount.

Result

Disulphides stain bluish reddish violet shade.

Rationale

First introduced by Barrnett and Seligman (1953, 1954). It is good for (1) disulphide linkages and sulphydryl group, (2) SH only, and (3) disulphide linkages. To demonstrate SH and SS linkages, thioglycolic acid is used. It splits the disulphide linkages, and SH groups result. Both these and pre-existing SH groups are then demonstrated. Both types of sulphydryl groups react with the DDD reagent and the primary reaction product is colourless. When it is coupled with the diazonium salt, Fast blue B, the colour of the final product is blue.

This method is in vogue and is considered as a specific method for the demonstration of disulphide linkages. For this it is necessary to use *N*-ethyl maleimide to block the already existing sulphydryl groups. It is better to use a weak reducing reagent to prevent reversing blocking of SH by *N*-ethyl maleimide. Barrnett and Seligman (1954) recommend 10 per cent potassium cyanide.

Performic Acid Alcian Blue for Disulphides
(Adams and Sloper, 1955, 1956) (Plate 12, Figures 5 and 6)

Fixative

Formalin, Carnoy, Susa

Reagents Required

Formic acid
Hydrogen peroxide
Sulphuric acid
Alcian blue

Preparation of Reagents

Solution 1 (Oxidizing solution) (Performic acid (Pearse, 1960))

98 per cent formic acid	40.0 ml
100 V of H_2O_2	4.0 ml
Sulphuric acid	0.5 ml

Solution 2 (Staining solution)

Alcian blue	1.0 g
98 per cent sulphuric acid	2.7 ml
Distilled water	47.2 ml

Procedure

1. Deparaffinize and hydrate slides to water.
2. Keep sections in solution 1 (oxidizing solution) for 5 minutes.
3. Wash in tap water.
4. Dry sections by gently heating to 60°C.
5. Rinse in tap water.
6. Transfer to solution 2 (alcian blue solution) at room temperature for 1 hour.
7. Wash in running tap water.
8. Counterstain if necessary.
9. Wash.
10. Dehydrate, clear and mount in DPX.

Result

Disulphides stain dark blue.

Rationale

Adams and Sloper (1955) were the first to introduce it, to demonstrate cysteine in paraffin sections. With cryostat and freeze-dried sections, this is an excellent technique which gives best results. Performic acid oxidizes the cystein to cysteine sulphuric acid, which stains with the basic dye alcian blue.

Remarks

Oxidizing solution should be fresh. Before use, allow it to stand for one hour.

Thioglycolate–Ferric Ferricyanide for SS (Adams, 1956)

Fixation

Trichloroacetic acid–ethanol, formol calcium, formalin.

Reagents Required

Sodium thioglycolate
Phenyl mercuric chloride
Butanol
N-ethyl maleimide
Potassium ferrocyanide
Ferric chloride

Preparation of Reagents

Solution 1 (Sodium thioglycolate)
    Sodium thioglycolate   2.5 g
    Distilled water   100.0 ml

Procedure

1. Hydrate slides to water (2 slides).
2. Immerse both slides in solution 1 for 30 minutes.
3. Wash in acidified distilled water for 3 minutes.
4. Wash in running water for 3 minutes.
5. Transfer the second slide to 0.1 M N-ethyl maleimide for 4 hours at 37°C.
6. Place both slides in freshly prepared solution containing 10 ml of 1 per cent freshly made potassium ferrocyanide and 30 ml of 1 per cent ferric chloride for 1½ minutes.
7. Wash in distilled water.
8. Dehydrate, clear and mount.

Result

Disulphides take prussian blue colour.

Sulphydryl Groups DDD Reaction (Barrnett and Seligman, 1952, 1954)

Reagents Required

2:2-dihydroxy 6:6'-dinapththyl disulphide
Veronol acetate
Fast blue B
Phosphate buffer

Preparation of Reagents

Solution 1 (Incubating solution)
    2:2′ dihydroxy-6:6′-dinaphthyl disulphide (DDD)
    0.1 M veronol acetate buffer pH 8.5 35.0 ml
    DDD reagent   25 mg dissolved in absolute alcohol
Solution 2
    Fast blue B   50 mg
    Phosphate buffer pH 7.4 (0.1 M)   50.0 ml

This solution is prepared just before use.

Procedure

1. Deparaffinize and bring slides to water.
2. Incubate sections in solution 1 (DDD reagent) at 50°C for 1 hour.
3. Allow to cool.
4. Wash sections in distilled water.
5. Rinse sections in distilled water acidified with acetic acid, pH 4.5, for 5 minutes.
6. Rinse with acidified distilled water for 5 minutes.
7. Rinse in 70 per cent, 95 per cent and alcohol distilled water for 2 minutes each.
8. Place sections in staining solution 2 (Fast blue B) for 2 minutes.
9. Wash sections in running tap water.
10. Dehydrate, clear and mount in DPX.

Result

SH group stains reddish purple.

Mercury Orange Method for Sulphydryl Group (Bennett and Watts, 1958)

Fixation

Carnoy (Lillie, 1965)

Reagents Required

Mercury orange
N',N'-dimethyl formamide

Preparation of Reagents

Mercury orange saturated in $N'-N'$-dimethyl formamide 40.0 ml

Procedure

1. Deparaffinize, bring sections to absolute alcohol.
2. Frozen sections can be placed directly in solution 1 (mercury orange solution) for 2 days.
3. Rinse rapidly in 2 changes of absolute alcohol.
4. Clear in xylene and mount in DPX.

Result

SH group stains pale orange to orange-red.

Rationale

Bennett and Watts (1958) first introduced this method. Mercury orange was originally called red sulphydryl reagent. This is a specific method for SH groups (Barka and Anderson, 1963). The SH groups react with mercury orange to form mercaptide which is coloured.

Ferric Ferricyanide Technique for SH (Plate 12, Figure 4)

Fixation

Any general fixative

Reagents Required

Potassium ferricyanide
Ferric chloride

Preparation of Reagents

Solution 1 (Potassium ferricyanide)
  Potassium ferricyanide   1.0 g
  Distilled water   100.0 ml

Solution 2 (Ferric chloride solution)
  Ferric chloride   1.0 g
  Distilled water   100.0 ml

Solution 3
  Working Solution:
  4 cc of solution 1 and 30 cc of solution 2 and 6 cc of distilled water.
  All these reagents should be prepared just before use.

Procedure

1.   Deparaffinize and bring slides to water.
2.   Immerse slides in solution 3 for 10 minutes.
3.   Wash in tap water.
4.   Dehydrate, clear and mount.

Result

Sulphydryls   Blue
Nuclei   Red

Nitroprusside Reaction for SH (Hammett and Chapman, 1938)

Fixation

Fresh slices—40–100 µm

Reagents Required

Ammonium hydroxide
Sodium nitroprusside
Ammonium sulphate

Preparation of Reagents

Solution 1 (Staining solution)
  27–29 per cent ammonium hydroxide   0.05 ml
  1 per cent sodium nitroprusside   0.05 ml (solution should not be
                                             more than 1 hour old)

Procedure

1. Cut fresh sections.
2. Place the section in a dry watch glass.
3. Flood section with solution 1.
4. Displace the section to one side and underline with 0.25 g crystalline ammonium Sulphate.

Result

A pink colour develops but it fades after a short time. SH groups stain pink.

p-NBAF Method for SH Groups (Gershstein, 1958)

Fixation

Carnoy, Bouin's

Reagents Required

p-NBAF
Veronol acetate
Stannous chloride
Concentrated HCl
Sodium nitrite
H-acid

Preparation of Reagents

Solution 1 (*p*-NBAF)

p-NBAF 56 mg	
95 per cent ethanol	20.0 ml
Veronol acetate buffer (pH 7.8)	8.0 ml

Solution 2 (Stannous chloride)

Stannous chloride	10.0 g
5 N HCl	100.0 ml
Sulphuric acid w/v (10 per cent)	5.0 ml

Procedure

1. Bring sections to 95 per cent alcohol.
2. Incubate in solution 1 for 1 hour at 22°C.
3. Wash in acetone.
4. Reduce in solution 2 for 30 minutes.
5. Wash in 5N HCl.
6. Diazotize with $NaNO_2$/HCl at 2°C for 30 minutes.
7. Wash in water followed by veronol acetate.
8. Couple with 2 per cent H-acid in veronol acetate buffer for 5 minutes.
9. Dehydrate, clear and mount.

Result

SH group proteins stain violet.

BLOCKING AND CONVERSION METHODS FOR PROTEIN END GROUPS

While the specific demonstration of many proteins is not yet possible with the methods available, a greater degree of specificity may be obtained by appropriate blocking methods for specific groups. Many of these are not universally accepted as being completely specific, but in combination with the above methods many proteins can be identified with a degree of certainty.

Amino Group Blockade

Deamination   This is achieved by treatment with nitrous acid alone. But the following deamination method is much more specific (Stoward, 1963).

The following is the procedure adopted.

1.   Hydrate slides to water.

2.   Immerse in fresh nitrous acid (1 g of sodium nitrite in 30 ml of 3 per cent H_2SO_4) for 48 hours in the refrigerator at 0–5°C in the dark or 24 hours at room temperature.

3.   Wash in distilled water.

4.   Treat for 4 hours at 60°C in either (a) water or (b) alcohol.

5.   Stain for protein by appropriate method. Absence of staining after treatment in a previously stained area is a reliable evidence of the presence of amino groups before treatment.

The fixation procedure adopted for lipids is by itself a deamination (formol calcium method post-chromation). In fact these sections can be used for lipid staining as well as deamination sections.

Persulphate Block for Tryptophan

The persulphate breaks the pyrrole ring and this blocks reactions for tryptophan.

The procedure is as follows:

1.   Hydrate slides to water.

2.   Incubate in 2.5 per cent potassium persulphate in 0.5 N KOH for 16–18 hours at room temperature.

N-Haloamide Bromination for Tyrosine or Tryptophan

N-haloamide bromination splits tyrosine and tryptophan and this is a specific method (Pearse, 1968).

1.   Bring sections to absolute alcohol.

2.   Transfer to 0.02 per cent *N*-bromosuccinamide $(CH_2CO)_2nBr$ in 50 per cent alcohol at pH 4.0.

Sulphydryl Group Block

Iodoacetate or maleimide methods block sulphydryl groups.

Iodoacetate method (Sulphydryl block)

1. Dewax and bring down slides to water.
2. Treat with 0.1 M (approximately 2 per cent) aqueous sodium iodoacetate at pH 8.0 (with NaOH) for 20 hours at 37°C.

Maleimide method

1. Dewax and bring down sections to water.
2. Treat with 0.1 M (1.25 per cent) *N*-ethyl maleimide in 0.1 M phosphate buffer at pH 7.4 for 4 hours at 37°C.
3. Rinse in 1 per cent acetic acid.
4. Rinse in tap water.

Disulphide Reduction Method (–SS to –SH)

Lillie (1965) suggests the use of a freshly prepared 10 per cent sodium thioglycolate solution, adjusted with sodium hydroxide to pH 9.5 for 10 minutes at room temperature. A freshly prepared 0.5 M thioglycolate acid solution at pH 8.0 (with 0.1 N NaOH) for 4 hours at 37°C is specified by Pearse (1968). In spite of precelloidinization, sections may fall off.

Mercaptide Block

Mercuric chloride and phenyl mercuric acetate are used for this reaction. The latter is used as a 0.001 M solution in *n*-propanol (72 hours at 20°C). Many other organic mercurials are also recommended.

Iodine Oxidation (SH)

Sections are treated for 4 hours at room temperature with an aqueous solution containing 38 mg iodine and 33 mg potassium iodide per 100 ml. This solution should be adjusted to pH 3.2 with 0.01 N-HCl before use.

Permolybdic and Pertungstic Acid

These acids can be prepared by adding 30 g orthophosphoric acid (H_2PO_4) and 3 ml 30 per cent H_2O_2 0.5 g of either ammonium molybdate or sodium tungstate to water (60 ml). Tissue sections are oxidized at room temperature for 4–16 hours.

Disulphide and Sulphydryls

Lillie (1965) stated that it is quite difficult to distinguish between the two groups. Dihydroxy dinaphthyl disulphide (DDD) method of Barnett and Seligman (1952), mercury orange method of Bennett and Watts (1958), performic acid alcian blue method of Adams and Sloper (1955, 1956) demonstrate the presence of disulphides. These sulphur-containing groups occur in cystine, cysteine and methionine.

p-DMAB–Nitrite Method for Tryptophan (Adams, 1957)

Fixative

Short fixation 6–12 hours in formalin; 70 per cent methanol.

Reagents Required

p-dimethyl aminobenzaldehyde
Concentrated hydrochloric acid
Sodium nitrite

Preparation of Reagents

Solution 1 Staining solution
    p-dimethylaminobenzaldehyde   5.0 g
    Concentrated hydrochloric acid 100.0 ml

Solution 2 Sodium nitrite solution
    Sodium nitrite   1.0 g
    Concentrated hydrochloric acid 100.0 ml

Solution 3 Acid alcohol
    Concentrated hydrochloric acid   1.0 ml
    70 per cent alcohol   99.0 ml

Procedure

1. Dewax and bring sections to alcohol.
2. Coat with celloidin (0.5 per cent).
3. Keep sections in solution 1 (DMAB) for 1 minutes.
4. Transfer sections to solution 2 (sodium nitrite) for 1 minutes.
5. Wash in tap water.
6. Rinse sections in solution 3 (acid alcohol) for 15 seconds.
7. Dehydrate, clear and mount.

Result

Tryptophan stains deep blue.

Rationale

Glenner (1957), Adams (1957) and Glenner and Lillie (1957a) gave details of the processing Adam's method is preferable. In this technique tryptophan reacts with *p*-dimethylaminobenzaldehyde and produces a compound β-carboline. This is oxidized by nitrite solution to produce a blue pigment.

The Rosindole Reaction of Indoles (Glenner, 1957)

Fixation

10 per cent calcium acetate-formalin for 3–6 hours.

Reagents Required

Perchloric acid
Acetic acid
Hydrochloric acid
p-dimethylaminobenzaldehyde
Sodium nitrite

Preparation of Reagents

Solution 1 (*p*-dimethylaminobenzaldehyde)

Perchloric acid	60.0 ml
Acetic acid	34.0 ml
Concentrated HCl	1.0 ml
p-dimethylaminobenzaldehyde	1.0 g

Solution 2 (Sodium nitrite)

Acetic acid	35.0 ml
Concentrated HCl	5.0 ml
Sodium nitrite	500 mg

Procedure

1. Bring sections to absolute alcohol.
2. Immerse in solution 1 for 3 minutes at 25°C.
3. Transfer to solution 2 for 1 minute.
4. Wash three times in acetic acid and place it in 50:50 acetic xylene to pure xylene.
5. Mount in a synthetic medium.

Result

Tryptophan-containing proteins stain blue.

Rationale

This technique is almost similar to DMAB–nitrite method. Either 2 pyrrole compounds condense with one aldehyde to give a red colour or one pyrrole condenses one aldehyde resulting in a red dye. DMAB nitrite method gives a brighter colour than rosindole reaction.

Xanthydrol Reaction for Indoles (Lillie, 1957b)

Fixation

10 per cent calcium acetate formalin for 48 hours
6 per cent glutaraldehyde

Reagents Required

Xanthydrol
Acetic acid
Concentrated HCl

Preparation of Reagents

Solution 1 (Xanthydrol)

Xanthydrol	2.5 g
Acetic acid	90.0 ml
Concentrated HCl	10.0 ml

Procedure

1. Bring sections to absolute alcohol.
2. Dry in air for 30 seconds.
3. Place sections in solution 1.
4. Wash in acetic acid.
5. Clear in 50:50 acetic xylene, clear in paraxylene and mount.

Result

Indole-containing tissue stains violet.

Rationale

This method was first described by Dickman and Crockett (1956). Pyrrole reacts with xanthydrol in acetic acid resulting in a purplish red product. This colour is more prominent than DNAB–nitrite reaction.

Tryptophan Method for Formalin-fixed Tissues (Adams, 1960)

Fixation

Formalin

Reagents Required

Glycerol
Ferric chloride
Sulphuric acid
Methyl alcohol
Glacial acetic acid

Preparation of Reagents

Solution 1 (Staining solution)

Glycerol	5.0 ml
60 per cent ferric chloride (60 g/100 ml water)	1.0 ml
Concentrated sulphuric acid	5.0 ml
Distilled water	9.0 ml
Methyl alcohol	80.0 ml

This solution will keep for several months.

Procedure

1. Deparaffinize and bring sections to absolute alcohol.
2. Transfer sections to solution 1.
3. Drain of excess fluid, hold slide with a forceps hold the slide on a flame.
4. Wash in absolute alcohol.
5. Rinse in glacial acetic acid.
6. Clear in xylene and mount.

Result

Tryptophan stains mauve colour.

Post-Coupled Benzylidene Reaction for Indoles (Glenner and Lillie, 1957a)

Fixation

10 per cent calcium acetate formalin for 6 hours

Reagents Required

p-dimethylaminobenzaldehyde
Hydrochloric acid
Acetic acid
S-acid
Sodium nitrite

Preparation of Reagents

Solution 1

p-dimethylaminobenzaldehyde	1.0 g
Concentrated hydrochloric acid	10.0 ml
Acetic acid	10.0 ml

Solution 2 (S-acid)

S-acid	240 mg
NHCl	3.0 ml
Distilled water	6.0 ml
N-Sodium nitrite	1.0 ml

Procedure

1. Bring sections to absolute alcohol and dry.
2. Transfer to solution 1 at 25°C for 5 minutes.
3. Differentiate in acetic acid.
4. Transfer sections to solution 2 for diazotization for 5 minutes.
5. Differentiate in acetic acid.
6. Clear in acetic xylene (50 per cent and 20 per cent).
7. Clear in xylene and mount.

Result

Indoles stain dark blue.

Naphthyl Ethylenediamine Method for Tryptophan (Bruemmer *et al.*, 1957)

Fixation

Carnoy, Zenker, formalin.

Reagents Required

Ethylenediamine
Hydrochloric acid
Concentrated sulphuric acid
Acetic acid
Sodium nitrite

Preparation of Reagents

Solution 1
Sodium nitrite   8.0 g
Distilled water   100.0 ml

Take 50 ml of the above solution and to this add 50 ml of 6 NHCl.

Solution 2
N-ethylenediamine   2.0 g
95 per cent ethanol   100.0 ml
Add few drops of N-HCl
(This must be a fresh solution)

Procedure

1.   Bring paraffin sections to ice cold 50 per cent alcohol.
2.   Transfer to solution 1 for 15 minutes.
3.   Wash in distilled water.
4.   Transfer to solution 2 for 15 minutes.
5.   Dehydrate in 70 per cent alcohol containing few drops of concentrated H_2SO_4 for 15 minutes.
6.   Repeat it at room temperature.
7.   Clear in xylene containing 2 ml acetic acid at room temperature.
8.   Mount.

Result

Tryptophan activity sites stain purple.

Sakaguchi Reaction for Arginine (Sakaguchi, 1925, Modified by Baker, 1947)

Fixative

Carnoy, acetic alcohol.

Reagents Required

Sodium hydroxide
α-naphthol
Sodium hypochlorite
Pyridine
Chloroform

Preparation of Reagents

Solution 1 (Sodium hydroxide solution)
Sodium hydroxide   1.0 g
Distilled water   100.0 ml

Solution 2 (Naphthol solution)
α-naphthol   1.0 g
70 per cent alcohol   100.0 ml

Solution 3 (Milton solution)
Sodium hypochlorite   1.0 ml
Distilled water   99.0 ml

Solution 4 (Incubating solution)
Solution 1   2.0 ml
Solution 2   2 drops
Solution 3   4 drops

Solution 5 (Pyridine chloroform solution)
Pyridine   30.0 ml
Chloroform   10.0 ml

Procedure

1. Deparaffinize and hydrate slides.
2. Rinse in 70 per cent alcohol.
3. Flood sections with solution 4 (incubating solution) for 15 minutes.
4. Drain blot and dry.
5. Transfer to solution 5 (pyridine chloroform solution) for 2 minutes.
6. Mount in pyridine–chloroform mixture and ring the coverslip.

Result

Arginine stains orange-red.

Rationale

α-naphthol reacts with arginine in the alkaline medium to produce a red colour. Baker (1947) modified this method. Other naphthols have been tried and a very stable reaction could be obtained. Among these mention may be made of 8-hydroxyquinone (Warren and McManus, 1951) and 2:4-dichloro α-naphthol. Satisfactory results can be obtained with 15 micron sections.

Sakaguchi Dichloronaphthol/Hypochlorite Technique for Arginine (Deitch, 1961)

Fixation

Acetic ethanol or Carnoy fluid

Reagents Required

Barium hydroxide
Sodium hypochlorite
2, 4-dichloro-α-naphthol

Solution 1 (Staining solution)
4 per cent barium hydroxide	25.0 ml
20 per cent commercial 5 per cent hypochlorite in distilled water (1:4)	5.0 ml
Dissolve 75 mg 2, 4-dichloro-α-naphthol in 5 ml tertiary butanol	5.0 ml

Solution 2 (Dehydrating solution)
5 per cent tributylamine in tertiary butanol (aniline in place of tributylamine is also good.

Solution 3 (Clearing agent)
5 per cent tributylamine in xylene (or 5 per cent aniline)

Solution 4 (Mountant)
Shillaber's oil plus 10 per cent tributylamine or cellulose caprate (Resin 50 g: 50 ml xylene) to which is added 10 per cent tributylamine (oramtine) (Lillie, 1965).

Procedure

1. Hydrate slides to water and dry.
2. Transfer sections to a coplin jar containing solution 1 (reactant) for 10 minutes at 22°C.
3. Immerse sections in solution 2 (dehydrant) giving 3 changes 5 seconds each.
4. Clear in clearing agent solution 3 giving 2 changes.
5. Drain and mount in solution 4 (Schillaber's oil or cellulose caprate).

Result

Arginine-containing sites stain orange-red.

Oxidized Tannin-Azo (OTA) Technique (Dixon, 1959, 1962)

Fixation

Carnoy's fluid gives excellent results.

Reagents Required

Tannic acid
Hydrochloric acid

Periodic acid
Fast blue B
Acetic acid
Sodium acetate
o-dianisidine

Preparation of Reagents

Solution 1 (Tannic–HCl)

Tannic acid	10.0 g
Distilled water	225.0 ml
1N hydrochloric acid	25.0 ml

Solution 2 (0.5 per cent aq. periodic acid, pH 4)

Solution 3

Buffered diazotized *o*-dianisidine (pH 4.0) ice-cold solution should be used.

Fast blue B	100 mg
0.2 M acetic acid	82.0 ml
0.2 M sodium acetate	18.0 ml

Procedure

1. Deparaffinize sections and hydrate slides.
2. Place sections in solution 1 (Tannic-HCl) and leave for 10 minutes.
3. Wash well in distilled water giving 3 changes.
4. Oxidize with solution 2 (periodic acid) for 5 minutes.
5. Wash in distilled water.
6. Place sections in ice-cold distilled water.
7. Place in solution 3 (ice-cold buffered diazotized *o*-dianisidine solution) in the refrigerator for 20 minutes.
8. Wash in running tap water.
9. Dehydrate, clear and mount.

Result

Tannophilic proteins are stained salmon pink.

Rationale

Tannic acid links with the protein in the tissue by reacting with amino groups. It is then oxidized by periodic acid to 1:2-quinone which is later coupled with diazotized *o*-dianisidine to produce a salmon red azo dye (azoquinone).

Oxidized Tannin Oxazine Technique (OTO) (Dixon, 1962)

Reagents Required

Tannic acid
Hydrochloric acid
Periodic acid
6-nitroso-3-dimethyl aminophenol
Glacial acetic acid

Preparation of Reagents

Tannic HCl solution as in previous technique
0.5 per cent aqueous periodic acid
Freshly prepared aminophenol solution

Solution 1 (Staining solution)
    6-nitroso-3-dimethyl aminophenol   500 mg
    Glacial acetic acid   50.0 ml

Dissolve 500 mg of salt in 50 ml acetic acid. The solution is dark orange. Cool the solution till crystals are formed and add excess zinc dust. Orange colour disappears and then the mixture is filtered. Measure the filtrate and add acetic acid to make up to 100 ml. Now the solution turns blue. It should be used immediately.

Procedure

1. Steps 1–5 are as for OTA technique.
2. Wash with 3 changes of glacial acetic acid.
3. Transfer to aminophenol solution at 37°C for 20 minutes.
4. Wash in glacial acetic acid.
5. Dehydrate, clear and mount.

Result

Tannophilic proteins are stained blue-grey.

Rationale

1 : 2 quinone reacts with freshly prepared 6-amino 3-dimethylamino phenol to produce blue-grey oxazine.

Mixed Anhydride Method for Protein-bound Side Chain COOH (Barrnett and Seligman, 1958)

Fixation

Carnoy, formalin.

Reagents Required

2-hydroxy-3-naphthoic acid hydrazide pyridine

Preparation of Reagents

NAH Reagents (Staining solution)
    2-hydroxy 3-naphthoic acid hydrazide   50 mg
    Glacial acetic acid   2.5 ml
    50 per cent alcohol   47.5 ml

First dissolve the hydrazide in warm glacial acetic acid and then add alcohol.

Procedure

1. Remove wax with petroleum.
2. Dry in air, transfer to glacial acetic acid for 2 minutes.

3. Incubate in equal parts of acetic anhydride and anhydrous pyridine for 1 hour at 60°C.
4. Rinse in glacial acetic acid and wash in absolute alcohol.
5. Incubate in NAH reagent for 2 hours at room temperature.
6. Wash in 50 per cent alcohol giving 3 changes of 10 minutes each.
7. Transfer to 0.5N HCl for 30 minutes.
8. Rinse in distilled water.
9. Rinse in 1 per cent sodium bicarbonate giving 3 changes.
10. Rinse in distilled water changing several times.
11. Transfer to a jar containing equal parts of absolute ethanol and 0.06 M phosphate buffer (pH 7.6) containing 1 mg/ml Fast blue B salt for 6 minutes.
12. Rinse in distilled water.
13. Dehydrate, clear and mount.

Result

Protein-bound side chain COOH group is stained red-purple.

Rationale

COOH groups are turned into amidoketones by acetic anhydride in pyridine. The ketones condense with 2-hydroxy-3-naphthoic acid hydrazide (NAH). This compound couples with Fast blue B to give a coloured substrate.

C-Terminal Carboxyl (COOH) Groups
(Stoward, 1963; Stoward and Burns, 1967)

Fixation

Formalin

Reagents Required

Acetic anhydride
Pyridine
Salicyloyl hydrazide
Zinc acetate
Pentacyanoamine ferroate

Preparation of Reagents

Solution 1 (Acetic anhydride/pyridine)
Equal parts of acetic anhydride and anhydrous pyridine (Redistilled over barium oxide)

Solution 2
1 per cent salicyloyl hydrazide in 5 per cent glacial acetic acid

Solution 3
PAF solution—A fresh dilute solution of pentacyanoamine ferroate (Trisodium salt)

Solution 4
1 per cent aqueous zinc acetate solution

Procedure

1. Remove wax with petroleum and dry.
2. Wash in glacial acetic acid for 2 minutes.
3. Treat with acetic anhydride pyridine at 60°C (solution 1).
4. Wash in 2 changes of 95 per cent alcohol.
5. Place in salicyloyl hydrazide for 30–40 minutes at room temperature (solution 2).
6. Rinse in distilled water.
7. Transfer to PAF for 2 minutes (solution 3).
8. Rinse in distilled water.
9. Treat with zinc acetate solution for 5–10 minutes (solution 4).
10. Rinse in distilled water.
11. Dehydrate, clear and mount in Uni mount or fluoromount.

Result

C-Terminal COOH groups give an intense blue fluorescence.

Rationale

COOH groups are converted into amido ketones by treating with anhydride in pyridine. Ketone interacts with salicyloyl hydrazide to give an intense blue fluorescence.

Fast Green Method for Histones
(Alfert and Geschwind, 1953) (Plate 12, Figure 3)

Fixation

10 per cent formalin for 3–6 hours, Zenker or Susa.

Reagents Required

Trichloroacetic acid
Fast green

Preparation of Reagents

Solution 1 (TCA)
    Trichloroacetic acid (TCA) 5.0 g
    Distilled water   100.0 ml

Solution 2 (Staining solution)
    Fast green FCF 100 mg
    Distilled water 100.0 ml

pH should be 8.0–8.1 with sodium hydroxide, should be prepared fresh.

Procedure

1. Deparaffinize and hydrate slides.
2. Immerse in solution 1 in boiling water bath.
3. Wash in 2 changes of 70 per cent alcohol for 10 minutes each.
4. Stain in solution 2 for 30 minutes.

5. Wash in distilled water for 5 minutes.
6. Transfer directly to 95 per cent alcohol.
7. Place in absolute alcohol, clear and mount.

Result

Histones are stained green.

Rationale

Basic groups are unmasked by treating with trichloroacetic acid. The basic groups bind to fast green.

Ammoniacal Silver Method (Black and Ansley, 1964)

Fixation

10 per cent neutral buffered formalin

Reagents Required

Silver nitrate
Ammonium hydroxide

Preparation of Reagents

Solution 1
Silver nitrate   10 g
Distilled water   100.0 ml

Add 38 ml of solution 1 drop by drop to 3 to 4 ml of concentrated ammonium hydroxide. A turbid fluid is formed. Constantly stir it.

Procedure

1. Dewax and hydrate slides to water.
2. Place in buffered formalin for 15 minutes.
3. Wash in distilled water giving several changes.
4. Transfer to solution 1 for 5–10 seconds.
5. Again wash in distilled water giving several changes.
6. Treat with 3 per cent neutral formalin for 2 minutes.
7. Wash again in distilled water giving several changes.
8. Dehydrate, clear and mount.

Result

Histones are stained black.
(For further details see black *et al*. (1964) and Vedal *et al*. (1971))

Table 10.1   Histochemical techniques applied for the demonstration of proteins

Technique	Fixative	Basic proteins	NH_2 groups	Phenyl (tyrosine)	Disulphides (S-S)
Mercury bromophenol blue	Any general fixative	Blue			
Acrolein Schiff	Susa	Red			
Biebrich Scarlet	Carnoy, buffered mercuric chloride, alcohol	Strongly			
Acid solochrome	Carnoy, Susa, Bauins cold formalin	Dark steel blue			
Ninhydrin/Schiff	Cryosat or Susa freeze-dried		Pink to red		
Hydroxynaphthaldehyde	Non-formalin fixed, Susa		Blue		
Chloramine T-Schiff	Carnoy or cold acetone		Pink or magenta		
Diacetylbenzene	Susa		Reddish		
Alkaline salicylaldehyde	Freeze-dried, Carnoy, alcohol		Green fluorescence		
Acid azide	Bouin's, Carnoy, alcohol-paraffin		Bright red		
Diazotization coupling method	Carnoy			Purple-red	
Millon's reaction	Formalin, alcohol			Red, pink or yellowish red	

Dinitrofluorobenzene (DNFB)	Carnoy, alcohol	Reddish purple
DDD reaction for S–S groups	Carnoy, formalin	Bluish reddish violet
Performic acid Alcian blue	Carnoy, formalin	Dark blue
Thioglycolate-ferric ferricyanide	Trichloroacetic acid–ethanol, formalin	Prussian blue

Table 10.2   Histochemical techniques applied for the demonstration of proteins

Technique	Fixative	Sulphydryls (SH)	Indoles (Tryptophan)	Arginine	Tannophilic	Histones	COOH group
DDD reaction	Veronol acetate	Reddish purple					
Mercury orange	Carnoy	Pale orange or orange-red					
Ferric ferricyanide	Any general fixative	Blue					
Nitroprusside reaction	Fresh or frozen	Pink					
p-NBAF method for SH	Carnoy or Bouin's	Violet					
p-dimethylamino-benzaldehyde-nitrite	Formalin or 70 per cent methanol		Deep blue				
Rosindole method	10 per cent calcium acetate-formalin		Blue				

Xanthydrol for tryptophan	10 per cent calcium acetate formalin	Violet
Tryptophan method for formalin-fixed tissue	Formalin	Mauve colour
Post coupled benzylidene for indoles	10 per cent calcium acetate formalin	Dark blue
Napthyl ethylene diamine for tryptophan	Carnoy, Zenker, formalin	Purple
Sakaguchi	Carnoy, acetic alcohol	Orange-red
Sakaguchi Dichloronaphthol/ hypochlorite method for arginine	Acetic ethanol or Carnoy	Orange-red

(Contd.)

Table 10.2 (Continued)

Technique	Fixative	Sulphydryls (SH)	Indoles (Tryptophan)	Arginine	Tannophilic	Histones	COOH group
Oxidized Tannin-Azo technique	Carnoy's fluid				Salmon pink		
Oxidized Tannin Oxazine	Tannic acid				Blue-grey		
Mixed anhydride method for protein-bound side chain	Carnoy, formalin						Red-purple
C-Terminal Carboxyl	Formalin						Blue flourescence
Ammoniacal Silver Method	10 per cent neutral buffered formalin					Black	
Fast Green Method	10 per cent formalin, Susa					Green	

AMYLOIDS

11

Amyloid has for long been considered as a mixture of carbohydrate and protein. Glenner *et al.*, (1968) regarded amyloid as a protein. Kidney, spleen, liver and adrenals are the most affected parts where amyloid is deposited extracellularly. Sometimes organs and tissues which do not appear to have a deposition of amyloid, still display it microscopically, such as the walls of small blood vessels. This is a very useful diagnostic characteristic for amyloids. Most of the patients suffering from amyloid deposition in various organs have demonstrable amyloid within the wall of the small blood vessel, in the rectal mucosa and submucosa. However, amyloid is usually widespread within the body (amyloid of the heart or amyloid deposition in the stroma of a type of cancer of the thyroid gland).

Amyloid is broadly classified into 1) Primary amyloid and 2) Secondary amyloid. Primary amyloid is not associated with any disease, whereas secondary amyloid is associated with disease. Some of the diseases with which amyloid may be associated are:

i. Chronic tuberculosis

ii. Rheumatoid arthritis

iii. Tumours of plasma cells

iv. Myeloma

In formalin-fixed paraffin sections, amyloid appears as a homogeneous material taking a pink shade with eosin and khaki colour with Van Geison. It could often be confused for hyaline or collagen. So few specific methods have been developed for amyloid.

1. Congo red method observed under fluoroscence microscope polarized light.

2. Thioflavine-T fluorescence method.

3. Methyl violet and methyl green method.

Amyloid is best demonstrable by electron microscope. It has ultrastructure with characteristic cylindrical fibrils. These are arranged in bundles of 2–8 fibrils arranged in a parallel manner. Each fibril is 75 Å in diameter comprising two 25 Å electron rods separated by a 25 Å wide interspace.

HISTOCHEMICAL METHODS FOR AMYLOIDS

Congo Red (Modified by Highman, 1946)

Fixation

All types, frozen sections recommended

Reagents Required

Congo red
Potassium hydroxide
Absolute alcohol

Preparation of Reagents

Solution 1 (Congo red)

Congo red	500 mg
Absolute alcohol	50.0 ml
Distilled water	50.0 ml

Solution 2 (Differentiator)

Potassium hydroxide	200 mg
Absolute alcohol	30.0 ml
Distilled water	20.0 ml

Procedure

1. Deparaffinize and hydrate slides to water.
2. Place in solution 1.
3. Wash in distilled water.
4. Transfer to solution 2.
5. Wash in tap water.
6. If desired, counterstain in haematoxylin.
7. Wash.
8. Dehydrate, clear and mount.

Result

Amyloid	Orange-red
Elastin	Orange

Rationale

This method was first described by Benhold (1922). The dye has a great affinity for amyloid. Differentiation is important. In Benhold's method, differentiation is too rapid resulting in collagen being over-stained. Highman used potassium hydroxide (0.2 per cent) to differentiate at a slower rate. In alkaline Congo red method, differentiation is not necessary.

Alkaline Congo Red Method (Puchtler *et al.*, 1962)

Fixation

Cryostat-pre-fixed, post-fixed
Frozen formalin-fixed paraffin

Reagents Required

Hydrochloric acid
Sodium chloride
Sodium hydroxide
Congo red
Absolute alcohol

Preparation of Reagents

Solution 1 (Acid alcohol)
Hydrochloric acid   1.0 ml
70 per cent alcohol   99.0 ml

Solution 2 (Stock solution of saturated sodium chloride)
80 per cent alcohol   100.0 ml

Add sodium chloride for saturation.

Solution 3 Alkaline alcohol
Solution 2   50.0 ml
1 per cent sodium hydroxide   0.5 ml

Use within fifteen minutes.

Solution 4 (Stock Congo red)
Solution 2 (Saturated with Congo red)   100.0 ml
This solution lasts for several months. Allow to stand for 24 hours.

Solution 5
Solution 4   50.0 ml
1 per cent aqueous sodium hydroxide   0.5 ml
Filter and use within 15 minutes.

Procedure

1. Deparaffinize and hydrate slides to water or bring frozen section to water.
2. Transfer to Mayer's haematoxylin for 10 minutes.
3. Wash in tap water.
4. Place in solution 1.
5. Wash.
6. Transfer to solution 3 in a closed jar for 20 minutes.
7. Keep in solution 5 in a closed jar for 20 minutes.
8. Dehydrate, clear and mount.

Results

Amyloid	Orange-red
Elastin	Orange-red
Nuclei	Blue

High pH Congo Red Technique (Eastwood and Cole, 1971)

Fixation

Any general fixative.

Preparation of Reagents

Solution 1 (Glycine buffer pH 10.0)
0.1 M glycine	30.0 ml
0.1 M sodium chloride	30.0 ml
0.1 M sodium hydroxide	40.0 ml

Solution 2 (Staining solution)
Equal parts of 0.5 per cent Congo red and glycine

Procedure

1. Hydrate slides to water.
2. Stain with alum haematoxylin for nuclear staining.
3. Treat with solution 2 for 10–20 minutes.
4. Rinse in 70 per cent alcohol until background is clear.
5. Dehydrate, clear and mount.

Result

Amyloid, elastic tissue	Red
Nucleus	Blue

Congo Red as Fluorescence Method for Amyloid (Cohen *et al.*, 1959; Puchtler and Sweat, 1965)

Fixation

Frozen

Reagents Required

Congo red
Potassium hydroxide

Preparation of Reagents

Solution 1 (Congo red)
Congo red	100 mg
Absolute alcohol	50.0 ml
Distilled water	50.0 ml

Solution 2 (Differentator)
Potassium hydroxide	200 mg

Absolute alcohol	80.0 ml
Distilled water	20.0 ml

Procedure

1. Deparaffinize and hydrate slides to water.
2. Place in solution 1 for 1 minute.
3. Wash in tap water.
4. Transfer to solution 2 (Differentator) (Sections should be colourless).
5. Wash.
6. Dehydrate, clear and mount.

Result

Amyloid stains with orange-red fluorescence.

Rationale

Amyloid fluoresces when stained with Congo red. Sometimes elastin and collagen also fluoresce. To minimize or avoid this, Cohen *et al.* (1959) modified the Congo red method. Accordingly this section is stained for a very short time and differentiated for a longer time so that the sections appear almost colourless when examined under microscope. Under such conditions, fluorescence microscopy reveals only amyloid crystals, other structures remain non-discernible.

Crystal Violet Method (Lieb, 1947)

Fixation

10 per cent formalin or alcohol

Reagents Required

Crystal violet

Preparation of Reagents

Solution 1 (Crystal violet stock solution)
Crystal violet   14.0 to 15.0 g

Crystal violet is saturated in 100.0 ml of 95 per cent alcohol.

Solution 2 (Working solution)

Solution 1	10.0 ml
Distilled water	300.0 ml
Concentrated HCl	1.0 ml

Procedure

1. Dewax and bring down slides to water or run frozen section.
2. Transfer to solution 2 for 5 minutes to 24 hours.
3. Wash in distilled water.
4. Mount in glycerine jelly.

Result

Amyloid	Purple
Background	Blue

Thioflavine-T Method (Vassar and Culling, 1959)

Fixation

Formalin paraffin, unfixed cryostat

Reagents Required

Thioflavine-T
Acetic acid
Haematoxylin

Preparation of Reagents

Solution 1 (Thioflavine-T)

Thioflavine-T	500 mg
Distilled water	50.0 ml

Solution 2 (Differentiator)

Acetic acid	1.0 ml
Distilled water	100.0 ml

Solution 3 (Cole's haematoxylin) (*see* Chapter 7)

Procedure

1. Hydrate slides to water.
2. Transfer sections to solution 3 for 2 minutes.
3. Wash in distilled water.
4. Transfer to solution 1 for 3 minutes.
5. Rinse in distilled water.
6. Place slides in solution 2 for 20 minutes.
7. Wash.
8. Dehydrate, clear in xylene and mount in fluoro-free mountant.

Results

Amyloid	Bright yellow
Mast cells	Yellow

Rationale

Thioflavine-T has a specific affinity for amyloid. This method was first suggested by Vassar and Culling (1959). When the section is examined under the microscope, amyloid shows a brilliant yellow fluorescence. Thioflavine-T has affinity for other structures like juxtaglomerular apparatus in kidney and cell nuclei which fluoresce. This can be minimized by staining the slides with haematoxylin which normally stains nuclei and obliterates fluorescence. Slides need to be dehydrated in fresh alcohol, and fresh xylene should be used for clearing.

Methyl Violet Method

Fixation

Any general fixative

Reagents Required

Methyl violet
Acetic acid

Preparation of Reagents

Solution 1 (Methyl violet)
    Methyl violet   250 mg
    Distilled water   100.0 ml

Solution 2 (Differentiator)
    Acetic acid   1.0 ml
    Distilled water   99.0 ml

Procedure

1. Hydrate slides to water.
2. Place sections in solution 1 for 10 minutes.
3. Wash in tap water.
4. Place in solution 2 for 1 minute.
5. Wash in tap water.
6. Mount in glycerine jelly.

Result

Amyloid   Red
Nuclei   Blue

Rationale

Methyl violet stains amyloid metachromatically, i.e., in a colour different from the colour of the dye itself. Amyloid stains in a pinkish violet shade while the background remains bluish violet. Frozen sections also display amyloid metachromatically (red). Satisfactory preparations can be obtained by differentiating in acid alcohol using acetic or formic acid.

Actually, methyl violet is an impure dye (Kramer and Windrum, 1955). Impurities impart colour to amyloid but the dye may not be metachromatic. Methyl violet leaks out of amyloid into other tissues. This can be avoided by using watery mountant like corn syrup or glycerine which is not a permanent preparation.

Methyl Green Method (Bancroft, 1967)

Fixation

Cryostat, paraffin

Reagents Required

Methyl green
Acetic acid

Preparation of Reagents

Solution 1 (Methyl green solution)
Methyl green   2.0 g
Distilled water   100.0 ml

Solution 2 (Differentiator) (For paraffin sections)
Acetic acid   1.0 ml
Distilled water   99.0 ml

Procedure

1. Fix cryostat sections in formol saline for 2 minutes.
2. Wash in tap water.
3. Transfer to solution 1 for 2 minutes.
4. Wash in tap water for 30 seconds.
5. Air-dry.
6. Rinse in triethyl phosphate for 10 seconds.
7. Transfer to xylol.
8. Mount in DPX.

Result

Amyloid   Pink
Nuclei   Green
Elastin   Blue

Rationale

Amyloid deposits could easily be demonstrated by this method in frozen sections. When frozen sections are treated with methyl green (which is not chloroform-extracted and contains traces of methyl violet) it takes a pink shade and the nuclei appear in green shade. Sections can be mounted in corn syrup. Permanent preparation can be made after drying the slide. It is then immersed in triethyl phosphate, soaked in xylol and mounted in DPX.

There are other stains like Sirius red and Alcian blue which though not specific can stain amyloid. Sirius red is similar to Congo red showing apple green birefringence. Alcian blue staining is not specific (Lendrum *et al.*, 1969).

Metachromatic Methods for Amyloid: Dahlia Method (Lendrum, 1951)

Fixation

Formalin, alcohol, fresh frozen alcohol fixed

Reagents Required

Methyl violet

Formalin
Sodium chloride

Preparation of Reagents

Solution 1 (1 per cent aq. methyl violet)
    Methyl violet   1.0 g
    Distilled water   100.0 ml

Solution 2 (70 per cent formalin)
    Formalin   7 parts
    Water   3 parts

Solution 3 (Saturated sodium chloride)

Procedure

1. Deparaffinize and hydrate slides to water.
2. Stain in solution 1 for 3 minutes.
3. Differentiate in solution 2.
4. Rinse and flood slides with solution 3 for 5 minutes.
5. Rinse and mount in corn syrup.

Result

Amyloid   Pink to red
Other structures   Violet

Modified Methyl Violet Method (Fernando, 1961)

Fixation

Formol saline, Carnoy

Reagents Required

Crystal violet
Formic acid
Dextrin
Sucrose
Sodium chloride
Sodium merthiolate

Preparation of Reagents

Solution 1 (1 per cent crystal violet)
    Crystal violet   1.0 g
    3 per cent formic acid   100.0 ml

Solution 2 (Mounting medium)
    Dextrin   16.7 g
    Sucrose   16.7 g
    Sodium chloride   10.0 g

Sodium merthiolate   10.0 g
Distilled water   100.0 ml
Heat until all are dissolved, cool and store.

Procedure

1. Deparaffinize and bring slides to 95 per cent alcohol.
2. Blot and stain in solution 1 for 10 minutes.
3. Blot excess stain.
4. Differentiate in 1 per cent formic acid.
5. Blot.
6. When amyloid appears bright pink, remove from formic acid.
7. Rinse in distilled water.
8. Blot carefully.
9. Mount in solution 2 and seal the edge of the coverslip.

Result

Amyloid   Bright pink
Other components   Blue

Modified Methyl Violet Method
(Bancroft, 1963) (Unfixed cryostat)

Fixation

Unfixed cryostat section

Procedure

1. Stain in 2 per cent aqueous methyl violet for 5 minutes.
2. Wash in distilled water.
3. Differentiate in 1 per cent acetic acid for 20 seconds.
4. Wash in distilled water.
5. Stain in methyl green (2 per cent aqueous).
6. Wash.
7. Mount in Apathy's medium.

Result

Amyloid   Red to pink
Mast cells   Purple
Nuclei   Green

Modified Congo Red Method

Fixation

Fresh, frozen, acetic ethanol preferred
All types of paraffin

Reagents Required

Potassium iodide
Congo red
Mayer's haematein

Preparation of Reagents

Solution 1 (1 per cent Congo red)
Congo red   1.0 g
Distilled water   100.0 ml

Solution 2 (1 per cent potassium iodide)
Potassium iodide 1.0 g
Distilled water   100.0 ml

Procedure

1.   Stain with solution 1 for 1–6 hours.
2.   Differentiate in solution 2 for 60 seconds.
3.   Differentiate in 70 per cent alcohol.
4.   Wash in water.
5.   Counterstain with Mayer's haematein (*see* Chapter 7).
6.   Differentiate in 1 per cent acid alcohol.
7.   Dehydrate, clear and mount.

Result

Amyloid   Brick red
Nuclei   Blue

Permanent Gold Stain Method for Amyloid (Lynch and Inwood, 1963)

Fixation

Formalin

Reagents Required

Iodine
Potassium iodide
Gold chloride
Hydrogen peroxide

Preparation of Reagents

Solution 1 (Iodide solution)
Iodine   1.0 g
Potassium iodide 2.0 g
Distilled water   100.0 ml

Solution 2 (Gold chloride solution)
Gold chloride   1.0 g
Distilled water   100.0 ml

Solution 3
   3 per cent hydrogen peroxide (freshly prepared from 30 per cent stock)

Procedure

1.   Deparaffinize and hydrate slides to water.
2.   Stain with solution 1 for 5 minutes.
3.   Rinse in distilled water giving 3 changes.
4.   Immerse in solution 2 for 5 minutes.
5.   Rinse in distilled water giving 3 changes.
6.   Transfer to solution 3 at 37°C for 3–16 hours.
7.   Dehydrate, clear and mount.

Result

Amyloid	Golden yellow
Nuclei	Brown
Connective tissue	Grey

Sweat–Puchtler Method for Amyloid (Sweat and Puchtler, 1965)

Fixation

10 per cent neutral buffered formalin

Reagents Required

Sodium hydroxide
Sirius red
Sodium chloride
Sodium borate
Haematoxylin

Preparation of Reagents

Solution 1 (Sodium hydroxide solution)
   Sodium hydroxide   1.0 g
   Distilled water   100.0 ml

Solution 2 (Alkaline alcohol solution)
   80 per cent alcohol   100.0 ml
   Solution 1   1.0 ml

Solution 3 (Staining solution)
   Sirius red   1.0 g
   Distilled water   100.0 ml
   Sodium chloride   500 mg

Dissolve sirius red in water and then add sodium chloride. Allow it to stand for 24 hours.

Solution 4
   0.2 M buffer (*see* Chapter 5)

Procedure

1. Dewax and hydrate slides to water.
2. Wash in distilled water.
3. Place in neutral formalin overnight.
4. Wash in water.
5. Transfer to solution 2 for 1 hour.
6. Rinse in distilled water.
7. Transfer to pre-heated solution 3 for 1½ hour.
8. Rinse in solution 4.
9. Wash in running water.
10. Counterstain in Mayer's haematoxylin.
11. Wash in running water.
12. Dehydrate, clear and mount.

Result

Amyloid	Pink to red
Nuclei	Blue
Background	Unstained
Elastin	Pink to red

Sirius Red Technique (Llewellyn, 1970) (A Modification of Puchtler and Sweat, 1965)

Fixation

Formol saline, formol sublimate

Preparation of Reagents

Solution 1

Sirius red F3F	500 mg
Distilled water	45 ml
Absolute alcohol	50.0 ml
1 per cent sodium hydroxide	1.0 ml

Stir well and add 4.0 ml of 20 per cent sodium chloride.

Procedure

1. Hydrate slides to water.
2. Treat with alum haematoxylin for nuclei.
3. Rinse in water and then in 70 per cent alcohol.
4. Treat with solution 1 for 1 hour.
5. Wash in tap water.
6. Dehydrate, clear and mount.

Results

Amyloid	Red
Nuclei	Blue

Toluidine Blue Technique (Wolman, 1971)

Fixation

Any fixative.

Reagents Required

1 per cent toluidine blue in 50 per cent isopropanol

Procedure

1.   Deparaffinize and hydrate slides to water.
2.   Treat with toluidine blue for 30 minutes at 37°C.
3.   Blot and immerse in absolute isopropanol.
4.   Clear and mount.

Result

Amyloid and many other tissue components stain an orthochromatic blue colour but when examined under polarized light, amyloid gives a dark red birefringence.

Digestion Technique: Peptic Digestion

Fixation

All types including fresh and fixed frozen sections

Reagents Required

Pepsin, hydrochloric acid, eosin

Preparation of Reagents

Solution 1 (Pepsin)

Pepsin	0.2 g
0.02N hydrochloric red	40.0 ml

The pH of final solution should be 1.6.

Solution 2 (0.02N HCl)

0.02N HCl	40.0 ml (*see* Chapter 5)

Solution 3 (Eosin solution)

Eosin	1.0 g
Distilled water	100.0 ml

Procedure

1.   Hydrate slides to water.
2.   Place experimental section in solution 1 at 37°C for 4 hours.
3.   Place control section in solution 2 at 37°C for 4 hours.

4. Wash both sections.
5. Transfer both to solution 3 for 2 minutes.
6. Wash in tap water.
7. Dehydrate, clear and mount in DPX.

Result

Amyloid   Pink
Other tissue   Digested
In control section, digestion should not take place.

Rationale

Amyloid is resistant to proteolytic enzymes. A section containing amyloid is treated with pepsin–hydrochloric acid mixture and control section is treated with hydrochloric acid alone. Only amyloid resists peptic digestion and can be demonstrated by eosin.

Techniques mentioned above give excellent results with frozen sections than paraffin sections.

Table 11.1 Histochemical techniques applied for the demonstration of amyloid

Technique	Fixative	Amyloid
Congo red method	All types, frozen recommended	Orange-red
Alkaline Congo red method	Cryostat prefixed, post-fixed, frozen formalin-fixed	Orange-red
Technique	Fixative	Amyloid
High pH Congo red	Any general	Red
Congo red fluorescence method	Frozen	Orange-red fluorescence
Crystal violet method	10 per cent formalin or alcohol	Purple
Thioflavine T-method	Formalin paraffin, unfixed cryostat	Bright yellow
Methyl violet method	Any general	Red
Methyl green method	Cryostat, paraffin	Pink
Metachromatic method	Formalin, alcohol, fresh frozen, alcohol fixed	Pink to red
Modified methyl violet (Fernando)	Formol saline, Carnoy	Bright pink
Modified methyl violet (Bancroft)	Unfixed crysotat	Red to pink
Modified Congo red method	Fresh frozen, acetic ethanol preferred	Brick red
Permanent gold stain method	Formalin	Golden yellow
Sweat and Puchtler's method	10 per cent neutral buffered formalin	Pink to red
Sirius red technique	Formol saline, formol sublimate	Red
Toluidine blue method	Any general	Orthochromatic blue colour

NUCLEIC ACIDS

12

As early as 1869, Friedric Miescher experimented with nuclei of cells and isolated a substance and named it as nuclein which today is deoxyribonucleic acid (DNA). Though Robert Feulgen in 1914 made his first test on DNA, only in 1924 did he publish his results where he described the method of staining the nucleic acids and stated that DNA is located in chromosomes in the nucleus. A number of investigators (biochemists, histochemists, pathologists, biologists, biophysicists) subsequently continued their research on DNA. It was later discovered that the molecule consists of a long unbranched chain which in turn is made up of a five-carbon sugar (deoxyribose) alternating with phosphate group, a nitrogenous base the purine (adenine and guanine) and pyrimidines (thymine and cytosine). This chain is subdivided into units called nucleotides made up of phosphate–sugar–base. The work of Watson and Crick (1953) has been an important milestone in the history of the discovery of nucleic acids. They provided the DNA model—two polynucleotide chains intertwined into a double helix held together by hydrogen bonds.

In RNA, the sugar is ribose instead of deoxyribose, and adenine, guanine, cytosine and uracil (instead of thymine) are the bases. These are linked by phosphate groups forming a polynucleotide chain. When nucleic acids are hydrolysed, the following products result, (a) phosphate groups, (b) five-carbon sugars, (c) nitrogen bases namely the purines and pyrimidines.

The ladder has side bars consisting of sugars alternating with phosphate and transverse bars of nitrogenous bases, the purines and pyrimidines, linked with each other and also with the side bars. The purines and pyrimidines are four (2 each). This is the basic structure of DNA as well as RNA. The two nucleic acids differ in the type of sugar and in the type of pyrimidines. In DNA, the 5-carbon sugar is deoxyribose and in RNA, it is ribose. Another difference is, purines in DNA are adenine and guanine and pyrimidines are thymine and cytosine whereas in RNA, purines are the same but in pyrimidines thymine is replaced by uracil.

A number of phosphate radicals occur in the nucleic acids. So basic dyes can colour them, e.g. haematoxylin. In some cells, cytoplasm does not contain sufficient RNA and as a result the cytoplasm is eosinophilic. On the other hand, in plasma cells there is large amount of RNA. Here the cytoplasm does not appear reddish in colour, instead it turns purplish due to superimposition of eosinophilia on haematoxylinophilia. The demonstration of nucleic acid depends on phosphate radical, nucleotide structure, sugar and nitrogenous base. Phosphate radical being acidic will combine with basic dyes and hence the nuclear staining by basic dyes. Backler and Alexander (1965) modified Turchini method. But Feulgen reaction for DNA and other methods for RNA are in practice in most of the laboratories.

For nitrogenous bases, there are as yet no specific reliable histochemical tests.

FIXATION

A nuclear fixative like Carnoy, or formol saline are good. For smears of tissues, methyl alcohol or Clarke's fixative are superior. When fixed in neutral buffered formalin at 4°C, DNA degradation by cell nuclease is prevented (Tokuda *et al.*, 1990).

Clarke's Fluid

     Absolute alcohol   75.0 ml
     Glacial acetic acid   25.0 ml

The penetration quality of this fixative is superior to others. It is good for cytoplasmic elements also.

When a decalcified tissue is demonstrated for nucleic acid, there is a chance of these nucleic acids being extracted since nucleic acids are extracted with acids.

EDTA decalcification of calcified tissue may give best results.

Nucleic acids exhibit basophilia but are not metachromatic with toluidine blue or azure-A.

DNA can be demonstrated by Feulgen method of Feulgen and Rossenbeck (1924), methyl green–pyronin technique, fluorescent methods using acridine orange (though not very reliable). Both DNA and RNA can be demonstrated with gallocyanin–chrome alum technique but this method does not separate the two nucleic acids.

HISTOCHEMICAL METHODS FOR NUCLEIC ACIDS

DNA Feulgen Nuclear Reaction (Feulgen and Rosenbeck, 1924)

Fixation

Any fixative

Reagents Required

Concentrated hydrochloric acid, Schiff's reagent (*see* Chapter 9)
Potassium metabisulphite

Preparation of Reagents

Solution 1 (HCl)
Concentrated HCl   8.5 ml
Distilled water   91.5 ml

Solution 2
Schiff's reagent

Solution 3 (Potassium metabisulphite solution)
10 per cent potassium metabisulphite   5.0 ml
(10 g/100 ml distilled water)
N-hydrochloric acid   5.0 ml
Distilled water   90.0 ml

Procedure

1. Hydrate slides to water.
2. Place sections in solution 1 at room temperature for 1 minute.
3. Transfer sections to solution 2 (Schiff's) for 45 minutes.
4. Rinse sections in solution 3 giving 3 changes of 2 minutes each.
5. Rinse in distilled water.
6. Counterstain with light green if necessary for 2 minutes.
7. Wash in water.
8. Dehydrate, clear and mount.

Result

DNA   Purple
Cytoplasm   Green

Remarks

Caution is to be taken for hydrolysis time. It depends upon the fixative. N-HCl should be pre-heated.

Rationale

First introduced by Feulgen and Rosenbeck (1924), it is a specific technique for DNA. During hydrolysis, only deoxyribose sugar reacts with N-HCl and ribose sugar does not interfere and is not hydrolysed. During hydrolysis, free aldehydes result which when treated with Schiff's reagent produce a coloured compound with DNA. Results depend upon the hydrolysis time which varies with fixative. If hydrolysis is prolonged and exceeds the time limit, intensity of stain becomes weaker and weaker and ultimately disappears. It is better to avoid Bouin's fixative because over-hydrolysis takes place during fixation. Formol saline and Carnoy are the best. Immersion in cold N/1 hydrochloric acid before and after treatment at 60°C is desirable.

Hydrolysis in N-HCl at 60°C (Bauer, 1932)

The duration of hydrolysis varies with the fixatives employed.

Fixative	Time in minutes
Bouin's	Not recommended
Helly	6
Carnoy	8
Flemings	8
Formalin	8–10
Zenker-formol	5
Newcomers	20
Regaud	14
Susa	18
Zenker	5
Champy	25

Pyronin–Methyl Green for Nucleic Acids (Elias, 1969)

Fixation

Carnoy

Reagents Required

Methyl green
Acetate buffer (Walpole)
Pyronin

Preparation of Reagents

Solution 1

Methyl green	500 mg
Acetate buffer	100.0 ml
Pyronin G or Y	200 mg

Procedure

1. Deparaffinize and hydrate slides to water.
2. Treat with solution 1 for 1 hour at 37°C.
3. Rinse in cold distilled water.

4. Rinse in butanol.
5. Dehydrate in butanol giving 2 changes of 5 minutes each.
6. Clear and mount.

Result

Nuclear and cytoplasmic basophilic substances are stained red.

Thionin Methyl Green for Nucleic Acids (Roque *et al.*, 1965)

Fixation

4 per cent formaldehyde in 1 per cent sodium acetate for 3 hours.

Reagents Required

Methyl green
Thionine
Citrate buffer

Preparation of Reagents

Solution 1

Methyl green	100 mg
Thionin	16 mg
Citrate buffer (pH 5.8)	100.0 ml

First dissolve thionin in a little water and then add buffer and methyl green. Shake well, filter. Methyl green is purified with chloroform extraction.

Procedure

1. Dewax and hydrate slides to water.
2. Treat with solution 1 for 30 minutes at 40°C.
3. Rinse in distilled water.
4. Dehydrate in a mixture containing 80.0 ml of 3-butyl alcohol and 20 ml of absolute alcohol giving 3 changes.
5. Rinse in absolute alcohol, clear and mount.

Result

Chromatin	Green or blue-green
Nuclear and cytoplasmic basophilic substance	Red

RNA, DNA–Methyl Green Pyronin Method (Pappenheim, 1899; Unna, 1902; Bancroft and Cook, 1994)

Fixation

All types, preferably freeze-dried

Reagents Required

0.1 M acetate buffer
Pyronin Y

Methyl green
Chloroform

Preparation of Reagents

Solution 1 (Methyl green)
Dissolve 2 g of methyl green in 100 ml of distilled water. Place this solution, while stirring, in a separating funnel. Add 100.0 ml chloroform and shake well. Discard contaminated chloroform. Repeated extraction with chloroform is necessary.

Solution 2 (Pyronin Y)
Pyronin Y　　　2.0 g
Distilled water　100.0 ml

Solution 3 (Staining solution)
Methyl green (Solution 1)　7.5 ml
Pyronin Y (Solution 2)　　12.5 ml
Acetic buffer (pH 4.8)　　30.0 ml

Procedure

1. Deparaffinize and hydrate slides to water.
2. Transfer slides to solution 3 for 4–10 minutes.
3. Blot and dry.
4. Dip rapidly in absolute acetone.
5. Dip rapidly in 10 per cent acetone in xylene.
6. Dip rapidly in 50 per cent acetone in xylene.
7. Transfer to xylene.
8. Transfer sections to fresh xylene and mount in DPX.

Result

DNA　Green
RNA　Red

Rationale

First introduced by Pappenheim (1899), this was later modified by Unna (1902), Taft (1951), Trevan and Sharrock (1951), Brachet (1953) and Kurnic (1955). Kurnic (1955) has suggested that while methyl green blends with DNA, two sites are involved and two phosphoric groups of the DNA combine with the amino group.

Methyl Green–Pyronin Y Method for RNA–DNA (Trevan and Sharrock, 1951) Modified

Fixative

Neutral fixatives, neutral buffered formalin.

Reagents Required

Acetate buffer (pH 4.8)

5 per cent pyronin Y
2 per cent methyl green

Preparation of Reagents

Solution 1 (Methyl green–pyronin Y solution)

2 per cent methyl green (chloroform-washed)	10.0 ml
5 per cent pyronin Y	17.5 ml
Distilled water	250.0 ml

Solution 2 (Acetic buffer)
Acetic buffer (pH 4.8)

Solution 3 (Working solution)

Solution 1	25.0 ml
Solution 2	25.0 ml

Procedure

1. Deparaffinize and bring down slides to water.
2. Dip in distilled water and then blot.
3. Transfer to solution 3 (working solution) for 20–30 minutes.
4. Rinse rapidly in distilled water.
5. Dehydrate, clear and mount.

Result

DNA	Green to bluish green
RNA	Red

Gallocyanin–Chrome Alum Method for RNA and DNA (Einarson, 1932, 1951)

Fixation

All types of fixatives

Reagents Required

Gallocyanin
Chrome alum

Preparation of Reagents

Solution 1 (Chrome alum–Gallocyanin)

Dissolve 5.0 g of chrome alum in 100.0 ml of distilled water and then add 150 mg of gallocyanin. Heat the solution to boiling for 5 minutes. Cool it and adjust it to 100.0 ml.

Procedure

1. Deparaffinize and hydrate slides to water.
2. Transfer slides to solution 1 for 18–48 hours.

3.   Wash thoroughly in tap water.

4.   Dehydrate, clear and mount.

Result

DNA and RNA are stained blue.

Rationale

This method was first introduced by Einarson (1932) for Nissl granules. Later in 1951 this technique was applied to demonstrate nucleic acids. In this technique, phosphoric acid residue molecules of nucleic acids combine with gallocyanin at pH 1.0. At pH 2.0 and above, other tissues stain. Thus it is important to maintain the pH at 1.0.

DNA–Naphthoic Acid Hydrazine–Feulgen Method (Pearse, 1951)

Fixation

All types of fixatives

Reagents Required

Veronol acetate buffer
2-hydroxy 3-naphthoic acid hydrazine
Acetic acid
Fast blue B
N-hydrochloric acid

Preparation of Reagents

Solution 1 (Acid water)
    Concentrated HCl   8.5 ml
    Distilled water   91.5 ml

Solution 2 (Fast Blue B Solution)
    Fast blue B   50 mg
    Veronol acetate buffer (pH 7.4)   50.0 ml

This buffer should be prepared before use.

Solution 3 (NAH Solution)
    2-hydroxy 3-naphthoic   50 mg
       acid hydrazide
    Absolute alcohol   47.0 ml
    Conc. acetic acid   3.0 ml

Procedure

1.   Deparaffinize and hydrate slides to water.

2.   Immerse briefly in solution 1 (room temperature).

3.   Transfer sections to solution 1 at 60°C.

4.   Again transfer to solution 1 at room temperature for 1 minute.

5.   Dip in distilled water for 1 minute.

6. Transfer to 50 per cent alcohol for 10 minutes.
7. Place in sections to solution 3 (NAH) at room temperature.
8. Transfer sections to 50 per cent alcohol for 10 minutes.
9. Place in distilled water for 1 minute.
10. Transfer sections to solution 2 (Fast blue B) for 3 minutes.
11. Dehydrate, clear and mount.

Result

DNA	Bluish purple
Protein material	Purple-red

Rationale

This method can also be used as a control for Feulgen reaction. When sections are hydrolysed at 60°C, free aldehydes are released and they combine with 2-hydroxy 3-naphthoic acid hydrazide. This is then coupled to Fast blue B producing a purplish blue colour at the site of coupling. Though dinitrophenyl hydrazine was used in the ratio 2:4 by Danielli (1947), best results are obtained with 2-hydroxy 3-naphthoic acid.

RNA–DNA Acridine Orange (Bertalanffy and Nagy, 1962)

Fixation

Freeze-dried, frozen cryostat
Paraffin—Best fixatives are acetic ethanol and 70 per cent alcohol

Reagents Required

0.2 M phosphate buffer (*see* Chapter 5)
Calcium chloride
Acridine orange
Acetic acid

Preparation of Reagents

Solution 1 (Acridine orange solution)
    Acridine orange   50 mg
    Distilled water   40.0 ml
The pH of the solution should be 6.0 with addition of phosphate buffer. The volume is made to 50 ml.

Solution 2 (Phosphate buffer) pH 6.0 (*see* Chapter 5)

Solution 3 (Calcium chloride solution)
    Calcium chloride   11.0 g
    Distilled water   50.0 ml

Procedure

1. Deparaffinize and hydrate slides to water.
2. Rapidly dip in 1 per cent acetic acid for 15 seconds.
3. Rinse in distilled water.

4. Transfer slides to solution 1 (acridine orange) for 2 minutes.
5. Place sections in solution 2 (phosphate buffer) for 1 minute.
6. Differentiate in solution 3 for 20 seconds.
7. Bring sections again to solution 2 (phosphate buffer).
8. Mount sections wet and examine under fluorescence microscope.

Result

RNA   Red
DNA   Light green

Rationale

Clarity and perfection of the technique depend on the fixative, concentration of acridine orange and pH of the working solution (6.0). If the concentration of the stain is more than what is required, red colour will overwhelm green colour.

Menzies Method (Menzies, 1963)

Fixation

10 per cent neutral buffered formalin.

Reagents Required

Tetrahydrofuran
Hydrochloric acid
Azure B
Basic fuchsin
Glacial acetic acid

Preparation of Reagents

Solution 1 (Hydrochloric acid tetrahydrofuran)
    Hydrochloric acid   10.0 g
    Tetrahydrofuran   90.0 ml

Solution 2 (1 per cent Azure B solution)
    Azure B   1.0 g
    Distilled water   100.0 ml

Solution 3 (0.3 per cent basic fuchsin solution stock)
    Basic fuchsin   100 mg
    Distilled water   100.0 ml

Solution 4 (Azure B basic fuchsin solution)
    Solution 2 (Azure B stock)   30.0 ml
    Solution 3 (Basic fuchsin stock) 8.0 ml
    Glacial acetic acid   2.0 ml

Procedure

1. Deparaffinize and hydrate slides to water.
2. Place in solution 1 at 37°C for 5 minutes.

3.   Directly transfer to solution 4 for 30 minutes.
4.   Rinse in acetone.
5.   Clear in xylene and mount.

Result

DNA   Red
RNA   Blue

Spicer's Method for Nucleic Acids (Spicer, 1961b)

Fixation

Bouin's solution

Reagents Required

Basic fuchsin
Sodium metabisulphite
Citric acid
Disodium phosphate
Methylene blue

Preparation of Reagents

Solution 1
Schiff's reagent (*see* Chapter 9)

Solution 2 (0.5 per cent sodium metabisulphite solution)
Sodium metabisulphite   500 mg
Distilled water   100.0 ml

Solution 3 (0.1 M citric acid solution)
Citric acid   19.21 g
Distilled water   1000.0 ml

Solution 4 (0.2 M disodium phosphate solution)
Disodium phosphate   28.40 g
Distilled water   1000.0 ml

Solution 5 (Methylene blue solution)
Methylene blue   10 mg
Solution 3 (Citric acid 0.2 M)   28.6 ml
Solution 4 (Disodium phosphate)   11.4 ml

Procedure

1.   Dewax and hydrate slides to water.
2.   Transfer to solution 1 for 10 minutes.
3.   Rinse in solution 2 giving three changes.
4.   Wash in running water.
5.   Transfer to solution 5 for 30 minutes.
6.   Dehydrate, clear and mount.

Result

DNA	Red
Chromosomes	Red
Chromatin	Red
Cytoplasmic RNA	Blue

DNA Extraction by Enzyme Deoxyribonuclease

Reagents Required

Deoxyribonuclease
0.3 M tris buffer (pH 7.6)

Preparation of Extraction Solution

0.2 M tris buffer (pH 7.6)	10.0 ml
Distilled water	50.0 ml
Deoxyribonuclease	10 mg

Procedure

1. Deparaffinize and hydrate slides to water.
2. Place sections in extraction solution for 4 hours at 37°C.
3. Wash in running water.
4. Stain as for Feulgen method.

Result

Test section: DNA stains negative.
Control: DNA stains red.

Remarks

Nucleic acids may be extracted not only by enzymes but also by using perchloric acid. Although enzyme digestion is the best for routine analyses, perchloric acid treatment is satisfactory.

Extraction of Nucleic Acid by Perchloric Acid

Reagents Required

Ribonuclease

Preparation of Reagents

Solution 1 (Perchloric acid)
    Perchloric acid   2.5 ml
    Distilled water   47.5 ml

Solution 2 (Perchloric acid)
    Perchloric acid   5.0 ml
    Distilled water   45.0 ml

Solution 3 (Sodium carbonate)
    Sodium carbonate 1.0 g
    Distilled water   100.0 ml

Procedure

1. Deparaffinize and hydrate slides to water.
2. Keep sections in solution 2 (10 per cent perchloric acid) at 4°C overnight.
3. Rinse in distilled water.
4. Transfer sections to solution 3 (sodium carbonate) for 5 minutes.
5. Wash.
6. Proceed for nucleic acid method.

Remarks

If both RNA and DNA have to be removed, place sections in 5 per cent perchloric acid (solution 1) at 60°C for 30 minutes at stage 2) Then continue the same procedure.

EXTRACTION OF NUCLEIC ACIDS

Trichloroacetic acid Extraction of Nucleic Acids

1. Hydrate slides to water after deparaffinizing.
2. Treat with 4 per cent trichloroacetic acid at 90°C for 15 minutes.
3. Wash.
4. Stain with toluidine blue.

Both RNA and DNA are extracted by this method.

Hydrochloric Acid Extraction for DNA and RNA

1. Hydrate slides to water after deparaffinizing.
2. Treat with 1 M HCl for 3 hours at 37°C.
3. Wash.
4. Stain with dilute solution of methylene blue at pH 5.7 for 24 hours.

Both DNA and RNA are removed.

Method for Removal of RNA by RNase

1. Treat with 5 per cent trichloroacetic acid for 30 minutes at 60°C.
2. Treat with 10 per cent perchloric acid for 12 hours at 4°C.
3. Treat with 1N hydrochloric acid for 12 minutes at 60°C.
4. Incubate in the enzyme ribonuclease for 4 hours at 40°C.
    Ribonuclease   1 mg
    Distilled water   1.0 ml

Nuclear and cytoplasmic RNA is extracted but chromosomal RNA is resistant to enzyme.

Method for Removal of DNA by Deoxyribonuclease Method (Daoust, 1964)

Incubate slides for 24 hours at 37°C in
DNase   50 mg
1 M tris maleate buffer (pH 6.5)
        containing ($MgSO_4 \cdot 7H_2O$ at a concentration of 0.2 M   1.0 ml

Tandler Method (1974)

1. Place slides from water into a mixture of 90 ml saturated aqueous picric acid and 10 ml concentrated formaldehyde for 12–24 hours.
2. Wash in distilled water (2 changes, 2 minutes each).
3. Place in 25 per cent acetic acid (2 changes, 2 minutes each).
4. Place in 10 per cent aniline blue in 25 per cent acetic acid for 1 hour.
5. Wash in 2 changes of 25 per cent acetic acid for 2 minutes.
6. Wash in distilled water and follow the usual procedure of staining.

LIPIDS

13

It has been known for long that animal and vegetable fats are combinations of glycerol with fatty acids. Bloor (1925–1926), Lisen (1936) and Cain (1950) contributed to the study of lipids. Lipids are naturally occurring fats and fat-like substances which are insoluble in water and soluble in solvents like chloroform, benzene, petroleum, ether and acetone. There are some exceptions. Lecithin is slightly soluble in water and insoluble in acetone, isolecithin is soluble in warm water and insoluble in ether and sphingomyelin and cerebrosides are insoluble in a range of fat solvents. Frozen sections are best to demonstrate lipids. Biochemical classification of lipids is at variance from histochemical classification because lipids do not occur in pure form. They are often linked to carbohydrates as glycolipids or to proteins as lipoproteins. As a result, both their physical and chemical properties are altered, which in turn fail to give precise results. Lipids as stored fats are common in all tissues which sometimes may be dangerous to life. Large amounts of stored lipids in some organs engender diseased conditions.

Examples of lipids are carotenoids, fatty acids, triglycerides (neutral fats), phospholipids, cerebrosides, sterols and lipid pigments (lipofuchsins).

CLASSIFICATION OF LIPIDS

Lipids are categorized as (1) simple lipids, (2) compound lipids and (3) derived lipids (Figure 13.1).

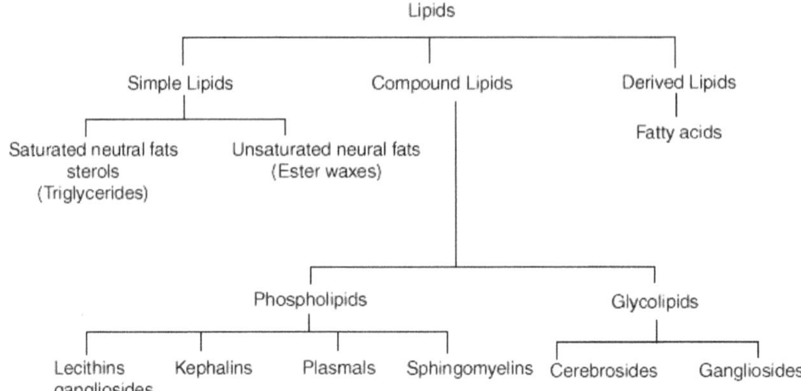

Figure 13.1   Classification of lipids

Simple Lipids

These are mostly esters of saturated and unsaturated fatty acids. Unsaturated fatty acids with long chains are neutral lipids. These fatty acids can be stearic, palmitic or oleic and they are insoluble in water and soluble in organic solvents. Neutral lipids are triglycerides. Ester waxes also come under simple lipids. Sudan dyes are the best for the demonstration of lipids in tissues. In Nile blue sulphate technique, the red oxazone component stains them red.

Compound Lipids

These lipids contain a non-lipid group and are categorized as (a) phospholipids and (b) glycolipids.

Phospholipids   These contain phosphoric acid which is bound to a nitrogenous base and alcohol which is glycerol. Heart, muscles, liver and brain are some of the organs sequestering high quantities of phospholipids known as esters of phosphatidic acid. They are usually termed as glycerophosphatides and are categorized as follows.

1. Lecithins
2. Kephalins
3. Plasmals
4. Sphingomyelins

Lecithins   These are major constituents of phospholipids, glycerol, saturated and unsaturated fatty acid residues and phosphoric acid, and choline is the base.

Kephalins   In chemical composition, they are similar to lecithins but differ in having the base as either serene or ethanolamine.

Plasmals   There are similar to lecithins and kephalins. In the place of fatty acids, they have acetals of fatty acid aldehydes, combined with acetal phosphatides and plasmalogens.

Sphingomyelins These contain fatty acid, phosphoric acid, choline and the complex alcohol sphingosine, but no glycerol. Except for this they are similar to cerebrosides. Sphingomyelins are PAS-negative whereas cerebrosides are PAS-positive. Sphingomyelins can be histochemically demonstrated by the NaOH OTAN method.

Glycolipids (Cerebrosides, galactolipids) These are lipids consisting of cerebrosides containing a single fatty acid chain (sphingoline) and one or more molecules of hexose sugars usually glucose or galactose and one fatty acid. Besides cerebrosides, there is a second group of gangliosides which are almost similar to cerebrosides except for containing neuraminic acid. It is difficult to demonstrate them histochemically except by PAS reaction.

Derived Lipids

As the name indicates, these are derived from simple and compound lipids. They contain fatty acid (either saturated or unsaturated). Saturated fatty acids are palmitic or stearic acids and they lack double bonds. Unsaturated fatty acid is oleic acid with double bonds. Sterols (the commonest being cholesterol), could be demonstrated histochemically. They are produced by hydrolysis of ester waxes, and histochemically cholesterol is the most important.

Lipids never occur as individual compounds, but in mixtures. Because of this quality they cannot act in their characteristic way. Lipids usually occur as granules or droplets and are occasionally bound to components within the cell. Staining methods and extraction methods with organic solvents are some methods by which lipid droplets can be demonstrated. Bound lipids on the other hand are intermingled or associated with other tissue components and as such it is difficult to demonstrate them histochemically. Complex groups like lipoproteins are mixtures of proteins and lipids.

There are several methods to identify lipids (Figure 13.2).

SOLUBILITY TEST

Keiling in 1944 used an extraction apparatus and concluded that lipids could be differentiated by their solubility in different fat solvents. The old method was to use Soxhlet apparatus for extraction. Fresh unfixed blocks of tissue were extracted for 24 hours with 3 changes of solvent. Sometimes extraction methods do not specify a particular lipid compound because they are bound to other tissue structures. Soxhlet method is still employed by some. Usually in the extraction procedure, fresh tissue is placed in a suitable solvent for 48 hours at 60°C.

After extraction, the block is fixed in formol saline and processed and thin sections are cut. These are stained with dyes along with sections taken from unextracted tissue.

Majority of the lipids are extracted in a mixture of hot chloroform methanol in the ratio 2 : 1. Cerebrosides could be extracted in hot acetone, triglycerides, cholesterol and cholesterol esters in cold acetone; and phospholipids in hot ether. Baker's (1946) hot pyridine method can be routinely used. Glycolipids and phospholipids are not soluble in cold acetone.

STAINING METHODS

Majority of lipids can be differentiated from other compounds with specific dyes and appropriate histochemical techniques. Compound lipids which include some water-soluble hydrophilic lipids, when fixed in formol calcium become insoluble, whereas simple lipids are soluble in organic solvents. Frozen sections are best for precise demonstration. Lipids generally never occur as independent compounds. They are always interlinked with either carbohydrates or proteins resulting in alteration of their physical and chemical characteristics. Majority of them are solid at room temperature and liquid at body temperature.

Most routine methods used in many histological and histopathological laboratories is the staining of neutral fats with fat-soluble dye.

Sudan dyes, Sudan black B, Sudan-III, Sudan-IV, Fettrot and Oil Red O have been in vogue to demonstrate lipids. Sudan black B is the most popular dye and relatively easy to deal with.

Sudan-III as a fat stain is popular, though other better dyes have come into practice. 70 per cent alcohol is the vehicle, though small quantities of lipid dissolves. A mixture of equal parts of acetone and 70 per cent alcohol is better.

Minimal removal of lipids could be engendered with oil Red O in isopropyl alcohol (Lillie and Ashburn, 1943). Best results could be obtained by using Sudan-IV or Fettrot in propylene glycol without loss of small quantities of lipids.

This is an intricate process involving a number of staining techniques and some extraction methods. When a particular tissue is suspected to be lipid material, two frozen sections are taken, and the lipid-soluble dye technique is applied to one section which will give if any lipid material is present or not. The other slide is viewed under a polarizing microscope, for cholesterol or cholesterol esters which can be detected by their birefringence. After confirming the presence of lipid material, the acid lipid is distinguished from non-acid lipid by applying the Nile blue technique. Blue shade indicates acid lipids and red shade non-acid lipids. For further classification of acid lipids (for phospholipids and fatty acids) acid haematein test is performed. Non-acidic lipids may be either unsaturated lipids or glycolipids, or sterols and esterases. Only sterols or triglycerides could be detected by applying performic acid Schiff technique or PAS method or PAN method or digitonin technique or copper rubeanic acid technique respectively. Identification of lipids also depends upon extraction methods which are used in combination with staining methods.

Caution to be taken while demonstrating lipids   Most of the solvents evaporate and leave precipitate on the sides. To avoid this, tight screw-capped jars should be used while staining.

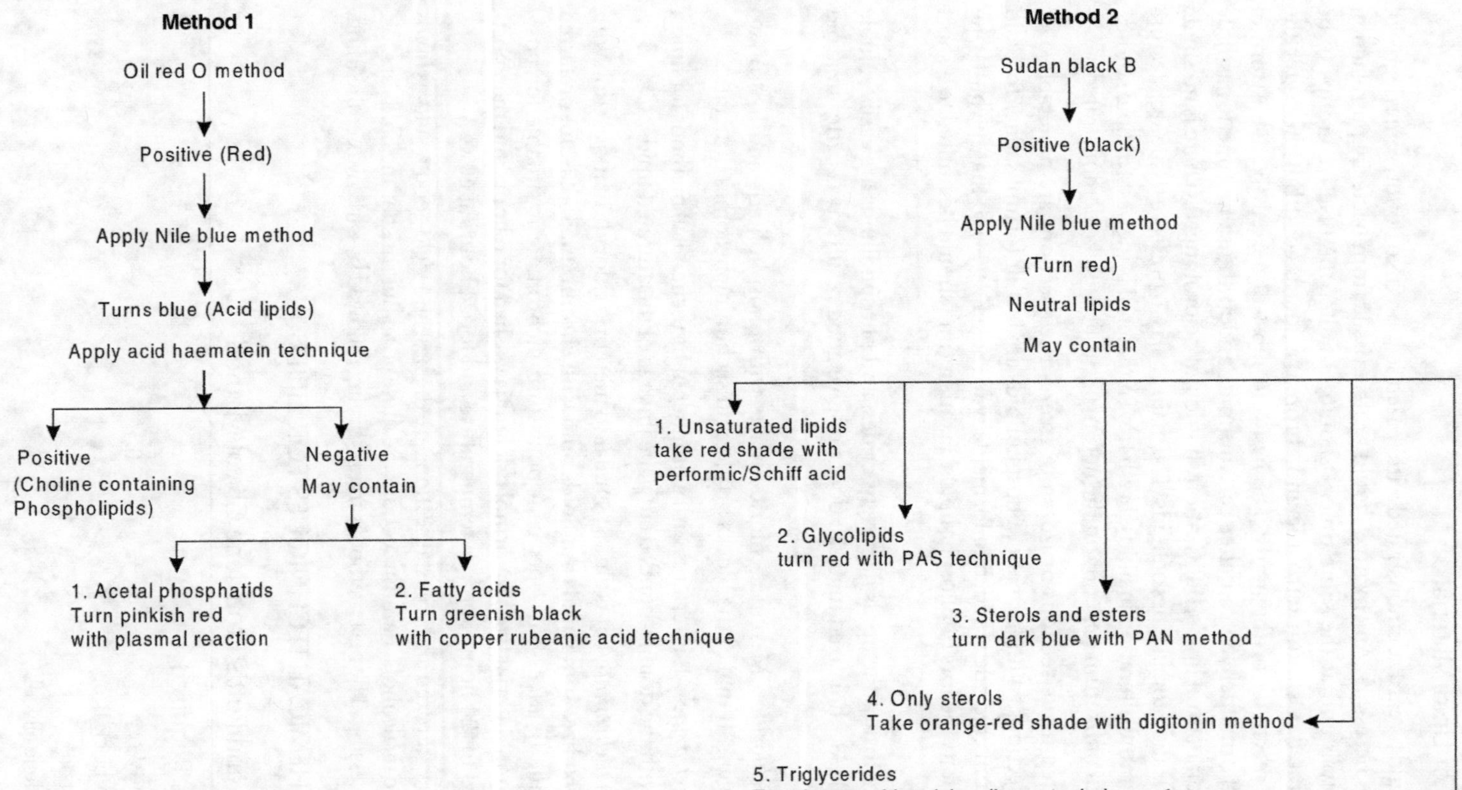

Figure 13.2   Methods for lipid identification

Fixation of Lipid Material

Lipids could be easily demonstrated with frozen sections. No doubt, some degree of fixation is necessary. Any fixative should protect tissue from bacterial putrefaction and autolysis and at the same time it should not alter the reactive groups to be demonstrated. Osmium tetroxide and chromic acid are good but they alter the chemical reactivity of the lipids. The best and most preferred fixative is formol calcium. Formaldehyde can alter certain lipids and at the same time it has little effect upon neutral lipids, but fixation should not be prolonged. There is no single mode of fixation for all types of lipids. The main goal is to preserve tissue architecture with minimum alteration in lipids. Fresh frozen material is the best for lipid studies. The length of fixation depends on the nature of the lipid to be demonstrated.

Unfixed tissue sections or frozen or freeze-dried sections are preferable over paraffin sections. With formalin, phospholipids are lost in formol saline. To avoid this, calcium chloride is added to formaldehyde (Baker, 1944) so that the buffering effect inhibits the loss of phospholipids. Fixation time should be limited to a short duration.

Alcohol   Most of the lipids like triglycerides, and phospholipids are easily soluble in alcohol. Alcohol-fixed material does not give good results. It is better to avoid alcohol.

Mercuric chloride   Mercuric chloride reacts with phospholipids and hydrolyses plasmalogens. It makes the blocks brittle, and sectioning becomes difficult.

Osmium tetroxide   It interferes with a number of histochemical methods for lipids. It reacts with the double-bonded ethylene linkage of unsaturated lipids.

Potassium dichromate   It is almost similar to mercuric chloride in its reaction with lipids. It oxidizes some lipids but lipids bind with chromium. So it is not preferred as a routine fixative. But 2 per cent potassium dichromate is suitable for post-chromation. Material fixed in formol calcium is washed for 6 hours and then post-chromated for 24 hours at room temperature and for 24 hours at 60°C.

Embedding   Frozen and cryostat sections are the best and yield good results as lipids are soluble in alcohols. Paraffin sections are quite good but most of the lipids are lost while processing. The best method is freeze-drying followed by carbowax embedding.

HISTOCHEMICAL TECHNIQUES FOR LIPIDS

Chiffelle and Putt's Propylene Glycol Method

Reagents Required

Sudan-IV or
Sudan black B or
Fettrot or
Propylene glycol

Preparation of Reagents

Solution 1 (Staining solution)

Sudan-IV or Sudan black	1.0 g
Propylene glycol	100.0 ml

Heat the solution for few minutes. Cool and filter through glass wool.

Procedure

1. Cut frozen sections and wash in several changes of water to remove formalin.
2. Immerse in pure propylene glycol giving 2 changes.
3. Transfer to solution 1 (Sudan-IV).
4. Immerse in 85 per cent propylene glycol for 2–3 minutes.
5. Rinse in 50 per cent propylene glycol.
6. Wash in distilled water.
7. Counterstain if desired with haematoxylin.
8. Wash in tap water and mount in glycerine jelly.

Results

Lipids	Red or black
Nuclei	Blue

Lillie and Ashburn's Isopropanol Oil Red O Method

Reagents Required

Oil red O
Isopropyl alcohol

Procedure

1. Cut frozen sections.
2. Wash in water.
3. Transfer to staining solution (saturated solution of Oil red O in isopropanol) for 10–15 minutes.
4. Differentiate in 70 per cent alcohol.
5. Wash in water.
6. Counterstain if desired with haematoxylin.
7. Wash and mount in glycerine jelly.

Result

Lipids	Bright red
Nuclei	Blue

Oil red O Method (Lillie and Ashburn, 1943)

Fixation

Formol calcium, frozen, cryostat post-fixed.

Reagents Required

Oil red O
Triethyl phosphate
Hematoxylin

Preparation of Reagents (Oil red O solution)

Oil red O   1.0 g
Triethyl phosphate   60.0 ml
Distilled water   40.0 ml

Add distilled water to triethyl phosphate and then add the dye. Heat the solution, stirring constantly, cool and filter. This forms the stock solution which should be filtered before use.

Procedure

1.   Wash sections in distilled water.
2.   Transfer sections to 60 per cent triethyl phosphate.
3.   Stain in Oil Red O solution at 20°C for 15 minutes.
4.   Wash sections in 60 per cent triethyl phosphate for 30 seconds.
5.   Wash in distilled water.
6.   Stain sections in haematoxylin for 1 minute.
7.   Wash and mount in glycerine jelly.

Result

Lipids   Red
Nuclei   Blue

Rationale

Oil Red is insoluble in water and is slightly soluble in organic solvents. Lipids absorb the dye from solvent and yield a bright red colour which is more intense than with other Sudan dyes.

Osmium Tetroxide (Mallory, 1944)

Fixation

10 per cent neutral buffered formalin.

Preparation of Reagent

Solution 1
    Osmium tetroxide   1.0-g ampoule
    Distilled water   100.0 ml

With a file make a deep constriction on the ampoule and drop into distilled water and shake vigorously so that the ampoule breaks and osmium tetroxide dissolves in water. This procedure prevents inhalation of frames.

Procedure

1.   Cut 10–15 µm frozen sections.
2.   Treat with osmium tetroxide for 24 hours.

3. Wash in distilled water giving several changes for 12 hours.
4. Treat with absolute alcohol for 5 hours.
5. Wash well in distilled water.
6. Mount in glycerine jelly.

Result

Lipids	Black
Background	Brown

Fluorescence Method (Metcalf and Patton, 1944; Peltier, 1954)

Fixation

Fresh tissue or 10 per cent neutral buffered formalin.

Preparation of Reagent

Solution 1

Phosphine 3R	0.1–1.0 g
Distilled water	100.0 ml

Procedure

1. Cut frozen sections 10 µm thick.
2. Wash in distilled water.
3. Treat with solution 1 for 5 minutes.
4. Rinse in distilled water.
5. Mount in glycerine jelly.

Result

Lipids emit white fluorescence.

Sudan Black B Method (Lison and Dagnelie, 1935)

(Plate 13, Figures 1–3; Plate 14, Figure 2)

Fixation

Formalin, frozen, cryostat post-fixed.

Reagents Required

Sudan black B, triethyl phosphate

Preparation of Reagents

Solution 1 (Staining of solution)

Sudan black B	1.0 g
Triethyl phosphate	60.0 ml
Distilled water	40.0 ml

To triethyl phosphate, add distilled water and then the dye. Boil the solution for 5 minutes stirring constantly. Allow it to cool and then filter. This is the stock solution and filter it before use.

Procedure

1. Wash sections in distilled water.
2. Transfer sections to 60 per cent triethyl phosphate.
3. Place sections in solution 1 (Sudan black B) at 20°C for 10 minutes.
4. Immerse sections in 60 per cent triethyl phosphate for 30 seconds.
5. Wash in distilled water.
6. Counterstain if desired in Mayer's carmalum for 3 minutes.
7. Wash in distilled water.
8. Mount in glycerine jelly.

Results

Lipids and phospholipids	Black
Nuclei	Red

Rationale

Sudan black B has amino groups and forms a basic dye unlike Oil red O. This basic dye combines with acidic groups in compound lipids, and as such phospholipids are also stained.

Sudan Black B Method for Bound Lipids in Paraffin Sections

Fixation

Formol calcium; post-treatment in 3 per cent potassium dichromate.

Preparation of the Reagent

Sudan black B saturated in 70 per cent alcohol

Procedure

1. Bring sections to 70 per cent alcohol.
2. Stain for 30 minutes in Sudan black B solution.
3. Remove excess by rinsing quickly in 70 per cent alcohol.
4. Wash in running water.
5. Counterstain in Mayer's haematein.
6. Wash in water.
7. Mount in glycerine jelly.

Result

Lipids stain black.

Remarks

Sudan black B is more soluble and has greater affinity to fats. So Sudan black B flows into the section from staining solution. Therefore it is not actually a histochemical binding. However, in conjunction with other methods like PAS staining (Hanumantha Rao, 1959a, 1959b), it is very reliable.

Bromine Sudan Black Method for Lipids (Bayliss and Adams, 1972)

Fixation

Cryostat, post-fixed for 1 hour in formol calcium.

Procedure

1. Dry sections mounted on slides.
2. Immerse sections in 2.5 per cent aqueous bromine for 30 minutes at room temperature inside a fume cupboard.
3. Wash thoroughly and immerse in 0.5 per cent sodium metabisulphite for 1 minute.
4. Wash in distilled water and as usual apply Sudan black B technique.

Result

Lipids stain black.

The Fettrot Method for Neutral Fat

Fixation

Frozen formalin

Procedure

1. Cut frozen sections, 10–15 μ thick. Mount on albuminized slides and dry thoroughly in air.
2. Stain in Mayer's haematein for 4–6 minutes.
3. Wash in running water for 30 minutes.
4. Rinse quickly in 50 per cent alcohol.
5. Stain in saturated solution of Fettrot for 10–15 minutes.
6. Wash in water.
7. Mount in glycerine jelly.

Results

Lipids   Pinkish red
Nuclei   Dark blue

Lillie's Sulphuric Nile Blue Technique for Fatty Acids

Reagents Required

Nile blue sulphate
Conc. sulphuric acid

Reagents Preparation

Solution 1 (Nile blue sulphate solution)

Nile blue sulphate	50 mg
Distilled water	99.0 ml
Conc. sulphuric acid	1.0 ml

Procedure

1.   Bring frozen sections or paraffin sections to water.
2.   Transfer to solution 1 (Nile blue sulphate solution) for 20 minutes.
3.   Wash in running water.
4.   Mount in glycerine jelly.

Result

Fatty acids   Dark blue
Neutral fats   Pinkish red

Remarks

Nile blue sulphate acts at low pH (0.09). In such low pH concentration carboxylic and phosphoric acid radicals do not get ionized, as such they are not able to bind the dye. Under such circumstances, fatty acids bind the dye and stain blue.

Fischler's Method for Fatty Acids

Reagents Required

Haematoxylin
Absolute alcohol
Potassium ferricyanide
Borax
Copper acetate

Preparation of Reagents

Solution 1 (Weigert's lithium haematoxylin)
    Solution A   10 g haematoxylin in 100 cc absolute alcohol
    Solution B   10 ml saturated lithium carbonate in 90.0 ml distilled water

Mix equal parts of A and B just before use.

Solution 2 (Weigert's borax ferricyanide)
    Borax   20.0 g
    Potassium ferricyanide   25.0 g
    Distilled water   1000.0 ml

Procedure

1.   Fix tissue in 10 per cent formalin saturated with calcium salicylate.
2.   Cut frozen sections 8–10 μ thick.
3.   Mordant the sections in saturated copper acetate solution for 12–24 hours at 37°C.
4.   Wash in water.
5.   Place sections in solution 1 (lithium haematoxylin) for 20 minutes.
6.   Differentiate in solution 2 (Borax ferricyanide).
7.   Wash in distilled water.
8.   Mount in glycerine jelly.

Result

Fatty acids	Dark blue
Neutral lipids	Red

Copper Rubeanic Acid Method for Fatty Acids (Holczinger, 1959)

Fixation

Cryostat-unfixed
Cryostat-pre-fixed
Formalin-fixed frozen sections

Reagents Required

Copper acetate
Rubeanic acid
Ethanol
Ethyl alcohol

Preparation of Reagents

0.005 per cent copper acetate

Solution 1 (Copper acetate solution)
Copper acetate   5 mg
Distilled water   50.0 ml

Solution 2 (0.1 per cent EDTA)
Ethylene diamine   50 mg
  tetra-acetic acid
Distilled water   50.0 ml

Solution 3 (1.0 per cent rubeanic acid)
Rubeanic acid   50 mg
Absolute alcohol   35.0 ml
Distilled water   15.0

Rubeanic acid is dissolved in absolute alcohol by warming slightly and to this distilled water is added.

Procedure

1. Keep sections in solution 1 (Copper acetate solution) for 3–5 hours.
2. Wash sections in solution 2 for 10 seconds.
3. Wash in distilled water for 10 minutes.
4. Transfer sections to solution 3 (rubeanic acid) for 30 minutes.
5. Wash sections in 10 per cent alcohol for 3 minutes.
6. Wash sections in running tap water.
7. Mount sections in glycerine jelly or dehydrate, clear and mount in DPX.

Results

Fatty acids are stained greenish black.

Rationale

Fischler (1904) first introduced this method. After repeated attempts, Holczinger (1959) introduced copper rubeanic acid method. In sections exposed to copper acetate for 3–5 hours after treatment with EDTA, non-specific absorbed copper is removed. Dilute rubeanic acid is applied to the sections and copper soaps are formed which indicate that fatty acids are present.

Acid Haematein Method for Phospholipids
(Baker, 1946) (Plate 14, Figure 1)

Fixation

Unfixed piece of tissue

Reagents Required

Potassium dichromate
Calcium chloride
Formalin
Haematein
Sodium iodate
Potassium ferricyanide
Sodium tetraborate

Preparation of Reagents

Solution 1 (Fixative)

Formalin	10.0 ml
Anhydrous calcium chloride	1.0 g
Distilled water	90 ml

Solution 2 (Post-chroming solution)

Potassium dichromate	5.0 g
Calcium chloride	1.0 g
Distilled water	100.0 ml

Solution 3 (Acid haematein solution)

Haematein	50 mg
1 per cent sodium iodate	1.0 ml
Distilled water	49.0 ml

Heat solution, cool and add 1 ml glacial acetic acid.

Solution 4 (Differentiator)

Potassium ferricyanide	250 mg
Sodium tetraborate	250 mg
Distilled water	100.0 ml

Procedure

1. Keep blocks in fixative (Solution 1) at 22°C for 6–12 hours.
2. Keep tissue in solution 2 (Post-chroming solution) at 22°C for 18 hours.

3. Transfer tissue to solution 2 at 60°C for 24 hours.
4. Wash in running tap water for 6 hours.
5. Cut frozen sections 10 μ thick.
6. Immerse sections in solution 2 (Post-chroming solution) at 37°C for 1 hour.
7. Wash in distilled water.
8. Stain in solution 3 (acid hematein) at 60°C for 5 hours.
9. Rinse in distilled water.
10. Place them in solution 4 (differentiating solution) at 37°C for 18 hours.
11. Wash in tap water.
12. Mount in glycerine jelly.

Results

Phospholipids stain dark blue.

Rationale

This technique was first introduced by Baker (1946) which is specific. With a control section extracted with pyridine, compounds other than phospholipids also give a blue colour. Unless pyridine extraction is done, phospholipids are not removed. Chrome ions in conjunction with phospholipids are not removed by this technique. On the other hand, they form a dye lake. Later, differentiation is carried out with borax–ferricyanide mixture. Only compounds containing phosphoric acid retain the dye and phospholipids contain phosphoric acid, hence the blue colour.

Sudan Black B Method for Phospholipids (Elftman, 1957)

Fixation

5 per cent aqueous $HgCl_2$ 100.0 ml
Potassium dichromate   2.5 g

pH should be 2.5 which can be adjusted with HCl. Fixation time is 3 days.

Preparation of Reagents

Solution 1
     Sudan black B   700 g
     Propylene glycol   100.0 ml

Add Sudan black B to glycol in small amounts, heat and stir it. Let the temperature be maintained at 100°C. Filter and cool.

Solution 2 (Ethylene glycol solution)
     Ethylene glycol   85.0 ml
     Distilled water   100.0 ml

Procedure

1. Dewax and bring the slides to absolute alcohol.
2. Transfer to solution 2.

3. Transfer to solution 1 for 30 minutes.
4. Differentiate in 95 per cent ethylene glycol for 3 minutes.
5. Wash in distilled water.
6. Mount in glycerine jelly.

Result

Phospholipids stain black.

Copper Phthalocyanin Method for Phospholipids
(Kluver and Barrera, 1953) (Plate 13, Figures 4–6; Plate 15, Figure 2)

Fixation

Formol calcium, frozen section

Reagents Required

Luxol fast blue G
Lithium carbonate
Neutral red

Preparation of Reagents

Solution 1 (Luxol fast blue solution)
    Luxol fast blue   10 mg
    95 per cent alcohol   100.0 ml

Solution 2 (Lithium carbonate solution)
    Lithium carbonate   50 mg
    Distilled water   100.0 ml

Solution 3 (Neutral red solution)
    Neutral red   1.0 g
    Distilled water   100.0 ml

Procedure

1. Bring sections to absolute alcohol.
2. Place sections in solution 1 for 6–18 hours at 60°C.
3. Rinse in 70 per cent alcohol, wash in water.
4. Differentiate in solution 2 for 30 minutes to 2 hours.
5. Rinse in water.
6. Counterstain in solution 3 (neutral red) for 10 minutes.
7. Rinse in water.
8. Dehydrate, clear and mount in Canada balsam.

Results

Phospholipids except sphingomyelin stain blue. A dark reddish blue shade denotes the presence of phospholipids.

Acid Haematein for Phospholipids (Hori, 1963) (Plate 14, Figure 1)

Fixation

Formalin calcium (10 per cent formalin in 1 per cent calcium chloride)
Formalin/calcium/cadmium (1 per cent cadmium chloride added to formalin calcium)
Glutaraldehyde calcium chloride (2 : 5 : 25)

Reagents Required

Haematoxylin
Sodium iodate
Glacial acetic acid
Borax
Potassium ferricyanide

Preparation of Reagents

Solution 1 (Acid haematein)

Haematoxylin	50 mg
0.01 per cent sodium iodate	50.0 ml

Heat to boil, cool and then add 1.0 ml of glacial acetic acid.

Solution 2

Borax	250 mg
Potassium ferricyanide	250 mg
Distilled water	100.0 ml

Store in refrigerator.

Procedure

1. Cut frozen sections of 10 μm thick.
2. Post-chromate sections in 5 per cent aqueous potassium dichromate at 60°C for 4 hours.
3. Wash in distilled water giving several changes.
4. Treat with solution 1 for 30 minutes at 37°C.
5. Wash in distilled water.
6. Differentiate in solution 2.
7. Dehydrate, clear and mount.

Result

Phospholipids stain blue to blue-grey.

Controlled Chromatin Procedure for Phospholipids (Elftman, 1954)

Reagents Required

Dichromate
Acetate buffer
Potassium ferricyanide
Haematoxylin

Preparation of Reagents

Solution 1 (Buffered dichromate fixative)
Potassium dichromate   2.5 g
Distilled water   100.0 ml

Final pH should be 3.5. Adjust with 0.2 M acetate buffer.

Solution 2 (Buffered haematoxylin solution)
0.2 M acetate buffer pH 3.0   100.0 ml
Potassium ferricyanide (0.005 per cent)   5 mg
Haematoxylin (0.1 per cent)   100 mg

Procedure

1. Fix fresh tissues in solution 1 for 18 hours at 56°C.
2. Wash in running water.
3. Dehydrate, clear and embed for sectioning.
4. Hydrate slides to water.
5. Transfer to solution 2 (buffered haematoxylin) (Pre-heated for 2 hours at 56°C).
6. Rinse in water, dehydrate, clear and mount in synthetic resin.

Result

Phospholipids stain dark blue.

Bromine–Acetone–Sudan Black Method for Phospholipids (Baylis High, 1981)

Procedure

1. Treat sections with 2.5 per cent aqueous bromine for 30 minutes at room temperature in a fume cupboard.
2. To remove excess bromine, treat with 0.5 per cent sodium metabisulphite.
3. Wash thoroughly.
4. Treat with anhydrous acetone for 20 minutes at 4°C to extract neutral lipids.
5. Follow the standard Sudan black method.

Result

Phospholipids are stained grey.

Pyridine Extraction Method for Phospholipids (Baker, 1946)

Fixation

Bouin's fluid

Reagents Required

Picric acid
Formalin
Glacial acetic acid

Distilled water
70 per cent alcohol
50 per cent alcohol
Pyridine
Distilled water

Procedure

1. Fix the tissue in dilute Bouin's fixative for 20 hours.
2. Wash in 70 per cent alcohol for 1 hour.
3. Wash in 50 per cent alcohol for 30 minutes.
4. Wash in running tap water for 30 minutes.
5. Keep in pyridine at 22°C for 1 hour.
6. Transfer tissue to fresh pyridine at 22°C for 1 hour.
7. Keep tissue in fresh pyridine at 60°C for 24 hours.
8. Wash in running tap water for 2 hours.
9. Transfer to post-chroming solution as in method 5.
10. Stain in acid haematein at 60°C for 5 hours.

Result

Phospholipids are stained negative.

Otan Method (Adams, 1959)

Fixation

Formol calcium, frozen sections

Reagents Required

Osmium tetroxide
Potassium perchlorate
α-naphthylamine

Preparation of Reagents

Solution 1 (Osmium tetroxide solution)
    1 per cent osmium tetroxide solution   1 part
    1 per cent potassium perchlorate   3 parts

Solution 2 α-naphthylamine solution
    Saturated solution in warm distilled water
    (β-naphthylamine is a carcinogen)

Procedure

1. Treat free-floating sections with solution 1 (osmium solution) in suitable containers (tightly stoppered) for 18 hours.
2. Wash in distilled water for 10 minutes.
3. Pick up on slides.
4. Treat with solution 2 (α-naphthylamine solution) at 37°C for 20 minutes.

5.   Wash sections in distilled water for 5 minutes.

6.   Mount in glycerine jelly.

Result

Phospholipids   Orange-red
Cholesterol esters   Black
Triglycerides   Black

Rationale

Osmium tetroxide which is soluble in all types of lipids is reduced by unsaturated fatty acids. The reduction takes place in two stages. First product is colourless. In the second stage, it is further reduced to form osmium dioxide. Simple lipids as already stated are not water-soluble and they are further reduced to black osmium dioxide. Compound lipids like phospholipids are partially soluble in water. Sodium perchlorate inhibits further reduction of osmium tetroxide. The resulting product is colourless which is chelated with a-naphthylamine to produce an orange-red colour. With this method, simple lipids like triglycerides, cholesterol and its esters and fatty acid could be detected as black spots and phospholipids as orange-red.

Ferric Haematoxylin Method for Phospholipids (Elledes and Lojda, 1973)

Fixation

Unfixed cryostat, frozen short fixed.

Preparation of Reagents

Solution 1

Distilled water	298.0 ml
Conc. HCl	2.0 ml
Ferric chloride ($FeCl_2.6H_2O$)	2.5 g
$FeSO_2\ 7H_2O$	4.5 g

This lasts for some weeks.

Solution 2

Distilled water   10.0 ml
Haematoxylin   100 mg

This solution should be prepared fresh before use. Haematoxylin dissolves with gentle heat.

Solution 3 (Working solution)
Solution 1   3 parts
Solution 2   1 part

This solution lasts for 1 hour.

Procedure

1.   Take two sets of sections. Treat one set with chloroform–methanol extract for 1 hour at room temperature. Treat the other set with dry acetone at 4°C for 15 minutes.

2. Fix both sections in formol calcium for 30 minutes.
3. Rinse in distilled water.
4. Wash in distilled water.
5. Immerse in 0.2 per cent HCl.
6. Wash in tap water.
7. Dehydrate in acetone, clear in xylene and mount in DPX.

Result

Phospholipids	Blue
Nuclei	Blue

Phosphomolybdic Acid Method (Landing *et al.*, 1952)

Fixation

Formalin, formol calcium, frozen sections

Procedure

1. Place sections on gelatinized slides, drain, blot and expose to formalin vapour.
2. Allow the sections to dry and dip in 50/50 acetone–ether.
3. Place slides in 1 per cent phosphomolybdic acid in 50/50 ethanol/chloroform for 15 minutes.
4. Rinse in ethanol–chloroform, then in chloroform and dry.
5. Rinse slide in 1 per cent aqueous stannous chloride in 3N HCl (freshly made).
6. Wash in water.
7. Counterstain with aqueous eosin.
8. Dehydrate, clear and mount in Canada balsam.

Result

Choline-staining lipids stain in molybdenum blue shades.

Nile Blue Method for Phospholipids (Menschik, 1953) (Plate 15, Figure 1)

Fixation

Formol calcium, frozen sections

Reagents Required

Gelatin
Nile blue sulphate
Sulphuric acid
Acetic acid
Hydrochloric acid

Procedure

1. Fix for 12 hours and cut frozen sections without embedding in gelatin or carbowax.
2. Stain in saturated aqueous Nile blue sulphate (500 ml) with 50 ml 0.5 per cent sulphuric acid for 1½ hours at 60°C.

3.   Rinse in distilled water.
4.   Transfer to pre-heated acetone at 50°C.
5.   Keep acetone at room temperature for 30 minutes.
6.   Differentiate in 5 per cent acetic acid for 30 minutes.
7.   Rinse in distilled water.
8.   Differentiate in 0.5 per cent HCl for 3 minutes.
9.   Wash in distilled water and mount in glycerine jelly.

Result

Phospholipids are stained blue.

Bromine–Silver Method for Unsaturated Lipids (Norton *et al.*, 1962)

Procedure

1.   Cut frozen or cryostat sections of formol calcium-fixed tissue.
2.   Treat with bromine potassium bromide (1 ml of bromine in 390 ml of 2 per cent potassium bromide) for 1 minute.
3.   Wash in water.
4.   Treat with 1 per cent sodium bisulphate for 5 minutes.
5.   Rinse in distilled water.
6.   Transfer to 1 per cent silver nitrate in 1N nitric acid for 18 hours.
7.   Rinse in distilled water (several times).
8.   Reduce in Kodak Dektol developer for 1 minute.
9.   Wash and mount in glycerine jelly.

Result

Unsaturated lipids are stained brown to black.

Remarks

While brominating unsaturated lipids, they react with silver to give silver bromide with a metallic lustre.

Performic Acid/Schiff Method for Unsaturated Lipids (Lillie, 1952)

Fixation

Cryostat post-fixed, cryostat pre-fixed, formalin-fixed frozen sections, freeze-dried, paraffin sections.

Preparation of Reagents

1.   Performic acid

90 per cent formic acid	40.0 ml
100 vol. or 30 per cent hydrogen peroxide	4.0 ml
Sulphuric acid concentrated	0.5 ml

2.   Schiff's reagent (*see* Chapter 9)

Procedure

1. Deparaffinize and bring sections down to water.
2. Transfer sections to solution 1 for 30 minutes.
3. Wash in tap water for 15 minutes.
4. Transfer to Schiff's reagent for 40 minutes.
5. Wash in running tap water.
6. Dehydrate, clear in xylene and mount in DPX.

Result

Unsaturated lipids including cerebrosides are stained red.

Rationale

This technique involves oxidation of unsaturated bonds to produce free aldehydes. Oxidation is mostly carried out by performic acid or peracetic acid but performic acid is preferred. While demonstrating, it is better to block by bromination since DNA and keratin also react.

Bromination of Unsaturated Lipids (Lillie, 1954a)

Fixation

Frozen sections

Reagents Required

Carbon tetrachloride
Sodium metabisulphate
Bromine water

Preparation of Reagents

Solution 1 (2.5 per cent bromine water)
Bromine water   1.0 ml
Distilled water   30.0 ml

Solution 2 (0.5 per cent sodium metabisulphate)
Sodium metabisulphate   500 mg
Distilled water   50.0 ml

Procedure

1. Place sections on slides and allow them to dry.
2. Place sections in solution 1 (2.5 per cent bromine water) for 1–6 hours.
3. Wash in tap water.
4. Transfer sections to solution 2 (0.5 per cent sodium metabisulphate) for 2 minutes.
5. Wash in running tap water.
6. Apply Sudan black method.

For Paraffin Sections

Preparation of Reagent

Solution 1 (Carbon tetrachloride solution)
Carbon tetrachloride   10.0 ml
Bromine water   1.0 ml

Procedure

1. Deparaffinize in xylol.
2. Place sections in carbon tetrachloride for 4 minutes giving 2 changes.
3. Immerse sections in bromine solution for 2–4 minutes.
4. Immerse sections in carbon tetrachloride for 4 minutes giving 2 changes.
5. Rinse sections in 60 per cent alcohol for 1 minute.
6. Rinse sections in 60 per cent alcohol.
7. Rinse sections in water.
8. Apply Sudan black B method.

Ultraviolet Schiff Method for Unsaturated Lipids (Belt and Hayes, 1956)

Fixation

Unfixed cryostat sections or short fixed frozen.

Reagents Required

Schiff's reagent   (*see* Chapter 9)

Procedure

1. Cut frozen or fresh sections and mount them on slide.
2. Expose the sections to ultraviolet light for 2 hours.
3. Treat with Schiff's reagent for 15 minutes.
4. Also treat non-irradiated control section with Schiff's reagent to exclude non-lipid aldehydes.
5. Wash in tap water and again in distilled water.
6. Mount sections in glycerine jelly.

Result

Unsaturated lipids are stained magenta.

Osmium Tetroxide Method for Unsaturated Lipids

Fixation

Unfixed, cryostat sections or short-fixed frozen sections.

Procedure

1. Cut frozen or unfixed sections and mount them on slide.
2. Transfer them to one per cent osmium tetroxide for 1 hour at room temperature.
3. Wash in distilled water and mount in glycerine jelly.

Result

Unsaturated lipids are stained brown or black.

Bismuth Trichloride Method for Differentiation between Cholesterol and Cholesterol Esters (Grundland *et al.*, 1949)

Reagents Required

Bismuth trichloride
Acetyl chloride
Nitrobenzene
Digitonin

Preparation of Reagents (Bismuth trichloride solution)

Bismuth trichloride	200 mg
Acetyl chloride	1.0 ml
Nitrobenzene (anhydrous)	100.0 ml

Procedure

1. Fix tissue in saturated solution of digitonin in 70 per cent alcohol for 36 hours.
2. Infiltrate with paraffin wax containing 5 per cent of glyceryl monostearate for 12–16 hours.
3. Embed in wax and cut 6–10 μ sections.
4. Without removing the wax, treat the sections with bismuth trichloride reagent for 15–45 minutes.
5. Rinse in 10 per cent acetyl chloride in nitrobenzene.
6. Wash in 75 per cent nitric acid in absolute alcohol.
7. Rinse rapidly in absolute alcohol.
8. Immerse in 20 per cent yellow ammonium sulphide for 3 seconds.
9. Wash in absolute alcohol.
10. Clear in xylol and mount in Canada balsam.

Result

Cholesterol	Dark brown
Cholesterol esters	Colourless

Cholesterol and Related Substances: Perchloric Acid Naphthoquinone Reaction (Adams, 1961)

Fixation

Formol saline – fixed frozen sections
Formol calcium-fixed, frozen sections, free-floating

Reagents Required

1,2-Naphthoquinone 4-sulphonic acid
Ethanol
Perchloric acid
Formaldehyde

Preparation of Reagents

Staining solution

1,2-Naphthoquinone 4-sulphonic acid	12 mg
Ethanol	6.0 ml
60 per cent perchloric acid	3.0 ml
Concentrated formaldehyde	0.3 ml
Distilled water	2.7 ml

The ethanol–perchloric acid–formaldehyde–water solution is prepared first and the reagent is dissolved in it.

Procedure

1. Cut frozen sections and float into formalin. Leave for 7 days.
2. Place sections on slides and dry at room temperature.
3. Transfer sections to staining solution.
4. Heat sections in reagent to 60–70°C for 10 minutes.
5. Mount sections in 60 per cent perchloric acid.

Result

Cholesterol and its esters stain dark blue.

Remarks

This blue colour does not last long during heating. Section should change colour from red to dark blue.

Rationale

Adams (1961) modified the method of Schultz (1924–1925). This method is called PAN's method. Perchloric acid first forms an insoluble perchlorate with cholesterol. This is later converted into cholesta-3–5 diene by the elimination of water (Adams, 1965). This finally reacts with 1 : 2 naphthoquinone-4-sulphonic acid to form a dark blue pigment. This pigment is stable for a few hours. As the colour lasts for a short duration, the slide should be examined immediately.

Filipin Method for Free Cholesterol (Kruth and Vaughan, 1980)

Fixation

Cryostat—Post-fixed, frozen

Preparation of Reagents

Solution 1 (Stock solution)

Filipin	25 mg
Dimethyl formamide	1.0 ml

Solution 2 (Staining solution)
    Solution 1   0.2 ml
    Phosphate buffer   10.0 ml

Procedure

1. Wash sections in phosphate buffer.
2. Treat with solution 2 for 30 minutes.
3. Again wash in phosphate buffer.
4. Mount in glycerine jelly.
5. Observe under fluorescence microscope.

Result

Free cholesterol shows silvery fluorescence.

Digitonin Method for Free Cholesterol (Adams and Baylis, 1974)

Fixation

Formalin-fixed frozen sections, free-floating. A control section is stained by the Oil Red O method.

Reagents Required

Ethyl alcohol
Digitonin

Preparation of Reagents

Solution 1 (Ethyl alcohol 50 per cent)
    Ethyl alcohol   100.0 ml
    Distilled water   100.0 ml

Solution 2 (Digitonin solution)
    Digitonin   500 mg
    Solution 1   100.0 ml

Procedure

1. Incubate sections in solution 2 for 3 hours at room temperature.
2. Differentiate sections in solution 1.
3. Float sections onto slides.
4. Mount in glycerine jelly.

Result

Digitonin section: Free cholesterol (birefringent)

Oil Red O section: Free cholesterol (birefringent). Cholesterol esters take Oil Red O stain.

Rationale

First introduced by Windaus (1910), it is a good method to distinguish cholesterol and cholesterol esters. Two slides are used in this method. One slide is stained with Oil Red O for lipid material. Another slide is immersed in digitonin. Free cholesterol on combining with digitonin precipitates a complex compound of birefringent crystals. Control sections stained with Oil Red O also display birefringent crystals of cholesterol but cholesterol esters take Oil Red O stain and are not birefringent.

Nile Blue Method for Acid Lipids
(Smith and Dietrich; Modified by Cain, 1947)

Fixation

Formalin, fixed frozen cryostat sections

Reagents Required

Nile blue
Acetic acid

Preparation of Reagents

Solution 1 (Nile blue solution 1)
    Nile Blue   500 mg
    Distilled water   50.0 ml

Section 2 (Nile blue solution 2)
    Nile blue   10 mg
    Distilled water   50.0 ml

Solution 3 (Differentiator)
    Concentrated acetic acid   0.5 ml
    Distilled water   50.0 ml

Three sections are used. Stain one section by Sudan black B or Oil red O methods (As stated previously). This is slide 3. With the other two sections the following procedure is adopted.

Procedure

1. Bring down sections 1 and 2 to water.
2. Place both sections in solution 1 (0.5 per cent Nile blue) for 5 minutes at 60°C.
3. Differentiate in solution 3 at 60°C for 30 seconds.
4. Wash in tap water.
5. Mount section 1 in glycerine jelly.
6. Immerse section 2 in solution 2 at 60°C for 5 minutes.
7. Wash in tap water.
8. Differentiate section 2 in solution 3 at 60°C for 30 seconds.
9. Wash in tap water.
10. Mount in glycerine jelly.

Result

Section 1   Blue—acidic lipid (If there is blue staining compare with section 3 and it is acidic lipid.)
Section 2   Red—non-acidic lipid (If there is red staining compare with section 3 and it is non-acidic lipid.)
Section 3   Control

NaOH Otan Method (Adams and Baylis, 1963)

Fixation

Formol calcium or frozen sections

Reagents Required

Osmium tetroxide
Naphthylamine
Sodium hydroxide
Potassium perchlorate

Preparation of Reagents

Solution 1 (Osmium tetroxide)
    1 per cent osmium tetroxide   1 part
    1 per cent potassium perchloride   3 parts

Solution 2 (α-naphthylene)
    Saturated solution of α-naphthylamine in warm distilled water.

Solution 3
    Sodium hydroxide 8.0 g
    Distilled water   100 ml

Procedure

1. Place free sections in solution 3 for 1 hour at 37°C.
2. Wash in distilled water.
3. Transfer to 1 per cent acetic acid.
4. Treat with solution 1 for 18 hours in a tightly stoppered bottle.
5. Wash in distilled water for 10 minutes.
6. Pick the sections onto a slide and treat with solution 2 for 20 minutes at 37°C.
7. Wash in distilled water.
8. Mount in glycerine jelly.

Result

Sphingomyelin is stained orange-red.

Rationale

This is a modified technique of OTAN method where lecithin is separated from sphingomyelin. Lecithin has ester linkages which on hydrolysis are broken. On the other hand, sphingolucin linkage (fatty acid) is not affected and as a result stains black.

Sodium Hydroxide/Ferric Haematoxylin Method for Sphingomyelin

Fixation

Unfixed cryostat, short-fixed frozen.

Procedure

1. Treat sections with 2 M sodium hydroxide for 1 hour at room temperature.
2. Wash thoroughly.
3. Rinse in 1 per cent acetic acid for 5 seconds.
4. Proceed for ferric haematoxylin method (*see* Chapter 7).

Result

Sphingomyelin is stained blue.

Acetal Phosphatides: The Plasmal Reaction
(Feulgen and Voit, 1924; Modified by Terner and Hayes, 1961)

Fixation

Frozen formalin fixed
Frozen unfixed
Cryostat post fixed
Cryostat pre-fixed

Reagents Required

Sodium chloride
Mercuric chloride
Hydrochloric acid
Schiff's reagent (*see* Chapter 9)
Sodium metabisulphite

Preparation of Reagents

Solution 1 (Sodium chloride solution)
    Sodium chloride 900 mg
    Distilled water 100.0 ml

Solution 2 (Mercuric chloride solution)
    Mercuric chloride 1.0 g
    Distilled water 100.0 ml

Solution 3 (Schiff's Reagent) (*see* Chapter 9)

Solution 4 (Sulphurous acid rinse)
    Sodium bisulphite 500 mg
    Distilled water 99.5 ml
    Conc. HCl 0.5 ml

Procedure

With this technique it is necessary to have a control section that is not placed in solution 2 (mercuric chloride). If section is unfixed, place in 3 changes of solution 1, otherwise wash fixed sections in distilled water.

All Sections

1. Immerse sections in solution 2 (mercuric chloride) for 7 minutes.
2. Stain sections in Schiff's reagent 3 for 10 minutes.

3. Keep them in sulphurous acid rinse for 2 minutes.
4. Wash in tap water.
5. Counterstain in 2 per cent methyl green for 3 minutes.
6. Wash in running tap water.
7. Mount in glycerine jelly.

Result

Acetal phosphate	Schiff-positive, reddish pink
Control section	Negative

Rationale

Feulgen and Voit (1924) introduced this method to demonstrate acetal phosphates. These are phospholipids which have a molecule of an aldehyde of fatty acids which combines with the glycerol group. The linkage between the two is of acetal type which is split by mercuric chloride. As a result, aldehydes are released which react with PAS method. Terner and Hayes (1961) later dealt with this method in detail. Pearse (1968) gave a better demonstration.

Caza Method for Choline Lipids (Boelsma–Van Houte, 1965)

Fixation

Fresh cryostat post-fixed formol calcium

Reagents Required

Sodium periodate
Toluene
Acetic anhydride
cis-aconitic anhydride
Cobalt chloride

Preparation of Reagents

Solution 1 (Cobalt chloride solution)
    Cobalt chloride   10.0 g
    Distilled water   100.0 ml
Solution 2 (Sodium periodate solution)
    Sodium periodate   1.0 g
    Distilled water   100.0 ml

Procedure

1. Place sections in 10 per cent aqueous cobalt chloride for 24 hours at room temperature.
2. Rinse thoroughly.
3. Keep in solution 2 (one per cent aqueous sodium periodate) for 1 hour at 37°C.
4. Upgrade through graded series of alcohol.
5. Remove ethanol by transferring to toluene.

6. Place in acetic anhydride/toluene (1 : 1 v/v).
7. Develop colour by treating for 30–60 minutes at room temperature with 2.5 per cent solution of *cis*-aconitic anhydride in acetic anhydride/toluene (2 : 3 v/v). This should be prepared 24 hours before use.
8. Remove excess reagent by rinsing in acetic anhydride/toluene (1 : 1 v/v)).
9. Clear in toluene and cover with a coverslip.

Result

Choline-containing lipids stain pink-red.

Modified Acid Haematein–Oil Red Method (Bourgeois and Hubbard, 1965)

Fixation

Fresh frozen cryostat

Reagents Required

Haematoxylin
Glacial acetic acid
Sodium iodate
Borax ferricyanide
Pyridine
Triethyl phosphate

Preparation of Reagents

Solution 1 Acid haematein

Haematoxylin	50 mg
1 per cent $NaIO_4$	1.0 ml
Distilled water	48.0 ml
Glacial acetic acid	1.0 ml

Procedure

1. Fix sections in formol calcium for 1–2 hours at room temperature.
2. Rinse and mordant for 12–16 hours at 60°C in dichromate.
3. Wash in distilled water.
4. Transfer to solution 1 at 37°C for 2 hours (solution should be fresh).
5. Rinse and differentiate in borax ferricyanide (refer acid haematein method).
6. Rinse and mount in glycerine. Proceed to stage 7 of acid haematein. Control sections before stage 1 are treated with pyridine for 2 hours at 60°C, washed in water and transferred to mordant.
7. Rinse and place in 60 per cent triethyl phosphate for 2 minutes.
8. Stain for 15 minutes at room temperature in 0.5 per cent Oil Red O in triethyl phosphate.
9. Immerse in 60 per cent triethyl phosphate.
10. Wash in distilled water.
11. Mount in glycerine.

Result

Choline-containing lipids	Black
Triglycerides	Red

Calcium Lipase Method for Triglycerides (Adams *et al.*, 1966)

Fixation

Cryostat, post-fixed in formol calcium, formol calcium-fixed frozen section

Reagents Required

Tris buffer
Calcium chlorides
Porcine pancreatic lipase

Preparation of Reagent

Solution 1 (Incubating medium)
Tris buffer at pH 8.0	15.0 ml
2 per cent calcium chloride	10.0 ml
(2 g/100 ml water)	
Distilled water	25.0 ml
Porcine pancreatic lipase	50 mg

Warm solution at 37°C and filter.

Procedure

1. Incubate floating frozen slide mounted in the lipase medium (solution 1) at 37°C for 3 hours.
2. Wash section and mount them on slide.
3. Treat this section along with untreated (not treated with lipase) section with 1 per cent lead nitrate solution.
4. Wash in distilled water.
5. Transfer to a dilute solution of ammonium sulphate (3 to 4 drops in a coplin jar full of water).
6. Wash in distilled water.
7. Mount in glycerine jelly.

Result

Triglycerides are stained brown.

Gold Hydroxamic Acid Method for Phosphoglycerides (Adams *et al.*, 1963)

Fixation

Cryostat post-fixed for 1 hour in formol calcium or short-fixed frozen.

Reagents Required

Hydroxylamine hydrochloride

Sodium hydroxide
Silver nitrate
Ammonium nitrate

Preparation of Reagents

Solution 1 (Hydroxylamine solution)
    Hydroxylamine hydrochloride 2.5 g
    Sodium hydroxide   6.0 g
    Distilled water   100.0 ml

Solution 2 (Silver solution)
    Silver nitrate   100 mg
    Ammonium nitrate   200 mg
    Distilled water   100.0 ml

Final pH should be 7.8 with dilute sodium hydroxide solution.

Procedure

1. Immerse slides or free floating section in solution 1 for 2 minutes.
2. Wash in 3 changes of distilled water for 5 minutes each.
3. Treat with solution 2 for 2 hours at room temperature.
4. Rinse in 1 per cent acetic acid.
5. Wash in distilled water.
6. Tone the section with 0.2 per cent gold chloride for 10 minutes.
7. Rinse in 5 per cent sodium thiosulphate for 5 minutes.
8. Wash, if floating sections, and mount them on slides.
9. Counterstain nuclei with one per cent methyl green for 5 minutes.
10. Mount in glycerine jelly or dehydrate, clear and mount in Canada balsam.

Result

Phosphoglycerides stain purple.

Bromine Silver Method (Mukherji *et al.,* 1960)

Fixation

Formalin, frozen sections

Reagents Required

Bromine
Sodium thiosulphate
Silver nitrate

Preparation of Reagents

Solution 1 (Sodium thiosulphate)
    Sodium thiosulphate 5.0 g
    Distilled water   100.0 ml

Solution 2

Silver nitrate	10.0 g
50 per cent ethanol	100.0 ml

Procedure

1. Wash well in running water.
2. Blot-dry and expose to bromine vapour in a closed jar for 2 hours at 37°C.
3. Transfer to 5 per cent sodium thiosulphate to remove yellow colour.
4. Wash in several changes of distilled water.
5. Transfer to a dark coloured glass vessel containing 10 per cent silver nitrate in 50 per cent ethanol at 37°C (until section turns yellow).
6. Wash in subdued light and reduce in five per cent methyl hydroquinone to the desired depth of colour.
7. Wash in water and fix in 5 per cent sodium thiosulphate for 2 minutes.
8. Wash in distilled water, dehydrate and mount.

Result

Unsaturated lipids are stained brown to black.

Toluidine Blue Method for Sulphatide (Bodian and Lake, 1963)

Fixation

Post-fixed cryostat, formol calcium-fixed frozen.

Preparation of Reagents

Toluidine blue	10 mg
Phosphate buffer (pH 4.7)	100.0 ml

(*see* Chapter 5)

Procedure

1. Mount sections on slides.
2. Transfer to working solution for 16–18 hours.
3. Wash in water.
4. Dehydrate with acetone for 5 minutes.
5. Mount in DPX.

Result

Sulphatides are stained metachromatic red/brown/yellow.

Acriflavin–DMAB Method for Sulphatide (Hollander, 1963)

Fixation

Post-fixed cryostat sections, formol calcium-fixed frozen sections

Preparation of Reagents

Solution 1 (Acriflavine stock)
Acriflavine   100.0 mg
Distilled water   20.0 ml
Heat and dissolve and cool at 4°C.

Solution 2 (Acriflavine working solution)
0.1 M acetate HCl buffer (pH 2.5)   99.0 ml
Solution 1   1.0 ml

Solution 3
p-dimethylaminobenzaldehyde   600 mg
20 per cent hydrochloric acid   30.0 ml
Isopropanol   70.0 ml

Procedure

1. Mount sections on slides.
2. Treat with solution 2 for 5 minutes.
3. Immerse (2 changes) in 70 per cent isopropanol.
4. Transfer to solution 3 for 40 seconds.
5. Rinse in distilled water.
6. Counterstain with Mayer's haemalum.
7. Blue in tap water and mount in glycerine jelly.

Result

Sulphatide is stained red.

Table 13.1 Histochemical techniques applied for the demonstration of lipids

Technique	Fixative	Lipids	Neutral fats	Fatty acids	Phospholipids
Chiffelle and Putt's propylene glycol method	Frozen formalin	Red or black			
Lillie's and Ashburn's isopropanol oil red O method	Formalin frozen	Bright red			
Oil red O method	Formol calcium, frozen, cryostat post-fixed	Red			
Osmium tetroxide method	10% neutral buffered formalin	Black			
Fluorescence method	Fresh tissue or 10% neutral buffered formalin	White fluorescence			
Sudan black B method	Formalin, frozen cryostat post-fixed	Black			Black
Sudan black B for bound lipids	Formol calcium Post-treatment in 3% potassium dichromate	Black			
Bromine Sudan black B	Cryostat post-fixed for 1 hour in formol calcium	Black			

(Contd.)

Table 13.1 (Continued)

Technique	Fixative	Lipids	Neutral fats	Fatty acids	Phospholipids
Lillie's sulphuric Nile blue method	Frozen, cryostat or paraffin		Pinkish red		
Fettrot method	Frozen formalin	Pinkish red			
Copper rubeanic acid technique	Cryostat-unfixed, formalin-fixed frozen			Greenish black	
Fischler's method	Formalin frozen sections			Dark blue	
Acid haematein method	Unfixed				Dark blue
Copper phthalocyanin method	Formol calcium, frozen				Blue (except sphingomyelin)
Bromine acetone Sudan black B	Formalin, Zenker, Carnoy, etc.				Grey
Otan method	Formol calcium-frozen				Orange-red
Ferric haematoxylin method	Unfixed cryostat, frozen				Blue
Nile blue method	Formol calcium-frozen				Blue
Pyridine extraction	Bouin's fluid				Negative

Table 13.2   Histochemical techniques applied for demonstration of lipids

Technique	Fixative	Choline-containing lipids	Unsaturated lipids	Cholesterol	Cholesterol esters	Sphingomyeline	Acid lipids
Phospho-molybdic acid method	Formol calcium, frozen	Molybdenum blue					
Bromine–Silver method	Formol frozen		Black				
Performic acid/Schiff	Cryostat pre-fixed post-fixed		Red				
Ultraviolet Schiff method	Unfixed cryostat or short-fixed		Magenta				
Osmium tetroxide method	Unfixed cryostat or short-fixed frozen sections		Brown or black				

(Contd.)

Table 13.2   (Continued)

Technique	Fixative	Choline-containing lipids	Unsaturated lipids	Cholesterol	Cholesterol esters	Sphingo-myeline	Acid lipids
Bismuth trichloride	Saturated solution of digitonin in 70% alcohol			Dark brown	Colourless		
Digitonin method	Formalin-fixed frozen			Birefringent			
Perchloric acid/naphthoquinone	Formol saline, frozen fixed, formol calcium fixed			Dark blue			
Filipin method	Frozen cryostat post-fixed			Silvery fluorescence			
Nile blue method	Formalin-fixed frozen						Blue
NaOH Otan method	Formol calcium or frozen					Orange-red	
NaOH and ferric haematoxylin	Unfixed cryostat, short-fixed frozen					Blue	
Caza method	Fresh cryostat post-fixed formol calcium	Pink					
Acid haematein oil red method	Fresh frozen cryostat	Black					

Table 13.3   Histochemical tehniques applied for demonstration of lipid

Technique	Fixative	Triglycerides	Phospho-glycerides	Phosphatides	Sulphatides
Plasmal reaction	Frozen formalin-fixed, frozen unfixed, cryostat fixed			Magenta	
Calcium lipase method	Cryostat post-fixed in formol calcium	Brown			
Gold hydroxamic acid	Cryostat post-fixed for 1 hour in formol calcium		Purple		
Acriflavine DMAB method	Post-fixed cryostat-formol calciumfixed frozen				Red
Toluidine blue	Post-fixed cryostat, formol calcium-fixed frozen				Metachromatic red, brown, yellow

PIGMENTS

14

Usually coloured pigments are evident without staining. But, staining improves recognition. Pigments occur in tissues either normal or pathological. They may be categorized into (1) Artefacts, (2) Endogenous including haematogenous and (3) Exogenous.

ARTEFACT PIGMENTS

When a tissue is put in a fixative, a chemical reaction takes place. Artefactual pigments are formaldehyde or mercury interaction. Formalin pigments are brown and black crystalline granules in and around blood, they are haematin derivatives. Such crystals develop at an acid pH but above 6.0, crystals do not develop. Such pigments can be identified by their distribution outside the cells. In the acid pH, formalin reacts with haemoglobin to form acid haemoglobin formaldehyde haematin. This is frequently seen in spleen. Formalin pigments are closely associated with haemorrhages and can be removed with saturated picric acid in two hours.

Mercury pigments In tissues fixed in solutions containing mercuric chloride, blackish brown pigment granules soluble in iodine develop.

Dichromate pigments In solutions containing potassium dichromate, a yellow pigment develops which could be removed on treatment with acid alcohol.

Malarial pigment Red blood cells display brownish black deposits of granules due to the presence of malarial parasite. This pigment like formalin pigment is easily removed with saturated alcoholic picric acid.

ENDOGENOUS PIGMENTS

Haemoglobin and haemosiderin   These arise in the organism from non-pigmented materials mostly derived from coloured constituent matter, the blood.

1. *Haemoglobin*   Haemoglobin contains iron. Formaldehyde-fixed tissues stain red with eosin. Haemoglobin occurs as droplets or granules (pathologically) which are yellow or yellow-brown in colour. It consists of two parts "Haem" (to which oxygen and iron are attached) and "Globin" (a colourless protein). Iron-containing haemoglobin can be broken up into iron-containing pigment, haematoidin. A number of histochemical techniques are available by which haemoglobin could be demonstrated—Benzidine or leucopatent blue, Dunn–Thompson method, modified Van Gieson method. Benzidine is carcinogenic so it is not preferred. The peroxidase technique or patent blue gives good results.

2. *Haemosiderin*   Haemosiderin is a breakdown product of haemoglobin and has ferric iron and protein. It occurs in the form of yellow-brown granules or golden brown granules, usually stored in bone marrow. It accumulates due to breakdown of blood in excess or due to the destruction of red blood cells. Haemosiderin contains iron and it can be best demonstrated by Prussian blue reaction. Ferrous iron is found in traces. Turnbull's blue reaction is specific and gives best results. Haemosiderin is insoluble in alkalis but soluble in strong acid solutions. Fixatives containing acids but no formalin can remove haemosiderin.

Bile pigment (Haematoidin)   This occurs as yellowish green granules and is mostly associated with old haemorrhages especially in spleen and brain. Bile contains bilirubin and haematoidin. Both are non-flourescent. Bile pigment is a mixture of biliverdin and both conjugated and unconjugated bilirubin. Good techniques like Gmelin technique, Stein's method, Fouchet's method are available. Gmelin and Glenner gave convincing results. Van Gieson's method is now popular. A detailed review of the histochemistry of bile pigments was given by Pearse (1985).

    i.   **Lipofuchsin**   This is a wear-and-tear pigment or brown atrophy pigment and is often found as yellowish brown granules in the liver cells and in cardiac cells. The granules concentrate around nucleus. Lipofuchsins are breakdown points of lipids and lipoproteins and exhibit fluorescence and are positive to PAS and performic acid Schiff (Pearse, 1972). Quite a number of methods for the demonstration of lipofuchsins are known. Among them, the important ones are Sudan black-B and Ziehl-Neelsen method. Lipofuchsin is PAS-positive and reduces silver. Schmorl's, Sudan black B staining gives good results.

        A type of lipofuchsin, the Dubin Johnson pigment, accumulates as yellow globules in phagocytes and liver cells. It is "Ceroid" and is capable of fluorescing. It emits a green flame which slowly gets faded when exposed to ultraviolet rays. It is positive with Masson's Fontana, methane-amine silver and Schmorl's reaction. It is soluble in strong acids and Sudan black B and is PAS-positive. However, many doubts remain.

ii. **Melanin** Melanin is a blackish brown pigment associated with skin, hair, eyes or other parts of body. It is formed from tyrosine by the enzyme tyrosinase (dihydroxyphenyl alanine DOPA). It stains with fast dyes and some basic aniline dyes such as fuchsin. It originates in the melanocytes of the skin. An insoluble pigment, it does not emit light (no fluoroscence is noticed). It reduces silver, and thus could easily be demonstrated by Fontana's method. Several enzymes may be associated especially non-specific esterases and acid phosphatase. It is metachromatic with methyl green and gives positive Schiff reaction. Melanin reduces ferric ferricyanide to give a positive Schmorl's reaction.

iii. **Chromaffin cells** These are dark brown cells normally containing small granules. The granules are highly labile to chromates and react to form brown granules. They are found in adrenal medulla. When these cells are fixed in fluids containing chromates, they lack colouration. Formol saline and other fixatives also exhibit the same results. They remain uncoloured. Adrenaline and noradrenaline precusors are found in chromaffin cells. Chromaffin cells contain large amounts of catecholamine and can reduce silver and ferric ferricyanide when they are fixed in formalin. Chromaffin cells become bottle green in colour when fixed in a fluid containing chromates.

iv. **Argentaffin cells** These occur frequently in the alimentary tract particularly concentrated at the base of the intestinal glands of the lower intestine. Argentaffin cells are capable of reducing silver salts and ferric ferricyanide with the aid or intervention of any other reducing agent. This reducing reaction is called argentaffin reaction and the cells are known as argentaffin cells. A high concentration of 5-hydroxytryptamine has been recorded in the argentaffin cells and perhaps the high content enables the cells to reduce silver. In addition they are also capable of reducing diazonium salts and ferric ferricyanide in an alkaline medium. This may also be due to high amine content.

EXOGENOUS PIGMENTS

These are substances which may gain entry into the body. Some of the important exogenous pigments or substances are chromolipids, carbon, silica, asbestos, silver, tattoo pigment and iron ore pigment.

HISTOCHEMICAL METHODS FOR PIGMENTS

Dunn-Thompson Method for Haemoglobin (Dunn-Thomspon, 1945)

Fixation

Any fixative

Reagents Required

Acid fuchsin
Haematoxylin
Iron alum
Picric acid

Preparation of Reagents

Solution 1 (Haematoxylin)
    Haematoxylin   250 mg
    Distilled water   100.0 ml

Solution 2 (Van Gieson stain)
    1 per cent acid fuchsin   13.0 ml
    Saturated picric acid   87.0 ml

Solution 3 (Alum)
    Iron alum   4.0 g
    Distilled water   100.0 ml

Procedure

1. Hydrate slides to water.
2. Place in solution 1 (aqueous haematoxylin) for 15 minutes.
3. Wash in water.
4. Mordant in solution 3 (iron alum) for 1 minute.
5. Transfer to 0.25 per cent aqueous haematoxylin.
6. Rinse in water.
7. Transfer to solution 2 (Van Gieson) for 15 minutes.
8. Rinse in 95 per cent alcohol for 3 minutes.
9. Rinse in 100 per cent alcohol, clear in xylene and mount in DPX.

Result

Sites of haemoglobin are stained greenish black.

Remarks

Any fixative with an acid ingredient should be avoided.

Leucopatent Blue Method for Haemoglobin (Dunn-Thompson, 1946)

Fixation

Formalin–paraffin sections

Reagents Required

Patent blue V
Powdered zinc
Glacial acetic acid
Hydrogen peroxide
Mayer's haemalum (see Chapter 7)

Preparation of Reagents

Solution 1 (Stock solution)

Patent blue V	1.0 g
Distilled water	100.0 ml
Powdered zinc	2.0 g
Glacial acetic acid	2.0 ml

In a round-bottomed flask, mix all the ingredients. Then boil the solution gently for about 15 minutes0 till the solution becomes straw coloured. Allow it to cool and then filter and store it.

Solution 2 (Working solution)

Solution 1	30.0 ml
Glacial acetic acid	6.0 ml
3 per cent (10 V) H_2O_2	3.0 ml

Procedure

1. Deparaffinize and hydrate slides to water.
2. Stain in solution 2 (working solution) for 5 minutes.
3. Rinse in water.
4. Counterstain in Mayer's haemalum for 1 minute.
5. Rinse in water.
6. Dehydrate, clear and mount.

Result

Haemoglobin	Dark blue-green
Eosinophil and neutrophil granules	Dark blue
Nuclei	Red

Rationale

It is based upon the demonstration of haemoglobin peroxidase and has replaced similar peroxidase methods such as carcinogenic benzidine method since hematoxylin is a peroxidase method. Otherwise oxidase-containing elements will also be determined.

Benzidine Method (Ralph, 1941; Glick, 1949)

Fixation

Formalin or alcohol or frozen sections of unfixed tissue

Reagents Required

Benzidine
Methyl alcohol
Peroxidase
Ethyl alcohol

Preparation of Reagents

Solution 1 (Benzidine reagent)
Benzidine	1.5 g
Absolute methyl alcohol	100.0 ml

Solution 2
3 per cent peroxidase	25.0 ml
70 per cent ethyl alcohol	75.0 ml

Procedure

1. Deparaffinize and hydrate slides to water.
2. Flood slides with solution 1 (benzidine) for 2 minutes.
3. Drain off excess and flood with solution 2 for 1–5 minutes.
4. Rinse in distilled water for 15 seconds.
5. Counterstain if necessary with any nuclear stain.
6. Dehydrate, clear and mount.

Results

Sites of haemoglobin are stained dark brown.

Rationale

Treatment with benzidine and then with peroxidase, activates the peroxidase system to produce oxidase in a rapid way to convert benzidine to an unstable blue compound. This blue compound undergoes auto oxidation to form a stable brown compound. Ericsson (1965) added 4.0 ml of benzidine to 60.0 ml of peroxidase, giving a staining time of 1 minute, whereas Putt (1951) combined thionine and benzidine which gives an olive green instead of brown colour to haemoglobin. Puchtler and Sweat (1962) used amido black for staining haemoglobin.

Cyanol Reaction (Dunn, 1946)

Fixation

Fixation in formalin buffered to pH 7 for 48 hours

Reagents Required

Cyanol
Zinc powder
Glacial acetic acid

Preparation of Reagents

Solution 1 (Cyanol reagent) (Stock solution)
Cyanol	1.0 g
Distilled water	100.0 ml
Zinc powder	10.0 g
Glacial acetic acid	2.0 ml

After adding all the ingredients, boil the mixture. After boiling, the solution becomes colourless. It lasts for several weeks.

Procedure

1. Deparaffinize and hydrate slides to water.
2. Stain the sections with solution 1 (cyanol reagent) for 3–5 minutes.
3. Rinse in distilled water.
4. Counterstain (if necessary) with any nuclear stain.
5. Dehydrate, clear and mount.

Result

Sites of haemoglobin are stained dark blue to bluish grey.

Puchtler's Method for Haemoglobin (Puchtler *et al.*, 1964)

Fixation

Zenker's solution

Reagents Required

Tannic acid
Phosphomolybdic acid
Buffalo black
Glacial acetic acid
Methyl alcohol

Preparation of Reagents

Solution 1 (5 per cent tannic acid solution)
Tannic acid   5.0 g
Distilled water   100.0 ml
This should be prepared 48 hours ahead.

Solution 2 (1 per cent phosphomolybdic acid)
Phosphomolybdic acid   1.0 g
Distilled water   100.0 ml

Solution 3 (Buffalo black NBR)
Buffalo black NBR   3.15 g
Glacial acetic acid   10.0 ml
Methyl alcohol   90.0 ml

Solution 4 (Methanol glacial acetic acid solution)
Glacial acetic acid   10.0 ml
Methyl alcohol   90.0 ml

Procedure

1. Deparaffinize and hydrate slides to water.
2. Transfer to solution 1 for 15 minutes.
3. Rinse in distilled water.
4. Place in solution 2 for 10 minutes.
5. Rinse in solution 4 for 5 minutes.

6. Transfer to solution 3 for 5 minutes.
7. Again rinse in solution 4.
8. Dehydrate, clear and mount.

Result

Erythrocytes	Dark blue
Haemoglobin	Dark blue
Other structures	Yellow

Ralph's Method for Haemoglobin (Ralph, 1941)

Fixation

Absolute alcohol or Carnoy in 10 per cent neutral buffered formalin.

Reagents Required

Benzidine
Methyl alcohol
Hydrogen peroxide
Nuclear fast red

Preparation of Reagents

Solution 1 (Benzidine solution)
    Benzidine   1.0 g
    Methyl alcohol   99.0 ml

Solution 2 (Peroxide solution)
    Hydrogen peroxide   25.0 ml
    70 per cent alcohol   75.0 ml

Procedure

1. Dewax and hydrate slides to water.
2. Flood slides with solution 1 for 1 minute.
3. Drain off excess stain.
4. Flood slides with solution 2 for 2 minutes.
5. Rinse in distilled water.
6. Place slides in solution 3 for 5 minutes.
7. Rinse in distilled water.
8. Dehydrate, clear and mount.

Result

Haemoglobin	Dark brown
Nuclei	Pink

Foetal Haemoglobin (Modified from Kiossoglou *et al.*, 1963)

Fixation

Air-dry blood smears for 1 hour.
Fix in absolute methanol for 5 minutes followed by 80 per cent ethanol for 5 minutes.

Reagents Required

Citrate buffer
Eosin Y
Ponceau S
Acetic acid

Preparation of Reagents

Solution 1 (Buffer)
McIlvaine buffer–Citric acid phosphate (pH 3.2–3.4) (*see* Chapter 5)

Solution 2 (Eosin)
Eosin Y   500 mg
Distilled water   100.0 ml

Solution 3 (Ponceau S)
Ponceau S   500 mg
1 per cent aqueous acetic acid   100.0 ml

Procedure

1. Rinse fixed slides in water (distilled).
2. Transfer to solution 1 for 5 minutes at 37°C.
3. Dip in water.
4. Transfer to either solution 2 or solution 3 for 3 minutes.
5. Rinse in water and dry.

Results

Sites of foetal haemoglobin   Deep pink to red
Adult cells   Colourless

Puchtler and Sweat Method for Haemoglobin and Haemosiderin (Puchtler and Sweat, 1963)

Fixation

Zenker formol

Reagents Required

Potassium ferrocyanide
Hydrochloric acid
Tannic acid

Phosphomolybdic acid
Phloxine B
Methanol
Glacial acetic acid

Preparation of Reagents

Solution 1 (2 per cent potassium ferrocyanide)
    Potassium ferrocyanide   2.0 g
    Distilled water   100.0 ml

Solution 2 (2 per cent hydrochloric acid (stock)
    Concentrated hydrochloric acid   2.0 ml
    Distilled water   100.0 ml

Solution 3 (Potassium ferrocyanide–Hydrochloric acid working solution)
    Solution 1   50.0 ml
    Solution 2   50.0 ml

Solution 4 (Tannic acid solution 5 g + 100 ml water)

Solution 5 (1 per cent phosphomolybdic acid) (1 g/100 ml water)

Solution 6 (Phloxine B solution)
    Phloxine B   5.0 g
    Glacial acetic acid   50.0 ml
    Methyl alcohol   90.0 ml

Solution 7 (Methanol glacial acetic acid solution)
    Methanol   95 ml
    Glacial acetic acid   5 ml

Procedure

1. Dewax and hydrate slides to water.
2. Place in solution 3 for 30 minutes.
3. Rinse in distilled water.
4. Transfer to solution 4 for 10 minutes.
5. Rinse in distilled water.
6. Place in solution 5 for 10 minutes.
7. Rinse in distilled water.
8. Transfer to solution 6 for 5 minutes.
9. Differentiate in solution 7.
10. Dehydrate, clear and mount.

Result

Haemoglobin   Red
Haemosiderin   Blue-green or dark blue

Gmelin Method for Bile Pigments

Fixation

All types of fixatives especially formalin

Reagents Required

Concentrated nitric acid
Absolute alcohol

Preparation of Reagents

Solution 1 (Nitric acid alcohol)
Concentrated nitric acid   5.0 ml
Absolute alcohol   5.0 ml

Procedure

1. Deparaffinize and hydrate slides to water.
2. Flood the slide with solution 1.
3. Drain off excess stain with filter paper.
4. Place the coverslip and ring the coverslip with paraffin.

Result

Bile pigment is stained reddish green.

Rationale

Tissues fixed in formalin can be used. Concentrated nitric acid damages tissue. Oxidation is completed in 3 stages producing 3 different colours—red, purple and green. Nitric acid is replaced by nitrous acid which on oxidation also exhibits 3 colours (green, purple and red). Both preparations are temporary.

Bile Pigment Staining (Bilirubin) (Glenner, 1957)

Fixation

Fresh frozen sections (Adamstone and Taylor, 1948) or formalin-fixed (fixation time 6 hours). For sectioning, use carbowax.

Reagents Required

Potassium dichromate
Potassium dihydrogen phosphate
Haematoxylin

Preparation of Reagents

Solution 1 (Potassium dichromate solution)
3 per cent potassium dichromate   25.0 ml
(3 g/100 ml water)
Buffer pH 2.2 (0.1N HCl)   8.0 ml
0.1N Potassium dihydrogen phosphate   17.0 ml

Procedure

1. Place sections in solution 1 for 15 minutes at room temperature.
2. Wash in running water for 5 minutes.

3. Counterstain in haematoxylin.
4. Dehydrate rapidly.
5. Clear and mount.

Result

Bilirubin is stained emerald green.

Glenner's Method for Bilirubin (Haemosiderin and Lipofuchsin, 1957)

Fixation

Frozen sections or cryostat sections

Reagents Required

Potassium dichromate
Acetic acid
Potassium ferrocyanide
Oil red O

Preparation of Reagents

Solution 1 (Potassium dichromate)
 Potassium dichromate 2.0 g
 Distilled water 100.0 ml

Solution 2 (Potassium ferrocyanide)
 Potassium ferrocyanide 2.0 g
 Distilled water 100.0 ml

Procedure

1. Cut frozen sections.
2. Immerse sections in solution 1 for 5 minutes.
3. Transfer sections to a solution containing equal parts of 5 per cent acetic acid and freshly prepared solution 2 for 20 minutes.
4. Rinse in running water and treat with buffered dichromate solution for 15 minutes.
5. Rinse in water.
6. Transfer sections to oil red O solution (saturated solution, i.e., 0.25–0.5 per cent in isopropyl alcohol) for 20 minutes.
7. Rinse in 70 per cent alcohol and remove excess stain.
8. Wash in running water.
9. Mount in Apathy's medium.

Result

Bilirubin Green
Haemosiderin Blue
Lipofuchsin Red

Stein's Technique for Bilirubin

Procedure

1. Hydrate slides to water.
2. Transfer sections to a mixture of 3 parts of Lugol's iodine and 1 part of tincture of iodine for 6–12 hours.
3. Decolorize with 5 per cent aqueous solution of sodium sulphate for 15–20 seconds.
4. Counterstain nuclei if desired in alum carmine for 1–3 hours.
5. Wash well.
6. Dehydrate in acetone, clear in xylene and mount in Canada balsam.

Result

Bilirubin   Green
Nuclei   Red

Hall's Method for Bilirubin (Hall, 1960)

Fixation

10 per cent buffered neutral formalin

Reagents Required

Trichloroacetic acid
Ferric chloride
Picric acid
Acid fuchsin

Preparation of Reagents

Solution 1 (Fouchet's solution)
    Trichloroacetic acid   25.0 g
    Distilled water   100.0 ml

To this solution, add 10 ml of 10 per cent ferric chloride.

    Ferric chloride solution
        Ferric chloride   10.0 g
        Distilled water   100.0 ml

Solution 2 (Van Gieson's solution) (*see* Chapter 19)

Procedure

1. Deparaffinize and hydrate slides to water.
2. Transfer to solution 1.
3. Wash in running water.
4. Place in solution 2 for 5 minutes.
5. Dehydrate, clear and mount.

Results

Bilirubin	Green
Collagen	Red
Muscle	Yellow

Ferric Iron Method for Bilirubin (Kutlik, 1957)

Fixation

Formalin, Carnoy

Reagents Required

Iron alum
Ferric chloride
Acetic acid
Brilliant yellow

Preparation of Reagents

Solution 1 (Alum)

Iron alum	5.0 g
Distilled water	100.0 ml

Solution 2 (Ferric chloride solution)

Ferric chloride	5.0 g
2 per cent acetic acid	100.0 ml

Procedure

1. Deparaffinize and hydrate slides to water.
2. Transfer slides to solution 1 for 10 minutes or solution 2.
3. Wash lightly in distilled water.
4. Rinse in 90 per cent alcohol.
5. Counterstain in brilliant yellow in 90 per cent alcohol.
6. Dehydrate, clear and mount.

Result

Bilirubin	Green
Background	Yellow

Azo-coupling Method

Fixation

10 per cent buffered formalin

Reagents Required

Sodium acetate
Sodium barbital
Conc. HCl
Fast red B

Preparation of Reagents

Solution 1 (Buffer solution)
Sodium acetate   1.17 g
Sodium barbital   2.94 g
Distilled water   100.0 ml

Solution 2
Conc. HCl   0.84 ml
Distilled water   100.0 ml

Solution 3 (Working solution) (pH 9.2)
Solution 1   5.0 ml
Solution 2   0.25 ml
Distilled water   17.75 ml

Solution 4 (Staining solution)
Fast red B   50 mg
Buffer (pH 9.2)   50.0 ml
(Solution 3)

Procedure

1. Dewax and hydrate sections to water.
2. Transfer to solution 4 for 30 seconds
3. Drain excess stain.
4. Counterstain in haematoxylin for 5 minutes.
5. Wash.
6. Dehydrate, clear and mount.

Results

Bile pigments   Orange-red
Nuclei   Blue

Trichrome Method for Bilirubin, Haemosiderin and Lipofuchsin (Glenner, 1957)

Fixation

Fresh sections

Reagents Required

Potassium ferricyanide
Acetic acid
Potassium dichromate
Oil red O

Preparation of Reagents

Solution 1 (Potassium ferricyanide)
Potassium ferricyanide   2.0 g
Distilled water   100.0 ml

Solution 2 (Acetic ferricyanide)
    Solution 1   50.0 ml
    5 per cent acetic acid   50.0 ml

Solution 3 (Dichromate solution)
    Potassium dichromate   3.0 g
    Distilled water   100.0 ml

Solution 4 (Oil red O)
    Freshly filtered oil red O in isopropanol

Procedure

1. Cut 10 μ sections and mount on slides.
2. Transfer to solution 1 for 5 minutes.
3. Place the slides in solution 2 for 20 minutes.
4. Rinse in running water.
5. Transfer to solution 3 for 15 minutes.
6. Directly transfer to solution 4 for 20 minutes.
7. Wash in distilled water.
8. Mount in Apathy's medium.

Result

Lipofuchsin	Dark orange-red
Haemosiderin	Blue
Bilirubin	Green

Iodine Method for Bile Pigments (Leibnitz, 1964)

Fixation

96 per cent ethanol

Procedure

1. Bring sections to methanol.
2. Place them in 1 per cent iodine in methanol.
3. Wash in distilled water.
4. Counterstain nuclei if desired in haematoxylin.
5. Dehydrate, clear and mount in synthetic resin.

Result

Bilirubin is stained green.

Long Ziehl-Neelsen Method for Lipofuchsin

Fixation

Any general fixative

Reagents Required

Basic fuchsin
Phenol
Absolute alcohol
Haematoxylin

Preparation of Reagents

Solution 1 (Carbol fuchsin)

Basic fuchsin	1.0 g
Phenol	500 mg
Absolute alcohol	10.0 ml
Distilled water	100.0 ml

Procedure

1. Deparaffinize and hydrate slides to water.
2. Immerse sections in solution 1 for 3 hours at 60°C.
3. Wash in tap water.
4. Differentiate in acid alcohol.
5. Wash in tap water.
6. Counterstain in haematoxylin.
7. Dehydrate, clear and mount.

Result

Lipofuchsin	Red
Nuclei	Blue

Rationale

Since lipofuchsin is lipid in nature, it exhibits the property of acid fastness due to tubercle bacilli. With increasing staining period, the colour becomes brighter.

Sudan black B and aldehyde fuchsin techniques have also been considered to be revealing lipofuchins (See Stevans and Chalk, 1996).

Schmorl's Method for Melanin, Lipofuchsin and Argentaffin

Fixation

Any fixative

Reagents Required

Ferric chloride
Potassium ferricyanide
Safranin
Acetic acid

Preparation of Reagents

Solution 1 (Ferric chloride solution)
Ferric chloride   50 mg
Distilled water   50.0 ml

Solution 2 (Potassium ferricyanide solution)
Potassium ferricyanide   500 mg
Distilled water   50.0 ml

Solution 3 (Staining solution)
Solution 1   37.5 ml
Solution 2   5.0 ml
Distilled water   7.5 ml

Procedure

1. Hydrate slides to water.
2. Immerse in solution 3 for 30 seconds and watch staining intensity. If the colour is intense, take out the slide.
3. Differentiate in 1 per cent acetic acid.
4. Wash in tap water.
5. Dehydrate, clear and mount.

Result

Melanin   Deep blue
Argentaffin granules   Deep blue
Lipofuchsin   Deep blue

Rationale

Ferric chloride and potassium ferricyanide reduce ferric ferricyanide to ferrocyanide when a blue precipitate develops. Staining time should be controlled. Melanin and lipofuchsin take the stain earlier than argentaffin.

Nile Blue Method for Melanin and Lipofuchsin (Lillie, 1956a, 1956b)

Fixation

Any general fixative

Reagents Required

Nile blue
Sulphuric acid

Preparation of Reagents

Solution 1 (Nile blue sulphate)
Nile blue sulphate   50 mg
Sulphuric acid 1 per cent   100.0 ml
(1 ml/100 cc water)

Procedure

1. Deparaffinize and hydrate slides to water.
2. Immerse in staining solution 1 (Nile blue) for 20 minutes.
3. Wash in running water for 20 minutes.
4. Mount in glycerine jelly.

Result

Lipofuchsin	Dark blue or blue-green
Melanin	Pale green
Erythrocytes	Greenish yellow
Myelin	Deep blue

Rationale

Nile blue sulphate stains free fatty acid, phospholipids and some lipids. Lipofuchsin stains with Nile blue at pH below 1.0 and an acid base operating above pH 3.0 when stained at acid base, lipofuchsins take a green stain but when they stain below pH 1.0 they get decolorized. But melanins stain with basic dyes at lower pH levels.

Indophenol Method (Alpert *et al.*, 1960)

Fixation

Fresh frozen sections or general fixation

Reagents Required

Sodium 2,6-dichlorophenolindophenol
Ethyl alcohol
Hydrochloric acid

Preparation of Reagents

Solution 1 (Staining solution)

Sodium 2,6-dichlorophenolindophenol	100 mg
50 per cent ethyl alcohol	100.0 ml

Prepare fresh, filter and then add hydrochloric acid (1 ml/99 ml water) until the solution becomes red.

Procedure

1. Deparaffinize and hydrate slides to water.
2. Immerse in staining solution 1 for 5 minutes.
3. Rinse in tap water giving several changes.
4. Mount in glycerine jelly.

Results

Lipofuchsins	Red
Erythrocytes	Blue

Chrome Alum Fuchsin Method for Lipofuchsins (Gomori, 1941a)

Fixation

Formalin

Reagents Required

Haematoxylin
Chrome alum
Potassium dichromate
Sulphuric acid
Potassium permanganate
Oxalic acid

Preparation of Reagents

Solution 1 (Potassium permanganate)
Potassium permanganate   500 mg
Distilled water   100.0 ml

Solution 2 (Oxalic acid solution)
Oxalic acid   1.0 g
Distilled water   100.0 ml

Solution 3 (Haematoxylin solution)
Haematoxylin   1.0 g
Distilled water   100.0 ml

Solution 4 (Alum)
Chrome alum   3.0 g
Distilled water   100.0 ml

Solution 5 (Dichromate solution)
Potassium dichromate   5.0 g
Distilled water   100.0 ml

Solution 6 (Working solution)
Solution 3   50.0 ml
Solution 4   50.0 ml
Solution 5   2.0 ml
5 per cent sulphuric acid   1.0 ml

Procedure

1. Deparaffinize and hydrate slides to water.
2. Oxidize sections in solution 1 for 5 minutes.
3. Wash.
4. Bleach in solution 2 for 2 minutes.
5. Wash.
6. Transfer slides to solution 6 for 10 minutes.
7. Differentiate in 1 per cent acid alcohol.
8. Counterstain in 1 per cent aqueous eosin.

9. Wash in running water.
10. Dehydrate, clear and mount.

Result

Lipofuchsin   Blue-black
Nuclei   Purple

Ferric Ferricyanide Method (Lillie, 1957c)

Fixation

All fixatives except liquids with chromate

Reagents Required

Ferrous sulphate
Potassium ferricyanide
Glacial acetic acid
Picro Ponceau

Preparation of Reagents

Solution 1 (Ferrous sulphate)
    Ferrous sulphate   2.5 g
    Distilled water   100.0 ml

Solution 2 (Ferricyanide solution)
    Potassium ferricyanide   1.0 g
    Distilled water   99.0 ml
    Glacial acetic acid   1.0 ml

Procedure

1. Hydrate slides to water.
2. Immerse in solution 1 for 1 hour.
3. Wash in distilled water giving 4 changes of 20 minutes each.
4. Place in solution 2 for 30 minutes.
5. Wash in 1 per cent acetic acid for 2 minutes.
6. Counterstain in picro Ponceau.
7. Wash, dehydrate, clear and mount.

Results

Melanin   Dark green
Collagen   Red
Cytoplasm   Yellow

Masson's Fontana Method for Melanin and Argentaffin Cells

Fixation

Formalin

Reagents Required

Silver nitrate
Ammonia

Preparation of Reagents

In an amber-coloured bottle take 20 ml of silver nitrate. To this, add concentrated ammonia drop by drop. A precipitate is formed. This precipitate dissolves after addition of few more drops. Now the solution is clear. Add 20 ml of distilled water. Then filter and store in a dark place. Filter every time before use.

Procedure

1.   Deparaffinize and hydrate slides to water.
2.   Place the slides in silver nitrate solution. Keep the coplin jar in a dark chamber.
3.   Wash in tap water.
4.   Counterstain if desired.
5.   Wash, dehydrate, clear and mount.

Results

Melanin stains black.

Rationale

Silver is reduced directly by the substance. Argentaffin cells are stained by a deposit of silver. The silver deposit is fixed by sodium thiosulphate (hypo) for a short time.

Method to Distinguish Melanins from Lipofuchsins (Hueck, 1912)

Fixation

Any general fixative

Reagents Required

Nile blue sulphate
Hydrogen peroxide

Preparation of Reagents

Solution 1 (Nile blue sulphate solution)
Saturated solution of Nile blue sulphate

Solution 2 (Hydrogen peroxide)
3 percent hydrogen peroxide or 10 per cent hydrogen peroxide

Procedure

1.   Deparaffinize and hydrate slides to water.
2.   Transfer the slides to freshly prepared solution 1 for 1½ hours.
3.   Rinse in distilled water giving 3 changes of 1 minute each.
4.   Transfer to solution 2 for 24 hours (10 per cent solution 2 is preferred).
5.   Rinse rapidly in distilled water.
6.   Mount in glycerine jelly.

Result

Melanin	Colourless
Lipofuchsins	Blue

Masson's Fontana Method for Melanin (with hexamine-silver variant)

Fixation

Any general fixative without chromate

Reagents Required

Silver nitrate
Ammonia
Hexamine
Phenolphthalein
Sodium hydroxide
Borax buffer

Preparation of Reagents

Solution 1 (Silver nitrate solution)

Silver nitrate	10.0 g
Distilled water	100.0 ml

Take 20 ml of this and add strong ammonia drop by drop till few granules of precipitate remain. Add 20 ml of distilled water. Keep it undisturbed for 24 hours. Decant.

Solution 2 (Hexamine solution)

Hexamine	3.0 g
Distilled water	100.0 ml

Solution 3 (Silver nitrate solution)

Silver nitrate	5.0 g
Distilled water	100.0 ml

Solution 4 (Working solution)

Solution 2	100.0 ml
Solution 3	5.0 ml

To this add 5 ml of borate buffer (pH 8.0)

Procedure

1. Hydrate slides to water.
2. Place in Gram's iodine for 10 minutes.
3. Wash in running water.
4. Transfer to either solution 1 or 4 for 24 hours and leave the jar in a dark chamber.
5. Rinse in distilled water.
6. Counterstain if desired in carbol safranin for 10 minutes.
7. Differentiate in 70 per cent alcohol.
8. Dehydrate, clear and mount.

Result

Melanin granules   Black
Nuclei   Red
Keratin   Orange

Diamine Silver Method for Melanin (Lillie, 1965)

Fixation

Formalin

Reagents Required

Ammonia
Silver nitrate

Preparation of Reagents

Solution 1 (Ammoniacal silver nitrate)
    28 per cent ammonia   2.0 ml
    5 per cent silver nitrate   35.0 ml

Shake well till last few granules of precipitate remain.

Procedure

1.   Hydrate slides to water.
2.   Transfer slides to solution 1 for 2 hours at 22°C in a dark chamber.
3.   Wash.
4.   Transfer to 0.2 per cent gold chloride for 5 minutes.
5.   Wash in distilled water.
6.   Mount in Apathy's medium.

Result

Melanin is stained black.

Ferrous Iron Method for Melanin (Lillie, 1957c)

Fixation

Formalin, Carnoy, chromate containing fixatives are not recommended

Reagents Required

Ferrous sulphate
Potassium ferricyanide
Acetic acid
Picric acid

Preparation of Reagents

Solution 1 (Ferrous sulphate solution)
    Ferrous sulphate   2.5 g
    Distilled water   100.0 ml

Solution 2 (Potassium ferricyanide–acetic acid)
    Potassium ferricyanide   1.0 g
    1 per cent acetic acid   100.0 ml

Procedure

1. Hydrate slides to water.
2. Transfer to solution 1 for 1 hour.
3. Wash in distilled water giving several changes.
4. Transfer to solution 2 for 30 minutes.
5. Wash in 1 per cent acetic acid.
6. Counterstain if necessary in Van Gieson.
7. Dehydrate, clear and mount.

Result

Melanin is stained dark green.

Chromaffin Reaction

Fixation

Regaud's or Orth's fluid

Procedure

1. Rinse tissues in distilled water.
2. Cut fresh frozen sections.
3. Stain in silver solution.

Results

Chromaffin cells are stained yellow-brown.

Modified Giemsa's for Chromaffin Cell Granules

Fixation

Any fixative containing chromates

Reagents Required

Giemsa stain
Acetic acid

Preparation of Reagents

Solution 1 (Giemsa stain)
    Giemsa's stain   2 ml
    Distilled water adjusted to
       pH 6.8 with 0.5 acetic acid   48.0 ml

Procedure

1. Deparaffinize and hydrate slides to water.
2. Rinse in distilled water.

3. Place in dilute solution 1 overnight.
4. Rinse in distilled water.
5. Wash in 0.5 per cent acetic acid.
6. Wash, dehydrate, clear and mount.

Result

Chromaffin granules	Greenish yellow
Nuclei	Blue

Alkaline Diazo Method for Argentaffin Cell Granules

Fixation

Formalin

Reagents Required

Diazonium salt Fast Red B
0.1 M Veronol acetate buffer (pH 9.2)
Fast red salt B
Haematoxylin

Preparation of Reagents

Solution 1 (Staining solution)
0.5 M veronol acetate buffer (pH 9.2) containing 1 mg or 1 ml diazotate of 5-nitroanisidine (Fast red salt B)

Solution 2 Mayer's haemalum (*see* Chapter 7)

Procedure

1. Deparaffinize and hydrate slides to water.
2. Immerse slides in staining solution 1 for 30 seconds at 4°C.
3. Wash.
4. Stain nuclei in solution 2 for 5 minutes.
5. Wash.
6. Dehydrate, clear and mount.

Result

Argentaffin	Orange-red
Nuclei	Blue

Rationale

At pH 9.2, Fast Red salt B added to veronol acetate engenders coupling between phenolic components of argentaffin granules and Fast Red salt B. The resulting compound is orange-red.

Noradrenaline Fluorescence Technique

Fixation

Formalin—frozen or freeze-dried

Procedure

1. Fix in formol saline for 24 hours.
2. Cut frozen sections of 10 µ thickness.
3. Mount in glycerine jelly.
4. Examine under fluorescence microscope.

For freeze dried sections

1. Freeze-dry.
2. Fix in formalin vapour.
3. Embed in paraffin wax.
4. Cut 8-µ thick sections.
5. Mount in light petroleum.
6. Examine under fluorescence microscope.

Result

Strong fluorescence (Yellowish green)

Chromaffin Reaction for Noradrenaline and Adrenaline (Hillarp and Hokfelt, 1955)

Fixation

Fresh sections

Reagents Required

Potassium dichromate
Potassium chromate

Preparation of Reagents

Solution 1 (Potassium dichromate solution)
Potassium dichromate   5.0 g
Distilled water   100.0 ml

Solution 2 (Potassium chromate solution)
Potassium chromate   5.0 g
Distilled water   100.0 ml

Solution 3 (Working solution)
Solution 1   10 vol. or 100.0 ml
Solution 2   1 vol. or 10.0 ml

Procedure

1. Place fresh slices in solution 3 for 16 hours at room temperature.
2. Wash in 3 changes of distilled water.
3. Mount in glycerine jelly.
4. Dehydrate, clear and mount.

Result

Adrenaline   Dark brown
Noradrenaline   Yellow-brown

Iodate Method for Noradrenaline (Hillarp and Hokfelt, 1955)

Fixation

Fresh sections

Reagents Required

Potassium iodate
Formalin

Preparation of Reagents

Solution 1 (Potassium iodate solution)
   Potassium iodate   10.0 g
   Distilled water   100.0 ml

Procedure

1. Keep fresh slices of tissue in solution 1 for 16 hours at room temperature.
2. Place them in 10 per cent formalin for 2 hours.
3. Cut sections (20 μ thick).
4. Wash and counterstain nuclei if desired.
5. Mount sections on slides, dehydrate, clear and mount.

Result

Noradrenaline is stained brown.

REMOVAL OF PIGMENTS

Pigments can be removed by the following methods.

1. Permanganate Method (Lillie, 1965)

1. Hydrate slides to water.
2. Transfer slides to 0.1 per cent potassium permanganate for 12–24 hours.
3. Wash in running water.
4. Transfer slides to 1 per cent oxalic acid for 1 minute.
5. Wash and proceed to stain.

2. Performic or Peracetic Acid Methods

Bleaching is done by immersing slides in either performic acid or peracetic acid for 1–2 hours.

Performic acid   Add 8 ml of 90 per cent formic to 31 ml of 30 per cent H_2O_2 and 0.22 ml of concentrated sulphuric acid, store it below temperature of 25°C. 4.7 per cent of performic acid is formed which will last only for few hours.

Peracetic acid   Add 95.6 ml of acetic acid to 259 ml of 30 per cent H_2O_2 and add 2.2 ml of concentrated H_2SO_4. Allow it to stand for 3 days. Add 40 mg of disodium phosphate and store at 1°C. It stands for several months.

3. Chlorate Method

Immerse sections in a mixture containing 50 per cent alcohol to which little of potassium chlorate and a few drops of HCl are added. Before staining, wash.

4. Bromine Method

Immerse slides in 1 per cent bromine water for 24 hours and wash.

5. Chromic Acid Method

Immerse slides in a mixture containing 1 per cent chromic acid and 5 per cent calcium chloride for 8–12 hours.

6. Peroxide Method

Immerse slides for 24–48 hours in 10 per cent H_2O_2 and wash well before staining. This is a specific method for melanin.

FORMALIN PIGMENT

Brown or black pigment in the form of crystalline granules are formed by formalin and they are considered as haematein derivatives.

The methods by which formalin pigment can be removed are given below.

Baker Method

One per cent potassium hydroxide is added to 80 per cent alcohol or picric acid until precipitate is removed.

Murdock Method (1945)

Place sections in a mixture of

3 per cent H_2O_2	25.0 ml
Acetone	25.0 ml
Ammonia	1 drop

Add ammonia drop by drop till precipitate dissolves.

Barrett Method (1944)

Place section in a saturated solution of picric acid in alcohol for 10 minutes to 2 hours.

Pearse Mehod (1960)

Immerse slides in 90 per cent formic acid.

MALARIAL PIGMENT

This is also in the form of brownish black pigments.

Gridley Method (1975)

1. Hydrate slides to water
2. Bleach for 5 minutes in a mixture of

Acetone	50.0 ml
3 per cent H_2O_2	50.0 ml
28–29 per cent ammonia	1.0 ml

  or overnight in 5 per cent ammonium sulphide. Wash well before staining.

HAEMOSIDERIN

This is in the form of yellowish brown or greenish brown pigment. Does not dissolve in acids and alkalis. It can be identified by Perl's test.

BILE PIGMENTS

This is in the form of yellowish green pigment. Bleaching is not effective and they are argentaffin-positive. It can be converted to biliverdin with H_2O_2, Lugol solution and nitrous acid.

Table 14.1   Histochemical techniques applied for the demonstration of pigments

Technique	Fixative	Haemoglobin	Bile pigments	Bilirubin	Haemosiderin	Lipofuchsin
Dunn–Thompson method	Any general fixative	Greenish black				
Leucopatent blue method	Formalin-paraffin sections	Dark blue-green				
Benzidine method	Formalin or alcohol frozen sections	Dark brown				
Cyanol reaction	Buffered formalin (pH 7)	Dark blue to bluish grey				
Puchtler's method	Zenker's solution	Dark blue				
Ralph's method	Absolute alcohol, Carnoy 10% neutral buffered formalin	Dark brown				
Foetal haemoglobin	Fix in absolute methanol for 5 minutes, followed by 80% ethanol for 5 minutes	Deep pink to red				
Puchtler and Sweat method	Zenker formol	Red			Blue green or dark blue	
Gmelin method	Any general fixative		Reddish green			

(Contd.)

Table 14.1 (Continued)

Technique	Fixative	Haemoglobin	Bile pigments	Bilirubin	Haemosiderin	Lipofuchsin
Bile pigment staining	Fresh frozen or formalin-fixed			Emerald green		
Glenner's method	Frozen sections or cryostat sections			Green	Blue	Red
Stein's technique	96% ethanol paraffin section			Green		
Hall's method	10% buffered neutral formalin			Green		
Ferric iron method	Formalin, Carnoy			Green		
Azo-coupling method	10% buffered formalin		Orange-red			
Trichrome method	Fresh sections			Green	Blue	Dark orange-red
Iodine method	96% ethyl alcohol			Green		
Long-Ziehl ethyl alcohol-Neelsen method	Any general fixative					Red

Table 14.2 Histochemical techniques applied for the demonstration of pigments

Technique	Fixative	Lipofuchsin	Argentaffin	Melanin	Chromaffin	Erythrocytes	Noradrenaline
Schmorl's reaction	Any general fixative	Deep blue	Deep blue	Deep blue			
Nile blue method	Any general fixative	Dark blue or blue-green		Pale green		Greenish yellow	
Indophenol method	Fresh frozen or general fixative	Red				Blue	
Chromealum Fuchsin method	Formalin	Blue-black					
Ferric ferricyanide method	All fixatives			Dark green			
Masson's Fontanas method	Formalin			Black			
Distinguishing melanins from lipofuchsins	Any general fixative	Blue		Colour-less			

Table 14.2 (Continued)

Technique	Fixative	Lipofuchsin	Argentaffin	Melanin	Chromaffin	Erythrocytes	Noradrenaline
Masson's Fontana method formelanin	Any general fixative without chromate			Black			
Diamine silver method	Formalin			Black			
Ferrous iron method	Formalin, Carnoy			Dark green			
Chromaffin reaction	Regaud's or Orth's fluid				Yellow-brown		
Modified Geimsa stain	Any fixative containing chromates				Greenish yellow		
Alkaline diazo method	Formalin		Orange-red				
Fluorescene technique for noradrenaline	Freeze-dried or formalin—frozen						Strong yellowish green fluorescence
Chromaffin technique for adrenaline and noradrenaline	Fresh sections						Yellow-brown
Iodate method for noradrenaline	Fresh						Brown

MINERALS

15

INTRODUCTION

Calcium, iron, sodium and potassium occur in tissues of both vertebrates and invertebrates. Iron and calcium are the commonly occurring minerals whereas sodium and potassium occur in high quantities in tissues. Other minerals like lead, aluminium, barium, carbon, etc., occur in low concentrations. Histochemical demonstration of most of these minerals is possible.

Iron occurs independently or in combination with protein to form haemosiderin, a golden brown pigment or in combination with protein as ferritin. Iron can be separated from protein by using reducing agents such as hydrosulphite. Pigments with combination of iron and protein are haemoglobin, myoglobin, etc.

All more or less calcified tissues contain great concentrations of calcium. Disease due to calcium deficiency and due to presence of excessive calcium are known. Osteoporosis and osteomalacia are the two most common diseases due to calcium disorders. Sometimes excessive calcium is deposited in tissues. Deposition of calcium takes place first in an insoluble form. Calcium deposition is a very common feature in several degenerative and chronic inflammatory diseases both in insoluble and soluble forms. The soluble form is calcium chloride or ionized calcium in combination with proteins. Calcium salts are usually monorefringent but calcium oxalate is birefringent. Various dyes have been used which form chelate complexes with calcium. With haematoxylin/eosin, calcium takes a blue colour. Alizarin, purpurin, naphthochrome green B and nuclear fast red are mostly used. Van kossa (1901) method is the classic one for calcium, with this phosphate and carbonate radicals react.

Aberrations in calcium metabolism lead to deposition of copper in liver or brain in traces. This is the reason why it is difficult to demonstrate calcium histochemically. Copper like many metallic ions forms a blue lake with Mallory's unripened haematoxylin. Rhodamine method is also good. The best method of demonstrating it is by rubeanic acid technique. Occasionally haematoxylin is also used. It forms blue lakes with calcium since many metals form blue lakes with haematoxylin. It is not a specific technique.

Workers in factories especially battery industry suffer from excessive deposition of lead in tissues. The silver sulphide method and Rhodizonate methods are known to detect lead histochemically. Many metals are positive to silver sulphate. Other methods include sulphide silver stated by Timm (1958).

Carbon is usually found in people working in coal mines. Most of the city dwellers inhale carbon particles, besides all smokers suffer from deposition of carbon in their respiratory tracts. It is common in lungs and lymph nodes of smokers. Smoke from industrial chimneys is the main source of carbon. Carbon occurs as black particles. The inhaled tobacco gets entry as particulate form and gets trapped by the mucous membrane of nose, trachea, lungs and bronchi. Of all, coal miners inhale the largest amount of carbon. Beryllium and aluminium gain entry into tissues of people working in fluorescent light tube. They cause serious diseases by settling on skin as particles. They can both be demonstrated by Solochrome azurine. Naphthochrome Green-B method is good for beryllium. Beryllium salts are fluorescent and can be easily identified by bluish white colour when exposed to ultraviolet light. Aluminium is positive with fluorescent Morin method (Pearse, 1985).

Silver also can gain entry into tissues of people working in industries. People working as goldsmiths or those who help them have slate grey skin due to silver grains on the skin. Skin sections display black or dark brown granules and when sections are subjected to Lugol's iodine treatment followed by sodium thiosulphate rinse, silver is removed. Ammonium sulphide converts silver deposits into silver sulphide. A common histochemical method to demonstrate silver is dimethyl-aminobenzaldehyde–Rhodamine method. Silica is usually found in mine workers who inhale large quantities of silica. Industrial workers who are involved in sand blasting and stone grinding suffer most. Silica is unreactive as such and it cannot be demonstrated with histological and histochemical methods.

Generally urates get deposited in large amounts. This disease is called "Gout". They get settled in tissues lining the joints and the synovial fluid. They are in the form of crystals that are floating. In acute cases they may get deposited elsewhere in the body of which kidney forms the important organ.

HISTOCHEMICAL METHODS FOR MINERALS

Perl's Prussian Blue Reaction for Iron (Perl, 1867) (Plate 16, Figure 1)

Fixation

Any general or frozen section

Reagents Required

Potassium ferricyanide

Hydrochloric acid

Neutral red

Preparation of Reagents

Solution 1 (Potassium ferricyanide solution)
Fresh potassium ferricyanide   2.0 g
Distilled water   100.0 ml

Solution 2 (Hydrochloric acid)
2 per cent hydrochloric acid   2.0 ml
Distilled water   49.0 ml

Solution 3 Staining solution
2 per cent hydrochloric acid   25.0 ml
 (Solution 2)
2 per cent potassium ferricyanide   25.0 ml
 (Solution 1)

Procedure

1. Deparaffinize and hydrate slides to water.
2. Transfer to freshly prepared solution 3 (Staining solution) for 30 minutes.
3. Wash in water.
4. Counterstain if desired in neutral red.
5. Wash rapidly.
6. Dehydrate, clear and mount.

Results

Ferric alum   Blue
Nuclei   Red

Rationale

First introduced by Perl (1867), both paraffin and frozen sections could be treated by this method. 2 per cent hydrochloric acid liberates iron and ferric ions. Later it reacts with potassium ferricyanide to form potassium ferric ferricyanide. The end product is an insoluble blue compound.

Iron Reaction: Dinitroresorcinol (Humphrey, 1935)

Fixation

Formalin
Any general fixative with acid ingredients. 10 per cent buffered formalin is also preferable.

Reagents Required

Dinitroresorcinol

Ammonium sulphate

Preparation of Reagents

Solution 1
30 per cent ammonium sulphide (analytical)

Solution 2
Saturated aqueous dinitroresorcinol or 3 per cent in 50 per cent alcohol.

Procedure

1. Deparaffinize and hydrate slides to water.
2. Place the slides in solution 1 (ammonium sulphide) for 1 minute.
3. Rinse in water.
4. Transfer to solution 2 (dinitroresorcinol) for 24 hours.
5. Wash in water.
6. Dehydrate, clear and mount.

Results

Iron	Dark green
Background	Brown

Remarks

An old solution seems to work well. Saturated solution of dinitroresorcinol should have excess salt.

Turnbull Blue Method for Ferrous Iron (Pearse, 1953b)

Fixation

10 per cent buffered formalin
Any other fixative without acid ingredient

Reagents Required

Ammonium sulphide
Potassium ferricyanide
Hydrochloric acid
Safranin O

Preparation of Reagents

Solution 1 (Saturated ammonium sulphide)
Saturated ammonium sulphide solution (Analytical)

Solution 2 (Potassium ferricyanide)
Potassium ferricyanide   20.0 g
Distilled water   100.0 ml

Solution 3 (Hydrochloric acid)
Hydrochloric acid (conc.)   1.0 ml
Distilled water   100.0 ml

Solution 4 (Safranin O solution)
    Safranin O   200 mg
    Distilled water   100.0 ml
    Glacial acetic acid 1.0 ml

Procedure

1. Deparaffinize and hydrate slides to water.
2. Wash in distilled water.
3. Transfer to solution 1 (yellow ammonium sulphide) for 1–3 hours.
4. Rinse in distilled water.
5. Transfer to a freshly prepared mixture containing equal parts of solution 2 and solution 3 for 20 minutes.
6. Rinse in distilled water.
7. Counterstain with solution 4 (safranin O) for 2–5 minutes.
8. Rinse in 70 per cent alcohol.
9. Dehydrate, clear and mount.

Result

Ferrous and ferric iron are stained deep blue.

Iron Reaction (Hutchinson, 1953)

Fixation

Sodium sulphate   12.0 g
Glacial acetic acid 33.0 ml
Formalin   40.0 ml
Distilled water   200.0 ml

Reagents

Potassium ferricyanide
Hydrochloric acid
Safranin

Preparation

Solution 1 (Potassium ferricyanide)
    Potassium ferricyanide 2.0 g
    Distilled water   50.0 ml

Solution 2 (Hydrochloric acid)
    Hydrochloric acid 2.0 ml
    Distilled water   50.0 ml

Solution 3 (Staining solution)

Mix solutions 1 and 2 freshly before use. Warm them and filter. Pre-heat at 56°C before use.

Procedure

1. Deparaffinize and hydrate slides to water.
2. Wash.
3. Transfer to staining solution 3 at 56°C for 10 minutes.
4. Rinse in distilled water giving several changes.
5. Counterstain in safranin for 2–5 minutes.
6. Rinse in 70 per cent alcohol.
7. Dehydrate, clear and mount.

Results

Iron pigment   Brilliant greenish blue
Nuclei   Red

Remarks

Warming the solution at 56°C is an important step to speed up the reaction and produce more intense colour.

Gomori's Method for Iron (Gomori, 1936)

Fixation

Absolute alcohol, 10 per cent neutral buffered formalin

Reagents Required

Concentrated hydrochloric acid
Potassium ferricyanide
Nuclear fast red

Preparation of Reagents

Solution 1 (Hydrochloric acid stock)
   Concentrated hydrochloric acid   20.0 ml
   Distilled water   80.0 ml

Solution 2 (Potassium ferricyanide)
   Potassium ferricyanide   10.0 g
   Distilled water   100.0 ml

Solution 3 (Working solution)
   Solution 1   50.0 ml
   Solution 2   50.0 ml

Mix just before use.

Procedure

1. Deparaffinize and bring slides to water.
2. Transfer slides to solution 3 for 30 minutes.
3. Rinse in distilled water.
4. Counterstain with nuclear fast red (*see* page 352) for 5 minutes.

5. Rinse in distilled water.
6. Dehydrate, clear and mount.

Result

Iron is stained light blue.

Hukil–Putt Method for Iron (Hukil and Putt, 1962)

Fixation

10 per cent neutral buffered formalin

Reagents Required

Batho phenanthroline
Glacial acetic acid
Thioglycolic acid

Preparation of Reagents

Solution 1 (Stock)
  Batho phenanthroline   100 mg
  3 per cent glacial acetic acid   100.0 ml

Solution 2 (Working solution)
  Solution 1   40.0 ml
  Thioglycolic acid   0.2 ml

Mix before use.

Solution 3 (Methylene blue solution)
  0.5 per cent methylene blue
  (500 mg + 100 ml water)

Procedure

1. Deparaffinize and hydrate slides to water.
2. Place in solution 2 for 2 hours.
3. Rinse in distilled water.
4. Counterstain in solution 3.
5. Rinse in distilled water.
6. Blot, and dry at 60°C.
7. Mount.

Result

Iron   Red
Nuclei Blue

Lillie's Method for Ferric and Ferrous Iron (Lillie, 1965)

Fixation

10 per cent neutral buffered formalin

Reagents Required

Potassium ferricyanide
Concentrated hydrochloric acid

Preparation of Reagents

Solution 1 (Potassium ferricyanide solution)
    Potassium ferricyanide   400 mg
    0.06N Hydrochloric acid   40.0 ml
    (1 ml + 199 ml water)

Solution 2 (Potassium ferricyanide solution)
    Potassium ferricyanide   400 mg
    0.6N Hydrochloric acid   40.0 ml

Solution 3 (Basic fuchsin solution)
    Basic fuchsin   500 mg
    Distilled water   100.0 ml
    Glacial acetic acid   1.0 ml

Procedure

1.   Deparaffinize and hydrate slides to water.
2.   Place in solution 2 for 1 hour.
3.   Wash in distilled water.
4.   Wash in 1 per cent glacial acetic acid.
5.   Place in solution 3 for 10 minutes.
6.   Rinse in distilled water.
7.   Dehydrate, clear and mount.

Result

Ferric iron   Dark Prussian blue
Ferrous iron   Dark turnbulls blue
Background   Red

Mallory's Method for Iron (Mallory and Wright, 1924)

Fixation

10 per cent neutral buffered formalin

Reagents Required

Hydrochloric acid
Potassium ferricyanide
Nuclear fast red

Preparation of Reagents

Solution 1 (HCl solution stock)

Solution 2 (5 per cent potassium ferricyanide stock)

Solution 3 (Working solution)
    Solution 1   50.0 ml
    Solution 2   50.0 ml

Solution 4 (Nuclear fast red solution) (*see* page 322)

Procedure

1. Deparaffinize and hydrate slides to water.
2. Keep in solution 3 for 10 minutes.
3. Rinse in distilled water.
4. Counterstain in solution 4 for 5 minutes.
5. Rinse, dehydrate, clear and mount.

Result

Iron pigments	Bright blue
Nuclei	Red
Cytoplasm	Pink

Alizarin Red S Method (Dahl, 1952) (Plate 16, Figures 2 and 3)

Fixation

Any neutral fixative. Acid-containing fixatives not preferred.

Reagents Required

Sodium alizarin sulphate
Hydrochloric acid
Ammonia

Preparation of Solutions

Solution 1 (Sodium alizarin sulphate)
    Sodium alizarin sulphate   500 mg
    Distilled water   45.0 ml

To this solution, add 28 per cent ammonia (1 part ammonia and water 99 parts) and stir well. The pH of the solution should be 6.3–6.5 or adjust it by using buffer.

Solution 2 (Differentiator)
    Hydrochloric acid   0.1 ml
    Distilled water   99.9 ml

Procedure

1. Deparaffinize and hydrate slides to water.
2. Stain in solution 1 (Alizarin red S) for 2 minutes.
3. Wash well in distilled water.
4. Differentiate in solution 2 (acid alcohol) for 1 second.
5. Dehydrate, clear and mount in cedar wood oil.

Result

Calcium deposits stain orange-red.

Rationale

This method was first introduced by Dahl (1952) and later modified by McGree and Russel (1958). A lake is formed in this technique. There are others like nuclear fast red and haematoxylin which also form lakes with calcium. But in this technique, the lake formed is that of calcium carbonate and calcium phosphate. Time schedule is important for staining.

Alizarin Red S (Puchtler *et al.*, 1969)

Fixation

Carnoy, avoid using formalin or Zenker formol

Reagents Required

Alizarin red S
Phosphate or barbital buffer (pH 9)

Preparation of Reagents

Solution 1

| Alizarin red S | 500 mg |
| Phosphate buffer (pH 9) | 100.0 ml |

Procedure

1. Dewax and hydrate slides to water.
2. Place in solution 1 for 30–60 seconds.
3. Differentiate in buffer (pH 9) for 5 seconds.
4. Dehydrate, clear and mount.

Result

Calcium stains orange-red.

Van Kossa Method for Calcium (Van Kossa, 1901)

Fixation

Neutral fixatives

Reagents Required

Silver nitrate
Sodium thiosulphate
Neutral red

Preparation of Reagents

Solution 1 (Silver nitrate solution)

| Silver nitrate | 750 mg |
| Distilled water | 50.0 ml |

Solution 2 (Sodium thiosulphate)
  Sodium thiosulphate 2.5 g
  Distilled water   50.0 ml

Procedure

1. Deparaffinize and hydrate slides to water.
2. Transfer slides to solution 1 (1.5 per cent silver nitrate) for 15–30 minutes.
3. Wash in distilled water.
4. Transfer to solution 2 (hypo solution) for 1 minute.
5. Wash in distilled water.
6. Counterstain in 2 per cent neutral red for 30 seconds.
7. Wash in distilled water.
8. Dehydrate, clear and mount in Canada balsam.

Results

Calcium deposits Brown to black
Nuclei   Red

Rationale

Calcium carbonate or calcium phosphate subjected to silver nitrate engenders calcium to be replaced by silver. With sunlight or ultraviolet light, the silver salt undergoes reduction to metallic silver. When such tissue sections are subjected to sodium thiosulphate solution, calcium deposits turn black.

Carr's Method for Calcium (Carr *et al.*, 1961)

Fixation

10 per cent neutral buffered formalin

Reagents Required

Sodium hydroxide
Chloranilic acid
Light green

Preparation of Solutions

Solution 1 (Staining solution)
  Distilled water   100.0 ml
  Sodium hydroxide 400 mg
  Chloranilic acid 1.0 g

After all acid has dissolved, filter.

Solution 2 (Counterstain)
  Light green   1.0 g
  Distilled water 100.0 ml

Procedure

1. Deparaffinize and bring down slides to water.
2. Place in solution 1 for 30 minutes.
3. Wash in running water for 15 minutes.
4. Transfer to solution 2 for 5 minutes.
5. Rinse in distilled water.
6. Dehydate, clear and mount.

Result

Calcium   Red-brown
Background   Green

Pizzalato's Method for Calcium Oxalate (Pizzalato, 1964)

Fixation

10 per cent neutral formalin

Reagents Required

Silver nitrate
Hydrogen peroxide

Preparation of Reagents

Solution 1 (Silver nitrate hydrogen peroxide)
    5 per cent silver nitrate   25.0 ml
    Hydrogen peroxide   25.0 ml

Mix before use.

Solution 2 (Nuclear fast red solution)
    Nuclear fast red   100 mg
    5 per cent aluminium sulphate   100.0 ml
        (5 g in 100 ml water)

Heat gently, cool, filter and add a grain of thymol.

Procedure

1. Deparaffinize and bring slides to water.
2. Flood slides with solution 1 for 30 minutes under 60-Watt bulb.
3. Rinse in distilled water.
4. Counterstain if desired with solution 2.
5. Rinse in distilled water.
6. Dehydate, clear and mount.

Result

Calcium   Red-brown
Background   Green

Calcium Red Method for Calcium (McGee–Russel, 1955)

Fixation

Formalin or alcohol

Reagents Required

Nuclear Fast red
Calcium red

Preparation of Reagents

Solution 1 (Staining solution)
Nuclear fast red   2.0 g
Distilled water   100.0 ml

Wash and take the residue. Mix 100 ml distilled water.

Procedure

1.   Deparaffinize and hydrate slides to water.
2.   Transfer to solution 1 for 10 minutes.
3.   Wash in distilled water.
4.   Dehydrate, clear and mount.

Result

Calcium deposits are stained red.

Phthalocyanin Method for Calcium

Fixation

Formalin or alcohol

Reagents Required

Durazol fast blue 8
Alcian blue 8 GS
Lithium carbonate
Hydrochloric acid

Preparation of Reagents

Solution 1 (Hydrochloric acid)
1 per cent aqueous hydrochloric acid

Solution 2 (Durazol/alcian blue solution)
Durazol fast blue G   100 mg
Distilled water   100.0 ml
Alcian blue 8 GS   100 mg

Solution 3 (Lithium carbonate solution)
Lithium carbonate   50 mg
Distilled water   100.0 ml

Procedure

1. Deparaffinize and bring slides to water.
2. Place one section in solution 1.
3. Place both sections in solution 2 for 30–60 minutes.
4. Wash in running water.
5. Differentiate in solution 3.
6. Dehydrate, clear and mount.

Result

In untreated sections (without step 2) calcium deposits are stained bright blue.

Eriochrome Black-T Method for Calcium (for invertebrates)

Fixation

Formalin

Reagents Required

Eriochrome black-T
Sodium hydroxide

Preparation of Reagents

Solution 1 (Staining solution)
Eriochrome black-T   500 mg
Distilled water   100.0 ml

pH should be 9.5 and adjusted by sodium hydroxide.

Procedure

1. Deparaffinize and hydrate slides to water.
2. Place in solution 1 for 5 minutes.
3. Wash in distilled water.
4. Mount in glycerine jelly.

Result

Sites with calcium deposits are stained red.

Silver Nitrate Method for Calcium Oxalate (Yasue, 1969)

Fixation

Formalin

Reagents Required

Acetic acid
Silver nitrate
Methylene blue
Rubeanic acid

Preparation of Reagents

Solution 1 (Silver nitrate solution)
    Silver nitrate   5.0 g
    Distilled water   100.0 ml

Solution 2 (Rubeanic acid solution)
    Rubeanic acid saturated in 70 per cent ethanol + 2 drops of ammonia

Solution 3 (Methylene blue solution)
    Methylene blue   2.0 g
    Distilled water   100.0 ml

Procedure

1. Hydrate slides to water.
2. Immerse in 5 per cent acetic acid for 30 minutes.
3. Wash in distilled water.
4. Transfer to solution 1 for 20 minutes.
5. Rinse in distilled water.
6. Transfer to solution 2 for 1 minute.
7. Rinse in 50 per cent alcohol.
8. Place slides in solution 3 for 2 minutes.
9. Dehydrate, clear and mount in synthetic resin.

Result

Calcium oxalate   Black
Nuclei   Green

O-Toluidine–Thiocyanate Method for Copper

Fixation

Cold formalin
Formol calcium-fixed frozen sections
Formol calcium-fixed paraffin sections

Reagents Required

O-toluidine
Acetone
Ammonium thiocyanate

Preparation of Reagents

Solution 1 (Working solution)
    O-Toluidine   50 mg
    Acetone   5.0 ml
    Distilled water   5.0 ml
    Ammonium thiocyanate   500 mg

Procedure

1. Deparaffinize and bring down slides to water.
2. Flood sections with solution 1. Frozen sections are placed in a small dish containing solution 1.
3. Rinse in 70 per cent alcohol.
4. Dehydrate, clear and mount.

Result

Sites of copper salts are stained bright blue.

Benzidine Method for Copper

Fixation

Formalin

Reagents Required

Benzidine hydrochloride
Ammonium thiocyanate
Neutral red

Preparation of Reagents

Solution 1 (Benzidine solution)

Benzidine hydrochloride	10 mg
Ammonium thiocyanate	30 mg
Distilled water	5.0 ml

Solution 2 (Counterstain)

Neutral red	1.0 g
Distilled water	100.0 ml

Procedure

1. Deparaffinize and hydrate slides to water.
2. Flood slides with solution 1 for 10 minutes.
3. Rinse in distilled water.
4. Counterstain with solution 2 for 15 seconds.
5. Rinse in water.
6. Dehydrate, clear and mount.

Result

Copper deposits are stained blue.

Mallory and Parker's Haematoxylin Method for Lead and Copper

Fixation

Formalin or alcohol

Reagents Required

Haematoxylin
Potassium hydrogen phosphate

Preparation of Reagents

Solution 1

Dissolve 10 mg of haematoxylin in few drops of 95 per cent alcohol. To this add 10 ml of 2 per cent potassium dihydrogen phosphate (filtered). Solution should be prepared fresh before use.

Procedure

1.   Deparaffinize and hydrate slides to water.
2.   Stain in freshly prepared solution 1 (haematoxylin) for 2–3 hours at 56–60°C.
3.   Wash in running water.
4.   Dehydrate.
5.   Clear in terpineol and mount in terpineol balsam.

Result

Lead   Dark grey-blue
Copper   Blue
Haemosiderin   Black

Rhodizonate Method for Lead Salts

Fixation

Neutral fixative

Reagents Required

Sodium rhodizonate
Acetic acid

Preparation of Reagents

Solution 1 (Staining solution)
Sodium rhodizonate   200 mg
Distilled water   99.0 ml
Acetic acid   1.0 ml

Procedure

1.   Deparaffinize and hydrate slides to water.
2.   Transfer to solution 1 (Rhodizonate solution) for 1 hour.
3.   Rinse in tap water.
4.   Stain in 0.1 per cent light green in 1 per cent acetic acid.
5.   Rinse in water.
6.   Mount in glycerine jelly.

Result

Lead salts   Red
Background   Green

Remarks

In this technique, sodium rhodizonate reacts with lead in the tissue.

Chromate Method for Lead Salts

Fixation

Cold microtome   Frozen
Formalin   paraffin

Reagents Required

Potassium dichromate
Acetic acid
Toluidine blue

Preparation of Reagents

Solution 1 (Potassium chromate solution)
   Potassium chromate   2.0 g
   1 per cent aqueous acetic acid   100.0 ml

Procedure

1. Fix fresh pieces in Regaud's solution for 2 days.
2. Deparaffinize paraffin sections and bring them to water.
3. Stain in solution 1 for few days.
4. Wash tissue in running water for 6 hours.
5. Dehydrate, clear and embed in paraffin wax.
6. Cut 10-μ thick sections.
7. Deparaffinize and bring down to water.
8. Stain with 0.5 per cent toluidine blue.
9. Dehydrate clear and mount.

Result

Sites of lead are stained yellow.

Solochrome Azurine Method for Aluminium and Beryllium

Fixation

Any general fixative

Reagents Required

Solochrome azurine

Preparation of Reagents

Solution 1 (Working solution)
Solochrome azurine   200 mg
Distilled water   100.0 ml

Procedure

1. Deparaffinize and hydrate slides to water.
2. Transfer slides to solution 1 (Solochrome azurine solution).
3. Rinse in distilled water.
4. Dehydrate, clear and mount in DPX.

Result

Aluminium   Deep blue
Beryllium   Deep blue

Remarks

Both aluminium and beryllium take deep blue colour. The two can be differentiated by pre-heating the slide with an alkali when aluminium is removed. Beryllium remains unaffected.

Naphthochrome Green B Method for Aluminium and Calcium (Denz, 1949; Pearse, 1960)

Fixation

Any general fixative

Preparation of Reagent

Solution 1 (Staining solution)
Naphthochrome green B   150 mg
Distilled water   100 ml

Procedure

1. Deparaffinize and hydrate slides to water.
2. Transfer slides to solution 1 (naphthochrome green B) for 5–10 minutes.
3. Wash in distilled water.
4. Dehydrate, clear and mount in Canada balsam.

Result

Aluminium   Deep green
Calcium   Brown, brownish red

Fluorescent (Solochrome) Method for Aluminium

Fixation

Paraffin sections fixed in formalin
Frozen sections fixed in formalin

Reagents Required

Solochrome black B
Hydrochloric acid

Preparation of Reagents

Solution 1 (Solochrome solution)
    Saturated solution of Solochrome black B in absolute alcohol

Add 2 per cent hydrochloric acid to this.

Procedure

1. Deparaffinize and hydrate slides to water.
2. Place sections in solution 1 for 20–60 minutes.
3. Rinse paraffin sections in absolute alcohol.
4. Rinse frozen sections in 2 per cent acid alcohol.
5. Wash frozen sections and mount in glycerine jelly.
6. Wash paraffin sections, dehydrate, clear and mount.

Result

Aluminium salts fluoresce from yellowish orange to brick red.

Acid Solochrome Method for Aluminium

Fixation

Formalin

Reagents Required

Solochrome cyanin R.S.
Hydrochloric acid

Preparation of Reagents

Solution 1 (Staining solution)

Solochrome cyanin R.S.	200 mg
1 per cent aqueous hydrochloric acid	100.0 ml

Procedure

1. Deparaffinize and hydrate slides to water.
2. Place in solution 1 for 15–30 minutes.
3. Rinse in running hot tap water till the sections turn pale grey-blue.
4. Dehydrate, clear and mount.

Result

Sites with aluminium are stained deep rose-red.

Aurine Method for Aluminium Hydroxide (Irwin, 1955)

Fixation

Formalin

Reagents Required

Ammonium chloride
Ammonium acetate
Aurine tricarboxylic acid
Ammonium carbonate

Preparation of Reagents

Solution 1 (Buffer solution)

5 M Ammonium chloride	60.0 ml
5 M Ammonium acetate	60.0 ml
6N Hydrochloric acid	10.0 ml

The pH should be 5.2.

Solution 2 (Working solution)

Aurine tricarboxylic acid	2.0 g
Solution 1 (Buffer)	5.0 ml
Distilled water	100.0 ml

Heat at 25°C, cool and filter.

Solution 3 (Differentiating solution)

1.6 M Ammonium carbonate	22.0 ml
Solution 1 (Buffer)	50.0 ml

The pH should be 7.2.

Procedure

1. Deparaffinize and hydrate slides to water.
2. Place in solution 2 at 75°C for 10 minutes.
3. Wash in distilled water.
4. Differentiate in solution 3.
5. Wash in distilled water.
6. Counterstain if desired in 1 per cent aqueous methylene blue.
7. Rinse in distilled water.
8. Dehydrate, clear and mount.

Result

Aluminium hydroxide stains bright red.

Naphthochrome Green B Method for Beryllium (Denz, 1949; Pearse, 1960)

Fixation

All types

Reagents Required

Naphthochrome green B
Phosphate buffer

Preparation of Reagents

Solution 1 (Staining solution)
Naphthochrome green B	50 mg
0.1 M phosphate buffer (pH 5.0)	50.0 ml

Procedure

1. Deparaffinize and hydrate slides to water.
2. Transfer slides to solution 1 (Naphthochrome green B) at 37°C for 30 minutes.
3. Wash sections in distilled water.
4. Immerse sections in 70 per cent alcohol.
5. Dehydrate, clear and mount in DPX.

Result

Beryllium stains light green.

Rationale

This method was first introduced by Denz (1949) and later modified by Pearse (1960). When metals like iron, calcium, aluminium and beryllium are treated with naphthochrome green B, lakes develop. Beryllium forms a green lake. By changing the pH, different types of metals can be demonstrated.

Alkaline Solochrome Azurine Method for Beryllium

Fixation

Formalin

Reagents Required

Solochrome azurine
Sodium hydroxide

Preparation of Reagents

Solution 1 (Stock solution)
Sodium hydroxide	2.0 g
Distilled water	100.0 ml

Solution 2 (Solochrome working solution)
Solochrome azurine	200 mg
Solution 1	100.0 ml

Procedure

1. Deparaffinize and hydrate slides to water.
2. Place in solution 2 for 30 minutes.

3. Rinse in distilled water.

4. Dehydrate, clear and mount.

Result

Sites of beryllium deposits are stained black.

Alkaline Quinalizarin Method for Beryllium and Calcium

Fixation

Paraffin fixed in formalin or alcohol
Frozen sections fixed in formalin

Reagents Required

Quinalizarin (1,2 5,8 tetrahydroxy anthraquinone)
Sodium hydroxide

Preparation of Reagents

Solution 1 (NaOH solution)
Sodium hydroxide   4.0 g
Distilled water   100.0 ml

Solution 2 (Quinalizarine solution)
Quinalizarine   200 mg
Solution 1   100.0 ml

Procedure

1. Deparaffinize and hydrate slides to water.

2. Place in solution 2 for 3–5 minutes.

3. Wash in distilled water.

4. If frozen, mount in glycerine jelly.

5. If paraffin, dehydrate, clear and mount. No counterstaining.

Result

Sites with beryllium and calcium deposits stain deep purple.

p-Dimethylaminobenzaldehyde Rhodamine Method for Silver (Okamoto and Otamura, 1938)

Reagents Required

p-dimethylaminobenzaldehyde
Ethanol
Nitric acid

Preparation of Reagent

Solution 1 (Staining solution)

p-dimethylaminobenzaldehyde rhodamine	200 mg
Distilled water	90 ml
Ethanol	10.0 ml
Nitric acid (Conc.)	0.1 ml

Procedure

1. Deparaffinize and hydrate slides to water.
2. Stain in saturated solution (3.5 ml), N-HNO$_3$ (3 ml) and distilled water (93.5 ml) for 24 hours at 37°C.
3. Wash in distilled water.
4. Mount in glycerine jelly.

Floating Sections

1. Cut frozen sections 15–20 μ thick float in staining solution for 2 hours at 37°C.
2. Rinse free-floating sections in 70 per cent alcohol.
3. Rinse in distilled water. Mount in glycerine jelly.

Results

Silver is stained red-violet or red-brown.

Rationale

Metal chelation produces reddish brown deposits of silver rhodonate.

Wachstein and Zak's Method for Bismuth

Fixation

Formalin or frozen sections

Reagents Required

Brucine sulphate
Sulphuric acid
Potassium iodide
Light green
Castel brucine reagent
Hydrogen peroxide

Preparation of Reagents

Solution 1 (Brucine solution)

Brucine sulphate	250 mg
Distilled water	100.0 ml

Add 3 drops of concentrated sulphuric acid and 2 g of potassium iodide. Store in a dark bottle. Filter before use.

Solution 2 (Light green solution)
    1 per cent light green 0.1 ml
        (1 g/100 ml water)
    Castel's reagent   10.0 ml

Procedure

1. Hydrate slides to water and air-dry.
2. Treat with 100 vol. (30 per cent) hydrogen peroxide for 15 seconds.
3. Wash in running water.
4. Treat with solution 2 (Castel's reagent) for 1 hour.
5. Mount in laevulose syrup.

Result

Bismuth stains orange-red.

Methenamine Silver for Urates (Gomori, 1951a)

Fixation

Any fixative

Reagents Required

Methenamine
Silver nitrate
Boric acid
Sodium tetraborate
Sodium thiosulphate

Preparation of Reagents

Solution 1 (Methenamine stock)
    5 per cent silver nitrate   5.0 ml
        (5 g/100 ml water)
    Methenamine   3 g
    Distilled water   100 ml

When 50 ml silver nitrate is mixed with 3 per cent methenamine, a precipitate is formed which slowly dissolves. Store the solution at 4°C (keep for 3 months).

Solution 2 (Buffer solution)
    Boric acid   800 mg
    Sodium tetraborate   650 mg
    Distilled water   100.0 ml

This solution should be stored at 4°C.

Solution 3 (Working solution)
    Methenamine stock 1   25.0 ml
    Buffer stock (solution 2)   5.0 ml
    Distilled water   20.0 ml

Procedure

1. Deparaffinize and bring slides to absolute alcohol.
2. Transfer slides to solution 3 (working solution) for 1 hour at 37°C.
3. Rinse in distilled water.
4. Treat with 2 per cent sodium thiosulphate for 15 minutes.
5. Rinse in tap water.
6. Counterstain if desired.
7. Dehydrate, clear and mount.

Results

Deposits of urates are stained black.

De Galantha's Method for Urate Crystals (De Galantha, 1935)

Fixation

10 per cent neutral buffered formalin (De Galantha, 1935)

Reagents Required

Silver nitrate
Gelatin
Hydroquinone

Preparation of Reagents

Solution 1 (20 per cent silver nitrate solution)
    Silver nitrate   20.0 g
    Distilled water   100.0 ml

Solution 2 (Developing solution)
    3 per cent gelatine in hot water   10.0 ml
    Solution 1   3.0 ml
    Hydroquinone   2.0 ml

Procedure

1. Dewax and place slides in absolute alcohol.
2. Treat slides with solution 1 exposing sections to strong sunlight for 3 hours.
3. Flood slides with solution 2 until sections turn black.
4. Wash.
5. Dehydrate, clear and mount.

Results

Urates   Black
Background   Yellow

Schmorl's Method for Reducing Substances
(Barka and Anderson, 1963)

Fixation

10 per cent neutral buffered solution

Reagents Required

Ferric chloride, potassium ferricyanide, mucicarmine, metanil yellow

Preparation of Reagents

Solution 1 (Stock)
    Ferric chloride   1.0 g
    Distilled water   100.0 ml

Solution 2 (Potassium ferricyanide solution)
    Potassium ferricyanide   100 mg
    Distilled water   100.0 ml

Solution 3 (Working solution)
    Solution 1   150.0 ml
    Solution 2   50.0 ml

Final pH should be 2.4 with few drops of dilute HCl.

Solution 4 (Mucicarmine solution)
    Carmine   1.0 g
    Aluminium chloride (anhydrous)   500 mg
    Distilled water   100.0 ml

Solution 5 (0.25 per cent metanil yellow)
    Metanil yellow   250 mg
    Distilled water   100.0 ml
    Glacial acetic acid   0.25 ml

Procedure

1. Dewax and hydrate slides to water.
2. Place slides in solution 4 for 7 minutes.
3. Rinse in distilled water.
4. Transfer to solution 3 for 1 hour.
5. Rinse rapidly in distilled water.
6. Counterstain in solution 5 for few seconds.
7. Rinse rapidly in distilled water.
8. Dehydrate, clear and mount.

Results

Argentaffin granules	Blue
Background	Yellowish green
Reducing substances	Blue
Goblet cells	Reddish pink

Copper Method for Arsenites and Arsenates (Castel, 1936)

Fixation

Formalin

Reagents Required

Formalin
Copper sulphate

Procedure

1. Fix tissues in 10 per cent formalin with 2.5 per cent copper sulphate for 5–6 days.
2. Wash in running water for 12–24 hours.
3. Dehydrate, clear and embed in paraffin.
4. Cut fairly thick sections.
5. Deparaffinize and mount sections.

Results

Sites with arsenic are stained green.

Rhodizonate Method for Barium Salts

Fixation

Formalin paraffin
Formalin frozen

Reagents Required

Sodium rhodizonate
Hydrochloric acid

Preparation of Reagents

Solution 1 (Sodium rhodizonate solution)
Sodium rhodizonate   200 mg
Distilled water   100.0 ml

Solution 2 (HCl)
20 per cent hydrochloric acid

Procedure

1. Deparaffinize and hydrate slides to water.
2. Place in solution 1 and incubate at 60°C for 2 hours or if frozen at 37°C.
3. Wash briefly.
4. If necessary, convert the black rhodizonate to red salt by immersion in solution 2.
5. Wash.
6. Dehydrate, clear and mount.

Result

Deposits of barium   Black
If treated with solution 2   Scarlet red

Dimethylglyoxime Method for Nickel Salts (Feigl, 1954)

Fixation

Formalin-fixed paraffin sections
Fresh frozen

Procedure

1. Stain sections with a saturated solution of dimethyl glyoxime in alcohol for 20 minutes at 37°C.
2. Rinse in alcohol.
3. Clear in xylene and mount in synthetic medium.

Results

Sites with nickel are stained pink.

Reduction Method for Inorganic Gold Salts (Christeller, 1927)

Fixation

Formalin, alcohol

Reagents Required

Stannous chloride
Hydrochloric acid

Preparation of Reagents

Solution 1 (Stannous chloride solution)
     Stannous chloride   5.0 g
     Distilled water   100.0 ml

Solution 2 (Working solution)
     Solution 1   40.6 ml
     Conc. hydrochloric acid   5.0 ml

Procedure

1. Deparaffinize and hydrate slides to water.
2. Incubate in solution 2 for 36 hours.
3. Wash in distilled water.
4. Dehydrate, clear and mount.

Results

Sites of gold deposits are stained black or brown.

Titan Yellow Method for Magnesium

Fixation

Fresh frozen sections
Formalin-fixed frozen
Paraffin

Reagents Required

Titan yellow
Sodium hydroxide

Preparation of Reagents

Solution 1 (Titan yellow solution)
  Titan yellow   200 mg
  Distilled water   100.0 ml

Solution 2 (Sodium hydroxide)
  2N Sodium hydroxide

Procedure

1. Flood the sections with solution 1.
2. Add a similar quantity of solution 2 for 40–60 minutes.
3. Wash.
4. Dehydrate, clear and mount in DPX.

Result

Magnesium salts are stained red. But colour fades away very soon.

Magneson Method for Magnesium

Fixation

Fresh frozen
Formalin-fixed frozen
Formalin-fixed paraffin

Reagents Required

Magneson II (*p*-nitrobenzene-azo-1-naphthol)
Sodium hydroxide

Preparation of Reagents

Solution 1
  Magneson II   1.0 g
  5 per cent sodium hydroxide   100.0 ml
  (Solution 2)

Solution 2 (Sodium hydroxide solution)
  Sodium hydroxide   5.0 g
  Distilled water   100.0 ml

Procedure

1.   If paraffin, hydrate slides to water.
2.   Incubate in solution 1 at 60°C for 30 minutes.
3.   Wash in dilute sodium hydroxide (Solution 2).
4.   Dehydrate, clear and mount in DPX.

Result

Magnesium salts stain bright blue.

Dithizone Method for Zinc (Mager *et al.*, 1953)

Fixation

Freeze-dried
Paraffin
Cold microtome

Reagents Required

Dithizone
Sodium thiosulphate
Sodium acetate
Potassium cyanide
Acetic acid
Carbon tetrachloride
Sodium potassium tartrate

Preparation of Reagents

Solution A (Stock solution)
    Dithizone   10 mg
    Absolute acetone 100.0 ml

Solution B (Buffer)
    Sodium thiosulphate   550.0 g
    Sodium acetate   90.0 g
    Potassium cyanide   10.0 g
    Distilled water   1000.0 ml

pH should be 5.5. Adjust with acetic acid.

Make up to 2000.0 ml. Extract with dithizone in CCl_4 to remove zinc.

Working Solutions
    A   Stock solution A   100.0 ml
           Distilled water   150.0 ml

    B   Stock solution A   24.0 ml
           Distilled water   18.0 ml

pH should be 3.7; adjust with N-acetic acid.

Add stock solution B   5.8 ml
    20 per cent sodium   0.2 ml
        potassium tartrate

Procedure

1. Take 2 pairs of sections.
2. Stain section in working solution A and B for 10 minutes.
3. Wash in chloroform.
4. Rinse quickly in water.
5. Mount in karo syrup.

Result

Sites of zinc are stained reddish purple.

Method of Gersh for Potassium

Fixation

Freeze-dried material

Reagents Required

Petroleum iet
Sodium cobalt nitrite
Glycerine

Procedure

All steps except first should be carried out at 0°C in a cold room.
1. Cut 15 µ thick paraffin section and place on coverslips.
2. Remove wax in petroleum and dry by burning excess petroleum.
3. Add a drop of 12 per cent sodium cobalt nitrite.
4. Allow to dry and mount in glycerine.

Result

Potassium crystals are stained yellow.

Modified Gersh Method for Potassium (Crout and Jennings, 1957)

Fixation

Freeze-dried

Reagents Required

Cobalt nitrite
Sodium nitrite
Acetic acid

Preparation of Reagents

Solution 1
Cobalt nitrite   25.0 g
Distilled water   500.0 ml

Solution 2

    Sodium nitrite   120.0 g

    Distilled water   100.0 ml

Add 210.0 ml of solution 2 to solution 1.

Procedure

1. Cut frozen sections and put in dilute petroleum at room temperature.
2. Mount sections on a dry slide. Remove petroleum and air-dry.
3. Transfer to cobalt nitrite solution.

   (Dissolve 25 g cobalt nitrite in 500 ml distilled water, add 12.5 ml acetic acid. Dissolve 120 g of sodium nitrite in 180.0 ml distilled water. Add 210 ml of this to cobalt nitrite solution.

4. Blot excess stain.
5. Dehydrate in absolute alcohol.
6. Clear and mount in synthetic resin.

Result

Potassium is stained as yellowish brown granules.

Table 15.1 Histochemical techniques applied for the demonstration of minerals

Technique	Fixative	Iron	Calcium	Calcium oxalate
Perl's Prussian blue method	Any general or frozen	Blue		
Dinitro Resorcinol method	Formalin	Dark Green		
Turnbull Blue method	10 per cent buffered formalin, any other fixative without acid ingredient	Deep blue		
Iron reaction	Sodium sulphate 12.0 g + Glacial acetic acid 33.0 ml + Formalin 40.0 ml + Distilled water 200 ml	Brilliant greenish blue		
Gomori's method	Absolute alcohol, 10 per cent neutral buffered formalin	Light blue		
Hukil and Putt method	10 per cent neutral buffered formalin	Red		
Lillies method	10 per cent neutral buffered formalin	Dark blue		

(Contd.)

Table 15.1   (Continued)

Technique	Fixative	Iron	Calcium	Calcium oxalate
Mallory's method	10% neutral buffered formalin	Bright blue		
Alizarin Red S method	Any neutral fixative		Orange -red	
Van Kossa method	Neutral fixative		Brown to black	
Carr's method	10 per cent neutral buffered formalin		Red-brown	
Pizzalato's method	10 per cent neutral buffered formalin		Red-brown	
Calcium Red Method	Formalin or alcohol		Red	
Phthalocyanin Method	Formalin or alcohol		Bright blue	
Eriochrome Black-T	Formalin		Red	
Silver NO_2 method	Formalin			Black

Table 15.2   Histochemical techniques applied for demonstration of minerals

Technique	Fixative	Calcium	Copper	Lead	Aluminium	Beryllium	Silver	Bismuth	Aluminum hydroxide
O-Toluidine–Thiocyanate	Cold formalin, formol calcium–fixed frozen		Bright blue						
Benzidine method	Formalin		Blue						
Mallory and Parker's haematoxylin	Formalin or alcohol		Blue	Dark grey-blue					
Rhodizonate method	Neutral fixative			Red					
Chromate method	Frozen formalin			Yellow					
Solochrome Azurine method	Any general				Deep blue	Deep blue			
Naphthochrome Green B method	Any general fixative	Brown			Deep green				

(*Contd.*)

Table 15.2 (Continued)

Technique	Fixative	Calcium	Copper	Lead	Aluminium	Beryllium	Silver	Bismuth	Aluminium hydroxide
Fluorescent method	Formalin-frozen, Formalin – paraffin				Fluore-scence				
Acid solochrome method	Formalin				Deep rose-red				
Aurine method	Formalin								Bright red
Naphthochrome green B method	Any general fixative					Light green			
Alkaline solochrome Azurine method	Formalin					Black			
Alkaline quinalizarin method	Formalin or alcohol formalin-fixed frozen	Deep Purple				Deep purple			
Dimethylaminoben zalidine rhodamine method	Fresh, cold knife or cold microtome or formalin fixed						Red-violet		
Wachstein and Zak's method	Formalin or forzed sections							Orange-Red	

Table 15.3   Histochemical techniques applied for demonstration of minerals

Technique	Fixative	Urates	Reducing substance	Arsenite	Barium	Gold salts	Magnesium	Zinc	Potassium	Nickel
Methenamine silver method	Any general fixative	Black								
De Galantha's method	10% neutral buffered formalin	Black								
Schmorl's method	10% neutral buffered formalin		Blue							
Copper method	Formalin			Green						
Rhodizonate	Formalin paraffin, Formalin frozen				Black					
Dimethyl glyoxine method	Formalin – paraffin, fresh frozen									Pink
Reduction method	Formalin, alcohol					Black or brownish				
Titan yellow method	Fresh frozen, formalin-fixed frozen paraffin						Red			

(Contd.)

Table 15.3   (Continued)

Technique	Fixative	Urates	Reducing substance	Arsenite	Barium	Gold salts	Magnesium	Zinc	Potassium	Nickel
Magneson method	Fresh-frozen, Formalin-fixed frozen, paraffin						Bright blue			
Dithizone method	Freeze-dried paraffin							Reddish purple		
Gersh method	Freeze-dried								Yellow	
Modified Gersh method	Freeze-dried								Yellowish brown granules	

MICROORGANISMS IN SECTIONS

16

BACTERIA

Bacteria are very small (ranging between 0.5 and 1.0 µm) and can be distinguished and diagnosed by their difference in shape, certain morphological features and their staining affinities to different stains. They react to single dyes, mixed dyes or polychrome dyes.

Most bacteria have one of three shapes. They may be spherical when they are termed as cocci, or cylindrical or rod-shaped when they are called bacilli and spiral when they are referred to as spirilla. Some may occur as short rods referred to as coccobacillus. A few others possess different shapes. Square cells were also reported from the salty pool of Senai Peninsula in Egypt in 1980.

Bacteria occur as independent cells attached to one another. Division in bacteria is by binary fission. Each of the resulting cells leads a life independent of all other cells. The plane of division of bacteria plays a major role in their arrangement. If bacteria divide in one plane, it results in chains, while those dividing in several planes form clusters.

Cocci are arranged in several characteristic arrangements depending on the plane of division and cocci are arranged in patterns whereas bacilli occur singly or in pairs and some form chains and long branched multinucleate filaments called hyphae. The bacteria that form long chains are termed as streptococci, those which occur as two cells and are bridged together are referred to as diplococci and those which occur in clusters are called staphylococci. Still others form cuboidal pockets with 8 or more cells. Long branched multinucleated filaments are seen in *Streptomyces*. Curved bacteria have a twist. This single twist gives them a vibrioid shape. Some have more than one twist which gives them a helical shape. For example spirilla are rigid helical bacteria. Some bacteria are pear shaped (*Pasteurella*), lobed spheres

(*Sulfolobus*), rods with square ends (*Bacillus anthracis*), some are discs arranged like stacks of coins (*Caryophanon*) and rods with sculptured (helically) surfaces (*Seliberia*).

Most bacteria have two structures, the cell wall and the cytoplasmic membrane, that surround the cell cytoplasm. Some bacteria have a third layer, the capsule or slime layer. All these are collectively called cell envelopes.

The bacterial cell wall is composed of subunits found nowhere else in nature. The cell wall and its component parts may produce symptoms of disease. The cell wall is the site of action of some of the effective antibiotics.

The chemical composition of the wall determines the gram-staining properties of the cell. The shape of the organism is also determined by the cell wall, for example, cylindrical shape is due to cylindrical wall and spherical wall gives the organism a spherical shape.

As already mentioned, different bacteria react differently when treated with the same dye. Based on this feature, most bacteria are separated into two groups, the gram-positive and gram-negative bacteria.

Gram-positive bacteria when treated with crystal violet retain the stain after they are treated with iodine and washed with alcohol, whereas in gram-negative bacteria, alcohol washes the crystal violet–iodine complex. Apparently the dye–iodine complex becomes trapped in alcohol in gram-positive bacteria. On the other hand the cell wall is permeable in gram-negative bacteria.

Flagellum is the major organ for movement. Flagella are differently arranged. In some bacteria there is a single flagellum at one end of the cell. This is the polar flagellum. Others have a tuft. In still others, along the sides of the body, flagellae are inserted at many points. This flagellar arrangement has a great bearing on identification of bacteria.

With this gross background let us see the most important bacteria.

Staining of Bacteria

As it is difficult to observe the transparent, rapidly moving organism under light microscope, cells are first killed and later treated with dyes. Entire organisms or some parts stain prominently in contrast to the unstained parts.

First a drop of liquid containing the organism is placed on a slide and fixed to the slide by passing the slide over a flame. The organism gets fixed and is later treated with the dye.

A number of staining procedures are in vogue. Some dyes will stain a particular component and some will stain only a particular group of organisms thus making it easy to classify the organisms based on their staining characteristics.

There are two varieties of stains, the positive stains which colour the components that have a strong attraction for the dyes and the negative stains that never penetrate the cell and as such are highly visible with a contrasting dark background. Negative stains are used to demonstrate surface structures which are not stained by positive stains.

The dye methylene blue is usually used on fixed bacteria. The dye stains the entire cell blue without staining the background material. Methylene blue is a basic dye (with a positive charge) which strongly binds the DNA and RNA of the cells (which have negative charge). Internal structures are not visible.

On the other hand, acidic dyes (having negative charges) such as eosin, acid fuchsin and Congo red stain basic compounds in the cell especially proteins carrying positive charges (composed of amino acids). Sudan black B is used to localize fat droplets in bacteria.

The most popular and widely applied stains for bacteria are Gram stain and acid-fast stain.

Gram Stain

In diagnostic microbiology, Gram stain is the most wanted and useful procedure. Dr. Hans Christian Gram developed this stain and introduced it in 1884. This procedure enabled him to distinguish bacteria which cause pneumonia and eukaryotic nuclei in infected mammalian tissue. The specimen should be either fixed or smeared, gram-stained and examined microscopically. On microscopic examination, the gram reaction resulting in purple-blue colour indicates gram-positive organisms and red, gram-negative. Morphology (shape—cocci, rods, fusiform or others) of bacteria should be noted. Gram-staining method is as follows:

1. Fix bacteria or smear by passing the slide over the flame.
2. Flood the slide with crystal violet.
3. Wash in water, allow it to dry.
4. Flood the slide with Gram's iodine.
5. Wash in water, allow it to dry.
6. Decolorize in acetate and alcohol while gently agitating.
7. Wash in water, allow it to dry.
8. Flood the slide with safranin for 30 seconds (25 per cent in 95 per cent ethanol).
9. Wash with water and allow it to dry.

This differential behaviour of bacteria in taking the stain is attributed to the chemical structure of the cell wall. The cell wall of gram-positive bacteria is different from that of gram-negative bacteria.

Acid-fast stain is also very popular especially in diagnostic laboratories engaged in detecting *Mycobacterium*. Ziehl–Neelsen acid-fast procedure is as follows:

1. As in the previous case, fix the organism or smear by heat.

2.   Flood the slide with carbol fuchsin or basic fuchsin and steam gently for 5 minutes.
3.   Wash with water.
4.   Decolorize in acid alcohol till a faint pink colour is retained.
5.   Wash with water.
6.   Counterstain with methylene blue for 15 seconds.
7.   Wash with water and allow it to dry.

The acid-fast nature is mainly due to the presence of mycolic acid (lipid) in their cell walls. This type of bacteria falls under the genus *Mycobacterium*. *Mycobacterium* species are causative agents of tuberculosis and leprosy. This lipid (mycolic acid) does not permit Gram stain to penetrate the cell wall.

SOME IMPORTANT BACTERIA

Staphylococcus aureus

These are spherical in shape measuring 1 μm in diameter and the characteristic feature is their arrangement in clusters resembling grape bunches (Figure 16.1). Sometimes they may be found singly or in pairs or in the form of chains having three or four cells. There is no motility or capsule formation. They are non-sporing and are gram-positive and can grow on any type of media.

Figure 16.1   *Staphylococcus aureus*

Almost 30 per cent of the people have staphylococci in the nose. Staphylococci may also come from infected domestic animals such as cows. Mode of transmission may be by direct contact or through fomites.

Streptococcus pyogenes

These are spherical to oval measuring 0.5 to 1.0 μm in diameter (Figure 16.2). These are arranged in chains due to the cocci dividing in one plane only and the daughter cells fail to separate. Length of the chains forms an important character in classification. These are also non-motile, non-sporing but capsulated. *Streptococcus* is responsible for respiratory infections, sore throat being the commonest, tonsillitis, scarlet fever which was once very prevalent in colder countries, skin infections such as erysipelas and impetigo, and general infections.

Figure 16.2   *Streptococcus pyogenes*

Pneumococcus

These are small elongated cocci with one broad end and one narrow end giving a lanceolate appearance (Figure 16.3). They occur in pairs with broad ends in apposition. They are capsulated, each capsule enclosing a pair. They stain with aniline dyes and are gram-positive.

In humans, *Pneumococcus* causes bronchopneumonia. Meningitis is the most serious pneumococcal infection. This disease is common in children.

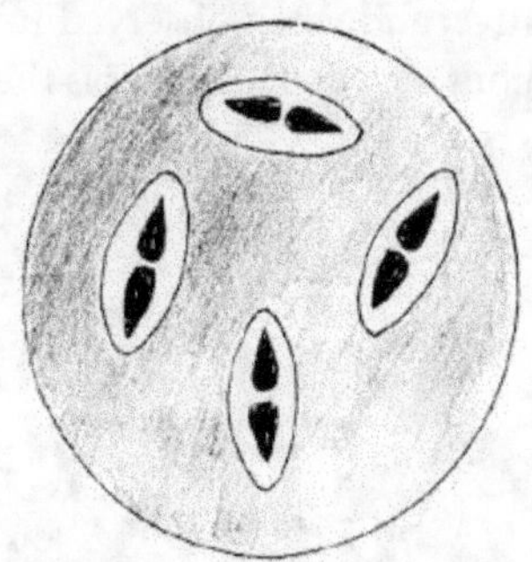

Figure 16.3   *Pneumococcus*

Neisseria gonorrhoeae

These bacteria resemble meningitidis in many aspects. The organism appears as a diplococcus with adjacent sides concave appearing kidney-shaped or pear-shaped (Figure 16.4). Gonococcus possesses pili on the surface which aid in adhesion of the cocci to the mucosal membrane. Piliate gonococci agglutinate human red blood cells. This is the causative agent of gonorrhoea which is acquired by sexual contact.

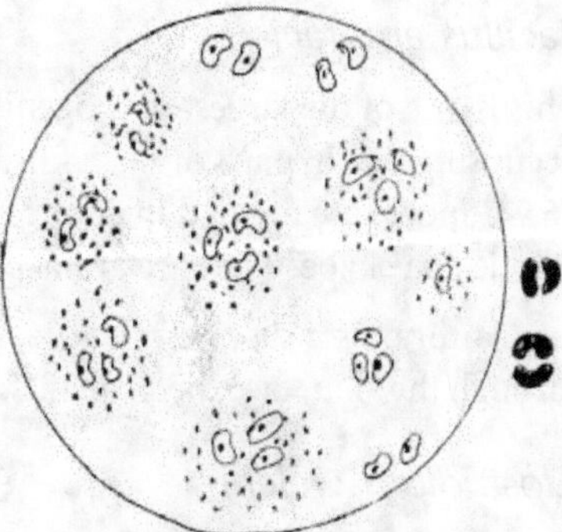

Figure 16.4   *Neisseria gonorrhoeae*. Enlarged view of kidney-shaped cells

Neisseria meningitidis

These are gram-negative bacteria either oval or spherical in shape (Figure 16.5), 0.6–0.8 μm in size arranged in pairs with one side flattened. The long axis is at right angles to the line joining the two cocci in pair. Size variation, shape variation and also variations in staining properties are noticed. These are non-motile. They are the causative agents of cerebrospinal meningitis and meningococcal septicaemia.

Corynebacterium diphtheriae

It is slender and rod-like measuring 3.6 × 0.6–0.8 μm (Figure 16.6). The bacilli are pleomorphic, non-motile and non-capsulated. Cells exhibit septa. It is gram-

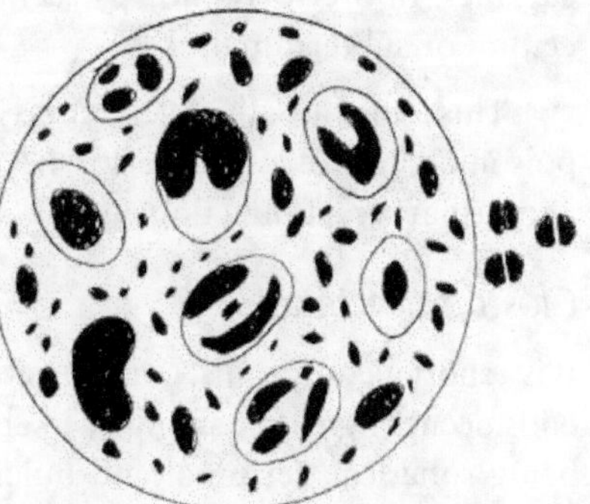

Figure 16.5   *Neisseria meningitidis* Enlarged view to show flat adjacent sides of cocci

positive, and when observed in smears, the bacilli are arranged in a characteristic fashion, seen as pairs. This is the causative agent of diphtheria.

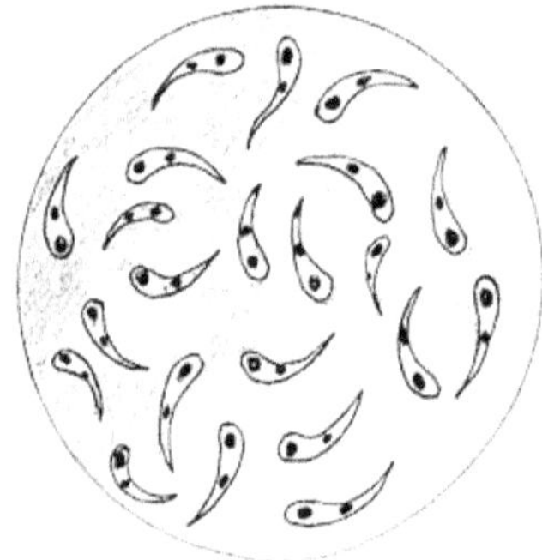

Figure 16.6   *Corynebacterium diphtheriae*

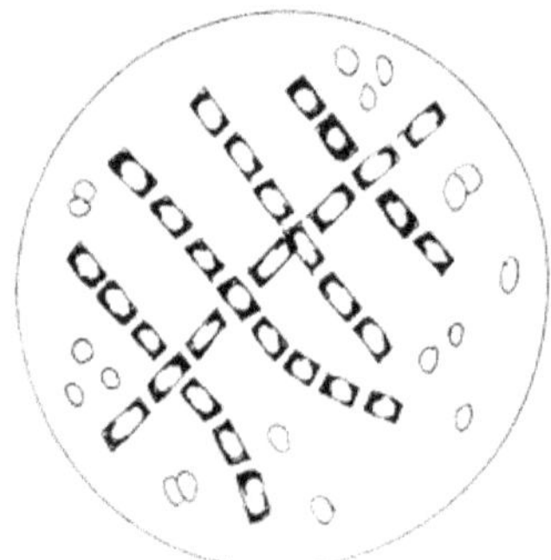

Figure 16.7   *Bacillus anthracis*

Bacillus anthracis

This is one of the largest pathogenic bacterium measuring 3–10 × 1–1.6 μm. It may occur singly or in pairs or in chains, the whole chain surrounded by a capsule (Figure 16.7). Spores are formed in cultures but never in the animal body. Spores are oval or elliptical in shape. They are gram-positive and non-acid-fast and non-motile.

Anthrax is a disease of cattle, sheep and horses. Human anthrax is contracted through the skin, face, neck, hands, arms and back being the usual sites.

Clostridium welchi

It is a gram-positive bacillus with rounded or truncated ends measuring 4–6 × 1 μm (Figure 16.8). They occur singly, in chains or in bundles. They are pleomorphic, and non-motile, and form capsules. Spore formation is seen, spores being central or subterminal.

This is the causative agent of gas gangrene, food poisoning, necrotizing enteritis, necrotizing colitis and urinary tract infections.

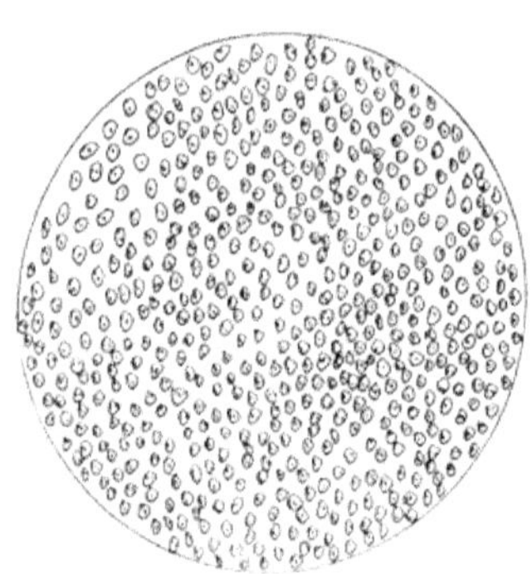

Figure 16.8
Clostridium welchi

Clostridium tetani

It is a short pleomorphic, gram-positive slender bacillus, 4.8 × 0.5 μm. It has rounded ends occurring either singly or in chains (Figure 16.9). It is spore-forming, spores being spherical, terminal and bulging and as such the bacillus has a drumstick appearance. It is the causative agent of tetanus.

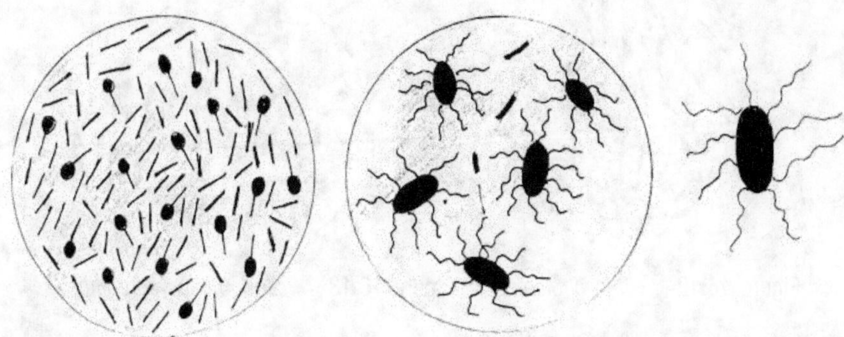

Figure 16.9   *Clostridium tetani*   Figure 16.10   *Escherichia coli*

Escherichia coli

It is gram-negative bacterium appearing as straight rods measuring 1.3 × 0.4–0.7 μm arranged singly or in pairs (Figure 16.10). It shows no spore formation, and is motile and capsulated. It is the causative agent of urinary tract infection, diarrhoea or gastroenteritis, pyogenic infections and septicaemia.

Shigella

These are rod-like, gram-positive bacteria 0.5 × 1.3 μm in size. They are non-motile, non-sporing and non-capsulated. They are the causative agents of bacillary dysentery.

Salmonella typhi

These are rod-like, gram-negative bacteria measuring 1.3 × 0.5–0.8 μm in diameter (Figure 16.11). They are motile, with no capsule or spore formation. These bacilli are the causative agents of enteric fever, septicaemia, gastroenteritis or food poisoning, enteric fever including typhoid fever.

Vibrio cholerae

These are curved cylindrical rods measuring 1.5 × 0.2–0.4 mm in diameter with rounded ends or slightly pointed ends (Figure 16.12). The cell is typically comma-shaped. They are motile with a single polar flagellum, gram-negative but stain with aniline dyes and are non-acid-fast. This is the causative agent of cholera.

Figure 16.11   *Salmonella typhi*

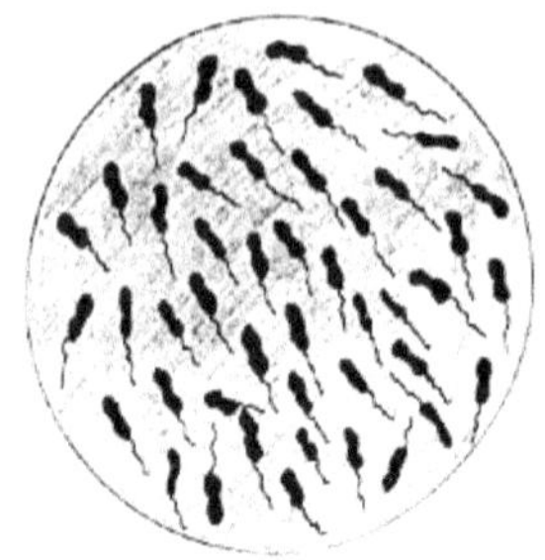

Figure 16.12   *Vibrio cholerae*   Figure 16.13   *Pseudomonas aeruginosa*

Pseudomonas aeruginosa

It is a slender bacillus (Figure 16.13) measuring 1.5–3.0 × 0.5 mm. It is motile with a polar filament, and is non-capsulated and gram-negative. It is the causative agent of infantile diarrhoea.

Yersinia pestis

This is the plague bacillus. It is a short plump bacillus, ovoid in shape, (Figure 16.14) and measures 1.5 × 0.7 m in size. Ends are rounded with convex sides, and it occurs in short chains or in small groups. Pleomorphism is common. It is gram-negative and is the causative agent of plague.

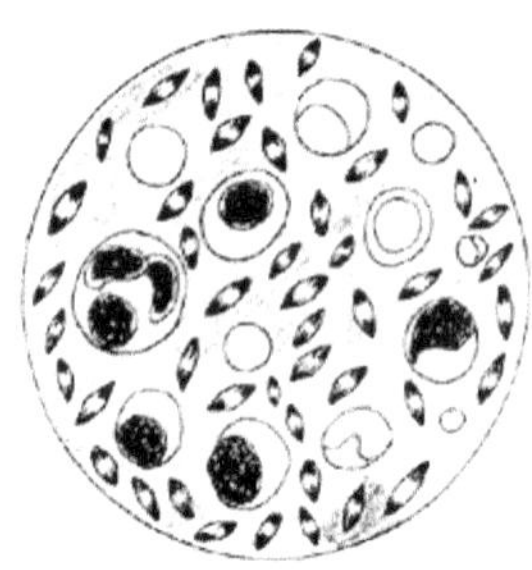

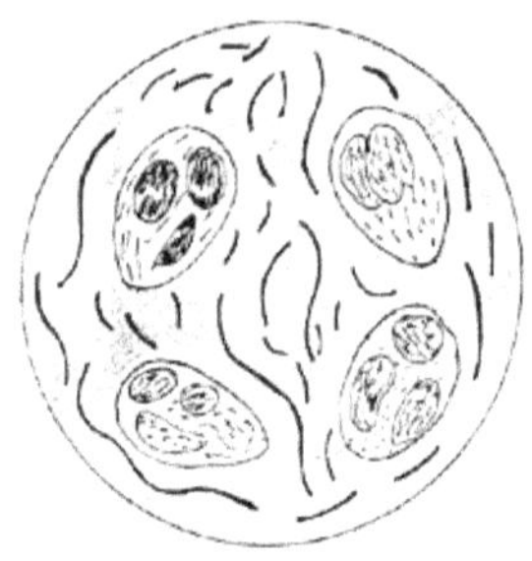

Figure 16.14   *Yersinia pestis*   Figure 16.15   *Haemophilus influenzae*

Haemophilus influenzae

It is small and measures 1.5 × 0.3 μm (Figure 16.15). It is non-motile, non-sporing and pleomorphic. It is gram-negative but stains with Loeffler's methylene blue or dilute carbol fuchsin. The most serious disease caused by *H. influenzae* is meningitis. 90 per cent of the people affected succumb to this disease.

Bordetella pertussis

This bacillus is small, ovoid and measures 0.5–0.7 × 0.6–1.5 μm in size (Figure 16.16). It is arranged either singly or in chains, is non-motile, non-capsulated and non-sporing, and is gram-negative. It is the causative agent of whooping cough in children, characterized by paroxysmal cough and respiratory distress.

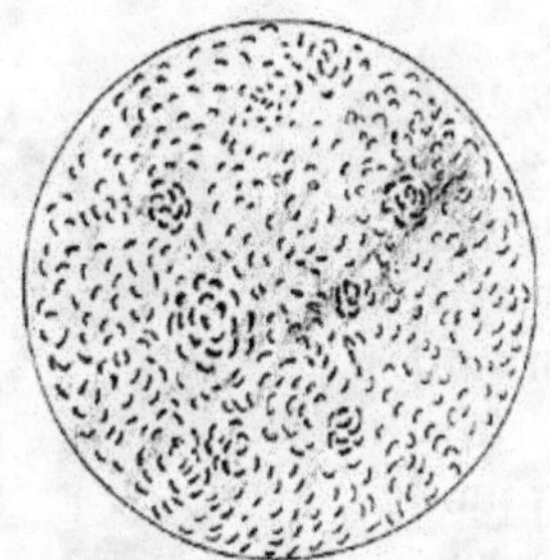

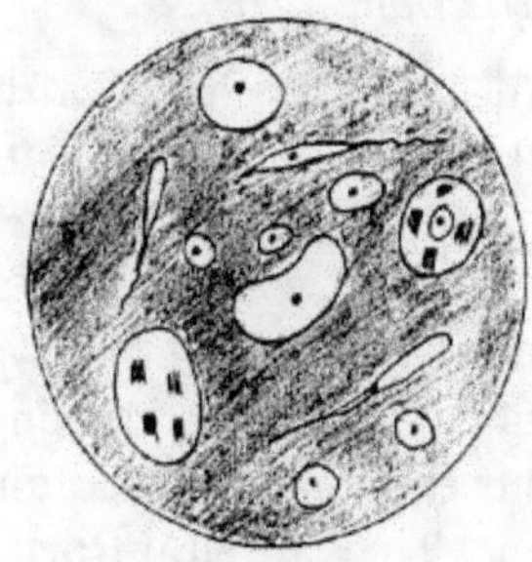

Figure 16.16   *Bordetella pertussis*   Figure 16.17   *Mycobacterium leprae*

Mycobacterium leprae

It is a straight or slightly curved rod measuring 1.8×0.2–0.5 µm size (Figure 16.17). Polar bodies may be present. Branching is observed. It is gram-positive and acid-fast. They occur either singly or in groups. This is the causative agent of leprosy.

Mycobacterium tuberculosis

It is either a straight or curved rod measuring 1.4×0.1–0.8 µm (Figure 16.18). It occurs either singly or in pairs or in clumps. Occasionally branching forms may be seen. It is non-motile, non-sporing and non-capsulated, gram-positive and with Ziehl–Nielsen or carbol fuchsin methods

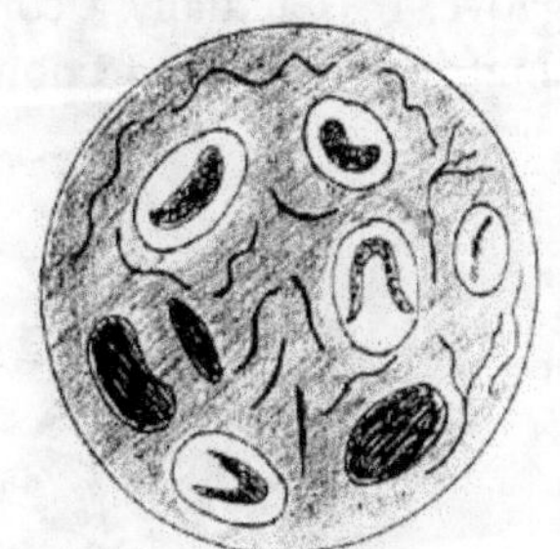

Figure 16.18
Mycobacterium tuberculosis

or fluorescent dyes. The bacilli resist decolorization. This is the causative agent of tuberculosis.

Spirochaetes

Spirochaetes have usually a helical shape but they are capable of twisting or controlling their shape. A special kind of flagellum is located between the outer membrane and the protoplasmic cylinder, i.e., they are located in the periplasmic space of the cell. Ultrastructural details of the periplasmic flagella are similar to that of ordinary flagella with a basal body with discs. Flagella help during swimming of the spirochaetes. Motility depends purely on ordinary flagella since periplasmic flagella cannot be extended outside the cell wall. The exact mechanism of motility is not clearly understood. Unlike bacteria which swim best in media of low viscosity, spirochaetes do it in media of high viscosities. In addition, spirochaetes are also capable of creeping and crawling when in contact with solid surface.

Spirochaetes are responsible for relapsing fever, syphilis. They divide by transverse fission.

Treponema pallidum

It is a thin delicate spirochaete with both ends pointed, about 10 μm in length, 0.1–0.2 μm in width (Figure 16.19). There are about ten angular spirals. They are motile, rotating around long axis. Interestingly they exhibit backward and forward movements simultaneously flexing the body. These movements could be observed under dark ground microscope. They are gram-negative. *Treponema pallidum* is the causative agent of syphilis.

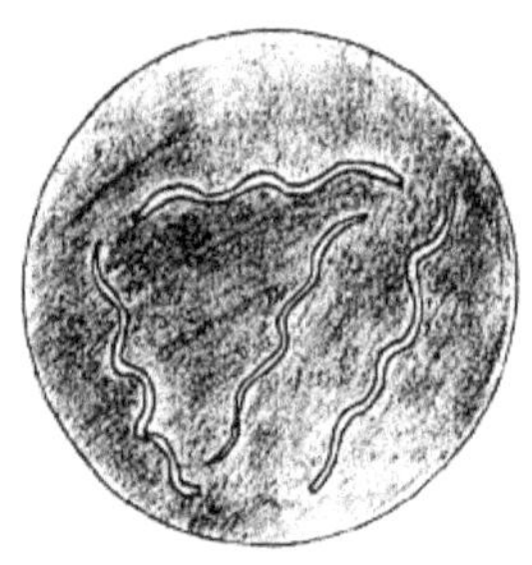

Figure 16.19   *Treponema pallidum*

Borrelia vincentii

It is a motile spirochaete measuring 5–20 μm long and 0.2–0.6 μm wide (Figure 16.20). It takes carbol fuchsin stain and it has 3–8 coils of varying sizes. It is gram-negative. It is actually a commensal but under pre-disposing conditions like malnutrition or viral infection, they may cause gingivostomatitis or oropharyngitis.

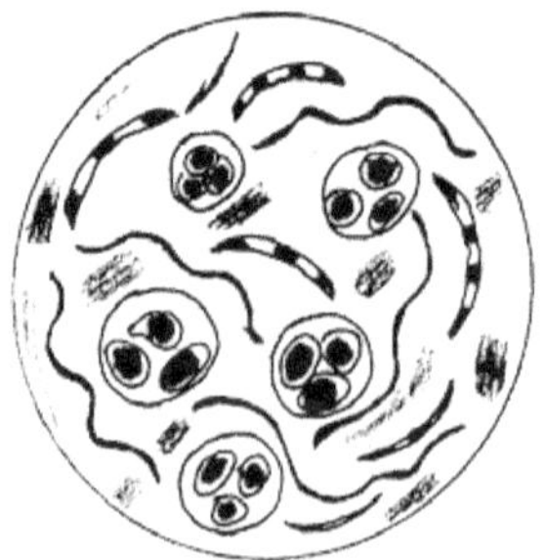

Figure 16.20   *Borrelia vincentii*

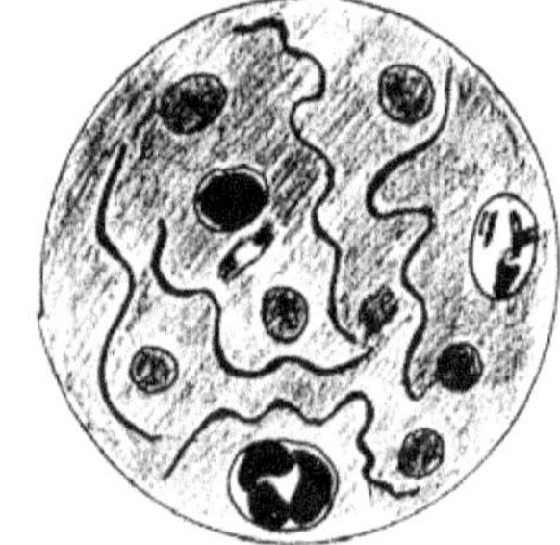

Figure 16.21   *Borrelia recurrentis*

Borrelia recurrentis

These are irregularly shaped with one or both ends pointed (Figure 16.21). It consists of five to ten loose spiral coils at intervals of about 2 μm and exhibit motility with lashing movements. It stains with Giemsa stain and is gram-negative. It is the causative agent of relapsing fever.

Leptospira

These are very delicate measuring 6–20 μm long and 0.1 μm thick (Figure 16.22). They have a number of coils all set together with great difficulty and can be observed under dark ground illumination or under electron microscope. They give the appearance of an umbrella handle because their ends are hooked. They stain with Giemsa but are gram-negative and are actively motile.

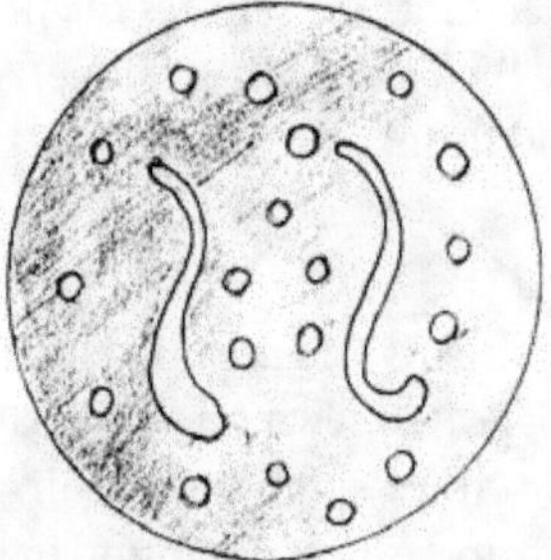

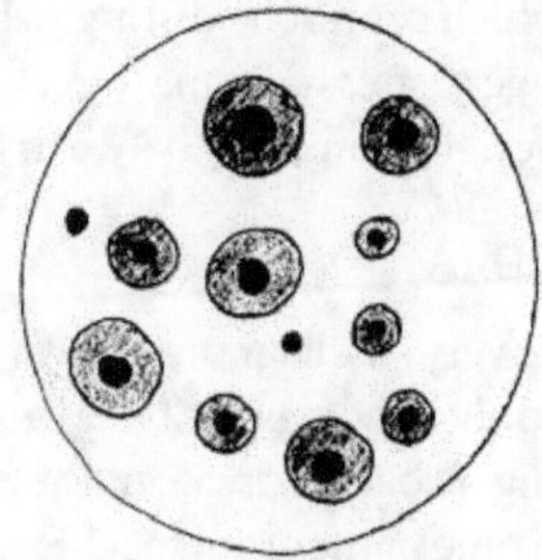

Figure 16.22   *Leptospira*   Figure 16.23   *Mycoplasma*

Mycoplasma

These are the minutest free-living microorganisms and are pleomorphic (Figure 16.23). They are in the form of granules and filaments differing in size. Granules vary in size from 125–250 μm and larger granules are about 500–1000 μm in size. These larger bodies are disc-, baloon- or star-shaped. There is no spore formation. They have no flagella, and are non-motile. Cell wall is absent. These are gram-negative but take Giemsa stain.

FUNGI

Without chlorophyll, fungi are eukaryotic chemoorganotrophic organisms. The thallus or body of the fungus is of a single cell as in yeasts and with filaments 5–10 μm across, and which are commonly branched. Some fungi are dimorphic, i.e., they exist in two forms. Some of the pathogenic fungi of man and other animals have a unicellular form in their hosts, but when they grow saprophytically either in laboratory culture medium or soil, they have a filamentous mould form, so a fungal colony is either a mass of yeast cells or it may be a filamentous mat of mould.

Like *Actinomycetes*, fungi also anastomose to form filaments and mycelia and produce sexual spores which are the result of the fusion of two cells and sexual spores are formed by the differentiation of spore-bearing cells of the spore-bearing hyphae without fusion. In some cases, the tips of the hyphae fragment and produce spores called conidia (*Pencillium*). Fungi can be stained easily.

A few fungi are pathogenic causing serious diseases like meningitis, European blastomycosis, and some infections of skin and mucous membrane.

Yeast cells in general are larger than bacteria. Their size varies from 1–5 μm in width and 5–30 μm in length. They are commonly egg-shaped, some are elongated and others are spherical. The yeast cell is surrounded by a true cell wall. They are non-cellular, uninucleated and are capable of budding and fission.

Actinomycetes have a system of anastomosing branching filaments which are known as hyphae. They intertwine and anastomose greatly to form a colony called mycelium. Sometimes these anastomosing filaments break up to form bacilli-like bodies which cannot be distinguished from bacteria. Sometimes the protoplasm of hyphae aggregates to form spore-like bodies which are gram-positive. According to some, they are transitory between bacteria and fungi.

RICKETTSIA

These are very small, gram-negative microorganisms and are obligate parasites able to grow only in host cells. They cause typhus, spotted fever and associated diseases. Appearing as bacilli or cocci, they may occur either singly or in pairs or in dense masses. Some of these are specific to cytoplasm and others are found in the nucleoli. Rickettsia could easily be demonstrated with Giemsa stain or with Machiavello's method. Some other diseases caused by them are classical typhus fever, murine typhus fever, rickettsial pox and scrub typhus.

VIRUSES

Viruses are too small and cannot be seen with naked eye. They are so small that they easily pass through filters. Viruses are responsible for causing serious diseases like yellow fever, poxes, influenza, measles, mumps, rabies, colds, infective hepatitis, encephalitis and poliomyelitis. Viruses have infectious particles called inclusion bodies of different sizes. These are found in the cytoplasm or nucleus of infected host cells. These inclusion bodies can be demonstrated by special staining methods. Many viruses like rabies, poliomyelitis and measles form inclusion bodies which are also seen in other infections (chlamydid) like psittacosis and trachoma.

STAINING PROCEDURES FOR BACTERIA

GRAM STAINING

With gram-staining technique, the gram-positive organisms take the colour of the dye. Iodine is the mordant and after mordanting, a precipitate is formed. This precipitate is soluble in water.

Gram Weigert Method (Krajian and Gradwohl, 1952)

Fixation

    Zenker's formol or 10 per cent formalin

Reagents Required

    Eosin
    Crystal violet
    Aniline oil

Preparation of Reagents

    Solution 1 (Eosin solution)
        Eosin   1.0 g
        Distilled water   100.0 ml

Solution 2 (Staining solution)

Crystal violet	5.0 g
(Gentian violet)	
95 per cent ethyl alcohol	10.0 ml
Aniline oil	2.0 ml
Distilled water	88.0 ml

First mix aniline oil in water. Then add crystal violet solution (dissolved in alcohol). This solution lasts for several months.

Solution 3 (Gram's iodine)

Iodine	1.0 g
Potassium iodide	2.0 g
Distilled water	300.0 ml

Dissolve potassium iodide in water and then add iodine.

Procedure

1. Deparaffinize and hydrate slides to water.
2. Transfer to solution 1 for 5 minutes.
3. Rinse in water.
4. Place in solution 2 (Gentian violet) for 10 minutes.
5. Wash and flood the slides with solution 3 (Gram's iodine).
6. Blot with filter paper.
7. Flood the slides with equal parts of aniline oil and xylene.
8. Clear in xylene and mount.

Result

Gram-positive bacteria and fungi	Violet
Gram-negative bacteria and fungi	Do not stain
Fibrin	Blue-black

Rationale

Differential staining depends on the permeability of the cell membrane (Bartholomew and Mittwer 1950, 1951). In gram-positive bacteria, the linkage is between the acidic groups of bacteria with alkaline group of the dye. Iodine forms a dye complex which is dissociated with alcohols. Cell membrane is permeable to alcohol. This decolorization occurs and the reaction is gram-negative. On the other hand if the membrane is not permeable, decolorization does not occur and the reaction is gram-positive. Others who dealt with gram-staining are Bartholomew and Finkelstein (1959) and Mittwer *et al.* (1950).

Brown and Brenn Method (1931)

Fixation

10 per cent formalin

Reagents Required

Haematoxylin
Gentian violet

Sodium bicarbonate
Basic fuchsin
Acetone
Picric acid

Preparation of Reagents

Solution 1 Mayer's haematoxylin (*see* Chapter 7)

Solution 2 Gentian violet
    Crystal violet   1.0 g
    Distilled water   100 ml

Solution 3 (Sodium bicarbonate)
    Sodium bicarbonate   5.0 g
    Distilled water   100.0 ml

Solution 4 Gram's iodine (*see* page 351)

Solution 5 Basic fuchsin
    Basic fuchsin   250 mg
    Distilled water   100.0 ml

Dilute this in the ratio 1:100 ml water.

Solution 6: Acetone picric acid
    Picric acid   100 mg
    Acetone   100.0 ml

Procedure

1. Deparaffinize and hydrate slides to water.
2. Transfer slides to solution 1 (Mayer's haematoxylin) for 3 minutes.
3. Wash in running water.
4. Flood slides with a solution containing 26 drops of solution 2 and 5 drops of solution 3 (sodium bicarbonate) for 2 minutes.
5. Rinse in water.
6. Transfer to solution 4 (Gram's iodine) for 1 minute.
7. Rinse in water.
8. Flood the slide with a solution containing 1 part ether and 3 parts acetone until no blue colour comes off.
9. Transfer to solution 5 (basic fuchsin) for 5 minutes.
10. Wash quickly.
11. Differentiate in solution 6 (acetone–picric acid mixture). Keep it till section turns yellowish pink.
12. Rinse in acetone, aceto-xylene (equal parts).
13. Transfer to xylene and mount.

Result

Gram-positive organisms	Deep violet or black
Gram-negative organisms	Bright red
Nuclei	Reddish brown

Cytoplasm   Yellow
Cartilage   Pink

Leaver *et al.* (1977) Substitute Method for Brown and Brenn

Fixation

10 per cent neutral buffered formalin

Reagents Required

Crystal violet
Gram's iodine (*see* page 351)
Malachite green
Pyronine Y

Preparation of Reagents

Solution 1 (Crystal violet)
    Crystal violet   1.0 g
    Distilled water   100.0 ml

Solution 2 (Gram's iodine) (*see* page 351)

Solution 3 (Sandiford stain)
    Malachite green   0.05 g
    Pyronin Y   0.15 g
    Distilled water   100.0 ml

Procedure

1. Dewax and hydrate slides to water.
2. Treat with solution 1 for 3 minutes.
3. Rinse in water.
4. Transfer to solution 2 for 3 minutes.
5. Rinse in water.
6. Differentiate in a mixture of equal parts of acetone and absolute alcohol.
7. Wash in water.
8. Counterstain with solution 3.
9. Rinse in distilled water.
10. Dehydrate clear and mount.

Result

Gram-positive bacteria   Purple-black
Gram-negative   Bacteria stain red
Background   Blue-green

Lillie's Quick Method (1928): Bartholomew's Modification (1962)

Fixation

Any general fixative

Reagents

1. Crystal violet
2. Ammonium oxalate
3. Lugol's iodine (*see* Chapter 1)
4. Safranin

Preparation

Solution 1 (Crystal violet solution)

Crystal violet	2.0 g
90 per cent ethyl alcohol	20.0 ml
1 per cent ammonium oxalate	80.0 ml
(1 g/100 ml water)	

Filter the solution. This solution lasts long.

Solution 2 (Lugol's iodine) (*see* Chapter 1)

Solution 3 (Safranin)

Safranin O	500 mg
Distilled water	100 ml

Procedure

1. Deparaffinize and hydrate slides to water.
2. Transfer to solution 1 (crystal violet solution) for 30 seconds.
3. Wash in running water.
4. Treat with solution 2 for 1 minutes.
5. Wash in running water.
6. Decolorize in propyl alcohol.
7. Wash.
8. Counterstain with solution 3.
9. Wash in running water.
10. Dehydrate, clear and mount.

Result

Gram-positive bacteria	Blue-black
Gram-negative bacteria	Red
Nuclei	Red
Cytoplasm, fibrin, and collagen	Pink

McCallum–Goodpasture Method for Gram-positive and Gram-negative Bacteria (McCallum, 1919)

Fixation

Any general fixative

Reagents Required

Basic fuchsin
Aniline

Phenol crystal
Gram's iodine
Gentian violet
Picric acid

Preparation of Reagents

Solution 1 (Goodpasture's solution)

Basic fuchsin	590 mg
Aniline oil	1.0 ml
Phenol crystals	1.0 ml
30 per cent alcohol	100.0 ml

Solution 2 (Gram's iodine (see page 351)

Solution 3 (Sterling's gentian violet solution)

Gentian violet	5.0 g
100 per cent alcohol	10.0 ml
Aniline	2.0 ml
Distilled water	88.0 ml

Solution 4 (Saturated picric acid solution)

Procedure

1. Deparaffinize and hydrate slides to water.
2. Place in solution 1 for 10 minutes.
3. Rinse in distilled water.
4. Differentiate in 40 per cent formalin.
5. Wash in running water.
6. Transfer to solution 4.
7. Rinse in distilled water.
8. Differentiate in 95 per cent alcohol.
9. Place in solution 3 for 3 minutes.
10. Rinse in distilled water.
11. Transfer to solution 2 for 1 minute.
12. Rinse in distilled water.
13. Dehydrate, clear and mount.

Result

Gram-positive bacteria	Blue
Gram-negative bacteria stain	Red
Other elements	Various shades of red to purple

Brown–Hopps Method for Gram-positive and Gram-negative Bacteria

Fixation

10 per cent neutral buffered formalin

Reagents Required

Crystal violet
Gram's iodine (*see* page 351)
Cellosolve
Basic fuchsin
Formalin
Glacial acetic acid

Preparation of Reagents

Solution 1 (1 per cent crystal violet solution)
Crystal violet   1.0 g
Distilled water   100.0 ml

Solution 2 (Gram's iodine solution (*see* page 351)

Solution 3 (Cellosolve–Ethylene glycol monoethyl ether)

Solution 4 (0.5 per cent basic fuchsin solution)
Basic fuchsin   500 mg
Distilled water   100.0 ml

Solution 5 (Gallego's differentiating solution)
Distilled water   50.0 ml
40 per cent formalin   1.0 ml
Glacial acetic acid   0.5 ml

Solution 6 (1.5 per cent tartrazine solution)
Tartrazine   1.5 g
Distilled water   100.0 ml

Procedure

1. Deparaffinize and hydrate slides to water.
2. Place slides in a staining rack and flood with solution 1 for 2 minutes.
3. Rinse in distilled water.
4. Mordant in solution 2 for 5 minutes.
5. Rinse in distilled water.
6. Differentiate in solution 3 for 10 seconds.
7. Rinse rapidly in distilled water.
8. Transfer to solution 4 for 5 minutes.
9. Rinse in distilled water.
10. Differentiate in solution 5 for 5 minutes.
11. Rinse in distilled water.
12. Transfer to solution 6 for 3 seconds.
13. Wash rapidly in distilled water.
14. Dehydrate, clear and mount.

Results

Gram-positive bacteria   Blue

Gram-negative bacteria   Red
Background   Yellow

Taylor's Method for Gram-positive and Gram-negative Bacteria (Taylor, 1966)

Fixation

10 per cent neutral buffered formalin

Reagents Required

Haematoxylin
Hydrochloric acid
Gentian violet
Ammonium oxalate
Basic fuchsin
Methyl alcohol

Preparation of Reagents

Solution 1 (Harris haematoxylin solution (*see* Chapter 7)

Solution 2 (Acid alcohol)

Solution 3 (Hucker's solution)

Gentian violet 10 per cent alcohol	2.0 ml
Distilled water	18.0 ml
1 per cent ammonium oxalate	80.0 ml

Mix well and filter.

Solution 4 (Basic fuchsin stock solution)

Basic fuchsin	100 mg
Methyl alcohol	95.0 ml
Distilled water	5.0 ml

Solution 5 (Basic fuchsin working solution)

Solution 4	5.0 ml
Distilled water	60.0 ml

Solution 6 (Ethyl ether–acetone solution)

Solution 7 (0.1 per cent picric acid–acetone solution)

Picric acid	100 mg
Distilled water	100.0 ml

Solution 8 (Acetone–xylene solution)

Acetone	1 part
Xylene	2 parts

Solution 9 (Acetone–xylene solution II)

Acetone	1 part
Xylene	3 parts

Procedure

1. Deparaffinize and bring down slides to water.
2. Place in solution 1 for 10 minutes.
3. Wash in running water.
4. Differentiate in solution 2.
5. Wash in running water.
6. Wash in saturated lithium carbonate for few seconds.
7. Wash in running water.
8. Transfer to solution 3 for 2 minutes.
9. Wash quickly in water.
10. Mordant in Gram's iodine (*see* page 351) for 1 minute.
11. Wash in running water.
12. Differentiate in solution 6 for 1 minute.
13. Transfer to solution 5 for 2 minutes.
14. Wash in water.
15. Pass on to solution 7.
16. Dehydrate and clear in solution 8 and 9.
17. Mount.

Result

Gram-positive bacteria	Blue or blue-black
Gram-negative bacteria	Bright red
Nuclei	Brownish red
Erythrocytes	Red
Connective tissue	Red

STAINING OF ACID-FAST BACTERIA

Ziehl–Neelsen (Putt's Modification 1951)

Fixation

Any general fixative, 10 per cent formalin preferred.

Reagents Required

New fuchsin
Phenol
Methyl alcohol
Glacial acetic acid
Methylene blue chloride
Lithium carbonate

Preparation of Reagents

Solution 1 (New fuchsin)

New fuchsin	1.0 g
Phenol	5.0 g

Ethyl or methyl alcohol	10.0 ml
Distilled water	100.0 ml

Solution 2 (Acetic alcohol)

Glacial acetic acid	5.0 ml
Absolute ethyl alcohol	95.0 ml

Solution 3 (Counterstain)

Methylene blue chloride	0.5 g
Absolute ethyl alcohol	100 ml

Procedure

1. Deparaffinize and hydrate slides to water.
2. Immerse in solution 1 for 3 minutes.
3. Treat with saturated lithium carbonate for 1 minute.
4. Differentiate in solution 2 (acetic alcohol). Tissue should take a pale pink colour.
5. Rinse in absolute alcohol for 2 minutes.
6. Couterstain in solution 3.
7. Wash, dehydrate, clear and mount.

Result

Acid-fast bacteria	Red
Mast cells stain deep	Blue
Other bacteria stain	Blue

Harada Method (1973)

Fixation

10 per cent neutral buffered formalin

Reagents Required

Basic fuchsin
Phenol
Methylene blue
Potassium hydroxide

Preparation of Reagents

Solution 1 (Carbol fuchsin)

Basic fuchsin (alcohol saturated)	10.0 ml
5 per cent aqueous phenol	90.0 ml

Solution 2 (Loeffler alkaline methylene blue)

Methylene blue	3.0 g
Absolute alcohol	30.0 ml
Potassium hydroxide (0.01 per cent aqueous)	100.0 ml

Procedure

1. Dewax and hydrate slides to water.
2. Place in 1 per cent potassium permanganate for oxidation for 1 hour.

3. Rinse in water.
4. Transfer to solution 1 for 5 minutes.
5. Bleach in 1 per cent oxalic acid for 3 minutes.
6. Rinse in distilled water.
7. Decolorize in 1 per cent HCl in 70 per cent alcohol for 20 seconds.
8. Wash in water.
9. Counterstain solution 2 (dilute in the ratio 1 : 9 with distilled water) for 15 seconds.
10. Rinse in tap water.
11. Dehydrate, clear and mount.

Result

Acid-fast bacteria	Red
Mast granules	Deep blue
Other bacteria	Blue
Nuclei	Blue

Kin Youn's Carbol Fuchsin Method (Marti and Johnson's Modification, 1951)

Fixation

Any general fixative
10 per cent formalin

Reagents

Basic fuchsin
Phenol
Nitric acid
Malachite green
Tergitol

Preparation of Reagents

Solution 1 (Carbol fuchsin)

Basic fuchsin	4.0 g
Phenol	8.0 g
95 per cent ethyl alcohol	20.0 ml
Distilled water	100.0 ml

Add 1 drop tergitol up to 7 to every 30 ml or above.

Solution 2 (Acid alcohol)

Conc. nitric acid	0.5 ml
95 per cent ethyl alcohol	95.0 ml

Solution 3 (Malachite green)

Malachite green oxalate	1.0 g
Distilled water	100.0 ml

Procedure

1. Deparaffinize and hydrate slides to water.
2. Place in solution 1 (carbol fuchsin) for 30 minutes.
3. Wash in running water for 5 minutes.
4. Decolorize in solution 2 (acid alcohol) for 3 minutes.
5. Wash in running water.
6. Rinse in 95 per cent alcohol.
7. Wash in running water.
8. Transfer to solution 3 (malachite green) for 30 seconds.
9. Wash in water.
10. Dehydrate, clear and mount.

Result

Acid-fast bacteria	Red
Tissue elements	Green

Fite–Formaldehyde Method (Wade's Modification, 1957)

Fixation

Preferably Zenker

Reagents Required

New fuchsin
Phenol
Methyl alcohol
Acid fuchsin
Picric acid
Potassium permanganate
Oxalic acid

Preparation of Reagents

Solution 1 (Phenol new fuchsin)

New fuchsin	500 mg
Phenol	5.0 g
Ethyl or methyl alcohol	10.0 ml
Distilled water	100.0 ml

Solution 2 (Van Gieson modified)

Acid fuchsin	10 mg
Picric acid	100 mg
Distilled water	100.0 ml

Procedure

1. Deparaffinize in turpentine–paraffin oil (2 : 1) giving 2 changes.
2. Drain and blot the excess fluid.

3. Stain in solution 1 (phenol fuchsin) overnight.
4. Wash in tap water.
5. Flood with formalin for 5 minutes.
6. Wash in water.
7. Treat with 5 per cent sulphuric acid (5 ml /95 ml water) for 5 minutes.
8. Wash.
9. Treat with 1 per cent $KMnO_4$ (1 g/100 ml water) for 3 minutes.
10. Wash.
11. Bleach in oxalic acid 2:5 per cent (2.5 g/100 ml water) for 30 seconds.
12. Wash.
13. Place in solution 2 (modified Van Gieson) for 3 minutes.
14. Rinse in 95 per cent alcohol.
15. Dehydrate, clear and mount.

Results

Acid-fast bacteria	Deep blue
Connective tissue	Red
Other tissue elements	Yellow

(Also refer Beamer and Firminger, 1955, Tilden and Tanaka 1945).

Truant's Fluorescent Method for Acid-fast Bacteria (Truant's 1962)

Fixation

10 per cent neutral buffered formalin

Reagents Required

Haematoxylin
Auramine
Rhodamine B
Phenol

Preparation of Reagents

Solution 1 (Weigert's haematoxylin solution (*see* Chapter 7)

Solution 2 (Auramine–Rhodamine B solution)

Auramine	1.5 g
Rhodamine B	750 mg
Glycerine	75.0 ml
Phenol crystals	10.0 ml
Distilled water	50.0 ml

Procedure

1. Dewax and hydrate slides to distilled water.
2. Place in solution 1 for 10 minutes.
3. Wash in running water.

4.   Transfer to solution 2 at 60°C for 10 minutes.

5.   Wash in distilled water.

6.   Differentiate in acid alcohol.

7.   Wash in distilled water.

8.   Dehydrate, clear and mount.

Results

Bacilli fluoresce reddish yellow.

Wade's Method for Acid-fast Bacteria (Wade, 1952)

Fixation

Zenker

Reagents Required

Terpentine
Potassium permanganate
Paraffin oil
Phenol crystals
New fuchsin
Oxalic acid
Sulphuric acid

Preparation

Solution 1 (Deparaffinizing solution)
    Turpentine   2 parts
    Paraffin oil   1 part

Solution 2 (Carbol new fuchsin)
    New fuchsin magenta III   500 mg
    Phenol crystals liquid   5.0 ml
    100 per cent alcohol   10.0 ml
    Distilled water   100.0 ml

Solution 3 (5 per cent sulphuric acid)
    Conc. sulphuric acid   5.0 ml
    Distilled water   95.0 ml

Solution 4 (1 per cent potassium permanganate)
    1 g/100 ml water

Solution 5 (2 per cent oxalic acid)
    2 g/100 ml water

Solution 6 (Modified Van Gieson)
    Acid fuchsin   10 mg
    Picric acid   1.0 g
    Distilled water   100.0 ml

Procedure

1. Deparaffinize in solution 1 and hydrate slides to water.
2. Place in solution 2 overnight.
3. Wash in distilled water.
4. Place in concentrated formalin until sections are blue.
5. Wash in running water.
6. Place in solution 3 for 1 minute.
7. Wash in running water.
8. Transfer to solution 4 for 3 minutes.
9. Rinse in tap water.
10. Transfer to solution 5 for 2 minutes.
11. Rinse in tap water.
12. Place in solution 6 for 5 minutes.
13. Wash.
14. Dehydrate and mount.

Result

Acid-fast bacilli	Deep blue
Connective tissue	Red
Background	Yellow

Wade's Modification of the Fite 1 Acid-fast Procedure (Wade, 1957; Fite, 1938; 1940)

Fixation

Zenkers'
10 per cent formalin

Reagents Required

New fuchsin
Picric acid
Acid fuchsin
Turpentine
Liquid paraffin
Formaldehyde
Sulphuric acid
Potassium permanganate
Oxalic acid

Preparation of Solutions

Solution 1 (Fite's new fuchsin)

Absolute alcohol	10.0 ml
Distilled water	100.0 ml
Phenol melted	5.0 ml
New fuchsin	500 mg

Solution 2 (Modified Giemsa solution)
    Picric acid   1.0 g
    Distilled water 100.0 ml
    Acid fuchsin   10 mg

Solution 3 (Sulphuric acid solution)
    Sulphuric acid   0.5 ml
    Distilled water   90.0 ml
Make up the volume to 100 ml with distilled water.

Solution 4 (Formaldehyde)

Solution 5 (Potassium permanganate solution)
    Potassium permanganate   1.0 g
    Distilled water   90.0 ml
Make up the volume to 100 ml with distilled water.

Solution 6 (Oxalic acid solution)
    Oxalic acid   2.0 g
    Distilled water 100.0 ml

Solution 7 (Deparaffinizing agent)
    Liquid paraffin 1 part
    Turpentine   2 parts

Procedure

1. Deparaffinize in solution 7.
2. Drain the slide, blot, dry and then place in water.
3. Place the slides in solution 1 overnight for 16–24 hours at room temperature.
4. Place them directly in solution 4 for 5 minutes.
5. Wash in tap water.
6. Decolorize in solution 3 and then wash in tap water.
7. Now place the slides in solution 5 for oxidation for 3 minutes.
8. Bleach in solution 6 for 50 seconds.
9. Transfer the slides to solution 2 for 3 minutes.
10. Dehydrate rapidly, clear and mount in permount.

Result

Acid-fast bacilli   Deep blue
Connective tissue   Red
Background   Yellow

Fluorescent Method (Richards *et al.*, 1941; Richards, 1941)

Fixation

10 per cent formalin preferred

Reagents Required

Auramine
Phenol

Conc. hydrochloric acid
Sodium chloride

Preparation of Reagents

Solution 1 (Auramine stain)

Auramine	300 mg
Distilled water	97.0 ml
Melted phenol	5.0 ml

Shake well and dissolve by gentle heat.

Solution 2 (Decolorizing solution)

Conc. hydrochloric acid	0.5 ml
70 per cent ethyl alcohol	100.0 ml
Sodium chloride	500 mg

Procedure

1.   Deparaffinize and hydrate slides to water.
2.   Keep in solution 1 (Auramine stain) for 2 minutes.
3.   Wash in water for 2–3 minutes.
4.   Decolorize in solution 2 for 1 minute.
5.   Wash.
6.   Dry and examine or mount sections in fluorescent mount (Harleco mountant).

Result

Bacilli stain golden yellow.

Cresyl Violet Acetate Method for *Helicobacter* sp.

Fixation

Formalin

Procedure

1.   Dewax and bring sections to water.
2.   Spread 0.1 per cent cresyl violet acetate on the slide for 5 minutes.
3.   Rinse in distilled water.
4.   Blot, dehydrate, clear and mount.

Result

Helicobacter sp. and nuclei stain blue-violet.

Gimenez Method for *Helicobacter* (McMullen *et al.*, 1987)

Preparation of Reagents

Solution 1 Buffer phosphate (*see* Chapter 5)

Solution 2 (Stock carbol fuchsin solution)

Basic fuchsin	1.0 g

    Absolute alcohol   10.0 ml
    5 per cent aqueous phenol   10.0 ml
Filter before use.

Solution 3 (Working carbol fuchsin solution)
    Solution 1   10.0 ml
    Solution 2   4.0 ml
Filter before use

Solution 4 (Malachite green solution)
    Malachite green   0.8 g
    Distilled water   100.0 ml

Procedure

1. Dewax and bring sections to water.
2. Treat with solution 3 for 2 minutes.
3. Wash in tap water.
4. Immerse in solution 4 for 20 seconds.
5. Wash in tap water.
6. Repeat the steps 4 and 5 until sections are blue-green.
7. Blot sections, air-dry.
8. Clear in xylene and mount.

Results

Helicobacter   Red-magenta
Background   Blue-green

CAPSULE STAINING

Hiss Method (Burrows, 1954)

Fixation

Any general fixative
10 per cent formalin

Reagents Required

Basic fuchsin
Crystal violet
Copper sulphate

Preparation

Solution 1 (Basic fuchsin solution)
    a) Basic fuchsin   0.15–0.3 g
        Distilled water   100.0 ml
                    or
    b) Crystal violet   50 to 100 mg
        Distilled water   100.0 ml

Solution 2 (Copper sulphate solution)
  Copper sulphate crystals 20.0 g
  Distilled water   100.0 ml

Procedure

1. Deparaffinize and hydrate slides to water.
2. Flood the slides with either solution 1a or 1b. Gentle heat is required.
3. Wash and blot.
4. Wash with solution 2 (copper sulphate).
5. Blot and dry.
6. Dehydrate, clear and mount.

Result

Capsule stains light pink with solution 1 and blue with solution 2.

SPIROCHAETE STAINING

Dieterle's Method (Modified by Beamer and Firminger, 1955)

Fixation

10 per cent formalin

Reagents Required

Gum mastic
Hydroquinone
Sodium sulphate
Formalin
Glycerol
Pyridine
Uranium nitrate
Silver nitrate

Preparation of Reagents

Solution 1 (Dilute gum mastic)
  Gum mastic saturated in absolute alcohol 30 drops
  95 per cent ethyl alcohol   40.0 ml

Solution 2 (Developing solution)
  Hydroquinone 500 mg
  Sodium sulphate 60 mg
  Distilled water 20.0 ml

Stir the mixture and add 4 ml formalin and 5 ml glycerol. When all ingredients are thoroughly mixed, add drop by drop.

Procedure

1. Deparaffinize and hydrate slides to water.
2. Transfer to 80 per cent alcohol for 3 minutes.

3. Wash in distilled water for 5 minutes.
4. Transfer to uranium nitrate, 2–3 per cent (2.3 g/100 ml water, pre-heated to 60°C).
5. Wash in distilled water.
6. Place in 95 per cent alcohol.
7. Place in solution 1 for 5 minutes.
8. Wash.
9. Impregnate with silver nitrate 2 per cent (2 g/100 ml water, pre-heated to 60°C) for 40 minutes.
10. Transfer to pre-heated (60°C) solution 2 (developing solution) until sections are pale brown.
11. Rinse in 95 per cent alcohol.
12. Treat with silver nitrate 2 per cent (2 g/100 ml water) for 2 minutes.
13. Wash.
14. Dehydrate, clear and mount.

Result

| Spirochaetes | Black |
| Background | Yellow |

Warthin-Starry Silver Method
(Kerr, 1938; Faulkner and Lillie, 1945a; Bridges and Luna, 1957)

Fixation

10 per cent formalin

Reagents Required

Citric acid
Silver nitrate
Hydroquinone
Gelatine
Triple distilled water

Preparation of Reagents

Solution 1 (Acid water)
    Triple distilled water   100.0 ml
    Citric acid   10.0 g

Solution 2 (2 per cent silver nitrate)
    Silver nitrate   2.0 g
    Acidulated water   100.0 ml

Solution 3 (1 per cent silver nitrate)
    Silver nitrate   1.0 g
    Acid water   100.0 ml

Solution 4 (0.15 per cent hydroquinone)
    Hydroquinone   150 mg
    Acidulated water   100.0 ml

Solution 5 (5 per cent gelatine)
 Gelatine (extra pure) 5.0 g
 Acidulated water 100.0 ml

Solution 6 (Developer)
 Solution 2 1.5 ml
 Solution 5 3.75 ml
 Solution 4 2.0 ml

Pre-heat solution to 55–60°C. Mix in order while staining.

Procedure

1. Deparaffinize and hydrate slides to solution 1.
2. Impregnate in solution 3 (1 per cent silver nitrate) at 55–60°C for 30 minutes.
3. Place slides in a coplin jar and pour warm solution 6 (developer) till the sections become golden brown or yellow.
4. Rinse with warm tap water.
5. Dehydrate, clear and mount.

Result

Spirochaetes	Black
Background	Yellow

Levaditi Method for Block Staining (Mallory, 1944)

Fixation

10 per cent formalin

Reagents Required

Silver nitrate
Pyrogallic acid
Formalin

Preparation of Reagents

Solution 1 (Silver nitrate)
 Silver nitrate 1.5–3.0 g
 Distilled water 100.0 ml

Solution 2 (Reducing solution)
 Pyrogallic acid 4.0 g
 Formalin   5.0 ml
 Distilled water 100.0 ml

Procedure

1. Rinse blocks of tissue in tap water.
2. Place in 95 per cent ethyl alcohol.
3. Transfer to distilled water.
4. Impregnate with solution 1 (silver nitrate) at 37°C for 3–5 days.
5. Wash in distilled water.

6. Keep in solution 2 and reduce at room temperature in the dark for 24–72 hours.
7. Wash in distilled water.
8. Dehydrate, clear in cedar wood oil and infiltrate with paraffin.
9. Embed, take 5-μm sections and mount on slide.
10. Dewax with xylene and mount.

Result

Spirochaetes	Black
Background	Brownish yellow

Remarks

All glassware should be cleaned with potassium dichromate and sulphuric acid. All solutions must be fresh. Coat forceps with paraffin and use only triple distilled water.

Krajian Silver Stain for Spirochaetes
(Krajian, 1939 Modified by Walter Smith *et al.*, 1969)

Fixation

10 per cent formalin for 24 hours

Reagents Required

Acetone
Glycerine
Uranium nitrate
Gum mastic
Silver nitrate
Hydroquinone
Sodium sulphite
Formaldehyde
Sodium thiosulphate

Preparation of Reagents

Solution 1 (Mordant solution)

Acetone	30.0 ml
90 per cent ethanol	30.0 ml
Uranium nitrate	3.0 g

Preparation should be done in a clean Erlenmeyer flask. After dissolving the ingredients, keep the solution in an amber-coloured bottle in refrigerator. It lasts long.

Solution 2 (Gum mastic)

Gum mastic	25.0 g
Absolute ethanol	35.0 ml

Put this in clean flask and shake frequently at least for 5 days till the solution is clear. Now the solution is stable and can be stored for a long time. Only the clear portion of the solution is used.

Solution 3 (Dilute gum mastic solution)
Every day add 70 drops of saturated alcoholic gum mastic, i.e., solution 2 to 100 ml of 95 per cent ethanol.

Solution 4 (Silver nitrate stock solution)
Silver nitrate	10.0 g
Triple distilled water	100.0 ml

Store in an amber-coloured bottle in a refrigerator.

Solution 5 (Developer)
Hydroquinone	620 mg
Sodium sulphate	200 mg
Formaldehyde	5.0 ml
Acetone	5.0 ml
Pyridine	5.0 ml
Saturated gum mastic (Solution 2)	5.0 ml
Triple distilled water	30.0 ml

Mix this in order and it should be prepared very fresh, i.e., just before immersing the slides in the solution. Heat it in a water bath at 60°C.

Solution 6
Sodium thiosulphate	5.0 g
Distilled water	100.0 ml

Procedure

1. Deparaffinize and bring down slides to triple distilled water (2 sets of slides, one as control).
2. Place the slides in solution 1 at 60°C in a water bath for 10 minutes.
3. Rinse in distilled water.
4. Transfer slides to solution 3 for 5 seconds.
5. Rinse in distilled water.
6. Take solution 4 in a coplin jar and place the slides in it. Heat in a water bath at 75°C for 7 minutes.
7. Now prepare the final mixing of solution 5 (developer) and heat at 60°C in a water bath.
8. Then expose the slides to light, again dip and expose 5–7 times.
9. Wash in distilled water.
10. Transfer the slides to solution 6 for 5 minutes.
11. Rinse in distilled water.
12. Dehydrate, clear and mount.

Result

Spirochaetes	Black
Background	Brown

STAINING OF FUNGI

Gomori's Methenamine Silver Nitrate Method
(Grocott's Adaptation, 1955; Mowry's Modification, 1959)

Fixation

10 per cent formalin

Reagents Required

Silver nitrate
Methenamine
Borax
Light green
Glacial acetic acid
Periodic acid
Chromic acid
Sodium bisulphite

Preparation of Reagents

Solution 1 (Methenamine silver nitrate stock solution)
    5 per cent silver nitrate   5.0 ml
    (5 g/100 ml water)
    3 per cent methenamine   100.0 ml
    (3 g/100 ml water)

Solution 2 (Working solution)
    5 per cent borax   2.0 ml
    (5 g/100 ml water)
    Distilled water   25.0 ml
    Methenamine stock   25.0 ml
    (Solution 1)

Solution 3 (Light green stock solution)
    Light green yellowish   200 mg
    Distilled water   100.0 ml
    Glacial acetic acid   0.2 ml

Solution 4 (Working solution)
    Solution 3   10.0 ml
    Distilled water   100.0 ml

Procedure

1. Deparaffinize and hydrate slides to water.
2. Oxidize in 0.5 per cent periodic acid for 45 minutes.
3. Wash in water.
4. Oxidize in 5 per cent chromic acid (5 g/100 ml water) for 45 minutes.
5. Wash.
6. Transfer to sodium bisulphite (2 per cent) (2 g/100 ml of water) for 5 minutes.

7. Wash in tap water.
8. Wash in distilled water.
9. Transfer to solution 2 (methenamine silver nitrate) 58°C for 30 minutes.
10. Wash, giving several changes.
11. Tone in gold chloride.
12. Rinse in distilled water.
13. Place in 5 per cent sodium thiosulphate (5 g/100 ml water) for 3 minutes.
14. Wash in running water.
15. Counterstain in solution 4 (Light green) for 30 seconds.
16. Dehydrate, clear and mount.

Results

Fungi	Black
Background	Light green

Metachromatic Method (Kelly *et al.*, 1962)

Fixation

10 per cent formalin

Reagents Required

Sulphuric acid
Toluidine blue
Acetic acid

Preparation of Reagents

Solution 1 (Sulphation reagent)

Concentrated sulphuric acid is added dropwise to an equal volume of acid diethyl ether.

Solution 2 (Toluidine blue)

Toluidine blue	10 mg
3 per cent aqueous acetic acid	100.0 ml

Procedure

1. Deparaffinize and hydrate slides to water.
2. Air-dry for 10 minutes.
3. Keep slides in solution 1.
4. Wash.
5. Transfer to solution 2.
6. Wash in 3 per cent acetic acid.
7. Dehydrate, clear and mount.

Result

Fungi	Red
Background	Pale blue or colourless

(see also Schneider, 1963.)

Gridley's Method for Fungi (Gridley, 1953)

Fixation

10 per cent neutral buffered formalin

Reagents Required

Chromic acid
Metanil yellow
Basic fuchsin
Paraldehyde
Hydrochloric acid
Potassium metabisulphate
Charcoal

Preparation of Reagents

Solution 1 (4 per cent chromic acid solution)
Chromic acid   4.0 g
Distilled water   100.0 ml

Solution 2 (Feulgen's reagent (*see* Chapter 12)

Solution 3 (Aldehyde fuchsin (*see* Chapter 9)

Solution 4 (0.25 per cent metanil yellow)
Metanil yellow   250 mg
Distilled water   100.0 ml
Glacial acetic acid   0.25 ml

Procedure

1. Dewax and bring down slides to water.
2. Oxidize in solution 1 for 1 hour.
3. Wash in running water till colourless.
4. Transfer to solution 2 for 15 minutes.
5. Wash in running water for 10 minutes.
6. Rinse in 70 per cent alcohol.
7. Transfer to solution 3 for 30 minutes.
8. Differentiate in 70 per cent alcohol.
9. Counterstain with solution 4 for 1 minute.
10. Rinse in distilled water.
11. Dehydrate, clear and mount.

Results

Mycelia   Deep purple
Conidia   Deep rose
Background   Yellow
Elastic fibres Purple

Grocott's Method for Fungi (Grocott, 1955)

Fixation

10 per cent neutral buffered formalin

Reagents Required

Chromic acid
Silver nitrate
Methenamine
Borax
Sodium bisulphite
Gold chloride

Preparation of Reagents

Solution 1 (4 per cent chromic acid solution)
    4 g/100 ml water

Solution 2 (5 per cent silver nitrate solution)
    5 g/100 ml water

Solution 3 (3 per cent methenamine solution)
    Hexamethylenetetramine   3.0 g
    Distilled water   100.0 ml

Solution 4 (5 per cent borax solution)
    Borax   5.0 g
    Distilled water   100.0 ml

Solution 5 (Methenamine–silver nitrate stock)
    Solution 2   5.0 ml
    Solution 3   100.0 ml

Solution 6 (Methenamine–silver nitrate working solution)
    Solution 5   25.0 ml
    Distilled water   25.0 ml
    Solution 4   2.0 ml

Solution 7 (1 per cent sodium bisulphite solution)
    Sodium bisulphite   1.0 g
    Distilled water   100.0 ml

Solution 8 (0.1 per cent gold chloride)
    Gold chloride water   100 mg
    Distilled water   100.0 ml

Solution 9 (2 per cent sodium thiosulphate solution)

Solution 10 (0.2 per cent light green solution)
    Light green   200 mg
    Distilled water   100.0 ml
    Glacial acetic acid 0.2 ml

Procedure

1. Dewax and bring down slides to water.
2. Place in solution 1 for oxidizing for 1 hour.
3. Wash in running water.
4. Place in solution 7 for 1 minute.
5. Wash in running water.
6. Transfer to solution 6 at 58°C for 1 hour.
7. Rinse in distilled water.
8. Differentiate in solution 9 for 1 minute.
9. Tone in solution 8 for 5 minutes.
10. Rinse in distilled water.
11. Counterstain with solution 10.
12. Rinse in distilled water.
13. Dehydrate, clear and mount.

Result

Fungi	Black
Mucin	Grey
Mycelia	Rose
Background	Green

Pickett's Fluorescence Method for Fungi (Pickett *et al.*, 1960)

Fixation

10 per cent neutral buffered formalin

Reagents Required

Weigert's haematoxylin (*see* Chapter 7)
Acridine orange

Preparation of Reagents

Solution 1 (Weigert's haematoxylin

Solution 2 (0.1 per cent acridine orange solution)
Acridine orange   100 mg
Distilled water   100.0 ml

Procedure

1. Dewax and bring down slides to water.
2. Place in solution 1 for 5 minutes.
3. Wash in running water.
4. Transfer to solution 2 for 2 minutes.
5. Rinse in distilled water.
6. Dehydrate, clear and mount.

Results

Fungi fluoresce brightly.

Schneider's Gram Stain for the Demonstration of Fungi (Schneider, 1963)

Fixation

10 per cent neutral buffered formalin

Reagents Required

Crystal violet chloride
Ammonium oxalate
Potassium iodide
Iodine
Picric acid

Preparation of Reagents

Solution 1 (Crystal violet stock solution)

Crystal violet chloride	2.0 g
95 per cent alcohol	20.0 ml
1 per cent ammonium oxalate	80.0 ml

Solution 2 (Burke's iodine)

Potassium iodide	2.0 g
Iodine	1.0 g
Distilled water	100.0 ml

Put potassium iodide in a mortar and then add iodine and grind it. Then add 1 ml distilled water and another 5 ml until all potassium iodide and iodine are mixed and then add 10 ml, and make up to 100 ml.

Solution 3 (Saturated picric acid)

Solution 4 (*N*-butanol)

Procedure

1. Deparaffinize sections and hydrate to water.
2. Place in solution 1 for 2 minutes.
3. Rinse in distilled water.
4. Transfer to solution 2 for 2 minutes.
5. Quickly rinse in distilled water.
6. Decolorize in 100 per cent acetone for 10 seconds.
7. Quickly rinse in distilled water.
8. Counterstain in solution 3 for 2 minutes.
9. Place sections in 50 per cent *n*-butanol.
10. Dehydrate in fresh butanol.
11. Clear in xylene.
12. Mount.

Result

Fungi like *Candida*, *Histoplasma*, *Sporothrix*, *Nocardia* and *Actinomycetes* are gram-positive.

STAINING OF RICKETTSIAE

Modified Pappenheim Stain (Castaneda, 1939)

Fixation

Regaud's fluid preferred

Reagents Required

Jenner's stain
Giemsa stain
Glycerol
Methyl alcohol

Preparation of Reagents

Solution 1 (Jenner stock solution)
  Jenner's stain   1.0 g
  Methyl alcohol   400.0 ml

Solution 2 (Stock Giemsa stain)
  Giemsa stain   1.0 g
  Glycerol   66.0 ml

Keep the mixture in an oven (60°C) for 2 hours and then add methyl alcohol (66.0 ml).

Solution 3 (Working solution A)
  Distilled water   100.0 ml
  Glacial acetic acid   1 drop
  Solution 1 (Jenner stock solution)   20.0 ml

Solution 4 (Working solution B)
  Distilled water   100.0 ml
  Glacial acetic acid   1 drop
  Solution 2 (Giemsa stock solution)   5.0 ml

Procedure

1. Deparaffinize and hydrate slides to water.
2. Stain in solution 3 at 37°C for 15 minutes.
3. Transfer directly to solution 4 at 37°C for 30–60 minutes.
4. Dehydrate, clear and mount.

Result

Rickettsiae stain blue to purplish blue.

Castaneda's Method (Gradwohl, 1963)

Fixation

Regaud's or any general fixative

Reagents Required

Dibasic sodium phosphate
Monobasic sodium phosphate
Methylene blue
Potassium hydroxide
Formalin
Safranin

Preparation of Reagents

Solution 1 (Buffer solutions)
　　Solution 1a
　　　　Dibasic sodium phosphate (Na_2HPO_4)　23.86 g
　　　　Distilled water　　　　　　　　　　　1000.0 ml

　　Solution 1b
　　　　Monobasic sodium phosphate (anhydrous)　11.34 g
　　　　　(NaH_2PO_4)
　　　　Distilled water　　　　　　　　　　　　1000.0 ml

Solution 2 (Working solution)
　　Solution 1a　88.0 ml
　　Solution 1b　12.0 ml
　　Formalin　　　0.2 ml

Solution 3 (Methylene blue solution)
　　Methylene blue　　　21.0 g
　　95 per cent alcohol　300.0 ml
　　Potassium hydroxide　0.1 g
　　Distilled water　　　1000.0 ml

Dissolve methylene blue in alcohol and potassium hydroxide in distilled water. Mix methylene blue solution and potassium hydroxide solution and allow to stand for 24 hours.

Solution 4 (Staining solution)
　　Mix solution 2 with solution 3.

Procedure

1. Hydrate slides to 50 per cent alcohol.
2. Place in solution 4 (methylene blue solution) for 2–3 minutes.
3. Wash in running water for 30 seconds.
4. Counterstain in 1 per cent safranin (1 g/100 ml water) for 2 minutes.
5. Rinse in 95 per cent alcohol.
6. Dehydrate, clear and mount.

Result

Rickettsiae stain light blue.

Ordway–MacChiavello Method (Gradwohl, 1963)

Fixation

Regaud's fixative

Reagents Required

Porrier's blue
Eosin bluish

Preparation of Reagents

Solution 1

Porrier's blue 1 per cent   10.0 ml
(1 g/100 ml water)
Eosin bluish 0.45 per cent   15.0 ml
(0.450 mg/100 ml water)

Solution 2
Add 25.0 ml distilled water to solution 1 while stirring.

Procedure

1.   Deparaffinize and hydrate slides to water.
2.   Place in staining solution 2 for 8 minutes.
3.   Decolorize in 95 per cent ethyl alcohol till slides are pale bluish pink.
4.   Dehydrate, clear and mount.

Result

Rickettsiae	Bright red
Inclusion bodies	Bright red

Wolback's Variation of the Giemsa Stain for Rickettsiae (Wolback *et al.*, 1922)

Fixation

Zenker's fluid

Reagents Required

Giemsa
Methanol
Rosin
Sodium carbonate

Preparation of Reagents

Solution 1 (Giemsa stain solution)
(Prepare stock solution of Giemsa as Good Burn and Marmion stain)

Giemsa stock	2.5 ml
Distilled water	100.0 ml
Absolute methanol	2.5 ml
0.5 per cent sodium carbonate	0.25 ml

Solution 2 (Colophonium alcohol)

Rosin (Colophonium)	10.0 g
95 per cent ethanol	100.0 ml

Solution 3 (Working solution)

Solution 2 (10 per cent rosin)	0.5 ml
85 per cent ethanol	100.0 ml

Procedure

1. Deparaffinize and bring down sections to water.
2. Immerse in solution 1 for 1 hour. Repeat it at 1-hour interval overnight.
3. Rinse in distilled water.
4. Differentiate in 95 per cent alcohol and later place in solution 3 and expose to sunlight.
5. Dehydrate, clear and mount.

Results

Rickettsiae and Chlamydiae	Reddish purple
Nuclei	Blue-purple
Cytoplasm	Blue

Pinkerton's Method for Rickettsiae (Simmons and Gentzkow, 1944)

Fixation

Zenker's, Regaud's

Reagents Required

Methylene blue
Basic fuchsin
Citric acid

Preparation of Reagents

Solution 1 (1 per cent methylene blue solution)

Methylene blue	1.0 g
Distilled water	100.0 ml

Solution 2 (0.25 per cent basic fuchsin)

Basic fuchsin	250 mg
Distilled water	100.0 ml

Solution 3 (0.5 per cent citric acid solution)

Citric acid	500 mg
Distilled water	100.0 ml

Procedure

1. Dewax and hydrate slides to water.
2. Place in solution 1 overnight.
3. Rinse in 95 per cent alcohol.
4. Rinse in distilled water.
5. Transfer to solution 2 for 30 minutes.
6. Decolorize rapidly in solution 3.
7. Dehydrate, clear and mount.

Result

Rickettsia   Bright red
Nuclei   Blue

STAINING METHODS FOR INCLUSION BODIES

Lendrum Method (Lendrum, 1947)

Fixation

Zenker's

Reagents Required

Mayer's haematoxylin (*see* Chapter 7)
Phloxine
Tartrazine

Preparation of Reagents

Solution 1 (Mayer's haematoxylin)

Solution 2 (0.5 per cent phloxine solution)

Phloxine	500 mg
70 per cent alcohol	200.0 ml
Calcium chloride	1.0 g

Solution 3 (2.5 per cent tartrazine solution)

Tartrazine	2.5 g
Ethylene glycol monoethyl ether	100.0 ml

Procedure

1. Dewax and hydrate slides to water.
2. Place in solution 1 for 10 minutes.
3. Wash in running water till sections are blue.
4. Place in solution 2 for 30 minutes.
5. Rinse in distilled water.
6. Place in solution 3 until inclusion bodies are red.
7. Dehydrate, clear and mount.

Result

Inclusion bodies	Red
Nuclei	Blue
Background	Yellow

Page–Green Method (Page and Green, 1942)

Fixation

Any general fixative

Reagents Required

Hydrochloric acid
Beibrich scarlet
Orange G
Fast green
Phosphotungstic acid
Phosphomolybdic acid
Acetic acid
Haematoxylin
Ammonia

Preparation of Reagents

Solution 1 (1 per cent acid alcohol)

Concentrated HCl	100.0 ml
Distilled water	100.0 ml

Solution 2 (S. Lorr's staining solution)

50 per cent alcohol	100.0 ml
Biebrich scarlet	1.0 g
Orange G	250 mg
Fast green FCF	75 mg
Phosphotungstic acid	500 mg
Phosphomolybdic acid	0.5 g
Glacial acetic acid	2.0 ml

Solution 3 (Harris haematoxylin (*see* Chapter 7)

Solution 4 (Ammonia water solution)

Tap water	1000.0 ml
Ammonium hydroxide 20 per cent	2.3 ml

Procedure

1. Dewax and bring down slides to water.
2. Place in solution 3 for 5 minutes.
3. Rinse in water.
4. Differentiate in solution 1 until no haematoxylin comes out.
5. Wash in water.
6. Blue sections in solution 4.
7. Wash in running water for 10 minutes.

8. Transfer to solution 2 for 1 minute.
9. Rinse in 95 per cent alcohol.
10. Dehydrate, clear and mount.

Result

Inclusion bodies	Brilliant red
Connective tissue	Light green
Elastic tissue	Purplish red
Keratin	Orange
Erythrocytes	Orange-red
Nuclei	Blue

Modified Gomori Method for Trachoma

Fixation

Any general fixative; Regaud's preferred

Reagents Required

Giemsa strain
Monobasic sodium phosphate
Dibasic sodium phosphate

Preparation of Reagents

Solution 1a

Dibasic sodium phosphate	9.5 g
Distilled water	1000.0 ml

Solution 1b

Monobasic sodium phosphate	9.2 g
Distilled water	1000.0 ml

Solution 2 (Working buffer)

Solution 1a	72.0 ml
Solution 1b	28.0 ml
Distilled water	900.0 ml

Solution 3 (Giemsa stock)

Giemsa stain	1.0 g
Glycerol	66.0 ml

Solution 4

Dilute 1 drop of Giemsa stock (solution 3) with 2 ml of solution 2 (buffer working).

Procedure

1. Hydrate slides to water (these include smears).
2. Transfer to solution 4 for 1 hour at 37°C.
3. Rinse in 95 per cent alcohol giving 2 changes.
4. Dehydrate, clear and mount.

Result

Inclusion bodies are stained blue to purplish blue.

Table 16.1 Histochemical techniques applied for the demonstration of microorganisms

Technique	Fixative	Gram-positive	Gram-negative	Acid-fast	*Helicobacter*	Capsule	Bacilli	Spirochaetes
Gram Weigert method	Zenker formol, 10% formalin	Violet	Unstained					
Brown and Brenn method	10% formalin	Deep violet	Bright red					
Leaver *et al.* method	10% neutral buffered formalin	Purple-black	Red					
Lillies quick method	Any general fixative	Blue-black	Red					
Brown–Hopps method	10% neutral buffered formalin	Blue	Red					
McCallum–Good Pasture method	Any general fixative	Blue	Red					

Taylor's method	10% neutral buffered formalin	Blue or blue-black	Bright red		
Ziehl–Neelsen method	Any general, 10% formalin			Red	
Harada's method	10% neutral buffered formalin			Red	
Kin Youn's carbol fuchsin method	10% formalin, any general			Red	
Fite-formaldehyde method	Zenker			Deep blue	
Traunt's Fluorescent method	10% neutral buffered formalin				Reddish yellow
Wade's method	Zenker			Deep blue	

(*Contd.*)

Table 16.1 (Continued)

Technique	Fixative	Gram-positive	Gram-negative	Acid-fast	*Helicobacter*	Capsule	Bacilli	Spirochaetes
Wade's modification of Fite	Zenker's 10% formalin			Deep blue				
Fluorescent method	10% formalin						Golden yellow	
Cresyl Violet Acetate method	Formalin				Blue-violet			
Hiss method	10% formalin					Light pink		
Dieterle's method	10% formalin							Black
Warthin-Starry Silver method	10% formalin							Black
Levaditi method	10% formalin							Black

Table 16.2   Histochemical techniques applied for the demonstration of microorganisms

Technique	Fixative	Spirochaetes	Fungi	Rickettsiae	Inclusion bodies	Mycelia	Conidia
Krajian Silver Sain	10% formalin	Black					
Gomori's Methenamine Silver Nitrate method	10% formalin		Black				
Metachromatic method	10% formalin		Red				
Gridley's method	10% neutral buffered formalin					Deep purple	Deep rose
Grocott's method	10% neutral buffered formalin		Black			Rose	
Pickett's Fluorescence method	10% neutral buffered formalin		Fluoresces brightly				
Schneider's Gram stain	10% neutral buffered formalin		Gram-positive				
Casteneda's method	Regaud's			Blue or purplish blue			
Casteneda's method	Regaud's			Light blue			

(Contd.)

Table 16.2   (Continued)

Technique	Fixative	Spirochaetes	Fungi	Rickettsiae	Inclusion bodies	Mycelia	Conidia
Ordway–Macchiavello method	Regaud's			Bright red	Bright red		
Wolback's variation	Zenker			Reddish purple			
Pinkerton's method	Zenker or Regaud			Bright red			
Lendrum's method	Zenker				Red		
Page–Green method	Any general				Brilliant red		
Modified Gomori's method	Regaud's or any general				Blue or purplish blue		

MICROORGANISMS IN SMEARS

17

There are a number of techniques applied to microorganisms, most of them involving the use of smears. The usual procedure of fixing the smear is to pass the slide through the flame of a burner. This is heat-fixation. This does not hold good for fine cytological details such as the nuclear apparatus of bacterial cells or for the demonstration of mycoplasma cells. Heat-fixed specimens could be stored for quite some time but not indefinitely.

HISTOCHEMICAL TECHNIQUES FOR BACTERIAL SMEARS

Wet and Dry Procedures (Bartholomew, 1962)

Reagents Required

Crystal violet chloride
Ammonium oxalate
Iodine
Potassium iodide
Safranin O

Preparation of Reagents

Solution 1 (Crystal violet)
Crystal violet chloride   2.0 g
95 per cent ethanol   20.0 ml

Solution 2 (Ammonium oxalate)
Ammonium oxalate   1.0 g
Distilled water   100.0 ml

Solution 3
To 20 ml of solution 1, add 80 ml of solution 2.

Solution 4 (KIO_3)
Potassium iodide   2.0 g
Iodine   1.0 g

Mix these in a mortar and make a paste by adding distilled water (5 ml), next add some more distilled water and transfer to a reagent bottle. Rinse mortar and pestle with distilled water and let the volume be 100 ml.

Solution 5 (Safranin solution)
  Safranin O   2.5 g
  95 per cent ethanol 20.0 ml

Solution 6 (Working solution)
  10 ml of solution 5 is mixed with 90 ml of 95 per cent ethanol. This solution lasts for several days.

Procedure

1. Take heat-fixed slide and flood it with solution 3.
2. Wash excess stain (5 seconds).
3. Again flood the slide with solution 4 for 1 minute.
4. Wash excess stain if it is a wet preparation.
5. Decolorize with 95 per cent ethanol. Flood ethanol dropwise with a drop bottle for 15 seconds.
6. Blot, air-dry, or slightly heat on a flame.
7. Decolorize with 95 per cent ethanol as in step 5.
8. Wash.
9. Flood with solution 6.
10. Blot, dry and examine.

Result

Gram-positive organisms	Blue-black
Gram-negative organisms	Red

ACID-FAST STAINING

Generally 3 types of staining techniques are in vogue 1) The Ziehl–Neelsen, 2) Fluorescent method, 3) Cold method. Of these, the most popular is the Ziehl–Neelsen's procedure which is less complicated. Fluorescent technique is also good provided fluorescent microscope is available. Cold procedures, though of great interest, did not get popularity as routine procedures.

The mechanism of acid fastness is not clearly understood but high lipid content of mycobacteria is suspected to be involved.

Ziehl–Neelsen Method (Ziehl, 1882; Neelsen, 1883)

Reagents Required

 Phenol crystals
 Basic fuchsin chloride
 Ethanol

Hydrochloric acid
Methylene blue

Preparation of Reagents

Solution 1 (Phenol solution)

Melt phenol crystals at a temperature of 45°C and later add warm distilled water. The concentration of phenol is 90 per cent.

Solution 2 (Basic fuchsin)

Basic fuchsin chloride   300 mg
95 per cent ethanol   10.0 ml

Solution 3 (Working solution)

Solution 2   10.0 ml
Solution 1   100.0 ml

Solution 3 should be prepared at least 2 weeks before it is used and it lasts for 2 months.

Solution 4 (HCl)

Concentrated HCl   3.0 ml
95 per cent ethanol   97.0 ml

Solution 5 (Methylene blue solution)

Methylene blue   300 mg
95 per cent ethanol   30.0 ml
1 per cent potassium hydroxide   100.0 ml

Procedure

1. Make a smear, 3–4 sq.cm.
2. Heat gently on the flame or on a hot plate adjusted to 65°C for 2 hours.
3. Cover the smear with a piece of filter paper and then flood the slide with Ziehl–Neelson carbol fuchsin (solution 3). Gently heat it for 5 minutes. See that the stain is not evaporated due to heating. Keep the filter paper always wet with stain.
4. Cool the slide and rinse in tap water.
5. Decolorize with solution 4 for 30 seconds.
6. Rinse in tap water.
7. Flood the slide with solution 5.
8. Rinse in tap water.
9. Blot, dry and examine.

Result

Acid-fast bacteria   Bright red
Background   Blue

Fluorescent Method (Truant *et al.*, 1962)

Reagents Required

Auramine O
Rhodamine B

Glycerol
Concentrated hydrochloric acid
Potassium permanganate

Preparation of Reagents

Solution 1 (Staining solution)

Auramine O	3.0 g
Rhodamine B	1.5 g
Glycerol	150.0 ml
Phenol (Melted)	20.0 ml
Distilled water	100.0 ml

Keep the solution in refrigerator.

Solution 2 (Acid alcohol)

Conc. hydrochloric acid	0.5 ml
70 per cent ethanol	99.5 ml

Solution 3 ($KMnO_4$ solution)

Potassium permanganate	500 mg
Distilled water	100.0 ml

Procedure

1. Take heat-free sections (as in the case of Ziehl–Neelsen method) and flood the smear with solution 1 and heat it at 60°C for 10 minutes or 37°C for 15 minutes.
2. Rinse in tap water.
3. Decolorize with solution 2 for 2 minutes.
4. Rinse in water.
5. Flood smear with solution 3 for 4 minutes.
6. Rinse, blot and dry.
7. Examine under fluorescence microscope.

Results

Acid-fast organisms fluoresce.
Non-acid-fast organisms and background show no fluorescence.

Acid Fast-staining at Room Temperature
[Aubert, 1950 modified by Gross (1952) and Desbordes *et al.* (1952)]

Reagents Required

Tween-80
Basic fuchsin
Phenol
Ethanol
Conc. nitric acid
Methylene blue

Preparation of Reagents

Solution 1 (Staining solution)

Basic fuchsin	4.0 g
Phenol	12.0 ml
Ethanol	25.0 ml
Tween-80	10 drops

Dissolve basic fuchsin in phenol and heat it. Continuously stir it and allow it to cool. Then add ethanol, make the solution up to 300 ml with distilled water. Let it stand for a couple of weeks and later filter through glass wool. Take 100 ml of the solution and add 10 drops of Tween-80 and filter. Before applying the stain, it has to be filtered.

Solution 2 (Acid alcohol)

Conc. nitric acid	5.0 ml
70 per cent ethanol	95.0 ml

Solution 3

Methylene blue solution

Methylene blue	2.1 g
95% alcohol	30 ml

Procedure

1. Take heat-fixed smear and flood it with solution 1 for 10 minutes.
2. Rinse in tap water.
3. Decolorize with solution 2 for 30 seconds.
4. Rinse in tap water.
5. Counterstain with solution 3 for 1 minute.
6. Rinse, blot-dry and examine.

Result

Acid-fast organisms	Red
Non-acid-fast organisms	Blue
Background	Blue

BACTERIAL CAPSULES

Wet Negative Capsule Stain (Duguid, 1951)

Reagents Required

Indian Ink

Procedure

1. Place Indian ink on the slide.
2. On the ink place a colony of cells; do not dilute the ink.
3. Cover the smear with a cover glass. Blot excess ink.
4. Observe under oil immersion microscope.

Results

Capsules are clear.
Background appears dark.

Negative Capsule Stain Using Congo Red (White, 1947)

Reagents Required

Congo red
Rabbit or horse serum
Concentrated hydrochloric acid
Methylene blue chloride

Preparation of Reagents

Solution 1 (Congo red solution)

Congo red	5.0 g
Distilled water	100.0 ml
Rabbit serum	11.0 ml

Mix Congo red with distilled water and add serum, mix thoroughly, filter and discard the residue.

Solution 2 (HCl)

Concentrated HCl	0.5 ml
Distilled water	95.5 ml

Solution 3 (Methylene blue solution)

Methylene blue chloride	1.0 g
Distilled water	100.0 ml
Glacial acetic acid	5 drops

Procedure

1. On a clean slide, place a drop of solution 1 and put the organisms on it.
2. Make a thick or thin smear and slightly heat it over flame to fix it.
3. Flood the slide with solution 2 and pour off excess.
4. Drain and then put solution 3 for 15 seconds.
5. Air-dry and examine.

Result

Cells	Blue
Background	Orange
Capsule	Colourless

Modification of Hiss Capsule Stain (Anthony, 1931)

Reagents Required

Crystal violet chloride
Copper sulphate

Preparation of Reagents

Solution 1 (Crystal violet solution)
Crystal violet chloride   1.0 g
Distilled water   100.0 ml

Solution 2 (Copper sulphate solution)
Copper sulphate   20.0 g
Distilled water   100.0 ml

Procedure

1. A thick smear without heat-fixing is required. Air-dry.
2. Flood the slide with solution 1 for 2 minutes.
3. Wash it with solution 2.
4. Blot, dry and examine.

Result

Capsules are stained light blue.

SPIROCHAETES IN SMEARS

Fontana–Tribondeau Silver Stain (Fontana, 1926)

Reagents Required

Glacial acetic acid
Formalin
Tannic acid
Phenol
Ammonium hydroxide
Silver nitrate

Preparation of Reagents

Solution 1 (Acetic acid formaldehyde)
Glacial acetic acid   1.0 ml
Formaldehyde   2.0 ml
Distilled water   100.0 ml

Solution 2 (Tannic acid)
Tannic acid   5.0 g
Distilled water   100.0 ml

Add a crystal of phenol.

Solution 3 (Ammonium hydroxide)
Ammonium hydroxide   10.0 ml
Distilled water   190.0 ml

Solution 4 (Silver nitrate solution)
Silver nitrate   1.0 g
Distilled water   100.0 ml

To this, add solution 3 dropwise until the solution turns chocolate colour; see that chocolate colour is retained.

Procedure

1.   Prepare a fairly thin smear and dry it.
2.   It is heat-dried.
3.   Now flood the slide with solution 1 for 1 minute.
4.   Wash in running water.
5.   Flood the slide with solution 2 for 20 seconds and gently heat it.
6.   Flood the slide with solution 4 and heat gently.
7.   Wash, dry and examine.

Result

Spirochaetes are stained dark brown.

Potassium Permanganate–Crystal Violet Stain (Harris, 1930)

Reagents Required

Potassium permanganate
Crystal violet

Preparation of Reagents

Solution 1 (Permanganate solution)
Potassium permanganate   1.0 g
Distilled water   100.0 ml

Solution 2 (Crystal violet solution)
Crystal violet   2.0 g
Distilled water   100.0 ml

Procedure

1.   Make a thin smear, air-dry and heat it over a flame to fix.
2.   Flood the smear with solution 1 for 10 minutes.
3.   Wash.
4.   Flood slide with solution 2 for 10 minutes.
5.   Wash, air-dry and examine.

Result

Spirochaetes are stained black.

Sodium Carbonate Basic Fuchsin Stain (Ryu, 1963)

Reagents Required

Sodium bicarbonate
Basic fuchsin
Ethanol
Formalin

Preparation of Reagents

Solution 1 (NaHCO$_3$ solution)
Sodium bicarbonate   5.0 g
Distilled water   100.0 ml

Solution 2 (Basic fuchsin solution)
Basic fuchsin   750 mg
95 per cent ethanol   25.0 ml

Just before use, mix 9 parts of distilled water to 1 part of solution 2.

Solution 3 (Formalin)
Formalin   1.0 ml
Distilled water   99.0 ml

Procedure

1. Make a thin smear and dry the slide.
2. Fix the smear in solution 3.
3. Flood the slide with solution 1 and add 10 drops of solution 2, stain for 3–5 minutes (a precipitate forms).
4. Wash, dry and examine.

Result

Spirochaetes are stained red.

Warthin and Starry Silver–Agar Stain

Reagents Required

Silver nitrate
Agar
Gelatin
Glycerol
Hydroquinone
Sodium thiosulphate
Hydrogen peroxide

Preparation of Reagents

Solution 1 (Silver nitrate solution)
Silver nitrate   2.0 g
Distilled water   100.0 ml

This solution should be kept in an amber-coloured bottle, tightly closed.

Solution 2
Take 1.5 g of agar in a clean flask and add 30.0 ml of distilled water. After few minutes, saturated solution of agar is ready. Pour off excess water and wash in distilled water (several changes). Then add 100 ml of distilled water to the agar and melt it by keeping in an oven adjusted to 55°C. Take this into a tightly stoppered bottle. When it cools to 45°C, a gel is formed. This gel should be broken by shaking vigorously. The final product is kept in an oven (55°C) and should be in a fluid manner.

Solution 3

Take a flask and put 3 ml of 2 per cent silver nitrate and to this add 5 ml of warm aqueous 10 per cent gelatin and 5 ml of warm glycerol (35–45°C) and 5 ml of solution 2 (agar). Just before use, while stirring, add 2 ml of 5 per cent hydroquinone (aqueous). This solution must be prepared fresh. Keep this solution in a cover glass staining dish.

Procedure

1. Prepare a smear on a clean cover glass, and air-dry. Do not heat-fix.
2. Immerse the smear in absolute alcohol for 5 minutes.
3. Wash in distilled water.
4. Rinse in 2 per cent silver nitrate (solution 1). Now cover it with another cover glass which has also been rinsed with 2 per cent silver nitrate. Now immerse the cover glasses of 2 per cent silver nitrate in a tightly stoppered bottle, place the bottle in an incubator adjusted to 37°C
5. The cover glass with smear is placed in a cover glass staining jar with solution 3 for 2 minutes. The smear slowly develops a brown colour.
6. Rinse in 5 per cent sodium thiosulphate for few seconds.
7. Rinse in distilled water.
8. Expose it to absolute ethanol, blot-dry or clean in xylene and mount.

Result

Spirochaetes are stained black.

Positive Capsule Stain (Moller, 1951)

Fixation

Lead acetate	9.0 g
Distilled water	280.0 ml
40 per cent formalin	20.0 ml

Reagents Required

Crystal violet chloride
Copper sulphate
Sodium chloride
Normal serum
Glucose

Preparation of Reagents

Solution 1 (Suspension solution)

Sodium chloride (0.85 per cent)	0.65 ml
Normal serum	0.2 ml
20 per cent glucose	0.15 ml

Solution 2 (Staining solution)

Crystal violet chloride	500 mg
95 per cent ethanol	10.0 ml
Distilled water	90.0 ml

Solution 3 (Saturated copper sulphate)
    Copper sulphate   31.6 g
    Distilled water   100.0 ml
Solution 4
    Lead acetate   9.0 g
    Distilled water   280.0 ml
    Formalin   20.0 ml

Procedure

1. Place a drop of solution 1 on a clean slide.
2. Add organisms on to the slide, spread and air-dry.
3. Keep the slide in a slanting position and add solution 4. Allow it to stand for 15 seconds and then blot and dry.
4. Put a drop of solution 2 on to the slide for 15 seconds and then carefully dry.
5. Rinse with saturated copper sulphate (solution 3) for 10 seconds. Drain, blot and dry.

Result

Bacterial cells stain dark bluish violet.

BACTERIAL SPORE STAINING

Generally bacterial spores do not permeate ordinary dyes at room temperature. Staining should be carried on for longer period by increasing the temperature or by increasing the concentration of the dye.

Schaeffer and Fulton Modification (Wirtz Spore Stain (1933))

Reagents Required

Malachite green chloride
Safranin O chloride

Preparation of Reagents

Solution 1
    Malachite green   5.0 g
    Distilled water   100.0 ml

Solution 2
    Safranin O   500 mg
    Distilled water   100.0 ml

Procedure

1. Make a smear of the organism, air-dry and heat-fix over a bunsen burner (passing the slides over the flame).
2. Flood the slide in the smear with solution 1, heat for 30 seconds and cool.
3. Wash in water for a few seconds.

4.   Again flood the slide with solution 2 for 30 seconds.
5.   Wash, blot, dry and examine.

Result

Spores   Green
Vegetative cells Red

Bartholomew and Mittwer Modification (1950)

This method is a modification of the previous procedure omitting the heat step.

Solution 1
      Malachite green   10.0 g
      Distilled water   100.0 ml

Solution 2
      Safranin O   250 mg
      Distilled water   100.0 ml

Procedure

1.   Prepare a smear, air-dry without heat.
2.   Heat-fix by passing the slide 20 times over a burner and allow it cool.
3.   Flood the slide with solution 1 for 10 minutes.
4.   Rinse in tap water.
5.   Flood the slide with solution 2 for 15 seconds.
6.   Rinse in tap water and blot-dry.

Result

Spores   Green
Vegetative cells Red

Fluorescent Spore Stain (without heat) (Bartholomew *et al.*, 1965)

Fixation

Heat-fixation

Reagents Required

Solution 1
      Auramine O chloride   100 mg
      Distilled water   100.0 ml

It is better to use freshly prepared solution.

Solution 2
      Safranin O   250 mg
      Distilled water   100.0 ml

Procedure

1.   Heat-fix the smears.
2.   Flood the slide with solution 1 at room temperature for 2 minutes.

3. Wash with tap water.
4. Flood the slide with solution 2 for 1 minute.
5. Wash lightly, blot, dry and examine.

Result

Spores should fluoresce brightly.
Vegetative cells do not show fluorescence.

Spore Staining Using Inorganic Acids (without heat) (Lechtman *et al.*, 1965)

Fixation

Heat-fixed smears

Reagents Required

Crystal violet chloride
Phenol
Safranin O
6N HNO_3
Sodium hydroxide

Preparation of Reagents

Solution 1 (Crystal violet solution)

Crystal violet	2.0 g
Distilled water	64.0 ml
Ethanol	26.0 ml

Dissolve crystal violet in ethanol and then add distilled water. To this, add 10 ml of 10 per cent phenol. Store it undisturbed for 2 days and then use.

Solution 2 (Safranin solution)

Safranin O	250 mg
Distilled water	100.0 ml

Solution 3 (Normal nitric acid)

Conc. HNO_3	30.0 ml
Distilled water	80.0 ml

Keep this in a coplin jar tightly closed.

Solution 4 (Sodium hydroxide solution)

Sodium hydroxide	4.0 g
Distilled water	100.0 ml

This is 1N NaOH. Keep this also in a tightly closed coplin jar.

Procedure

1. First keep the heat-fixed slide in solution 3 at room temperature for 10 seconds.
2. Wash in tap water.
3. Transfer the slide to solution 4 for 30 seconds.

4. Wash in tap water.
5. Flood the slide with solution 1 for 2 minutes at room temperature.
6. Wash in tap water.
7. Transfer to solution 2 for 1 minute.
8. Wash lightly in tap water and blot-dry.

Result

Spores	Blue
Vegetative cells	Red

STAINING THE BACTERIAL CELL WALL

Till now there is no procedure to pinpoint the bacterial cell wall material. From electron microscopic studies the bacterial wall thickness is 0.01 to 0.08 mm. So such a thin flimsy structure when fully stained will not be visible.

Cell wall staining procedures are of 2 types. In one type using tannic acid as a mordant, newly completed transverse septa and the developing cross wall regions are stained and the ultrathin cell wall appears obscure. In the second method, a thick cell wall could be seen because of a precipitate on the surface of the cell and developing cross wall and newly formed transverse septa remain unstained. This is with Dyar cell wall stain.

Neither of these are specific stains for cell wall.

Dyar Cell Wall Stain (Dyar, 1947)

Fixation

Heat fixation

Reagents Required

Cetyl pyridinium chloride
Congo red
Methylene blue

Preparation of Reagents

Solution 1 (Cetyl pyridinium chloride solution)

Cetyl pyridinium chloride	340 mg
Distilled water	100.0 ml

Solution 2

Saturated solution of Congo red

Solution 3 (Methylene blue solution)

Methylene blue	500 mg
Distilled water	100.0 ml

Procedure

1. Cover heat-fixed slide with 2–3 drops of solution 1.
2. Add one drop of solution 2 and mix both.
3. Wash in tap water.
4. Flood with solution 3 for few seconds.
5. Wash, blot and dry.

Results

Cell wall	Red
Cytoplasm	Blue
Developing cross wall	Remains unstained or very poorly stained

Remarks

When Congo red is added to cetyl pyridinium chloride, a precipitate is formed on the outer surface of the cell. A red ring of precipitate can be seen.

BACTERIAL NUCLEAR MATERIAL

Bacteria have a double-stranded DNA molecule 800–1200 mm in length with both ends joined. The nuclear material is not associated either with a nuclear membrane or with a histone-type protein. Ultrastructurally the nuclear structure is fibrous with the DNA molecule arranged like strands of yarn. Depending upon the fixative, hydrolysis is carried out with 1N HCl at 60°C which unmasks the DNA region of the cell by removing RNA from the cell. After this staining can be done with 0.1 per cent solution of any basic dye or dilute Giemsa stain (1:10). An ordinary optical microscope cannot give full details but DNA-concentrated areas can be made out.

Cassel and Hutchinson Method (1955)

Fixation

Acetaldehyde	4.0 ml
Distilled water	95.0 ml
Formic acid	1.0 ml

Keep this in a tightly closed jar. Acetaldehyde has a low boiling point and is liquid at ordinary temperature. It should be stored in the refrigerator (20°C).

Reagents Required

Basic fuchsin
Hydrochloric acid

Culture

Brain–heart infusion agar is prepared in a petri dish. The whole surface is inoculated with organisms like *Escherichia coli* or *Bacillus cereus* using a sterile glass spreader and the culture is incubated at 37°C for a period of 3 hours.

Procedure

1. Cut 1-cm^2 blocks from the agar culture and press culture side down on the surface of 3 clear glass slides. Press firmly down. Then remove the agar block and discard these. These impression smears should be air-dried.
2. Keep three slides in fixative for 30 minutes.
3. Wash the slides in tap water.
4. Do not air-dry but immerse in 1N HCl at 60°C (hydrolysis).
5. Remove the slides, one after 8 minutes, 1 after 10 minutes and the third one after 12 minutes.
6. Wash in tap water.
7. Flood all slides with 0.1 per cent basic fuchsin for 5 seconds.
8. Wash, blot and dry.

Result

DNA-rich areas	Red
Cytoplasm	Colourless

It is not known at what time digestion takes place and this is the reason why the three slides are hydrolysed at different timings.

STAINS FOR BACTERIAL FLAGELLA

Bacterial flagella are too thin (0.01 μm) to be detected under optical microscope. Some of the stains used to detect flagella form a precipitate at the surface of the flagella so that the width of the flagella increases and can be seen under optical microscope. These stains also enable us to see the polar and lateral distribution of the flagella.

Clean the slides by immersing in cleaning fluid (sulphuric acid–potassium dichromate) at 70–80°C for 12 hours. Then the slides should be thoroughly washed and dried in a hot-air oven at 200°C for several hours as recommended by Gray (1926).

Culture

Cultures should be 12–18 hours old (Leifson, 1960) and incubation should be at room temperature. All cultures should be checked in wet mount to ascertain their mobility before staining. If mobile cultures do not show flagella after staining that means the procedure is at fault.

Cells from a solid medium are suspended in sterile distilled water in a test tube to a light turbidity. This is left at room temperature for 20–30 mm. During the period, cells are checked for motility. Procedures should be carried out in a gentle way. Do not place again. Centrifuge down the cells and carefully resuspend to a similar volume in 10 per cent formalin 1 : 1 part of 40 per cent formaldehyde to 9 parts of distilled water. Cells in the liquid medium should first be killed by adding 1 part of formalin to 9 parts of culture. The cells should be centrifuged and the supernatant decanted. Resuspend the cells in 2 ml of distilled water, then dilute to the original volume. Recentrifuge the cells and decant the supernatant, then resuspend the cells in distilled water to a light turbidity.

Smears

A large loop of the suspension is placed on the slide at one end with a narrow strip of lens paper. One end of the paper is over the suspension and drawn gently towards the other end. A thin film is formed, air-dried or dried in a hot-air oven.

Another method is to take a clean slide and mark a line with wax pencil across the middle. A drop of cell suspension is placed on this distal end of the slide and tilted so that the suspension runs along the slide to the centre line. It is then air-dried.

Leifson's Bacterial Flagella Stain (Leifson, 1960)

Reagents Required

Basic fuchsin
Ethanol
Tannic acid
Sodium chloride

Preparation of Reagents

Solution 1 (Basic fuchsin solution)
Basic fuchsin   1.2 g
95 per cent ethanol   100.0 ml

Solution 2 (Tannic acid solution)
Tannic acid   3.0 g
Distilled water   100.0 ml

This solution should be light yellow. To avoid moulds, add 0.1 ml liquid phenol.

Solution 3 (Sodium chloride solution)
Sodium chloride   1.5 g
Distilled water   100.0 ml

Solution 4 (Staining solution)
Solution 1   10 parts
Solution 2   10 parts
Solution 3   10 parts

Keep this in a tightly stoppered bottle. This stain lasts for a week and later a precipitate is formed.

Procedure

1. Make a smear as directed above.
2. Take 1 ml staining solution 4 and flood the smear; the stain must stand over the smear and not be running from the slide to any point. Leave for 15 minutes. A colloidal precipitate is formed which will settle on to the flagella.

3. Once the precipitate is formed, wash the slide in tap water.
4. Drain off the water and blot-dry.
5. If desired, counter stain with 1 : 10 dilution of methylene blue for 1 minute.

Result

Flagella are stained violet.
With counterstain, vegetative cells stain blue.

Gray's Modification of the Muir Flagella Stain (Gray, 1926)

Reagents Required

Mercuric chloride
Aluminium potassium sulphate
Basic fuchsin
Ethanol

Preparation of Reagents (Saturated potash alum)

Solution 1

a) Potash alum   11.4 g
    Distilled water   100.0 ml

b) Mercuric chloride   6.9 g
    Distilled water   100.0 ml

c) Tannic acid   20.0 g
    Distilled water   100.0 ml

Solution 2 (Mordant)
Solution 1a   5.0 ml
Solution 1b   2.0 ml
Solution 1c   2.00 ml

Solution 3 (Saturated basic fuchsin)
Basic fuchsin   6.0 g
Ethanol   100.0 ml

Solution 4 (Staining solution)
Solution 2   9.0 ml
Solution 3   0.4 ml

This solution should be used within 24 hours.

Procedure

1. Make a smear as detailed above.
2. Flood the smear with solution 4 (0.5 ml) for 10 minutes.
3. Wash in distilled water.
4. If desired, counterstain with carbol fuchsin for 5–10 minutes. (This stain is used in Ziehl–Neelsen acid-fast stain).
5. Wash in tap water, blot and dry.

Results

Flagella should be visible.
Vegetative cell stains red.

STAINING OF RICKETTSIAE, COXIELLA AND CHLAMYDIAE

From ultrastructural studies, it has been confirmed that rickettsiae and chlamydiae are small bacteria since their prokaryotic cells are possessing both DNA and RNA and they reproduce by binary fusion. They are different from all other bacteria by being obligate intracellular parasites. The chlamydiae differ from rickettsiae in being completely host-dependent for oxidative phosphorylation and in possessing a definite life cycle from the infective elementary body to the reproductive form called the initial body and it also differs in the ability of the elementary body to pass through bacteriological filters.

It is very difficult to demonstrate the chlamydiae and rickettsiae with optical microscope since they are very small (0.2–0.3 μm wide and 0.2 to 1.0 μm long). In addition to this, they are gram-negative and as such it is difficult to differentiate them from the cellular material.

Macchiavello Stain for Rickettsiae (Macchiavello, 1937)

Reagents Required

0.1 M phosphate buffer (*see* Chapter 5)
Basic fuchsin
Citric acid
Methylene blue
Phenol

Preparation of Reagents

Solution 1

0.1 M Phosphate buffer	100.0 ml (*see* Chapter 5)
Basic fuchsin	250 mg

Solution must be prepared fresh every day. A precipitate forms which should be removed by filtration through a coarse filter paper.

Solution 2

Citric acid	500 mg
Distilled water	100.0 ml

Solution 3 (Counterstain stock)

95 per cent ethanol	10.0 ml
Methylene blue	1.0 g
Distilled water	100.0 ml
Phenol	5.0 ml

Procedure

1. Prepare a smear on a clean slide from yolk sac material, dry it and heat gently.
2. Cover the smear with solution 1 for 3 minutes. Drain off excess dye.
3. Rinse in solution 2 for 2 seconds.
4. Rinse in tap water.
5. Place in solution 3 (diluted in the ratio 1 : 10) for 2 seconds.
6. Rinse with tap water, blot, dry and examine.

Result

Rickettsiae and Chlamydiae	Brilliant red
Cellular material	deep blue

Gimenez stain for Rickettsiae and Psittacosis Agents in Yolk Sac Cultures

Reagents Required

Basic fuchsin
Ethanol
0.1 M phosphate buffer (*see* Chapter 5)
Malachite green
Ferric nitrate
Fast green

Preparation of Solutions

Solution 1 (Stock solution)

Basic fuchsin	1.0 g
95 per cent ethanol	10.0 ml
Distilled water	65.0 ml
4 per cent phenol	25.0 ml

Stir the solution until basic fuchsin dissolves. Maintain at 37°C for 48 hours.

Solution 2

0.1 M phosphate buffer (pH 7.45)	4.0 ml
(*see* Chapter 5)	
Solution 1	10.0 ml

A precipitate is formed and it should be filtered immediately. Solution should be prepared fresh every time before use.

Solution 3 (Malachite green solution)

Malachite green	800 mg
Distilled water	100.0 ml

Solution 4 (Ferric nitrate solution)

Ferric nitrate	4.0 g
Distilled water	100.0 ml

Procedure (for all species of *Rickettsia* except *R. tsutsugamushi*)

1. Prepare a thin smear from yolk sac tissue. Heat-fix.
2. Flood the slide with solution 2 for 2 minutes.
3. Wash with tap water.
4. Flood again with solution 3 for 6–9 seconds.
5. Blot, dry and examine.

Procedure (for *R. tsutsugamushi*)

1–3 are the same as in previous procedure.
4. Flood slide with solution 4 for 6–9 seconds.
5. Cover with 0.5 per cent fast green for 30 seconds and wash.
6. Blot, dry and examine.

Result (for all species of *Rickettsia* except *R. tsutsugamushi*)

Rickettsia and Psittacosis organisms stain red.
Cells take a greenish blue colour.

Result (for *R. tsutsugamushi*)

Rickettsiae	Reddish black
Background	Green

STAINING OF MYCOPLASMA

These are very small bacteria ranging from 0.125–0.15 µm without cell walls. They are capable of free-living existence and are highly pleomorphic because of the absence of cell walls. They are so small that they can easily pass through bacteriological filters. It is highly difficult to demonstrate them with an optical microscope. They are capable of growing on artificial culture media on which they form small colonies (10–600 µm). Colonies are also small and transparent and it is difficult to see without the aid of staining. Through electron microscopic studies, the morphology of individual cells could be studied.

L-phase variants of bacteria are produced by a variety of bacterial species and their variants are also without cell walls. But these variants are slightly larger than the mycoplasma cells. The variants also grow on artificial culture media and also produce colonies which are definitely larger than the colonies of mycoplasma.

Dienes Stain for L-phase Variants of Bacteria and Mycoplasma (Madoff, 1960)

Reagents Required

Methylene blue
Azure II
Maltase
Sodium carbonate
Benzoic acid

Preparation of Reagents

Solution 1

Methylene blue	2.5 g
Distilled water	100.0 ml
Azure II	1.25 g
Maltase	10.0 g
Sodium carbonate	250 mg
Benzoic acid	200 mg

Procedure

1. Keep clean microscope cover glasses ready.
2. With a sterile applicator, make a thin film of the stain on the cover glass (on one side only). Allow to dry.
3. Make a 6–8 mm agar block from fresh agar culture of the organism. Place the block with culture side up on a clean slide.
4. Now cover the block with the cover glass with thin film of stain.
5. The space between cover glass and agar block is filled with a mixture of paraffin and 10 per cent vaseline.
6. Examine the surface of the agar for stained colonies.

Results

Mycoplasma colony stains deep blue in the centre and light black at edge.

Bacterial colonies stain uniformly under high magnification. Pleomorphic mycoplasmal cells can be seen. Bacterial cells appear with constant shape and size.

Klieneberger–Noble Agar Fixation Stain for Colonies of Mycoplasma and L-phase Variants (Klieneberger–Noble,, 1962)

Reagents Required

Bouin's fixative (*see* Chapter 1)
Geimsa powder
Glycerol
Methanol

Preparation of Reagents

Solution 1

Bouin's fixative (*see* Chapter 1)

Solution 2 (Giemsa stock)

Glycerol	33.0 ml
Giemsa	500 mg
Methanol	33.0 ml

Put glycerol in a beaker and start heating in a water bath adjusted to 55°C. Now slowly add Giemsa powder while stirring. Retain this temperature for 2 hours. Now add 33 ml of methanol. Cool and store in a tight stoppered bottle. Store it for 2 weeks before use. Filter before use.

Solution 3
   Phosphate buffer 0.1 M (pH 6.8) (*see* Chapter 5)

Procedure

1. Take an agar block with colonies and place in with culture side down.
2. Keep the cover glass in a watch glass with agar block up and add solution 1 overnight.
3. Peel off the agar and throw it. Wash the cover glass carefully until yellow colour disappears.
4. Now keep the cover glass in a coplin jar with a mixture of 1 ml of solution 2, 47 ml of distilled water and 2 ml of solution 3.
5. Drain off excess dye.
6. Pass the cover glass through
    i.   Acetone 19 ml + xylene 1 ml
    ii. Acetone 14 ml + xylene 6 ml
    iii. Acetone 6 ml + xylene 14 ml
    iv. 2 or 3 changes of xylene
7. Mount in Canada balsam.

Results

The individual cells are preserved and the individual cells to colony appearance can be seen. These mounts are permanent.

Dienes Agar Fixation Stains for Colonies of Mycoplasma and L-phase Variants (Dienes, 1967)

Reagents Required

Toluidine blue O
Azure II
Tartaric acid

Preparation of Solutions

Solution 1 (Toluidine blue solution)
   Toluidine blue   2.0 g
   Distilled water   100.0
        Or
   Azure II   2.0 g
   Distilled water   100.0 ml

Solution 2
   Tartaric acid   20.0 g
   Distilled water   100.0 ml

Procedure

1. Take a 3-mm agar block and inoculate with colonies. Incubate for 2 days.
2. Place the block on a filter paper and add few drops of formalin. Keep the block with filter paper in a closed container with the oven off.

3. Take a cover glass and put a loop each of solution 1 and solution 2. Mix and spread evenly.
4. Wet the thin agar block from formalin-fixed plate and place the block on a clean cover glass, culture side up.
5. Staining should be carried out at 50–60°C temperature. Block should not dry.
6. Place the newly cut 1 mm thick agar block with a film of stain, the culture side against the stain.
7. Place the cover glass, blocks upon the surface.
8. Place the preparation the lid of an inverted petri dish containing culture medium at 50°C, then place the petri dish in an oven at 50°C for 1 hour.
9. Remove the preparation from the petri dish. Allow the agar block to dry.
10. Mount in Balsam.

Result

Mycoplasma, L-phase variant, normal bacteria, fungi, are all stained. Observe, under oil immersion microscope.

ENZYMES

18

Enzymes are biological catalysts that either activate or accelerate a reaction at the appropriate temperature. As a manipulator, they occupy a position of "Bystander" initiating a deliberation between two participants without getting themselves involved. The etymology of an enzyme is based on its action. The enzyme that hydrolyses sucrose into monosaccharide is a sucrase. The suffix "ase" indicates an enzyme. The earliest enzymes which were mostly hydrolases however were not named using the suffix "ase", for example, the enzymes of the digestive tract pepsin, trypsin and chemotrypsin which hydrolyse proteins. Enzymes are essentially proteins with specific action and therefore they are defined as protein catalysts formed by living cells.

Enzymes play important and significant roles in the life of an organism. All physiological processes of an organism are regulated, controlled, and maintained in an orderly balance. More than 800 enzymes are known (Dixon and Webb, 1958). However only a few enzymes have so far been amenable to histochemical detection to date.

Preservation of an enzyme for subsequent demonstration is much more difficult than other cellular constituents. Unless maximum amount of enzyme is preserved intact, it is difficult to demonstrate cytochemically. Enzymes may be bioenzymes (e.g. glucuronidase) which dissolve in the cytoplasm and are likely to diffuse out as "desmoenzymes" (e.g. leucine aminopeptidase) bound to cytoplasmic constituents.

Tissue for cytochemical localization of enzymes should be compulsorily fresh. All procedures should be done in cold temperature (4°C), especially fixation (Seligman *et al.*, 1951; Nicholas *et al.*, 1956; Holt, 1959; Pearse, 1968). Acid and alkaline phosphatases and esterases are fixative-resistant. Fixation is followed by freezing the tissue in 0.88 M sucrose (30 per cent) in 1 per cent gum acacia. Ice-cold (0–4°C) gum sucrose treatment is good for cryostat cutting for enzyme localization.

USE OF CONTROLS

As usual, controls are essential for the demonstration of enzymes. Substrates or their components break down with time. Control sections therefore should be maintained simultaneously with test section. A control which is positive will demonstrate that nothing is wrong with the reagents and that they are in working condition. A negative control prepared by destroying the enzyme in the test section by immersion in boiling water for 15 minutes should be used. Enzyme activity is affected by factors like temperature, pH, inhibitors, activators, etc.

1. *Temperatures*   Optimal temperature for activity of many enzymes is 37°C. Lower temperatures between 4°C and –30°C are useful for employing enzyme technique. At higher temperatures rapid denaturation of protein impairs activity.

2. *pH*   Most enzymes have a pH at which the rate of the reaction is optimal pH of 7.0 is alright for a large number of enzymes. However, optimal pH for alkaline phosphatase is 9.2 and for acid phosphatase is 5.0.

3. *Inhibitors*   Enzyme activity can be destroyed by various inhibitors. Three types of inhibitors are known.

    1. Specific inhibitors, e.g. eserine for the cholinesterase.
    2. Non-specific inhibitors, e.g. heat for all enzymes.
    3. Competitive inhibitors

        Specific inhibitors destroy the enzyme molecule site whereas non-specific inhibitors affect the enzyme reaction by denaturing the protein. Competitive inhibitors are those which compete with the substrate for the active enzyme sites.

4. *Activators*   These are employed to promote enzyme activity. For example, magnesium ions are activators in the Gomori metal precipitation for alkaline phosphatase.

METHODS OF DEMONSTRATION

Any enzyme in the tissue is presented with its own specific substrate in the incubation medium and the reaction is usually undiscernible and, therefore, is allowed to couple with another substance so that an insoluble and visible final reaction product results at the site of enzyme activity. The coupling substance to be used differs from enzyme to enzyme—diazonium salts for phosphatase (Azo dye method), tetrazolium salt for dehydrogenases and organic chemicals like ammonium sulphite in the metal precipitation technique.

Five important techniques are available for enzymes.

1. Metal Precipitation

This has been a common technique for the demonstration of phosphate. In this technique ultimately there is formation of metal precipitate. With enzyme activity of the substrate, phosphate ions that are released combine with metallic ions. These phosphatases are not discernible as such but they can be made out by conversion to black sulphides by treating with ammonium sulphide.

2. Simultaneous Coupling

In this method, the enzyme hydrolyses a substrate which contains a diazonium salt combining with the coupler which is later released from the substrate by the enzyme. This is the primary resulting product (PRP). This PRP couples with diazonium salt to produce the final resulting product (FRP). Azo-coupling reaction or diazo technique is applied for alkaline phosphatase. In this method, the pH at which the enzyme exhibits its efficient activity and the pH at which the diazonium salt exhibits its maximum activity are very important.

3. Post-incubation or Post-coupling Reaction

This reaction involves the production of a colourless PRP which is insoluble. It is then coupled with a coloured substrate. No diazonium salt is used. According to Pearse (1968), diazonium salts are supposed to interfere with enzymes, and their long exposure to acid or alkaline solution results in non-specific staining (Nachlas *et al.*, 1957). This is the advantage of not using diazonium salts.

4. Self Coloured Substrate

A water-soluble coloured dye should be involved in this method. This soluble dye should be made insoluble by removing the solubilizing group without interfering with the colour. A coloured precipitate results at the site of enzyme activity. Fluorescence technique of Burstone (1960) for alkaline phosphatase is an example.

5. Intermolecular Rearrangement

This method involves rearrangement of molecular structure of a colourless substrate to give an insoluble coloured precipitate at the sites of enzyme activity. A clear-cut localization is however not possible because the final resulting products are not fully insoluble. This method has been described by Nachlas *et al.* (1957) for carboxylic acid esterase.

ENZYME FIXATION

There are some controversies in the fixation of enzyme. According to some, unfixed frozen tissues lose enzymes into the incubating medium and there will be lot of cell damage. Some of the enzymes can tolerate fixatives for a short duration say 15 minutes at 5–10°C.

Some of the fixatives suggested for enzyme histochemistry are listed below.

Formol Calcium

Formol calcium (Baker, 1944) is recommended for many enzymes like acid phosphatase, esterase and lipase.

Calcium chloride (anhydrous)	1.0 g
Distilled water	60.0 ml
Formaldehyde	10.0 ml

Adjust pH to 7.0–7.2 with 1N NaOH. Make up to 100 ml with distilled water. Check pH before use.

Glutaraldehyde

Sabatini *et al.* (1964) used 2–6 per cent glutaraldehyde in phosphate buffer for alkaline phosphatase and esterase.

50 per cent solution of glutaraldehyde	4–12 ml
0.1–0.2 M Phosphate buffer	100.0 ml

The final pH should be 7.4. If shrinkage is suspected, use 4 ml of glutaraldehyde in 100 ml of buffer. 1 per cent sucrose is recommended as a preservative. Store in a refrigerator.

Cacodylate buffer (pH 7.2)	
1 M Cacodylic acid (137.99 g/1000 ml)	54.6 ml
1N NaOH (40 g/1000 ml)	50.0 ml
Distilled water	895.4 ml

After glutaraldehyde fixation, tissue is post-fixed in 1 per cent osmium tetroxide buffered with 0.1 M phosphate buffer and glucose or veronol acetate-sucrose after cacodylate glutaraldehyde.

Physiological buffer
Solution 1   2.26 per cent $NaH_2PO_4H_2O$
Solution 2   2.52 per cent NaOH
Solution 3   5.4 per cent glucose
Solution 4   Isotonic disodium phosphate
   Solution 1   41.5 ml
   Solution 2   8.5 ml
Adjust pH to 7.4–7.6 with solution 2.

Phosphate-buffered Fixative

Solution 4	45.0 ml
Osmium tetroxide	500 mg
Solution 3	5.0 ml

Lasts for several weeks at 4°C. Wash tissues overnight in buffer.

Acetone

Acetone (cold 4°C) is used for some enzymes like alkaline phosphatase, esterase, peroxidase and DOPA oxidase.

Acetone	300.0 ml
0.03 M sodium citrate	32.0 ml
0.03 M citric acid	168.0 ml

Final pH should be 4.2. Store in refrigerator.

Paraformaldehyde

Solution 1 (0.2 MS Collidine buffer)

S-Collidine (2,4,6-trimethylpyridine)	2.64 ml
Distilled water	47.36 ml

Solution 2

0.1N HCl	9.0 ml
Distilled water	41.0 ml

Solution 3 (Working solution)

Equal parts of solution 1 and 2, final pH should be 7.4–7.5.

Solution 4 (Paraformaldehyde solution)

Paraformaldehyde	4 mg
Distilled water	66.0 ml

Take this into a graduated cylindrical jar and add distilled water till the total is 66.0 ml. Warm in a flask in a water bath then depolymerize by adding 0.1N NaOH drop by drop until the solution is clear.

Solution 5 (Working solution)

Solution 4	66.0 ml
Solution 3	33.0 ml
0.5 M calcium chloride	1.0 ml
(5.55 g/100 ml)	

Final pH should be 7.4.

Storage of Tissues

To obtain good results, fix 1 to 2 mm thick tissues in calcium formalin or glutaraldehyde in the refrigerator overnight. They can also be stored in gum sucrose or glycerine.

Gum sucrose (Holt *et al.*, 1960)

Sucrose	30.2 g
Distilled water	100.0 ml
Gum acacia	1.0 g

Prepare only small quantity.

Glycerine (Turchini and Malet, 1965)

Glycerine	50.0 ml
Distilled water	50.0 ml

DEHYDRATION AND EMBEDDING

Freeze-drying Method

Many expensive as well as cheaper freeze-drying apparatuses are now available. The principle behind freeze-drying is that the frozen tissue is kept under high vacuum until all water molecules are removed and condensed into cold surface or collected by a desiccant. 1-mm pieces are frozen rapidly to prevent large crystal formation which will disrupt the cells. After freezing, the tissue is dehydrated in a drying apparatus in vacuum at a temperature of $-30°C$ to $-40°C$ and the ice is sublimed into water vapour and removed. As there is no liquid phase, there is no diffusion of enzymes.

When the material is dry, allow it to rise to room temperature, infiltrate with paraffin and embed. Sections should not be floated on water and can be applied directly to warm albuminized slides.

Freeze-substitution Method

In this procedure, the tissue is rapidly frozen, ice formed and slowly dissolved in a fluid solvent like ethyl or methyl alcohol. When the tissue is free of ice and permeated by the cold solvent, it is brought to room temperature and embedded.

Freeze-substitution Technique
(Feder and Sidman, 1958; Hancox, 1957; Patten and Brown, 1958)

Procedure

1. Take a beaker containing liquid propane–isopentane (3:1) or SUVA-MP66 and cool to -170 to $-175°C$ with liquid nitrogen.
2. Take a small piece (1.5 to 3 mm) of tissue on a thin strip of aluminium foil and plunge into cold propane–isopentane. Stir vigorously.
3. Now transfer frozen tissue to ice solvent previously chilled to -60 to $-70°C$. After a few minutes, the tissue is loosened from the foil. Store in a dry-ice chest for a couple of weeks.
4. Remove fluid and tissue to refrigerator and wash for 12 hours in 3 changes of absolute alcohol in the refrigerator.
5. Transfer to chloroform for 6–12 hours in refrigerator.
6. Place in cold xylene ($-20°C$) and remove to room temperature for 10 minutes.
7. Transfer to fresh xylene at room temperature for 10 minutes.
8. Transfer to xylene–paraffin (50:50) 56°C – Vacuum 15 minutes.
9. Infiltrate paraffin no. 1 in vacuum for 15 minutes.

10. Transfer paraffin no. 2 in vacuum for 15 minutes.
11. Transfer paraffin no. 3 in vacuum for 15 minutes.
12. Embed.

Freeze-substitution of Sections (Chang and Hori, 1961; Patten and Brown, 1958)

Procedure

1. Quench 1 to 2 mm slices in liquid nitrogen or in isopentane chilled in a bath of liquid nitrogen.
2. Cut sections at 15°C in cryostat.
3. Slowly bring sections into screw cap vial of acetone pre-chilled in dry ice.
4. After some time keep vial in dry ice and complete dehydration.

CRYOSTAT SECTIONING

Enzymes and other similar tissue constituents must be sectioned at low temperatures of –25°C to –20°C. Since embedding is impractical at this temperature, tissue is frozen in isopentane or SUVA-MP66 (–16°C) or in dry ice–ethyl alcohol mixture (–70°C). The tissue must be kept at least at dry ice temperature (–20°C) and sectioned and mounted at this temperature. This is done in a cold chamber cryostat (–25° to –20°C). The object holder is pre-cooled and the tissue is frozen on it with a layer of embedding component. This medium supports and protects the tissue and facilitates sectioning. The embedding medium melts, leaves a thin transparent film against the slide and does not interfere with enzymatic reaction. If these slides are to be used for autoradiography, remove the embedding compound by washing slides in distilled water before dipping in emulsion.

The knife for sectioning must be dry and sharp. If the embedding compound is used and sections are to be mounted on slide, cut through the component and through the tissue just until the knife edge reaches the block along the opposite edge of the tissue. This means that the cut is complete through the tissue but not through the embedding compound. Stop sectioning. Most of the microtomes are now equipped with at least one device that prevents the rolling of sections as they are cut. Flatten the section with the brush and finish cutting through the blade. With the cold brush, lift the sections to a cold slide. If serial sections are required it is easy to collect several sections on to the blade and then brush them on to a cold slide in correct sequence. Thaw sections by putting the undersurface of the glass against a finger. Cold slides can be used to pick up tissue sections. Place a finger on the back of the slide to warm slightly and the tissue sections will adhere to the slide.

When thawed, the sections are ready for processing. Frozen tissue can be mounted on cover glasses in the same way as on slides. In some cases the section can be picked up by touching it with the cover glass. If the tissues have been suspended in gum sucrose, 6–10 μm sections are easily cut and dropped into the proper substrate for incubation or mounted on cover glasses or slides.

SECTIONING WITHOUT A CRYOSTAT

In laboratories where cryostat is not available, frozen sections can be made on the chemical freezing microtome or by the method developed by Adamstone-Taylor (1948).

Cold Knife Technique (Adamstone-Taylor, 1948)

Procedure

1. Freeze fresh tissue (2 mm) on block of microtome or on liquid nitrogen, isopentane.
2. Store in dry ice (–75°C) or in deep freeze (–25°C).
3. Chill knife by fastening solid block of dry ice on each end. Thin sheets of metal foil work well.
4. When the knife is chilled, freeze the frozen block of tissue in place on freezing head.
5. Cut sections, hold them flat on the knife with a camel's hair brush. Do not allow them to thaw.
6. Remove and transfer section to a small metal spatula kept cold with a few bits of ice. Transfer the section to cold slide directly from the knife.
7. Press the section in place and as it begins to soften but before it completely thaws, immerse slide in fixative.

TYPES OF ENZYMES AND THEIR NOMENCLATURE

Oxidoreductases

This forms a large group which were originally known as oxidases and dehydrogenases and together called as oxidative enzymes. Oxidoreductases include:

1. Oxidases which in the presence of oxygen catalyse oxidation of substrate.
2. Peroxidases remove hydrogen and catalyse oxidation of a substrate and hydrogen combines with hydrogen peroxide.
3. Dehydrogenases remove hydrogen to catalyse oxidation of substrate.
4. Diaphorases remove hydrogen and thus catalyse oxidation of NADH and NADPH.

Other groups of importance are:

Transferases

There are a number of them. They are neither involved in the uptake of water or lose water, catalyse the transfer of radicals of two compounds.

Hydrolases Though in some instances they remove water, in many an occasion they catalyse the introduction of water or its elements into specific substrate bonds.

These include:

Esterases

Lipases

Phosphatases

Glycosidases

Peptidases

Pyrophosphatases

i. *Lyases* These remove groups from substrates by a mechanism other than hydrolases resulting in the formation of carbon double bonds.

The above-mentioned classification is biochemical and there is no clear-cut classification in histochemistry, due to varied reasons. For a long time, a haphazard method was followed which landed in great confusion. Different workers gave different names to the same enzyme and conversely the same name was used for two or more enzymes leading to utter confusion. As mentioned above, the enzyme was named after the substrate on which it acted by adding the suffix "ase", but in recent times it became necessary that the enzymes are named not only on the name of the substrate but also on the type of reaction. Sometimes two enzymes catalyse similar reactions under different conditions, e.g. (a) acid phosphatase, (b) alkaline phosphatase (Bancroft and Hand, 1989).

Hydrolytic Enzymes

Hydrolytic enzymes hydrolyse organic phosphate esters, optional pH level plays an important role in the classification of these enzymes.

Enzymes which exhibit maximum activity at pH 9.0 are alkaline phosphatases.

Enzymes showing peak activity around pH 5.0 are acid phosphatases.

These enzymes catalyse the hydrolysis of organic phosphate esters, but on the other hand there are a few enzymes which act upon a specific substrate (glucose 6-phosphatase which dephosphorylates glucose 6-phosphate at pH 6.5). Such a type of enzyme is considered a specific phosphatase.

Thus, phosphatases are classified into

● Alkaline phosphatase

● Acid phosphatase

● Specific phosphatase

PHOSPHATASES

Many animal and plant tissues contain phosphatases. Phosphatases are mainly responsible for hydrolysis of organic phosphate esters. A specific phosphatase like adenosine triphosphatase acts specifically on a single substrate. Other phosphatases do not have this specificity or in other words are less specific and could be divided into 2 categories. Category-1 exhibits optional specificity at low pH values (alkaline phosphatases) and category-2 exhibits optimal activity at high pH value (acid phosphatase).

Alkaline phosphatases are active between a pH range of 9.0–9.6, i.e., pH should be alkaline. Manganese, magnesium and cobalt activate these phosphatases and cyanide and cysteine do the reverse action. These cyanides and cysteine could be incorporated in the incubating medium to provide control.

HISTOCHEMICAL METHODS FOR ALKALINE PHOSPHATASES

General Method

Preparation of Fixative

Solution 1 (4 per cent formol calcium)
Formaldehyde   4.0 ml
Distilled water   96.0 ml
$CaCl_2$ 1.0 g
Add calcium chloride until pH is 7.

Solution 2 (Gum sucrose)
Gum acacia   2.0 g
Sucrose   60.0 g
Distilled water   200.0 ml

Allow to dissolve and store at 4°C.

Solution 3 (Formaldehyde–Gelatin mixture)
1 per cent gelatin (1 g/100 ml water)   25.0 ml
2 per cent formaldehyde   25.0 ml

Procedure for Fixation and Sectioning

1. Cut small blocks of tissue.
2. Pre-cool in solution 1 (formol calcium) and keep the tissue at 4°C overnight.
3. Blot, dry, wash in tap water for 2 minutes and blot-dry.
4. Transfer to solution 2 (gum sucrose solution) at 4°C for 24 hours.
5. Blot-dry.
6. Blocks have to be attached to cryostat holder and frozen.
7. Sectioning is at –10°C. 8–10 µ thick sections are to be taken.
8. Pick sections upon slides pre-coated with solution 3 (formaldehyde–gelatin mixture) and allow to dry.

Modified Gomori's Method for Alkaline Phosphatase (1952b)

Frozen, paraffin, freeze-dried, cryostat pre- and post-fixed sections are preferable.

Reagents Required

2 per cent sodium β-glycerophosphate
2 per cent sodium veronol
2 per cent calcium nitrate
1 per cent magnesium chloride
2 per cent cobalt nitrate

Ammonium sulphate
2 per cent methyl green
Glycerine jelly

Preparation of Reagents

Solution 1 (Incubating medium)

2 per cent sodium β-glycerophosphate	2.5 ml
2 per cent sodium veronol	2.5 ml
2 per cent calcium nitrate	5.0 ml
1 per cent magnesium chloride	0.25 ml
Distilled water	1.25 ml

This incubating medium should have a pH of 9.0–9.4.

Procedure

1. Hydrate slides to water. After fixation, incubate in solution 1 for 45 minutes to 6 hours at 37°C.
2. Wash in distilled water.
3. Transfer sections to cobalt nitrate for 3 minutes.
4. Wash in distilled water.
5. Place sections in 1 per cent ammonium sulphate for 2 minutes.
6. Wash in distilled water.
7. Counterstain in 2 per cent methyl green.
8. Wash in distilled water and mount in glycerine jelly.

Result

Alkaline phosphatase	Brownish black
Nuclei	Green

Rationale

β-glycerophosphate combines with the calcium nitrate activated by magnesium chloride to form calcium phosphate at the sites of enzyme activity. Alkaline phosphatase releases phosphate from sodium β-glycerophosphate. The phosphate then combines with calcium ions to form calcium phosphate. Cobalt nitrate converts it to cobalt phosphate. With ammonium sulphide, black deposits of cobalt phosphides are formed at the sites of enzyme activity.

Remarks

Incubating medium should be prepared afresh whenever a reaction is to be started. pH values are important and should be lower than 9.0. Incubation should be carried out at 37°C.

Gomori Method (Modified by Fraenkel and Peters, 1964)

Fixation

Fix thin slices in cold acetone giving 3 changes 24–48 hours.

Freezing cryostat sections to be fixed in 95 per cent or absolute alcohol or neutral buffered formalin.

Reagents Required

Paranitrophenyl phosphate
Sodium barbital
Calcium chloride
Magnesium sulphate

Preparation of Reagents

Solution 1 (Incubating medium)

0.8 per cent paranitrophenyl phosphate	2.0 ml
2 per cent sodium barbital	2.0 ml
Distilled water	100.0 ml
2 per cent calcium chloride	4.0 ml
5 per cent magnesium sulphate	0.2 ml

Infiltration and Embedding

Chloroform or petroleum ether is used for paraffin sections.
Wax with melting point at 52°C is used.
Slides are dried not in usual manner but at 37°C.

Procedure

1. Dewax in light petroleum or chloroform for 1 hour.
2. Hydrate slides in absolute acetone, 80 per cent acetone water and 50 per cent acetone water.
3. Rinse in water.
4. Transfer slides to solution 1 (incubating medium) at 37°C for 45 minutes.
5. Wash in distilled water.
6. Place in 2 per cent cobalt nitrate for 5 minutes.
7. Rinse in distilled water.
8. Place in 2 per cent yellow ammonium sulphide for 2 minutes.
9. Counterstain with either haematoxylin or safranin, and wash.
10. Dehydrate, clear and mount.

Results

Enzyme sites	Black
Nucleus	Grey

Remarks

Paranitrophenyl phosphate is split by alkaline phosphate at a faster rate than other phosphates like phenyl phosphate, glycerophosphate and phenolphthalein phosphate. Reaction takes place in 30 minutes and artefact formation is minimized.

Kaplow Method Modified (1963)

Fixation

Same as in previous procedure

Reagents Required

Propanediol
2-amino-2-methyl-1,1,3-propanediol
Hydrochloric acid
Naphthol AS-MX phosphate
N-N-dimethyl formamide (DMF)
Fast red violet salt LB

Preparation of Reagents

Solution 1 (Propanediol stock solution)
2-amino-2-methyl-1,1,3-propanediol   10.5 g
Distilled water   500.0 ml

Solution 2 (Buffer working solution pH 9.7)
Propanediol solution (Stock)   25.0 ml
0.1 HCl   5.0 ml
Distilled water   100.0 ml

Solution 3 (Incubating solution)
Naphthol AS-MX phosphate   5 mg
N-N-dimethyl formamide   0.2–0.3 ml
Solution 2   60.0 ml
Fast red violet salt LB   30–40 mg

Procedure

1. Dewax in chloroform for 1 minute.
2. Bring down to 70 per cent acetone water, and then 50 per cent acetone water.
3. Place in solution 3 (incubating medium) for 30 minutes at room temperature.
4. Wash in running water.
5. Treat with Mayer's haematoxylin for 3 minutes.
6. Wash in running water.
7. Blue in Scott solution for a few seconds.
8. Wash in running water.
9. Treat with 0.5 per cent aqueous Wool green S for 30 seconds.
10. Rinse in distilled water.
11. Dehydrate in 2 changes of 95 per cent ethyl alcohol for a few seconds.
12. Dehydrate, clear and mount.

Results

Sites of alkaline phosphatase   Red
Nuclei   Blue

Azo Dye Coupling Method for Alkaline Phosphatase

Reagents Required

0.1 M tris buffer stock solution (*see* Chapter 5)

Sodium α-naphthyl phosphate
Diazonium salt (Fast red TR)

Preparation of Reagents

Solution 1 (Incubating medium)

Sodium naphthyl phosphate	10 mg
0.1 M tris buffer (pH 8.0)	10.0 ml
Diazonium salt	10 mg

pH of incubating medium should be adjusted to 9.0–9.4. Initially sodium naphthyl phosphate is dissolved in the buffer. Diazonium salt is now added and the solution is mixed well. This has to be used immediately.

Sections

Cryostat post-fixed, pre-fixed, freeze-dried, or paraffin sections.

Procedure

1. Bring sections to water, incubate in solution 1 at room temperature for 10–60 minutes.
2. Wash in distilled water.
3. Counterstain in 2 per cent methyl green.
4. Wash in distilled water.
5. Mount in glycerine jelly.

Result

Sites of alkaline phosphatase activity	Reddish brown
Nuclei	Green

Rationale

First described by Menton *et al.* (1944), it was later modified by Gomori (1951). The incubating medium should be at pH 9.2. From the substrate, the enzyme liberates naphthyl which couples with diazonium salt to form an insoluble azo dye at the sites of enzyme activity. The choice of the diazonium salt and pH of the incubating medium determine the quality of staining.

Remarks

pH of the incubating medium must be 9.2. Paraffin sections require more time for incubation.

Naphthol AS-BI Method for Alkaline Phosphatase (Simultaneous coupling with substituted naphthols)

Reagents Required

Naphthol AS-BI phosphate
$N:N$ dimethyl formamide
Sodium carbonate
Tris buffer

Preparation of Reagents

Solution 1 (Stock solution)

Naphthol AS-BI phosphate	25 mg
$N:N$ dimethyl formamide	10.0 ml
Distilled water	10.0 ml
Molar sodium carbonate	2–6 drops

Reagents have to be mixed in the order given above. Addition of sufficient molar sodium carbonate is necessary to obtain a pH of 8.0.

Distilled water	300.0 ml
0.2 M tris buffer (pH 8.3)	180.0 ml

The opalescent solution lasts for a long time.

Solution 2 (Incubating solution)

Stock naphthol AS-BI	10.0 ml
Fast red TR	10 mg

Shake well, filter and use immediately.

Sections

Cryostat post- and pre-fixed, frozen, paraffin, or freeze-dried sections.

Procedure

1. Bring sections to water. Incubate in solution 2 at room temperature for 5–15 minutes.
2. Wash in water.
3. Counterstain in 2 per cent methyl green.
4. Wash in running water.
5. Mount in glycerine jelly.

Results

Alkaline phosphatase	Red
Nuclei	Green

Rationale

Gomori (1952a) described this method. Burstone (1958) modified it. Four naphthol esters namely naphthol AS-BI phosphate, naphthol-AS-MX phosphate, naphthol AS-CL phosphate and naphthol AS-TR phosphate were substituted. With these esters, hydrolysis by alkaline phosphatase is quick. The resulting product is an extremely insoluble naphthol derivative. These naphthol derivatives react with diazonium salts to produce an insoluble azo dye at the sites of enzyme activity.

Remarks

Substituted naphthol phosphates have to be dissolved in dimethyl formamide (DMF). Distilled water is to be added and buffered to pH 8.3. These stock solutions last for several months in a refrigerator. For reactions, it is necessary to add diazonium salts such as Fast Red TR or Fast Blue RR in the ratio 1 mg/1 ml to the required amount of incubating solution. The incubating medium is to be prepared just before use and the staining has to be carried out at 37°C for 30 minutes.

Calcium Cobalt Method for Alkaline Phosphatase (Gomori, 1941c) (Plate 17, Figure 1)

Fixation

Cold acetone paraffin, Cold formalin frozen

Reagents Required

Sodium β-glycerophosphate
Sodium diethylbarbiturate
Calcium chloride
Magnesium sulphate
Cobalt nitrate
Yellow ammonium sulphide

Preparation of the Reagents

Solution 1 (Incubating medium)

3 per cent sodium β-glycerophosphate	10.0 ml
2 per cent sodium diethyl barbiturate	10.0 ml
Distilled water	5.0 ml
2 per cent calcium chloride	20.0 ml
5 per cent magnesium sulphate	1.0 ml

Solution 2 (Cobalt solution)

Cobalt nitrate or acetate	2.0 g
Distilled water	100.0 ml

Procedure

1. For paraffin sections dewax with light petroleum.
2. Bring them down to water via acetone.
3. Place the sections in solution 1 for 16 hours at 37°C.
4. Rinse in running water.
5. Transfer to solution 2 for 3–5 minutes.
6. Rinse in distilled water.
7. Flood slides with dilute yellow ammonium sulphide.
8. Wash in water.
9. Dehydrate, clear and mount.

Result

Sites of alkaline phosphatase activity are stained black.

Azo Dye Method for Leucocyte Phosphatase (Kaplow, Monis, Hayhoe)

Procedure

1. Take smear from venous blood and dry in air.
2. Fix the smear in ice-cold 10 per cent formalin–methanol (40 per cent formaldehyde 10 ml + methanol 90 ml) for 30 seconds.
3. Incubate the slides in the following medium at 22°C for 10 minutes.

0.1 M tris buffer (pH 9.2)	25.0 ml
Sodium naphthol phosphate	20 mg
Fast garnet GBC salt	30 mg

4. Wash in running water.
5. Counterstain with Mayer's haemalum for 3 minutes.
6. Wash in water.
7. Mount in glycerine jelly.

Result

Sites of alkaline phosphatase activity are stained brown.

Naphthol AS-Phosphate Azo Dye Method (Burstone, 1957a)

Fixation

Freeze-dried or cold acetone paraffin sections

Reagents Required

Diazonium salts

Procedure

1. Bring down sections to water via xylene and acetone.
2. Transfer to incubating medium at 22°C for 30 minutes.

   40 mg ASD salt + 100.0 ml distilled water

   or

   any diazonium salt such as

   Fast red
   Violet LB salt
   Fast blue RR
   Red RC
   Red TR

   are recommended.
3. Wash rapidly in water.
4. Mount in PVP medium.

Result

Sites of alkaline phosphatase activity are stained red or blue.

Gomori Red Method (Metal precipitation)

Fixation

Cryostat pre-fixed, post-fixed, frozen or paraffin sections.

Reagents Required

0.5 M veronol acetate buffer (pH 5.0)
Sodium β-glycerophosphate
Lead nitrate
1 per cent ammonium sulphide

Preparation of Reagents

Solution 1 (Incubating medium)

0.5 M veronol acetate buffer, pH 5.0	10.0 ml
Sodium β-glycerophosphate	32 mg
Lead nitrate	20 mg

Dissolve lead nitrate first in the buffer and then add β-glycerophosphate. Maintain pH at 5.0.

Procedure

1. Place sections in solution 1 incubating medium at 37°C for ½–2 hours.
2. Wash in distilled water.
3. Transfer to 1 per cent ammonium sulphide (prepared fresh) for 2 minutes.
4. Wash well in distilled water.
5. Counterstain in 2 per cent methyl green.

Result

Sites of acid phosphatase activity	Black
Nuclei	Green

Rationale

First introduced by Gomori (1941b). From the substrate, phosphate ions are split off by the enzyme and an insoluble precipitate of lead phosphatase results at the sites of enzyme activity. Treatment with ammonium sulphide converts the precipitate into lead sulphide. For unfixed frozen sections, see Bitensky (1963) for modification of Gomori method.

Azo Dye Coupling Method for Acid Phosphatase (Simultaneous Coupling)

Reagents Required

0.1N acetate buffer (pH 5.0)
Sodium naphthyl phosphate
Diazonium salt
Fast garnet GBC

Preparation of Reagents

Solution 1 (Incubating medium)

Sodium naphthyl phosphate	10 mg
0.1 M acetate buffer (pH 5.0)	10.0 ml
Fast garnet GBC	10 mg

Procedure

1. Incubate sections in solution 1 at 37°C for 15–60 minutes.
2. Wash in distilled water.
3. Counterstain in 2 per cent methyl green.
4. Wash in running tap water.
5. Mount in glycerine jelly.

Result

Sites of acid phosphatase activity	Red
Nuclei	Green

Rationale

α-naphthyl released at the sites of enzyme activity couples with a diazonium salt. A suitable diazonium salt which can couple satisfactorily should be selected at low pH. Gregg and Pearse (1952) carried out a number of reactions with a variety of diazonium salts and finally came to a conclusion that among all diazonium salts, Fast Garnet GBC was the best.

Azo Coupling Method (Kaplow and Burstone, 1964)

This is almost similar to the alkaline phosphatase method.

Reagents Required

Naphthol AS-MS phosphate
N-N-dimethyl formamide
Acetate buffer
Fast red violet salt LB

Preparation of Reagents

Solution 1

Naphthol AS-MS phosphate	5 mg
N-N-dimethyl formamide	0.2–0.3 ml
Distilled water	25.0 ml
0.2 M Acetate buffer (*see* Chapter 5)	25.0 ml
Fast red violet salt LB	40 mg

Shake well. Instead of acetate buffer, citrate buffer can also be used.

Procedure

1. Deparaffinize sections in chloroform or petroleum ether and hydrate through acetone–water combinations as in Gomori's method (Frozen sections can directly be placed in substrate).

2.   Transfer to solution 1 (incubating medium) for 2 hours at 37°C.
3.   Wash in running water.
4.   For unfixed sections, wash in formalin for 30 seconds. Wash in running water.
5.   Counter stain in Mayer's haematoxylin.
6.   Wash in running water.
7.   Mount in glycerine jelly.

Result

Sites of acid phosphatase	Red
Nuclei	Blue

Naphthol AS-BI Phosphate Method for Acid Phosphatase (Burstone, 1958 Modified by Barka, 1960)

Fixation

Cryostat post-fixed, pre-fixed, freeze-dried and frozen sections.

Reagents Required

Veronol acetate buffer
Naphthol AS-BI phosphate
N-N-dimethyl formamide
Pararosaniline hydrochloride
Hydrochloric acid
Sodium nitrite
0.1N sodium hydroxide

Preparation of Reagents

Solution 1 (Substrate solution)
    Naphthol AS-BI phosphate   50 mg
    Dimethyl formamide   5.0 ml

Solution 2 (Buffer solution)
    Veronol acetate buffer stock A (*see* Chapter 5)

Solution 3 (Sodium nitrite solution)
    Sodium nitrite   400 mg
    Distilled water   10.0 ml

Solution 4
    Pararosaniline stock (*see* Chapter 5)

Solution 5
    Distilled water

Solution 6 (Incubating medium)
    Solution 1   0.5 ml
    Solution 2   2.5 ml
    Solution 3 and 4   0.8 ml and 0.4 ml of solution 3 and 4 mixed before adding to the incubating solution
    Solution 5   6.5 ml
The pH of the solution should be 4.7–5.0, it is adjusted with 0.1N NaOH.

Procedure

1. Sections are placed in solution 6 (incubating medium) and incubated at 37°C for 15–60 minutes.
2. Wash in distilled water.
3. Counterstain with 2 per cent methyl green.
4. Wash in distilled water.
5. Mount in glycerine jelly or dehydrate quickly in alcohol and mount in DPX.

Result

Acid phosphatase sites Red
Nuclei   Green

Remarks

Solution 3 sodium nitrate should be fresh. pH should be 4.7–5.0. Solution should be filtered.

Rationale

Burstone (1958) recommended AS-BI phosphate instead of naphthol as the best. Barka (1960) suggested the use of hexazonium pararosaniline as the diazonium salt in the simultaneous coupling method. This diazonium along with AS-BI phosphate gives accurate localization and he recommended this.

Post-coupling Technique for Acid Phosphatase (Ruttenberg and Seligman, 1955)

Reagents Required

Sodium 6-benzoyl 2-naphthyl phosphate
Walpole acetate buffer, pH 5.0 (*see* Chapter 5)
Sodium chloride
Fast blue B or Garnet GBC

Preparation of Reagents

Solution 1 (Substrate preparation)

Sodium 6-benzoyl 2-naphthyl phosphate	25 mg
Distilled water	80.0 ml
Walpole acetate buffer	20.0 ml
Sodium chloride	2.0 g

Solution 2 (Diazonium salt solution)

Fast blue B or Garnet GBC	50 mg
Distilled water	50.0 ml

This solution is turned alkaline with sodium bicarbonate.

Procedure

1. Fresh frozen sections are treated with 0.85 per cent, 1 per cent and 2 per cent sodium chloride solution, each lasting 3 minutes.

2. Transfer sections to substrate at room temperature for 9 hours. If fresh sections are used keep them for 1 hour. If fixed, keep them for 2 hours.
3. Wash fresh section in cold saline, fixed in water.
4. Transfer to freshly prepared diazonium salt (solution 2).
5. Wash in cold saline or water.
6. Mount in glycerine jelly.

Results

Sites of enzyme activity are stained blue to red.

Lead Nitrate Method for Acid Phosphatase (Gomori, 1956)

Fixation

Cold acetone or cold formalin-fixed paraffin or cryostat frozen sections.

Procedure

1. For formalin-fixed frozen sections, they should be dried.
2. Transfer the slides to the incubating medium 0.01 M sodium β-glycerophosphate in 0.05 M acetate buffer (pH 5.0) containing 0.004 M lead nitrate. Incubating time is 4–16 hours at 37°C.
3. Wash briefly and place the slides in dilute yellow ammonium sulphide for 2 minutes.
4. Wash, counterstain in 1 per cent aqueous eosin for 5 minutes.
5. Wash.
6. Mount in glycerine jelly.

Result

Sites of acid phosphatase activity are stained black.

Lead Method for 5′-Nucleotidase (Wachstein and Meisel, 1957)

Reagents Required

Adenosine 5-phosphate
0.2 M tris buffer (pH 7.2)
2 per cent lead nitrite
Magnesium sulphate

Preparation of Reagents

Solution 1 (incubating medium)

1.25 per cent adenosine 5-phosphate	4.0 ml
0.2 M tris buffer (pH 7.2)	4.0 ml
2 per cent lead nitrite	0.6 ml
0.1 M magnesium sulphate	1.0 ml
Distilled water	0.5 ml

Sections

Sections fixed in cold formol calcium (4°C), cryostat

Procedure

1. Incubate slides in solution 1 at 37°C for 30 minutes to 1 hour.
2. Fix in formol saline if sections are unfixed.
3. Carefully transfer sections to distilled water with glass rod.
4. Wash again.
5. Immerse in 1 per cent ammonium sulphide for 3 minutes.
6. Wash in distilled water.
7. Repeat wash.
8. Mount on the slide, allow it to dry.
9. Mount in glycerine jelly.

Result

Sites of 5′ nucleotidase are stained blackish brown deposits.

Rationale

The reaction resembles that of acid and alkaline phosphatase. Magnesium ions are sequestered by the activated enzyme and lead nitrate helps to develop a precipitate. 5′-nucleotidase hydrolyses the substrate and the released phosphate ions precipitate as lead phosphate at the site of enzyme activity. On treatment with ammonium sulphide, black lead sulphide becomes evident.

Lead Method for Glucose 6-phosphatase (Wachstein and Miesel, 1956)

Reagents Required

Glucose 6-phosphate (potassium salt)
Tris maleate (pH 6.7)
2 per cent lead nitrate
Ammonium sulphide

Preparation of Solution 1 (Incubation medium)

0.12 per cent glucose 6-phosphate (aq.)	4.0 ml
0.2 M tris maleate (pH 6.7)	4.0 ml
2 per cent lead nitrate	0.6 ml
Distilled water	1.4 ml

Sections

Cryostat unfixed, Frozen unfixed.

Procedure

1. Incubate unfixed sections in solution 1 for 5–20 minutes at 37°C.
2. Wash in distilled water giving 3 changes.
3. Transfer sections to 1 per cent ammonium sulphide for 2 minutes.

4.   Wash in distilled water.
5.   Again fix sections in 10 per cent formaldehyde for 15–30 minutes.
6.   Wash in distilled water.
7.   Mount in glycerine jelly.

Result

Sites of glucose 6-phosphatase activity are stained brownish black.

Rationale

Chiquonine (1954) described the method and modified it in 1955. The present technique was first described by Wachstein and Meisel (1956). In glucose 6-phosphate, the potassium salt is the substrate, lead nitrate in the incubating medium produces a precipitate with the released phosphate. The lead precipitate on treatment with ammonium sulphide becomes the discernible lead sulphide.

Since glucose 6-phosphate is a sensitive enzyme, the reaction is advocated to be done with unfixed sections. Formol saline destroys it. Confusing as it may appear, the same method is applicable to both acid and alkaline phosphatases. By stringent control sections, confusion can be avoided.

Lead Method for Adenosine Triphosphatase (Wachstein and Meisel, 1960)

Reagents Required

Adenosine triphosphate
Tris buffer (pH 7.2)
2 per cent lead nitrate
2.5 per cent magnesium nitrate
Ammonium sulphide

Preparations of Reagents

Solution 1 (Incubating medium)

0.125 per cent adenosine triphosphate	4.0 ml
Tris buffer	4.0 ml
2 per cent lead nitrate	0.6 ml
2 per cent magnesium nitrate	1.0 ml
Distilled water	0.4 ml

Sections

Frozen fixed
Cryostat unfixed
Cryostat pre-fixed

Procedure

1.   Carefully incubate free-floating sections in solution 1 for 10–60 minutes at 37°C.
2.   Wash in distilled water.
3.   Transfer to 1 per cent yellow ammonium sulphide for 2 minutes.

4. Wash in tap water for 2 minutes.
5. Mount in glycerine jelly.

Result

Sites having adenosine triphosphatase (ATPase) turn brownish black.

Rationale

The method resembles Gomori's method for phosphatase. The enzyme is vulnerable to many fixations. Fixation is the most vexatious problem. In unfixed cryostat and frozen sections, enzyme diffuses. Formalin fixation (4°C) destroys the enzyme. To overcome this, perform the technique both on fixed and unfixed sections and compare. The enzyme splits the phosphate from the ATP. Lead phosphate is formed as a precipitate at the sites of enzyme activity. As usual, it is brownish black converted to lead sulphide with ammonium sulphide.

Calcium Method for Adenosine Triphosphatase (Maengwyn Davies *et al.*, 1952)

Fixation

Unfixed frozen

Reagents Required

Sodium acetate
Potassium hydroxide
Glycerine
Potassium chloride
Sodium adenosine triphosphate
Sodium phosphate
Calcium chloride
Yellow ammonium sulphide

Preparation of the Reagents

Solution 1 (Incubating medium)

0.01 M glycine	
0.4 M Potassium chloride in saturated sodium acetate	12.0 ml
Saturated sodium acetate	12.0 ml
0.36 M calcium chloride	3.6 ml
Potassium hydroxide	0.6 ml
0.04 M sodium adenosine triphosphate	6.0 ml
Distilled water	13.8 ml
Saturated sodium phosphate	0.3 ml

Solution 2 (Calcium chloride solution 1 per cent)

Calcium chloride	1.0 g
75 per cent ethanol	100.0 ml

Solution 3 (Calcium chloride solution 2 per cent)

Calcium chloride	2.0 g
Distilled water	100.0 ml

Procedure

1. Incubate floating sections in solution 1 for 5 minutes to 3 hours.
2. Wash in solution 2 giving 3 changes.
3. Immerse in solution 3 for 3 minutes.
4. Flood with yellow ammonium sulphide.
5. Wash, dehydrate and mount.

Result

Sites of ATPase activity are stained blackish brown.

Calcium Method for Adenosine Triphosphatase (Padykula and Herman, 1955)

Fixation

Unfixed frozen
Free-floating

Reagents Required

Sodium barbiturate
Calcium chloride
Adenosine triphosphate
Sodium hydroxide
Yellow ammonium sulphide

Preparation of Reagents

Solution 1 (Incubating medium)

0.1 M sodium barbiturate	20.0 ml
(2.062 g/100 ml)	
0.18 M calcium chloride	10.0 ml
(1.998 g/100 ml)	
Distilled water	30.0 ml
Adenosine triphosphate	152 mg

This should be prepared just before use.
Adjust the pH of the solution with 0.1 M NaOH and make up to 100 ml with distilled water.

Solution 2 (1 per cent calcium chloride solution)

Calcium chloride	1.0 g
Distilled water	100.0 ml

Solution 3 (2 per cent calcium chloride solution)

Calcium chloride	2.0 g
Distilled water	100.0 ml

Procedure

1. Place the slides in incubating medium at 37°C.
2. Wash in solution 2 giving 3 changes.

3. Place in solution 3 for 3 minutes.
4. Wash in distilled water.
5. Treat with yellow ammonium sulphide.
6. Wash, dehydrate, clear and mount in synthetic medium.

Result

Sites of ATPase activity are seen as black deposits.

Lead Nitrate Method for Phosphoamidase (Gomori, 1948)

Fixation

Cold acetone, double-embedded sections

Reagents Required

p-chloroanilidophosphonic acid
NH_4OH
Acetic acid
Maleic acid
N-NaOH
Lead nitrate
Manganese chloride

Preparation of Reagents

Solution 1 (Stock substrate)
Dissolve *p*-chloroanilidophosphonic acid to make 0.1 M solution in 10 per cent NH_4OH. pH should be 8 (with acetic acid) and make up the volume with distilled water.

Solution 2 (Maleate buffer)
Maleic acid	5.8 g
Distilled water	500 ml
N-NaOH	60 ml

Make up to 1000 ml. Final pH should be 5.6.

Solution 3 (0.1 M lead nitrate)

Solution 4
Magnesium chloride	10.0 g
Distilled water	100 ml

Solution 5 (Incubating medium)
Solution 1	2 ml
Solution 2	50 ml
Solution 3	1.5 ml
Solution 4	Few drops

Procedure

1. Keep frozen slides in solution 5 for 4 hours.
2. Rinse in distilled water.
3. Rinse in 0.1 M citrate or acetate buffer (pH 4.5).

4.   Rinse in distilled water.
5.   Keep in yellow ammonium sulphide solution for 1–2 minutes.
6.   Wash in running water.
7.   Counterstain with 1 per cent eosin for 5 minutes.
8.   Wash and mount in glycerine jelly.

Result

Phosphamidase sites are stained black.

Table 18.1   Histochemical techniques applied for the demonstration of enzymes

Technique	Fixative	Alkaline phosphatase	Acid phosphatase	5'-Nucleotidase	Glucose 6-phosphatase	Adenosine triphosphatase	Phospho amidase
Modified Gomori's method	Cryostat, frozen	Brownish black					
Gomori's method	Cold acetone	Black					
Kaplow method	Cold acetone	Red					
Azo Dye coupling method	Cold acetone	Reddish brown					
Naphthol AS-BI method	Cryostat post- and pre-fixed, freeze-dried frozen	Red					
Calcium cobalt method	Cold acetone, paraffin, cold formalin frozen	Black					
Azo Dye method	Frozen	Brown					
Naphthol AS-phosphate Azo Dye	Freeze-dried, cold acetone	Red or blue					

(Contd.)

Table 18.1 (Continued)

Technique	Fixative	Alkaline phosphatase	Acid phosphatase	5'-Nucleotidase	Glucose 6-phosphatase	Adenosine triphosphatase	Phospha-midase
Gomori's red method	Cryostat post-fixed, pre-fixed freeze-dried, frozen		Black				
Azo-Dye coupling method	Cryostat post-fixed, pre-fixed freeze-dried, frozen		Red				
Naphthol AS-BI phosphate method	Cryostat post-fixed, pre-fixed freeze-dried, frozen		Red				
Post-coupling technique	Fresh frozen		Blue to red				
Lead Nitrate method	Cold acetone, cold formalin frozen		Black				
Lead method (5'- nucleotidase)	Formol calcium (4°C) fixed			Blackish brown			
Lead method (Glucose 6-phosphatase)	Cryostat unfixed, frozen unfixed				Brownish black		

Method	Preparation	Result	
Lead method (ATP)	Cryostat pre-fixed, cryostat post-fixed frozen	Brownish black	
Calcium method	Unfixed frozen	Blackish brown	
Calcium method	Unfixed frozen	Black	
Lead nitrate method for phosphamidase	Cold acetone		Black

ESTERASE

Esterases are capable of hydrolysing carboxylic acid. A number of enzymes can hydrolyse carboxylic acid which can be demonstrated histochemically. The pH range of these enzymes lies between 5.0 and 9.0. Each enzyme is capable of hydrolysing a number of different types of substrates and many different enzymes can hydrolyse the same substrate.

Among esterase enzymes, there are two types 1) non-specific and 2) specific. If the substrate is a simple ester such as α-naphthyl acetate the hydrolysing enzyme is called non-specific. Specific esterases are of two types—acetylcholine esterase and choline esterase. Acetylcholine esterase is capable of hydrolysing acetylthiocholine whereas choline esterase will hydrolyse esters of choline other than thiocholine. Both can hydrolyse simple esters.

Non-specific esterases include:

1. Carboxyl esterases
2. Aryl esterases
3. Acetyl esterases

Specific esterases include:

1. Acetylcholine esterase
2. Choline esterase
3. Lipases

Non-specific esterase classification depends on the most suitable substrate that the enzyme will hydrolyse. According to Pearse (1972), the terminology adopted for non-specific esterases is

A-esters (aryl esterases)

B-esters (carboxyl esterases)

C-esters (acetyl esterases)

Specific esterases are cholinesterases comprising two types—acetylcholine esterase and cholinesterase. Acetylthiocholine esterase hydrolyses acetylthiocholine, whereas choline esterase hydrolyses esters of choline rapidly and efficiently. There is a difference between specific and non-specific esterases. Specific choline esterase has a great capacity to hydrolyse choline esters. This capacity can be destroyed by employing specific inhibitors, viz. eserine (10^5 M). Non-specific esterases remain unaffected by eserine.

Inhibitors and their Use

Inhibitors are necessary to separate different esterase enzymes. This allows precise identification of the specific enzymes. Some of the useful esterase inhibitors are the organophosphorous compounds such as dicropropyl fluoro phosphate (DFP) and diethyl-*p*-nitrophenyl phosphate (E600) and some aromatic mercuric compounds like *p*-chloromercuric benzoate.

Pearse (1972) has adopted one scheme to demonstrate whether esterase activity is due to either 'A', 'B' and 'C' esterases. According to this scheme, sections are incubated in a 10 mM solution of E600 in buffer at pH 5.3 for an hour at 37°C. This is followed by 'B' esterase being inhibited. Now esterase activity is due either to 'A' or 'C' esterases. These can be further distinguished by incubating a section in a solution of ρ-chloromercuric benzoate (PCMB) with a concentration of 100 mM. This will inhibit 'A' esterase but 'C' esterase is activated.

Non-specific Esterase Methods

Suitable controls and inhibitors should be used to identify the precise non-specific esterase because the substrate used is hydrolysed by a number of enzymes.

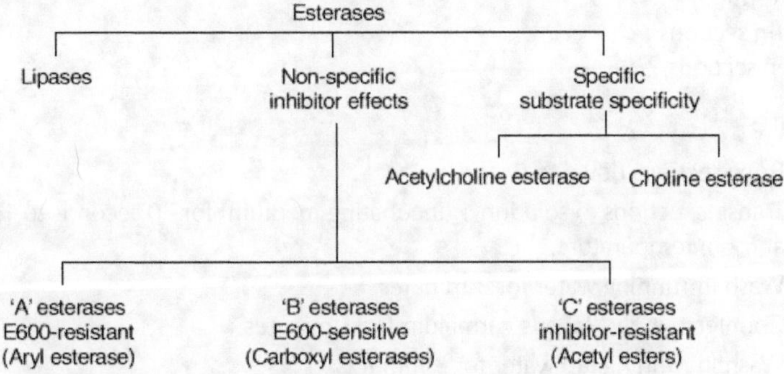

Some histochemical methods:

1. α-naphthyl acetate method with Fast Blue B for non-specific esterase.
2. α-naphthyl acetate method with hexazonium pararosaniline for non-specific esterase.
3. Indoxylacetate method for specific esterases, E 600-resistant esterase.
4. Naphthol AS-LC acetate method for non-specific esterase.
5. Tween method for lipase
6. Myristoyl choline for cholinesterase
7. Thiocholine

Non-specific Esterase: α-Naphthyl Acetate Method Using Fast Blue B (Gomori, 1950a)

Reagents Required

α-naphthyl acetate
Acetone
0.2 M phosphate buffer, (pH 7.4)
Fast blue B

Preparation of Reagent

Solution 1 (Incubating medium)

α-naphthyl acetate	5 mg
Acetone	0.1 ml
0.2 M phosphate buffer, (pH 7.4)	10.0 ml
Fast blue B	30 mg

α-naphthyl acetate is first dissolved in acetone and to this, phosphate buffer is added and mixed thoroughly. Fast blue B is then added and the solution is filtered.

Sections

Cryostat pre-fixed
Cryostat post-fixed
Freeze-dried
Paraffin sections
Frozen sections

Procedure

1. Bring sections down to water.
2. Transfer sections to solution 1 (incubating medium) for 30 seconds to 15 minutes at room temperature.
3. Wash in running water for 3 minutes.
4. Counterstain in Mayer's carmalum for 3 minutes.
5. Wash in running tap water for 3 minutes.
6. Mount in glycerine jelly.

Result

Sites of esterase activity	Reddish brown
Nuclei	Red

Rationale

All types of esterase activity may be revealed. The pH should be 7.4. The simultaneous coupling was first described by Nachlas and Seligman (1949). β-naphthyl acetate was first used with Diazo Fast Blue B as a coupling agent. Gomori (1952a) used α-naphthyl instead of β-naphthyl acetate because the resulting azo dye was not soluble in water.

Esterase in the tissue attacks α-naphthyl acetate and releases α-naphthol which combines with Fast Blue B to produce an insoluble azo dye at the site of enzyme activity. Cholinesterases may be involved here, but can be inhibited by using eserine (10^{-5} M).

Non-specific Esterase: α-Naphthyl Acetate Method using Hexazotized Pararosaniline (Gomori, 1950a; Davis and Ornstein, 1959)

Reagents Required

α-naphthyl acetate
0.2 M phosphate buffer (stock solution A)

Acetone
Pararosaniline hydrochloride
HCl
Sodium nitrate

Preparation of Reagents

Solution 1 (Substrate solution)
    α-naphthyl acetate   50 mg
    Acetone   5 ml

Solution 2 (Buffer solution)
    0.2 M phosphate buffer, stock solution A (*see* Chapter 5)

Solution 3 (Sodium nitrate solution)
    Sodium nitrate   400 mg
    Distilled water   10 ml

Solution 4 (Pararosaniline–HCl stock)
    Pararosaniline hydrochloride 2 mg
    2-N-Hydrochloric acid   50 ml
    Heat and cool to room temperature.

Solution 5 (Incubating medium)
    Solution 1   0.25 ml
    Solution 2   7.25 ml
    Solution 3   0.8 ml (0.4 ml of solution 3 and 4 are mixed
                     before being adding to the mixture)
    Solution 4   2.5 ml

Final pH should be 7.4.

Procedure

1. Bring sections to water.
2. Keep slides in incubating solution 5 for 2–20 minutes at 37°C.
3. Wash in running water.
4. Counterstain with 2 per cent methyl green.
5. Wash.
6. Dehydrate rapidly through alcohols, clear in xylene and mount.

Result

Esterase   Reddish brown
Nuclei   Green

Rationale

This is similar to the previous method except that hexazonium pararosaniline is used instead of Fast Blue B. This coupling agent for esterase was used by Davis and Ornstein (1959). It is better than the previous method because sections can be dehydrated through alcohol series to xylene and mounted in a synthetic mounting medium.

Non-specific Esterase: Indoxyl Acetate Method
(Holt and Withers1954), (Plate 17, Figure 2)

Reagents Required

5-Bromo-indoxyl acetate
Ethanol
0.1 M tris buffer, pH 6–8
Potassium ferricyanide
Potassium ferrocyanide
Calcium chloride

Preparation of Reagents

Solution 1 (Incubation medium)

5-Bromo-indoxyl acetate	1 mg
Ethanol	0.1 ml
Tris buffer (0.1 M) (pH 6–8)	2.0 ml
Potassium ferricyanide (0.05 M) (1.6 per cent)	1.0 ml
Potassium ferrocyanide (0.05 M) (2.1 per cent)	1.0 ml
Calcium chloride (0.1 M) (2.1 per cent)	1.0 ml
Distilled water	5.0 ml

The 5-bromo-indoxyl acetate is dissolved in ethanol, and the buffer is then added. The remaining chemicals are dissolved in distilled water and the solution is mixed. The solution should be freshly prepared.

Sections

Cryostat pre- and post-fixed
Freeze-dried
Frozen
Paraffin

Procedure

1. Bring sections to water.
2. Keep slides in solution 1 and incubate at 37°C for 15 minutes to 1 hour.
3. Rinse in tap water.
4. Counterstain in Mayer's carmalum for 5 minutes.
5. Rinse in tap water.
6. Mount in glycerine jelly or dehydrate quickly and mount in DPX.

Result

Esterase activity	Black
Nuclei	Red

Rationale

This technique was first described by Barnett and Seligman (1951) and Holt and Withers (1952). It is a specific method for non-specific esterase. Indoxyl acetate was recommended as a substrate. Holt (1954) introduced 4-chloro 5-bromo-indoxyl acetate as a substrate and this gives superior results to the earlier substrate. The incubating medium comprises 4-chloro 5-bromo-indoxyl acetate, potassium ferricyanide, potassium ferrocyanide, calcium chloride and tris buffer. The esterase in tissues hydrolyses the 5-bromo-indoxyl. This is then oxidized by potassium ferricyanide to an insoluble indigo dye. Calcium chloride in the medium acts as an enzyme activator.

Non-specific Esterase: Napthyl AS-LC Acetate Method (Gomori, 1952a; Burstone, 1957b)

Reagents Required

Naphthol AS-LC acetate
Dimethyl formamide
Ethyl cellulose
Tris buffer, pH 7.1
Fast garnet GBC

Preparation of Reagent

Solution 1 (Incubation medium)

Naphthol AS-LC acetate	5 mg
Dimethyl formamide	0.4 ml
Ethyl cellulose	1.0 ml
Tris buffer (pH 7.1)	2.0 ml
Distilled water	6.6 ml
Fast garnet GBC	10 mg

The substrate naphthol AS-LC acetate is dissolved in solutions dimethyl formamide and ethyl cellulose. The remaining tris buffer and distilled water are added. The Fast Garnet GBC is added finally. The solution is mixed thoroughly, filtered and used immediately.

Sections

Cryostat pre-fixed
Cryostat post-fixed
Frozen

Procedure

1. Bring sections to water.
2. Place sections in solution 1 (incubating medium) at 20°C for 10–30 minutes.
3. Rinse in tap water.
4. Counterstain in 2 per cent methyl green.
5. Wash in tap water.
6. Mount in glycerine jelly.

Result

Sites of esterase activity	Reddish brown
Nuclei	Green

Rationale

Introduced by Gomori (1952a), the technique was modified by Pearse (1953b) and Burstone (1957b). Burstone recommended the use of naphthyl AS-LC acetate. Pearse (1960) recommended this method especially for paraffin-embedded material. But for formalin-fixed frozen sections, α-naphthyl acetate method was preferred by him. Naphthol AS-LC acetate is included in the incubating medium. This is hydrolysed by the non-specific esterase to produce a highly insoluble product. This is coupled with a diazonium salt (Fast Garnet GBC) to give an insoluble azo dye at the site of enzyme activity.

α-Naphthyl Acetate Method for Esterase (Gomori, 1952a)

Fixation

Cold formalin
Cold acetone
Frozen
Paraffin
Cold microtome
Formalin post-fixed

Reagents Required

α-naphthyl acetate
Acetone
Phosphate buffer
Fast B salt

Preparation of Reagent

Incubating Medium

α-naphthyl acetate	10 mg
Acetone	0.25 ml
0.1 M phosphate buffer (pH 7.4)	20 ml

Mix and shake thoroughly until turbidity disappears. To this, add 50–100 mg Fast B salt. Shake and filter.

Procedure

1. Cut frozen sections of 10–15 μ thickness. Dry in air.
2. Incubate in the medium for 1–15 minutes.
3. Wash in running water.
4. Counterstain in Mayer's haemalum for 4 minutes.
5. Wash in running water for 30 minutes.
6. Mount in glycerine jelly.

Result

Sites of esterase activity are stained black.

Burstone's Azo-coupling Technique

Reagents Required

Naphthol AS-LC acetate
Acetone
0.1 M phosphate buffer (pH 7.2)
Fast blue RR (or fast garnet GBC)

Preparation

Solution 1 (Naphthol AS-LC solution)

Naphthol AS-LC acetate	3 mg
Acetone	0.3 ml
Distilled water	15.0 ml
0.1 M phosphate buffer (pH 7.2)	15.0 ml
Fast blue RR	15 mg

Dissolve naphthol acetate in acetone. Then mix the remaining in the given order. Stir well when Fast blue is added. Filter and use immediately.

Procedure

1. Cut frozen unfixed or acetone-fixed paraffin.
2. Place the free-floating or mounted section in solution 1 (substrate) for 30 minutes. Blue colour should develop.
3. Wash and mount in aqueous mountant.

Result

Sites of enzyme activity are stained red or blue.

Rationale

Naphthol from naphthol acetate is released and it couples with fast blue RR to form an insoluble azo dye.

LIPASES

These are a group of enzymes which are capable of hydrolysing long chain esters with more than seven carbon bonds which contain saturated fatty acids. Liver, pancreas and adrenals are the sources of lipases. Here again there is an overlap between lipases and non-specific esterases because both can hydrolyse simple esters. Lipase method is purely based on the enzyme hydrolysing the substrate Tween 60 to produce calcium ions to combine with fatty acids to form insoluble calcium soaps which when treated with lead ions and finally ammonium sulphide, form dark brown or black deposits of lead sulphide at the sites of enzyme activity.

HISTOCHEMICAL METHODS FOR LIPASES

Lipase: Tween Method (Gomori, 1952a)

Reagents Required

Tween 40, 60 or 80 (Tween 80 preferred)
Tris buffer (pH 7.2)
Calcium chloride
Thymol
Lead nitrate
Ammonium sulphide

Preparation of Reagents

Solution 1
    Tris buffer (pH 7.2)

Solution 2 (Tween)
    Tween 40, 60 or 80   5.0 g
    Tris buffer (pH 7.2)   100.0 ml
    Thymol   A few small crystals

Solution 3 (Calcium chloride solution)
    Calcium chloride   200 mg
    Distilled water   10.0 ml

Solution 4 (Lead nitrate solution)
    Lead nitrate   1.0 g
    Distilled water   50.0 ml

Solution 5 (Preparation of incubation medium)
    Solution 1   9.0 ml
    Solution 2   0.6 ml
    Solution 3   0.3 ml

Sections

Formalin-fixed, free-floating
Cryostat pre-fixed
Paraffin sections

Procedure

1. Bring sections to water.
2. Incubate sections at 37°C for 2–8 hours in solution 5. Paraffin sections require 24 hours.
3. Rinse sections in 3 changes of distilled water.
4. Place sections in pre-heated solution 4 (lead nitrate solution) at 55°C for 10 minutes.
5. Rinse sections in distilled water.
6. Transfer sections to one percent ammonium sulphide for 3 minutes.

7. Rinse in distilled water.
8. Counterstain in Mayer's carmalum for 5 minutes.
9. Wash in tap water for 1 minute.
10. Mount in glycerine jelly.

Result

Lipase activity	Yellow brown black
Nuclei	Red

Rationale

First described by Gomori (1945), it is not a specific test because tweens are hydrolysed also by some non-specific esterases. Gomori (1952a) modified it. Tween 60 and 80 are used besides lead nitrate and calcium chloride. The lipase in the tissue hydrolyses tween to release fatty acids. Fatty acids combine with lead nitrate to release lead which when treated with dilute ammonium sulphide precipitates black lead sulphide at the site of enzyme activity.

Cholinesterase: Myristoyl Choline Method (Gomori, 1948)

Reagents Required

Myristoyl choline
0.1 M veronol acetate buffer
Cobalt acetate
Calcium chloride
Magnesium chloride
Ammonium sulphide
Thymol

Preparation of Reagents

Solution 1

Cobalt acetate	400 mg
Distilled water	80.0 ml
0.1 M veronol acetate (pH 7.6)	20.0 ml
Magnesium chloride	1 mg
Calcium chloride	1 mg
Manganese chloride	1 mg
Thymol	A few small crystals

Solution 2 (Myristoyl solution)

Myristoyl chlorine	70 mg
Distilled water	10.0 ml
Thymol	A few small crystals

This solution should be in refrigerator (4°C).

Solution 3 (Incubating medium)

Solution 1	10.0 ml
Solution 2	0.2 ml

Sections

Cryostat pre-fixed
Formalin fixed
Cryostat post-fixed
Paraffin

Procedure

1.   Bring down sections to water.
2.   Incubate sections for 1–4 hours for 37°C. Paraffin sections require 2–16 hours.
3.   Wash in distilled water for 2 minutes.
4.   Transfer sections to 2 per cent ammonium sulphide.
5.   Wash in distilled water.
6.   Counterstain in 0.05 per cent eosin for 30 seconds.
7.   Wash in tap water.
8.   Mount in glycerine jelly.

Result

Sites of cholinesterase activity are stained dark brown to black.

Rationale

Gomori (1948) used myristoyl as a substrate. The incubating medium includes cobalt acetate in veronol acetate medium, calcium, manganese and magnesium are activators. Cholinesterase in the tissue splits myristoyl choline and releases fatty acids, which combine with cobalt acetate to form cobalt salt. On adding dilute ammonium sulphide, a precipitate of cobalt sulphide develops at the sites of enzyme activity.

Cholinesterase:Thiocholine Method (Gerebtzoff, 1959) (Modified by Filipe and Lake, 1983)

Reagents Required

Acetyl thiocholine iodide
0.1 M acetate buffer (pH 5.0–6.2)
Cupric sulphate
Glycerine
Ammonium sulphide

Preparation of Reagents

Solution 1 (Buffer)
    0.1 M acetate buffer (pH 5.0–6.2)

Solution 2 (Staining solution)

Acetyl thiocholine iodide	15 mg
Cupric sulphate	7 mg
Distilled water	1.4 ml

The solution is centrifuged at 4,000 rpm for 15 minutes and the supernatant is used.

Solution 3 (Glycerine)

    Glycerine   375 mg

    Distilled water   10.0 ml

Solution 4 (Cupric sulphate solution)

    Cupric sulphate 250 mg

    Distilled water   10.0 ml

Solution 5 (Incubating medium)

Solution 1	5.0 ml
Solution 2	0.8 ml
Solution 3	0.2 ml
Solution 4	0.2 ml
Distilled water	3.8 ml

The pH of the incubating medium varies with the type of tissue. Tissues with high cholinesterase activity are incubated at pH 5.0 and others at 6.2.

Sections

Cryostat pre-fixed

Formalin-fixed, free-floating

Cryostat post-fixed

Paraffin sections

Procedure

1. Place sections in incubating medium for 10–90 minutes at 37°C.
2. Rinse in 2 changes of distilled water.
3. Transfer sections to 2 per cent ammonium sulphide.
4. Wash well in distilled water.
5. Counterstain if necessary.
6. Wash in tap water.
7. Mount in glycerine jelly.

Result

Sites of cholinesterase activity are stained brown.

Rationale

Koelle and Friedenwald (1949) described the method, which is extensively used for both types of cholinesterase.

This incubating medium combines acetyl thiocholine iodide, copper sulphate and acetate buffer. Cholinesterases in the tissues hydrolyse the acteyl thiocholine to produce thiocholine which combines with copper ions to form copper thiocholine. The precipitate on treatment with 2 per cent ammonium sulphide is converted to copper sulphide visible as a precipitate.

Thiolactic Acid Method for Cholinesterase (Crevier and Belanger, 1955)

Fixation

Formalin fixed
Frozen
Fresh

Preparation of Reagents

Solution 1

0.12 M thiolacetic acid	0.75 mg
0.001 M lead nitrate	27 mg
0.1 M Na_2SO_4	83 ml
McIlvaine buffer (pH 6.2)	17.0 ml

Procedure

1.   Incubate in solution 1 at 22°C for ½ to 1 hour.
2.   Wash in ice-cold water.
3.   If desired, counterstain in 0.02 per cent basic fuchsin.
4.   Wash in ice-cold water.
5.   Dehydrate, clear in xylene.
6.   Mount in permount.

Result

Cholinesterase activity sites are revealed as black deposits.

Table 18.2   Histochemical techniques applied for the demonstration of esterase

Technique	Fixative	Non-specific esterase	Lipase	Cholinesterase
α-naphthyl acetate using fast blue B	Freeze-dried, frozen	Reddish brown		
α-naphthyl acetate method using Hexazotized pararosaniline	Cryostat post-fixed, cryostat pre-fixed, frozen-dried, frozen	Reddish brown		
Indoxyl acetate method	Freeze-dried, frozen	Black		
Naphthyl AS-LC acetate method	Cryostat pre-fixed, cryostat post-fixed, frozen	Reddish brown		
Burstone's azo-coupling technique	Cryostat prefixed, cryostat post-fixed, frozen	Red or blue		
α-naphthyl acetate method	Cold formalin, cold acetone frozen, paraffin, cold microtome, formalin post-fixed	Black		
Tween method	Formalin fixed, cryosat pre-fixed, cryostat post-fixed		Yellow brown black	
Miristoyl choline method	Formalin fixed, crysotat pre-fixed, cryostat post-fixed			Dark brown to black
Thiocholine method	Fresh or formalin-fixed, cryostat			Brown
Thiolactic acid method	Fresh or formalin-fixed, frozen			Black

OXIDOREDUCTASES

Oxidases

The oxidases comprise enzymes capable of hydrolysing certain appropriate substrates. Two ways of oxidation occur, 1) by acting as catalysts, catalysing between the substrate and atmospheric oxygen (oxidase) or 2) by electron transfer reaction (removing hydrogen from substrate to a hydrogen acceptor). These include dehydrogenases.

Oxidases occur in animal and plant tissues. They have copper and iron. Oxidases oxidize substrates by catalysing reaction between oxygen and substrate. These enzymes can be demonstrated histochemically. These enable to distinguish DOPA oxidase and peroxidases. Davenport (1960) stated that an easy way of distinguishing oxidases from dehydrogenases is that oxidases do not function anaerobically. Oxidases are categorized as follows:

1. **Cytochrome oxidase (Iron- and copper-containing)** Kidney, liver and muscle are abound with the enzyme. Lemberg (1959) from his experiments has shown that this is attached to cell mitochondria. It contains both haem groups and copper atom. Cytochromes are also known as indophenol oxidase.

2. **Peroxidase (Iron-containing)** Peroxidases break down peroxides. A special type of peroxidase is catalase. This breaks down hydrogen peroxide. Peroxidases are frequently found in leucocytes, haemoglobin, spleen and liver.

3. **DOPA oxidase (Copper-containing)** Also known as tyrosinase or catechol oxidase, it catalyses the oxidation of tyrosine to dihydroxyphenylalanine (DOPA) and the DOPA is finally oxidized to melanin pigments. It can also oxidize diphenol compounds like catechol and adrenaline, etc.

4. **Monoamine oxidase (Copper co-factor)** This acts on short-chain monoamines and diamines. Weiner (1960) stated that monoamine oxidase shows rapid action on tyramine and dopamine. It has a role in the breakdown of adrenaline and 5-hydroxytryptamine.

HISTOCHEMICAL METHODS FOR OXIDASES

DAB Technique for Peroxidase (Graham and Karnovsky, 1966)

Reagents Required

3,3´-diaminobenzidine tetrahydrochloride
Tris HCl buffer (pH 7.6)
1 per cent hydrogen peroxide

Preparation of Reagents

Solution 1 (Incubating medium)

3,3'-diaminobenzidine tetrahydrochloride	5 mg
Tris HCl buffer (pH 7.60)	10.0 ml
1 per cent hydrogen peroxide	1.2 ml

Sections

Fixation is very important. Graham and Karnovsky (1966) used 5 per cent glutaraldehyde and formaldehyde in cacodylate buffer (*see* Chatper 5)

Tissues should be pre-fixed in this solution.

Incubating Method

1.   Rinse fixed sections in water.
2.   Transfer to solution 1 (incubating medium) for 5 minutes.
3.   Wash in distilled water.
4.   Counterstain in 2 per cent methyl green for 5 minutes.
5.   Wash in distilled water.
6.   Dehydrate through graded alcohols.
7.   Clear in xylene and mount in DPX.

Result

Peroxidase activity	Brown
Nuclei	Green

Rationale

Graff (1916) introduced a histochemical method which was modified later. Subsequently a number of techniques emerged. Demonstration of peroxidases is somewhat difficult because many varieties of peroxidase enzymes abound in tissues. Most of the substrates of peroxidase produce weak results upon oxidation. A popular method to demonstrate peroxidase was the benzidine method. Benzidine is a carcinogenic substrate. The best technique in vogue is 3,3'-diaminobenzidine (DAB) (Graham and Karnovsky, 1966).

Benzidine Reaction for Peroxidase (De Robertis and Grasso, 1946)

Fixation

Cold microtome
Frozen fresh sections

Reagents Required

Ammonium molybdate
Benzidine
Saline
Hydrogen peroxide

Preparation of Reagents

Solution 1 (Incubating medium)
    Ammonium molybdate   100 mg
    Saline (0.85 per cent)   100.0 ml

Solution 2
    Hydrogen peroxide   1 drop
    Saturated benzidine in   1 drop
       0.85 per cent saline

To saturated benzidine, 20 vol. H_2O_2 per 2 ml is added just before use.

Procedure

1. Cut 20–40 μ thick section from fresh tissues at 4°C.
2. Immerse in solution 1 for 5 minutes.
3. Transfer directly to solution 2 for 2 minutes.
4. Wash in 0.85 per cent saline.
5. Mount in glycerine jelly.

Result

A blue precipitate indicates sites of peroxidase activity.

Straus Method (1964)

Fixation

10 per cent neutral buffered formalin + 30 per cent sucrose at 4°C for 18 hours.

Reagents Required

Benzidine
Hydrogen peroxide
Sodium nitroprusside
Methanol
Acetate buffer

Preparation of Reagents

Solution 1 (Benzidine solution)
    70 per cent ethanol   50.0 ml
    Benzidine   200 mg
    Hydrogen peroxide   0.015 ml

Solution 2
    Sodium nitroprusside 15.0 ml
    70 per cent ethanol   50.0 ml

Solution 3
    Sodium nitroprusside 15.0 ml
    Absolute methanol   35.0 ml
    0.2 M acetate buffer (pH 5) 2.5 ml
    (*see* Chapter 5)

Procedure

1. Cut frozen sections, 4–6 μm thick. Allow the slides to dry for one hour.
2. Place in solution 1 until peroxidase turns blue.
3. Place in freezing solution 2 for few seconds.
4. Place in freezing solution 3 for 1 to 2 hours.
5. Dehydrate through 3 changes of 70 per cent alcohol, clear and mount.

Results

Peroxidase sites are shown as grey-green to black.

DOPA Oxidase (tyrosinase): DOPA Reaction (Becker *et al.*, 1935)

Reagents Required

D.L. 3,4-dihydroxyphenylalanine (DOPA)
Phosphate buffer
Formol saline

Preparation of Reagents

Solution 1 Medium
D.L. 3:4-dihydroxyphenylalanine 100 mg
0.1 M phosphate buffer (pH 7.4)   100.0 ml

Section

Frozen

Procedure

1. Fix small pieces of tissue in formol saline at room temperature for 4 hours.
2. Cut thin sections (3 mm) from surface of blocks and wash in tap water for 3 minutes.
3. Place section in solution 1 (incubating medium for 1 hour at 37°C).
4. Place sections in fresh solution 1 (37°C overnight).
5. Wash in running water.
6. Blocks of tissue are fixed in 10 per cent formol saline overnight.
7. Dehydrate blocks through alcohols to chloroform.
8. Cut paraffin sections 8 μm thick.
9. Counterstain in Mayer's carmalum.
10. Dehydrate, clear and mount in DPX.

Result

Sites of DOPA oxidase activity are stained as blackish brown granules.

Rationale

Bloch (1917) introduced the first histochemical method for DOPA oxidase. Becker *et al.* (1935) modified it to be applied for paraffin sections also. Dihydroxyphenylalanine

(DOPA) of the incubating medium is oxidized by tyrosinase (DOPA oxidase) in the section. The pigment which results after oxidation is brownish black found in cells producing melanin. An element of doubt here is that peroxidase can catalyse the oxidation of tyrosine and DOPA (Okun, 1967; Okun *et al.*, 1970).

DOPA Oxidase (tyrosinase): DOPA Reaction (Becker *et al.*, 1935, Modified)

Reagents Required

D.L. 3:4 dihydroxyphenylalanine (DOPA)
0.1 M phosphate buffer (pH 7.4)
Formol saline

Preparation of Reagents

Solution 1 (Incubation solution)
D.L. 3:4 Dihydroxyphenylalanine   100 mg
0.1 M phosphate buffer (pH 7.4)   100.0 ml

Sections

Cryostat post-fixed (formalin)
Cryostat pre-fixed (formalin)
Frozen sections

Procedure

1. Keep sections in solution 1 (incubating medium) at 37°C for 45 minutes.
2. Wash in running tap water for 2 minutes.
3. Transfer sections to 10 per cent formol saline for 30 minutes.
4. Counterstain if necessary in Mayer's haemalum.
5. Wash in distilled water.
6. Mount in glycerine jelly.

Result

Sites of DOPA oxidase activity are stained bluish black.

Rationale

Bloch (1917) first introduced it which was later modified by Laidlaw and Blackberg (1932). Subsequently Becker *et al.* (1935) modified the method which could be applied to paraffin sections. DOPA is oxidized by tyrosinase which results in the formation of a brownish black pigment especially in melanin-forming cells.

Laidlaw DOPA oxidase method (Glick, 1949)

Fixation

Fresh tissue or 5 per cent formalin for 3 hours

Reagents Required

DL-β-(3,4-dioxyphenyl alanine) phenyl tyrosine
Disodium hydrogen phosphate
Potassium dihydrogen phosphate

Preparation of Reagents

Solution 1 (DOPA Stock)

DL-β-(3,4-dioxyphenylalanine) phenyl tyrosine	300 mg
Distilled water	300 ml

Store in refrigerator. It is not usable when it turns red.

Solution 2 (Buffer A)

Disodium hydrogen phosphate	11.0 g
Distilled water	1000.0 ml

Solution 3 (Buffer B)

Potassium dihydrogen phosphate	9.0 g
Distilled water	1000.0 ml

Solution 4 (Working solution) (Buffer DOPA solution)

Solution 1	25.0 ml
Solution 2	6.0 ml
Solution 3	2.0 ml

Filter. Final pH should be 7.4.

Solution 5 (Buffered control solution)

Solution 2	6.0 ml
Solution 3	2.0 ml
Distilled water	2.0 ml

Solution 6 (Cresyl violet)

Cresyl violet	500 mg
Distilled water	100.0 ml

Procedure

1. Frozen tissue is cut to 10 μm thick sections.
2. Wash in distilled water.
3. Place in solution 4 at 37°C. Solution turns red after 2 hours and finally to brown in 3–4 hours. Then remove the sections and place in fresh solution that is the control (Solution 5).
4. Wash in distilled water.
5. Counterstain with solution 6.
6. Dehydrate, clear and mount.

Results

DOPA oxidase	Black
Leucocytes	Black

Cytochrome Oxidase: Metal Chelation (Burstone, 1959)

Reagents Required

1-hydroxy 2-naphthoic acid
N-phenyl p-phenylenediamine
Absolute alcohol
Distilled water
Tris buffer 0.2 M (pH 7.4)
Cobalt acetate

Preparation of Reagents

Solution 1

1-hydroxy 2-naphthoic acid	10 mg
N-phenyl-p-phenylenediamine	10 mg
Absolute alcohol	0.5 ml
Distilled water	35.0 ml
0.2 M tris buffer, pH 7.4	15.0 ml

First two reagents are dissolved in absolute alcohol. Later distilled water and buffer are added.

Solution 2

Cobalt acetate	500 mg
Conc. formaldehyde	5.0 ml
Distilled water	45.0 ml

Procedure

1. Place sections in incubating solution for 15 minutes to 2 hours.
2. Place sections directly in solution 2 for 1 hour.
3. Rinse in distilled water.
4. Mount in glycerine jelly.

Result

Sites of cytochrome oxidase activity are stained blue-black.
[Further details on oxidases can be obtained from the work of Burstone (1962), Lillie and Burtner (1953) and Wachstein and Meisel (1964).]

Monoamine Oxidase: Tetrazolium Method (Glenner *et al.*, 1957)

Reagents Required

Tryptamine hydrochloride
Sodium sulphate
Tetranitro-blue tetrazolium
0.1 M phosphate buffer (pH 7.6)
Distilled water
10 per cent formol saline

Preparation of Reagents

Solution 1 (Incubating solution)

Tryptamine hydrochoride	25 mg
Sodium sulphate	4 mg
Tetranitro-blue tetrazolium	5 mg
0.1 M phosphate buffer (pH 7.6)	5.0 ml
Distilled water	15.0 ml

Sections

Cryostat unfixed 150 μm thick.

Procedure

1. Keep the sections in solution 1 (incubating medium) for 15–45 minutes at 37°C.
2. Wash in running water.
3. Treat sections with 10 per cent formol saline.
4. Wash in running water.
5. Mount in glycerine jelly.

Result

Sites of monoamine oxidase activity stain bluish black.

Rationale

This method was first introduced by Oster and Schlossman (1942) with aldehyde fuchsin PAS method. Francis (1953) introduced the use of tetrazolium salt. Tyramine was used as the substrate and neotetrazolium as tetrazolium salt. Glenner *et al.* (1957) used tryptamine as the substrate and nitro-blue tetrazolium as tetrazolium salt.

Although the reaction is not clearly understood, monoamine oxidase in the tissue oxidizes tryptamine in the incubating medium. The resulting product produces tetrazolium salt which in turn forms a formazon deposit.

Cytochrome Oxidase: Metachelation (Burstone, 1959)

Fixation

Unfixed cryostat

Reagents Required

1-hydroxy 2-naphthoic acid
N-phenyl-*p*-phenylenediamine
0.2 M tris buffer (pH 7.4)
Absolute alcohol
Cobalt acetate

Preparation of Reagents

Solution 1

1-hydroxy 2-naphthoic acid	10 mg
N-phenyl *p*-phenylenediamine	10 mg

Absolute alcohol	0.5 ml
Distilled water	35.0 ml
0.2 M tris buffer (pH 7.4)	15.0 ml

First dissolve the reagents in alcohol, then add distilled water and finally add buffer.

Solution 2

Cobalt acetate	500 mg
Conc. formaldehyde	5.0 ml
Distilled water	45.0 ml

Procedure

1. Place sections in solution 1 (incubating medium) for 15 minutes to 2 hours.
2. Place them directly in solution 2 for 1 hour.
3. Wash in distilled water.
4. Mount in glycerine jelly.

Result

Sites of cytochrome oxidase activity are stained blue-black.

Rationale

Graff (1916) first described this method. Many methods have by now come into vogue including Nadi type of reaction modified by Nachlas *et al.* (1958) and the metal chelation of Burstone (1959–60). The latter gives convincing results. Cytochrome oxidase in the tissue sections couples with 1-hydroxy 2-naphthoic acid. A naphthol compound is formed which in turn couples with free radical amine from *N*-phenyl-para phenylenediamine. This in turn is chelated to cobalt ions. Other methods are Thiazol blue method (Thielle, 1967) and that of Seligman *et al.* (1968).

Graff's G-Nadi Reaction

Reagents Required

α-naphthol
Dimethyl-*p*-phenylenediamine hydrochloride
Phosphate buffer

Preparation of Reagents

Solution 1 (α-naphthol solution)

α-naphthol	100 mg
Alcohol	1.0 ml
Distilled water	99.0 ml

Solution 2 (Oxidase reagent)

Dimethyl-*p*-phenylenediamine hydrochloride	120 mg
Distilled water	100.0 ml

This solution should be colourless. Store in a dark brown bottle.

Solution 3 (Nadi reagent)

Mix 20 ml of solution 1 and 20 ml of solution 2 and then add 8 ml of 0.1 M phosphate buffer at pH 7.5.

Procedure

1. Cut frozen section or unfixed tissue.
2. Incubate in solution 3 at 37°C for 1 hour.
3. Place in normal saline solution.
4. Counterstain nuclei.
5. Float on to slides. Drain—Mount in 20 per cent potassium acetate.
6. Place coverslip, ring it with paraffin wax.

Result

Sites of cytochrome oxidase are stained blue-violet.

Remarks

Cytochrome oxidase acts as a catalyst during oxidation of the mixture of solution 1 and 2. The resulting product is indophenol blue.

G-Nadi Reaction for Cytochrome Oxidase (Moog, 1943)

Fixation

Unfixed fresh

Reagents Required

Dimethyl-*p*-phenylenediamine
α-Naphthol
Sodium chloride
Phosphate buffer
Sodium nitrite (NaNO$_3$)
Phenyl urethine

Preparation of Reagents

Solution 1 (Nadi reagent)
    0.066 M phosphate buffer   100.0 ml

Just before use, combine equal parts of 0.1 M dimethyl-*p*-phenylenediamine and 0.01 M α-naphthyl in 1 per cent sodium chloride in 0.055 M phosphate buffer.

Solution 2 (Sodium azide)
    0.005 M sodium azide in physiological saline

Solution 3 (Solution 1 + solution 2)

Solution 4 (Phenyl urethine)
    0.02 M solution of phenyl urethine in physiological saline

Solution 5 (Solution 1 and solution 4)

Procedure

1. Cut fresh sections of tissue, 150 μm thick.
2. Incubate in solution 1 for 5 minutes at 37°C.

3. Wash another set of slides (control) in solution 2 and solution 4, incubate in solution 3 and in solution 5.
4. Transfer them first to solution 3 and solution 5 for 5 minutes.
5. Wash in physiological saline.
6. Counterstain in carmalum for 30 minutes.
7. Mount in 5 per cent potassium acetate and ring the coverslip with paraffin wax.

Result

Active sites of cytochrome oxidase stain blue or bluish violet. Sections incubated in the presence of sodium azide should be negative.

ADN Method for Cytochrome Oxidase (Nachlas *et al.*, 1958)

Fixation

Cold microtome
Fresh frozen

Reagents Required

Phosphate buffer
α-Naphthol
Cytochrome
Catalase
ADN

Preparation of Reagents

Solution 1 (Incubating medium)
0.1 M phosphate buffer (pH 7.4)	3.0 ml
α-Naphthol (1 mg/1 ml)	4.0 ml
Cytochrome (5 mg/1 ml)	3.0 ml
Catalase (30 mg/ml)	1.0 ml

Immediately before use add 4.0 ml of ADN (2 mg/ml).

Procedure

1. Cut fresh sections.
2. Incubate in solution 1 for 10–30 minutes at room temperature.
3. Mount in glycerine jelly.

Results

Sites of cytochrome oxidase stain purple.

An Improved Benzidine-Peroxide Reaction (Van Duijn, 1953)

Fixation

Cold microtome
Fresh frozen
Smear

Reagents Required

Ammonium chloride
EDTA
Benzidine
Hydrogen peroxide

Preparation of Reagents

Solution 1 (Saturated aqueous solution of benzidine)
    Benzidine   50 mg
    Distilled water   200.0 ml

Boil at 80°C, cool and filter.

Solution 2 (3 per cent hydrogen peroxide)

Solution 3 (Saturated ammonium chloride)
    Ammonium chloride   40.0 g
    Hot distilled water   100.0 ml

Solution 4
    EDTA   5.0 g
    Sodium hydroxide (pH 6.0)   100.0 ml

Solution 5 (Incubating medium)
    Solution 3   1.0 ml
    Solution 4   1.0 ml
    Solution 1   9.0 ml
    Solution 2   1 drop

Procedure

1. Incubate smears or slides in solution 5 for 5 to 10 minutes.
2. Rinse.
3. Mount in polyvinyl alcohol (PVA)–Fructose medium.
   Polyvinyl alcohol 18 g
   Cadmium iodide 34 g
   Fructose   8 g
   Distilled water   40 ml

Result

Peroxidase activity shows blue crystals.

Table 18.3   Histochemical techniques applied for the demonstration of oxidases

Technique	Fixative	Cytochrome oxidase	Peroxidase	DOPA oxidase	Monoamine oxidase
DAB technique	5% glutaraldehyde and formaldehyde in cacodylate		Brown		
Benzidine reaction	Cold microtome		Blue		
Straus method	Frozen, 10% neutral buffered formalin + 30% sucrose		Grey green to black		
DOPA reaction	Frozen			Blackish brown	
DOPA oxidase modified	Cryostat pre-fixed, post-fixed frozen			Bluish black	
Laidlaw's method	Fresh or 5% formalin (3 hours)			Black	
Metal chelation	Unfixed cryosat	Blue-black			
Tetrazolium method	Unfixed, cryostat				Bluish black
G-Nadi reaction	Frozen	Blue-Violet			
G-Nadi reaction (Moog, 1943)	Unfixed fresh	Bluish violet			
ADN method	Fresh frozen	Purple			
Benzidine peroxide	Fresh frozen or smears		Blue		

DEHYDROGENASES

Dehydrogenases remove hydrogen from a substrate and transfer it to another substance. Dehydrogenases carry out electron transfer and are oxidative in action. Contrary to oxidases, dehydrogenases act anaerobically.

The hydrogen receptor is either nicotinamide adenine dinucleotide (NAD) or nicotinamide adenine dinucleotide phosphate (NADP) or a flavoprotein. These are also called coenzyme 1 and coenzyme 2. With hydrogen, they become NADH and NADPH.

Occasionally dehydrogenase by itself acts as a hydrogen acceptor.

Histochemical demonstration of dehydrogenases involves colourless tetrazolium salts soluble in water. Hydrogen released during enzyme action is accepted and the brightly coloured tetrazolium salts, insoluble in water are the formazan compounds.

A number of tetrazolium salts are frequently used in the histochemical demonstration of oxidases and dehydrogenases.

Two types of tetrazolium salts, the monotetrazolium and ditetrazolium salts are known. Nachalas *et al.* (1957) for the first time used ditetrazolium chloride-nitro BT (NBT). The formazan produced in this reaction is brightly coloured and is insoluble.

Pearse (1957) first introduced monotetrazolium. The salt is 3 (4:5) dimethyl thiazolyl 2:5 diphenyl tetrazolium bromide (MTT). The formazon produced is chelated to cobalt ions present in the incubating medium to form brightly coloured granules soluble in lipids. A number of dehydrogenase enzymes can be histochemically demonstrated (Scarpelli *et al.*, 1958; Hess *et al.*, 1958; Pearse, 1972).

HISTOCHEMICAL METHODS FOR DEHYDROGENASES

Succinate Dehydrogenase (Pearse, 1972)

Reagents Required

3(4:5-dimethyl thiazolyl-2)5-diphenyl tetrazolium or nitro-blue tetrazolium (NBT) bromide (MTT)
Tris buffer (pH 7.4)
Cobalt chloride (if MTT is used)
Sodium succinate
Formol saline
Methyl green
0.05 M magnesium chloride

Preparation of Reagents

Solution 1 (Sodium succinate substrate solution)
Sodium succinate   6.75 g

Distilled water	8.0 ml
N-HCl	0.05 ml

Add sodium succinate to water and stir and then add NHCl. Final pH should be 7.1 and make up the volume to 10 ml keeping it at very low temperature (freezing).

Solution 2 (Stock tetrazolium solution)

MTT (1 mg per 1 ml distilled water)	2.5 ml
Tris buffer (pH 7.4)	2.5 ml
0.5 M cobalt chloride	0.5 ml
0.05 magnesium chloride	1.0 ml
Distilled water	2.5 ml

Final pH is 7.0. Adjust with tris buffer N-HCl keeping at freezing temperature.

Solution 3 (Formol saline)

10 per cent formol saline

Solution 4 (Methyl green solution)

2 per cent methyl green

Solution 5 (Incubating medium)

Solution 2 stock tetrazolium solution	0.9 ml
Solution 1 succinate substrate solution	0.1 ml

Sections

Cryostat unfixed
Frozen unfixed

Procedure

1. Flood sections with solution 5 (incubating solution) for 30 minutes at 37°C.
2. Pass them on to solution 3 (formol saline) for 10–15 minutes.
3. Wash in running water for 2 minutes.
4. Counterstain with solution 4 (methyl green) for 5 minutes.
5. Rinse in distilled water.
6. Mount in glycerine jelly.

Result

Succinate dehydrogenase	Black
Nuclei	Green

Rationale

From the substrate (sodium succinate), dehydrogenase releases hydrogen, which reduces the tetrazolium salt to yield formazan. Formazan then is chelated to cobalt ions resulting in a brightly coloured insoluble deposit.

In NBT technique, hydrogen is released by succinate dehydrogenase from the substrate. This in turn reduces the nitro-BT to produce a coloured, water-insoluble formazan.

Glucose 6-phosphate Dehydrogenase (Pearse, 1972)

Reagents Required

3(4:5)-dimethyl thiazolyl 2-5-diphenyl tetrazolium bromide
(MTT) or nitro-blue tetrazolium (NBT)
Tris buffer (pH 7.4)
Cobalt chloride (MTT is used)
Glucose 6-phosphate
Formol saline
2 per cent methyl green
Nicotinamide adenine dinucleotide phosphate (NADP)
0.8 M magnesium chloride

Preparation of Solution

Solution 1 (Glucose 6-phosphate substrate solution)

Glucose 6-phosphate	300 mg
Distilled water	0.8 ml
N-HCl	0.06 ml

Glucose 6-phosphate is dissolved in distilled water, then pH should be 7.1, neutralize it with N-HCl. The total volume should be 1 ml. Keep the solution in frozen condition.

Solution 2 (Stock MTT or NBT tetrazolium)

Solution 3 (10 per cent formol saline)

Solution 4 (2 per cent methyl green)

Solution 5

Solution 2 (Stock tetrazolium solution)	0.9 ml
Solution 1 (Glucose 6-phosphate substrate solution)	0.1 ml
Co-enzymes NADP	2 mg

Sections

Frozen unfixed
Cryostat unfixed

Procedure

1. Flood sections with solution 5 incubating medium for 30 minutes to 1 hour at 3°C.
2. Transfer sections to solution 3 (10 per cent formol saline) for 10–15 minutes.
3. Wash well in tap water.
4. Counterstain in solution 4 (2 per cent methyl green).
5. Wash in tap water.
6. Mount in glycerine jelly.

Result

Glucose 6-phosphate dehydogenase	Black formazon deposit with MTT
Nuclei	Green

Glucose 6-phosphatase Method (Wachstein and Meisel, 1956)

Fixation

Unfixed sections

Reagents Required

Potassium 6-phosphate
Tris buffer
2 per cent lead nitrate

Preparation of Reagent

Substrate

Potassium glucose-6-phosphate (0.125 per cent solution)	20 ml
0.2 M Tris buffer (pH 6.7)	20 ml
2 per cent lead nitrate	3 ml
Distilled water	7 ml

Procedure

1. Incubate cryostat or frozen sections in the substrate at 32°C for 5–15 minutes.
2. Wash in distilled water.
3. Treat with dilute ammonium sulphide for 1 minute.
4. Wash in water.
5. Post-fix in 10 per cent formalin for 2 minutes.
6. Mount in glycerine jelly.

Result

Sites of enzyme activity are stained brownish black.

Methods for Glucose 6-phosphatase

Fixation

Cold microtome

Reagents Required

Potassium glucose 6-phosphate
Tris buffer
Lead nitrate

Preparation of Reagents

Solution 1 (Incubating medium)

0.125 per cent potassium glucose 6-phosphate	20.0 ml
0.2 M tris buffer (*see* Chapter 5)	20.0 ml
2 per cent lead nitrate	3.0 ml
Distilled water	7.0 ml

Procedure

1. Cut 10–15 µm frozen sections.
2. Treat them with incubating medium for 5–15 minutes at 32°C.
3. Wash in distilled water.
4. Treat with yellow ammonium sulphide.
5. Wash in water.
6. Post-fix in 6 per cent neutral formaldehyde.
7. Mount in glycerine and ring the coverslip with nail polish.

Result

Sites of glucose 6-phosphatase activity stain brownish black.

NADH Diaphorase: MTT Method (Pearse, 1972)

Reagents Required

3(4:5-dimethyl thiazolyl-2)-5-diphenyl tetrazolium bromide
Tris buffer (pH 7.4)
Cobalt chloride
NADH
Formol saline
Methyl green
Distilled water
0.05 M magnesium chloride

Preparation of Reagents

Solution 1
    Stock MTT tetrazolium solution

Solution 2
    Coenzyme NADH

Solution 3
    10 per cent formol saline

Solution 4
    2 per cent methyl green

Solution 5 (Incubating medium)

Solution 1	0.9 ml
Distilled water	0.1 ml
Solution 2 Coenzyme NADH	2 mg

Co-enzyme is added to 1 ml of the medium: pH is adjusted with N-HCl.

Sections

Cryostat unfixed
Frozen unfixed

Procedure

1. Flood sections with solution 5 (incubating medium) for 30–40 minutes at 3°C.
2. Immerse section in solution 3 (10 per cent formol saline) for 15 minutes.
3. Wash.
4. Counterstain with solution 4 (2 per cent methyl green).
5. Wash.
6. Mount in glycerine jelly.

Result

NADH diaphorase   Black formazon deposits
Nuclei   Green

NADPH Diaphorase Using NBT (Pearse, 1972)

Reagents Required

Nitro-blue tetrazolium
Tris or phosphate buffer (pH 7.4)
0.05 M magnesium chloride
NADPH
0.1N HCl
Distilled water

Preparation of Reagents

Solution 1 (Stock NBT tetrazolium)
Solution 2
    Coenzyme NADPH 2 mg
Solution 3 (10 per cent formol saline)
Solution 4

Solution 1	0.9 ml
Distilled water	0.1 ml
Solution 2 (NADPH)	2.0 g

The co-enzyme is to be added just before use. Check pH and adjust to 7.1.

Sections

Cryostat unfixed
Frozen unfixed

Procedure

1. Flood sections with solution 4 incubating medium at 37°C for 30–60 minutes.
2. Transfer section to solution 3 (10 per cent formol saline) for 15 minutes.
3. Rinse in water.
4. Dehydrate, clear and mount in DPX.

Result

Enzyme activity sites are revealed as purple formazan deposits.

Table 18.4 Histochemical techniques applied for the demonstration of dehydrogenases

Technique	Fixative	Succinate dehydrogenase	Glucose 6-phosphate dehydrogenase	NADH diaphorase	NADPH diaphorase
Succinate dehydrogenase	Cryostat unfixed, frozen unfixed	Black			
Glucose 6-phosphate dehydrogenase	Cryostat unfixed		Black		
Method for glucose 6-phosphate	Cold microtome		Brownish black		
NADH diaphorase MTT method	Cryostat unfixed, frozen unfixed			Black	
NADPH diaphorase NBT method	Cryostat unfixed				Purple

OTHER ENZYMES

Enzymes not dealt with so far are considered here. Phosphorylases, β-glucuronidase, sulphatase and leucineaminopeptidase are some of the enzymes considered here.

Phosphorylases   Could be demonstrated by a number of techniques. A satisfactory method was first described by Tekeuchi and Kuruaki (1955). Fresh frozen sections which can be directly transferred to the incubating medium are chosen. Eranko and Palkama (1961) and Lake (1970, 1974) have modified the method of Tekeuchi and Kuruaki.

β-glucuronidase   This enzyme occurs in mammalian tissues such as spleen, liver and kidney. Three methods are available to demonstrate the enzyme—Azo dye method of Seligman *et al.* (1954) Fishman and Baker's (1956) ferric hydroquinoline method, Pugh and Walker's (1961) naphthol AS method. Of these, the last method (naphthol AS) is supposed to give good results. A later superior method developed by Hyashi *et al.*, (1964) involved AS BI glucuronide.

Sulphatase   Histochemical demonstration of the sulphatases is not easy. Five types of sulphatases (Pearse, 1972) are known arbitrarily, (a) aryl sulphatase, (b) sterol sulphatase, (c) glucosulphatase, (d) chondrosulphatase, and (e) myrosulphatase.

Aryl sulphatase can be demonstrated histochemically. Other sulphatases cannot be histochemically delineated. Aryl sulphatase post-coupling method of Ruttenberg *et al.* (1952) and naphthol AS sulphate and pararosaniline hydrochloride method are commonly performed.

Leucine Aminopeptidase   This enzyme hydrolyses free amino groups. Burstone and Folk (1956) suggested freeze-drying and double-embedding or cryostat sections. Enzyme inhibition is done by potassium cyanide or EDTA (ethylene diamine tetra acetic acid). Two techniques can easily be applied. Nachlas *et al.* (1957) method, i.e., metal chelation diazo coupling method and Burstone and Folk's (1956) method, i.e., the diazo coupling method.

HISTOCHEMICAL METHODS FOR OTHER ENZYMES

Leucine Aminopeptidase (Nachlas *et al.*, 1957)

Reagents

L-leucyl 4-methoxy β-naphthylamide
Ethyl alcohol
Distilled water
0.1 M acetate buffer
Potassium cyanide
Copper sulphate
Sodium chloride
Fast blue B salt

Preparation of Reagents

Solution 1 (Substrate solution)
    L-leucyl 4-methoxy β-naphthylamide   4 mg
    Ethyl alcohol   0.1 ml
    Distilled water   4.9 ml
Dissolve the salt in alcohol and then add alcohol.

Solution 2 (Sodium chloride)
    Sodium chloride   425 mg
    Distilled water   50.0 ml

Solution 3 (Copper sulphate 0.1 M)
    Copper sulphate   790 mg
    Distilled water   50.0 ml

Solution 4 (Potassium cyanide)
    Potassium cyanide   65 mg
    Distilled water   50.0 ml

Solution 5 (0.1 M acetate buffer, pH 6.5)

Solution 6 (Incubating medium)
    Solution 1   0.5 ml
    Solution 5   5.0 ml
    Solution 2   4.0 ml
    Solution 4   0.5 ml
    Fast blue B salt 5 mg

Sections

Cryostat post-fixed
Cryostat pre-fixed
Freeze-dried

Procedure

1. Immerse sections in solution 6 (incubating medium) for 15 minutes to 2 hours.
2. Rinse in saline solution 2 for 2 minutes.
3. Transfer to solution 3 (copper sulphate solution) for 2 minutes.
4. Rinse in saline.
5. Dehydrate, clear and mount in DPX.

Result

Leucine aminopeptidase activity site is stained red.

Rationale

Leucine aminopeptidase splits β-naphthylamine from the substrate L-leucyl 4-methoxy β-naphthylamine. β-naphthylamine then combines with the diazonium salt Fast blue B and forms an insoluble azo dye. When treated with a dilute solution of copper sulphate after a brief rinse in saline, the copper ions chelate with the azo dye. A purplish red final product at the site of enzyme activity reveals the enzyme location.

Burstone and Folk Method (1956)

Reagents Required

L-leucyl β-naphthylamide
Tris aminomethane

Preparation of Reagents

Solution 1 (Substrate stock)

L-leucyl β-naphthylamide	1.0 g
Distilled water	100.0 ml

Solution 2 (Buffer)

0.2 M tris hydroxymethine aminomethane	25.0 ml
0.1N HCl	45.0 ml

Dilute it to 100 ml with distilled water.

Solution 3 (Working solution)

Solution 1	1.0 ml
Distilled water	40.0 ml
Solution 2	10.0 ml
Garnet GBC	30 mg

Shake and filter.

Procedure

1. Dewax in chloroform or petroleum ether.
2. Immerse in absolute acetone.

3. Immerse in 95 per cent acetone for 1 minute.
4. Immerse in 85 per cent acetone for 1 minute.
5. Several dips in distilled water.
6. Transfer to solution 3 for 15 minutes to 1 hour at 37°C.
7. Wash.
8. Counterstain in haematoxylin for 3 minutes.
9. Wash in tap water.
10. Mount in glycerine jelly.

Result

Enzyme sites stain red.

(Details could be obtained in the investigation of Burstone (1962), Pearson *et al.* (1963) and Zugibe (1970).)

Aryl Sulphatase: Post-coupling Method (Ruttenberg *et al.*, 1952)

Reagents Required

6-Benzoyl 2-naphthyl sulphate
Sodium chloride
0.5 M acetate buffer (pH 6)
0.05 M phosphate buffer (pH 7.6)
Fast blue B salt
Distilled water

Preparation of Reagents

Solution 1 (0.05 per cent sodium chloride)
    Sodium chloride 850 mg
    Distilled water   100.0 ml

Solution 2 (1 per cent sodium chloride)
    Sodium chloride 1.0 g
    Distilled water   100.0 ml

Solution 3 (2 per cent sodium chloride)
    Sodium chloride 2.0 g
    Distilled water   100.0 ml

Solution 4 (Buffer)
    0.1 M acetate buffer (*see* Chapter 5)

Solution 5 (Substrate solution)

6-Benzoyl 2-naphthyl sulphate	250 mg
0.85 per cent hot sodium chloride (solution 1)	80.0 ml
Acetate buffer (pH 6.4)	20.0 ml
Sodium chloride	2.6 mg

Solution 6 (Fast Blue B)

Fast blue B	50 mg
0.05 M phosphate buffer (pH 7.6)	50.0 ml

Sections

Neutral formalin fixed
Frozen sections

Procedure

1. Immerse sections in solution 1 for 1 minute.
2. Immerse sections in solution 2 for 1 minute.
3. Immerse sections in solution 3 for 1 minute.
4. Place sections in solution 5 (substrate solution) for 2–20 hours.
5. Transfer sections to 2 per cent sodium chloride for 1 minute.
6. Transfer sections to 1 per cent sodium chloride for 1 minute.
7. Transfer sections to 0.85 per cent sodium chloride for 1 minute.
    Steps 5, 6, 7 for unfixed frozen sections only at 4°C.
8. Wash in distilled water at 4°C.
9. Place sections in Fast blue B for 5 minutes.
10. Wash in 0.85 per cent sodium chloride (Solution 1) at 4°C.
11. Mount in glycerine jelly.

Result

Aryl sulphatase activity sites stain red-blue.

Rationale

The substrate is 6-benzoyl 2-naphthyl sulphate. The technique was modified by Rutenburg *et al.* (1952). The substrate is split by the sulphatase enzyme and the resulting product is coupled with Fast blue B to form an azo dye which is insoluble. It is red or blue in colour.

Aryl Sulphatase-naphthol AS Sulphate Method (Woohsmann and Hartrodt, 1965) (Modified by Wachtler and Pearse ,1966)

Reagents Required

Naphthol AS sulphate (potassium salt)
Sodium chloride
0.2 M acetate buffer stock
Sodium nitrate
Pararosaniline hydrochloride

Preparation of Reagents

Solution 1 (Buffer stock)
    0.2 M acetate buffer 'A' solution (*see* Chapter 5)

Solution 2 (Sodium chloride (0.85 per cent))
    Sodium chloride 850 mg
    Distilled water   100.0 ml

Solution 3 (Naphthol AS sulphate)
    Naphthol AS sulphate   20 mg
    Sodium chloride (solution 2) 8.0 ml
    Acetate buffer (solution 1)   2.0 ml

Solution 4 (Hexazonium pararosaniline)
    Pararosanilin hydrochloride   0.3 ml
    Fresh 4 per cent sodium nitrate 0.3 ml

Solution 5 (Incubating solution)
    Solution 3   10.0 ml
    Sodium chloride 260 mg
    Solution 4   0.6 ml

Final pH is between 6 and 7.

Cryostat Pre-fixed Sections

Snail intestine sections serve as controls.

Procedure

1. Keep sections in solution 5 (incubating medium) at 37°C for ½–2 hours.
2. Wash in distilled water.
3. Counterstain in 2 per cent methyl green.
4. Wash in running water for 30 seconds.
5. Dehydrate, clear and mount in DPX.

Result

Aryl sulphatase   Red
Nuclei   Green

Phosphorylase: Iodine Method (Tekeuchi and Kuriaki, 1955) (Modified by Eranko and Palkama ,1961)

Reagents Required

Glucose 1-phosphate
Adenosine 5-phosphate
Glycogen
Sodium fluoride
Polyvinyl pyrrolidone
Insulin 40 I.U.
Ethyl alcohol
0.1 M acetate buffer
Sucrose
Grams' iodine
Glycerol

Preparation of Solutions

Solution 1 (0.32 M Sucrose)
Sucrose   11 mg
Distilled water   100.0 ml

Solution 2 (Gram's iodine sucrose)
Iodine   330 mg
Potassium iodide   660 mg
Sucrose   11 gm
Distilled water   100.0 ml

Solution 3 (40 per cent alcohol)
Ethyl alcohol   40.0 ml
Distilled water   60.0 ml

Solution 4 (Acetate buffer) (*see* Chapter 5)

Solution 5 (Incubating medium)
Glucose 1-phosphate   100 mg
Adenosine 5-phosphate   10 mg
Glycogen   7 mg
Sodium fluoride   180 mg
Polyvinyl pyrrolidone   900 mg
Insulin 40 IU   1 drop
Absolute alcohol   2.0 ml

Procedure

1. Place sections immediately after sectioning into the incubating medium at 37°C.
2. Dry sections for 15 minutes to 3 hours.
3. Immerse in solution 3 (40 per cent alcohol) for 2 minutes.
4. Allow sections to dry.
5. Transfer to solution 1 (sucrose solution) for 5 minutes.
6. Transfer to solution 2 (iodine sucrose) for 5 minutes.
7. Mount in glycerol–iodine mixture.

Result

Sites of phosphorylase activity are stained dark brown.

Best results could be obtained if sections are treated with solution 5 immediately after drying (Bancroft, 1975).

Rationale

Glucose 1-phosphate is converted into a polysaccharide. The reaction is initiated by the addition of glycogen. On treatment with iodine, a dark brown product results.

Phosphorylase (Lake, 1970)

Unfixed cryostat sections

Reagents Required

Magnesium chloride
Glucose 1-phosphate (dipotassium salt)
Glycogen
Adenosine 5-triphosphate
Sodium fluoride
Ethyl alcohol
Polyvinyl pyrrolidone
Acetate buffer
Lugol's iodine
Glycerol

Preparation of Incubating Medium

0.1 M acetate buffer pH 5.9	10 ml (*see* Chapter 5)
0.1 M magnesium chloride	1 ml
Glucose 1-phosphate	100 mg
Glycogen (rabbit or oyster)	2 mg
Adenosine 5-triphosphate	2 mg
Sodium fluoride	180 mg
Ethyl alcohol	2 ml
Polyvinyl pyrrolidone	900 mg

Add the reagents in the sequence given above. Chill at 20°C.

Procedure

1. Keep unfixed cryostat sections in incubating medium for 1 hour at 37°C.
2. Rinse in 40 per cent alcohol for 5 seconds.
3. Dry.
4. Fix in absolute alcohol and later dry.
5. Transfer to Lugol's iodine (1 part iodine + 30 parts water) for 5 minutes.
6. Mount in Lugol's iodine (1 part iodine + 9 parts glycerol).

Result

Sites of phosphatase activity are stained bluish black.

β-Glucuronidase: Naphthol AS–BI Method (Hayashi *et al.*, 1964)

Fixation

Pre-fixed cryostat

Reagents Required

Naphthol AS–BI glucuronide
Acetate buffer (pH 5)
Sodium bicarbonate
Hexazolium–pararosaniline
Sodium nitrate

Distilled water
1N sodium hydroxide

Preparation

Solution 1 (Sodium bicarbonate)
Sodium bicarbonate 210 mg
Distilled water 50.0 ml

Solution 2 (Substrate solution)
Naphthol AS-BI glucuronide 14 mg
Solution 1 0.6 ml
0.1 M acetate buffer (pH 5.0) 50.0 ml

Solution 3 (Hexazonium pararosaniline)
Pararosaniline hydrochloride 0.3 ml
4 per cent freshly prepared sodium nitrate 0.3 ml

Solution 4 (Incubating solution)
Solution 2 5 ml
Solution 3 0.3 ml
Distilled water 5 ml

Solution 3 is added just before use and pH (5.2) is adjusted with 1N NaOH.

Procedure

1. Keep sections in solution 4 (incubating medium) at 37°C for 20–40 minutes.
2. Wash for 2 minutes.
3. Counterstain in 2 per cent methyl green for 4 minutes.
4. Wash rapidly.
5. Dehydrate, clear and mount in DPX.

Result

Glucuronidase activity Red
Nuclei Green

Rationale

β-glucuronidase splits the substrate and releases naphthol AS-BI. The diazonium salt in the incubating medium combines with naphthol to produce an insoluble azo dye at the site of enzyme activity.

8-Hydroxy Quinoline Coupling Azo Dye Method

Fixation

Cold, formalin, frozen sections

Reagents Required

Hydroxy quinoline glucuronide
Acetate buffer (*see* Chapter 5)
4-benzoylamino 5-dimethoxyaniline

Preparation of Reagents

Solution 1 (Incubating medium)

    0.003 M – 8 hydroxyquinoline   10.0 ml
      in 0.1 M acetate buffer (pH 5.2)
    Diazotate of 4-benzoylamino   10 mg
      2 : 5 dimethoxyaniline

Shake the mixture and filter.

Procedure

1. Cut frozen sections 15 μ thick. Mount them on clean glass slides. Ring the sections with paraffin wax.
2. Incubate in solution 1 for 2–4 hours in a closed dish to avoid evaporation.
3. Wash in running water.
4. Remove the wax.
5. Counterstain if desired in Mayer's haemalum for 5 minutes.
6. Wash in running water.
7. Mount in glycerine.

Result

β-glucuronidase turns into an orange precipitate.

Post-coupling Method for β-glucuronidase (after Seligman *et al.*, 1954)

Fixation

Cold microtome – post-fixed in neutral formalin

Reagents Required

6-bromo 2-naphthyl-β D-glucopyranoside (glucuronide)
Methanol
Phosphate citrate buffer
Fast blue B

Preparation of Reagents

Solution 1 (Incubating medium)

6-bromo 2-naphthyl-β D-glucopyranoside	30 mg
Absolute methanol	5.0 ml
Phosphate citrate buffer (pH 4.95)	20.0 ml
Distilled water	75.0 ml

Procedure

1. Cut frozen sections.
2. Incubate sections for 4–6 hours at 37°C in solution 1.
3. Place sections in a solution containing 10 mg of Fast blue B in 10 ml of 0.02 M phosphate buffer (pH 7.5) at 4°C for 2 minutes.
4. Rinse in cold distilled water giving 2 changes.

5. Rinse in 1 per cent acetic acid.
6. Mount in glycerine jelly.

Result

β-glucuronidase becomes blue or purple.

Post-coupling Method for β-galactosidase (Ruttenberg *et al.*, 1958)

Fixation

Cold formalin
Frozen sections

Reagents Required

6-bromo 2-naphthyl *p*-D-galactopyranoside
Methanol
Phosphate citrate buffer
Fast blue B

Preparation of Reagents

Solution 1 (Incubating medium)

6-bromo 2-naphthyl *p*-D-galactopyranoside	100 mg
Methanol	15.0 ml
Phosphate citrate buffer	85.0 ml
Distilled water	200.0 ml

First dissolve the salt and then add phosphate citrate buffer after it cools. Add a further 100 ml of distilled water.

Procedure

1. Cut frozen sections.
2. Place slides in solution 1 at 37°C for 6 hours.
3. Rinse in water giving 3 changes.
4. Transfer sections to a solution containing 10 mg of Fast blue B and 10.0 ml of 0.02 M phosphate buffer at 4°C for 5 minutes. The final pH should be 7.4–7.0.
5. Wash in cold water giving 3 changes.
6. Mount in glycerine jelly.

Result

Sites of β-galactosidase activity stain blue or purple.

Nitro BT Method for DPN (NAD) Diaphorase (Nachlas *et al.*, 1958)

Reagents Required

Sodium lactate (must be stored in deep freeze)
Lactic dehydrogenase
DPN (NAD)

Nitro BT
0.2 M phosphate buffer

Preparation of Reagents

Solution 1 (Incubating medium)

0.5 M sodium lactate	0.6 ml
1.5 per cent aqueous lactic dehydrogenase	0.2 ml
DPN (NAD) (5 mg/ml)	0.3 ml
Nitro BT (5 mg/ml)	0.3 ml
0.2 M phosphate buffer (pH 7.4)	1.0 ml
Distilled water	0.6 ml

Procedure

1. Cut frozen sections and mount on slides.
2. Place on each section, 0.2 ml of substrate and incubate for 30 minutes.
3. Rinse in saline.
4. Fix in formol saline (10 per cent) for 10 minutes.
5. Wash in 15 per cent alcohol for 30 minutes.
6. Rinse in distilled water.
7. Mount in glycerine jelly.

Result

Sites of enzyme activity stain blue.

Nitro BT Method for TPN (NADP) Diaphorase (Nachlas *et al.*, 1958)

Reagents Required

Sodium DL-isocitrate (1.1 M)
Sodium L-malate (2.5 M)
Manganese chloride
TPN (NADP)
Nitro BT
Veronol acetate buffer

Preparation of Reagents

Sodium DL-isocitrate (1.1 M)	0.6 ml
Sodium L-malate (2.5 M)	0.5 ml
Manganese chloride (0.005 M)	0.3 ml
TPN (NADP) (5 mg/ml)	0.2 ml
Nitro BT (5 mg/ml)	0.3 ml
Veronol acetate buffer (pH 7.4)	1.1 ml

Keep the solution in deep freeze TPN (NADP). The remaining stock solution is adjusted to pH 7.4 and stored in the refrigerator.

Remaining procedure is same as for nitro BT method for DPN (NAD) diaphorase.

Result

Sites of enzyme activity stain blue.

MTT Method for DPN and TPN Diaphorase (Hess *et al.*, 1958)

Reagents Required

MTT
Cobaltous chloride
Tris buffer
Polyvinyl pyrrolidone

Preparation of Reagents

Solution 1 (Stock solution)

MTT 1 mg/1 ml	2.5 ml
Cobaltous chloride (0.5 M)	0.9 ml
0.2 M tris buffer (pH 8.0)	2.5 ml
Distilled water	4.1 ml
Polyvinyl pyrrolidone	770 mg

Cobaltous chloride and buffer are first mixed and the resulting precipitate is removed by filtration. The remaining constituents are slowly added and final pH is 7.2 adjusted with pH 10.4 tris buffer. This solution lasts for 4 weeks.

Substrate

Stock solution	1.0 ml
DPN or TPN H	6 mg (adjust pH with tris) buffer

Procedure

1. Cut frozen section.
2. Keep in 0.2 ml of substrate on each section and incubate at 37°C for 30 minutes.
3. Fix in formol calcium or formol saline for 10 minutes.
4. Rinse in distilled water.
5. Mount in glycerine jelly.

Results

Sites of enzyme activity DPN (NAD) or TPN (NADP) diaphorase stain black.

Table 18.5   Histochemical techniques applied for the demonstration of other enzymes

Technique	Fixative	Leucine amino peptidase	Aryl sulphatase	Phosphorylase	β-glucuronidase	β-galactosidase	DPN	TPN	DPN and TPN
Nachlas *et al.* method	Cryostat post-fixed, pre-fixed frozen	Red							
Burstone and Folk method	Cold acetone (10 hours)	Red							
Post-coupling method	Frozen, neutral formalin		Red-blue						
Naphthol AS sulphate method	Cryostat pre-fixed		Red						
Iodine method	Unfixed cryostat			Dark brown					
Lake method	Unfixed cryostat			Bluish black					
Naphthol AS-BI method	Cryostat pre-fixed				Red				
8-Hydroxy-quinoline coupling azo dye	Cold, formalin, frozen				Orange				

Post-coupling method	Cold microtome post-fixed in neutral formalin	Blue or purple		
Post-coupling method	Cold formalin frozen section		Blue or purple	
Nitro BT method for DPN diaphorase	Frozen			Blue
Nitro BT method for TPN diaphorase	Frozen			Blue
MTT method for DPN and TPN diaphorase	Frozen			Black

CONNECTIVE TISSUE

19

Connective and epithelial tissues are important components of an organism. Epithelial layer/tissue can be the outermost one with protective function or it could be the innermost one as that of the intestine. Wherever it is situated, it is a secreting and absorbing tissue. Epithelial tissue is attached to the connective tissue by a PAS-positive basement membrane. It is a thin flimsy layer of inter cellular substance. Cells may aggregate to develop a horny layer called keratin. Secretions of epithelium come from associated gland cells.

TYPES OF CONNECTIVE TISSUE

It is a tissue of varying nature depending on one of a variety of situations in the body with specific functions notably binding.

1. Aerolar or loose tissue
2. Adipose tissue
3. Bony tissue
4. Dense white fibrous tissue
5. Cartilage
6. Haemopoietic

Loose connective tissue or alveolar tissue   The cells are loosely arranged. It includes fibroblasts, macrophages, mast cells, plasma cells and fat cells.

1. Fibroblasts are found all over the connective tissue. They are elongated, fusiform, and numerous, and are responsible for the formation of collagen fibres.
2. Macrophages, also known as histiocytes are numerous, elongated and fusiform. They can be identified by any of the vital stains. Macrophages are also called wandering cells and are numerous in alveolar tissue.

3. Mast cells are numerous in all types of connective tissue especially in alveolar tissue. They have a granular texture because of metachromatic granules in the cytoplasm. These mast cells are responsible for the production of heparin and histamine in the body. They are supposed to contain 5-hydroxytryptamine or serotonin. The best technique to demonstrate is Czaba's alcian blue safranin method or aldehyde fuchsin method.

4. Plasma cells are numerous in haemopoietic tissue. These cells have abundant cytoplasm with eccentric nucleus. Cytoplasm has abundant RNA and hence stains crimson red with acridine orange. The chromatin in the nucleus aggregates into small masses around the periphery. These plasma cells are known to be responsible for secreting antibodies and could be demonstrated with Coon technique.

5. Fat cells are also numerous.

Adipose tissue   This is made up of only lipid or fat storing cells (without fibres). It has characteristic flattened nuclei. In short, adipose tissue is packed with fat cells. With fat solvents, these cells display a ringed appearance.

Fibrous tissue   It is full of tendons and ligaments. These are present in areas where great tensile strength is in demand. Membranes in this tissue usually consist of collagen fibres.

Bone and cartilage   Fibrillar elements, amorphous intercellular substance are present. In between fibrillar elements, there is deposition of mineral substance. The intercellular substance of bone is permeated by a system of five canals.

Composition

Connective tissue elements are composed of matrix or ground substance, fibres, and cells.

Ground substance or matrix   This is mainly composed of mucopolysaccharides. When fresh tissue is treated with silver nitrate, a dark background is seen with light brown spaces, indicating the presence of cells.

Fibres   These are of various types.

Collagenic fibres (Collagen)   This is the main fibrous protein of connective tissue, and may be tendinous. Fibres may be single or may occur in bundles. They may be straight or wavy and spread in all directions. When collagen is boiled, it yields gelatin which is digested by pepsin.

Yellow elastic fibres   These occur as bundles which are profusely branched to join other fibres, and are highly refractile. These elastic bundles are long, narrow, homogeneous and contain the protein substance, elastin. They are elastic and can be stretched to any extent. They are located in places where elasticity is necessary (walls of blood vessels and respiratory organs). Boiling has no effect. They are digested by trypsin.

Reticular fibres   According to some, they are immature collagen fibres but they are chemically and physically different from collagen. Even histologically they could be easily distinguished from collagen. When silver technique is applied, reticular fibres turn black whereas collagen fibres turn brown. Another important difference is that reticular fibres are gram-negative and isotropic when stained, whereas collagen fibres are gram-positive and unisotropic. Reticular fibres are delicate and connected to collagen fibres. They form a network with supporting cells, capillaries, nerve fibres and other tissue units. Reticular fibres form a sort of basement membrane containing fibres of collagen.

Hyalin   It is a transparent glossy refractile substance.

Fibrin   It is formed from fibrinogen (a type of protein in blood plasma). It forms a mesh of fibres at the site of cut blood vessels.

Cells

The most important cells found in connective tissue are fibroblasts, histocytes; mast cells and plasma cells.

Fibroblasts   These are found in all connective tissue proper and playa role in the formation of collagen fibres.

Histocytes   Also called as macrophages are numerous in areolar tissue. They act as scavengers in removing dye particles.

Mast cells   Their cytoplasm has basophilic and metachromatic granules with an indented nucleus. They are known to produce heparin and bulk of histamine in the body. They also contain serotonin. They are best demonstrated by Czaba's alcian blue-safranin method or aldehyde fuchsin technique.

Plasma cells   They are larger than lymphocytes with large quantities of cytoplasm and an eccentric nucleus. Chromatin in nucleus is aggregated into small masses. The cytoplasm takes a crimpson with acridine orange or Unna's Pappenheim's stain due to the presence of high concentration of RNA. They contain ribosomes (seen under electron microscope) and these play a very important role in the production of antibodies.

Muscle layers

There are three groups of muscles

a)   Smooth (non-striped non-voluntary)

b)   Striated (striped, voluntary)

c)   Striated voluntary (cardiac)

a)   Smooth muscles are involuntary either in the form of single fibre (fusiform) or in bundles with centrally placed (i.e., in the widest part) nucleus. Each muscle is ensheathed by reticular fibres.

b)   Striated muscles are voluntary, striped, much larger than smooth muscle with more than two nuclei ensheathed by a membrane, the sarcolemma.

The striations are diagonal or cross-alternating with one dark and one light material.

c)   Cardiac muscle   Though striated, striations are not distinct with dark and light bands.

MALLORY'S TRIPLE CONNECTIVE TISSUE STAIN

Various investigators have employed a number of modifications to this method. For connective tissue staining, phosphomolybdic acid is very important (Holde and Isler, 1958). Molybdic acid diminishes the background staining so that connective tissue stains with aniline blue, showing off very well. According to Baker (1958), phosphomolybdic acid is a colourless acid dye in the tissue chiefly on collagen. It also acts as a dye excluder on acid dyes such as acid fuchsin. Aniline blue stains collagen. Phosphomolybdic acid is not a mordant since it also opposes the action of aniline blue. Without phosphomolybdic acid, aniline blue stains strongly without selectivity. There are a number of dyes which form lakes with phosphomolybdic acid. Puchtler and Sweat (1964b) stated that such compounds were mostly used in textile dyeing. The interaction between acid groups of phosphomolybdic acid and basic groups of the dye bind some basic dyes and amphoteric dyes (Smith *et al.*, 1966). There is lot of controversy regarding the mordanting action of phosphotungstic acid and phosphomolybdic acid. After a series of techniques established, Everett and Miller (1974) concluded that most of the histological structures take a deep shade of aniline blue but treatment with phosphotungstic acid allows only connective tissue to take the stain. Phosphomolybdic acid and phosphotungstic acid are not mordants but they block all tissues except connective tissue from staining. Quantarelli *et al.* (1971) and Scott (1976) stated that phosphotungstic acid combined with cationic groups electrostatically, provided the pH is optimum and a salt type linkage is involved between organic cations and acid.

An interesting factor is that certain tissue elements react more specifically to certain metals, for example neuroglia and connective tissue to tungsten and collagen to molybdenum. The intensity of staining of fibre depends on phosphomolybdic acid and that of plasma on phosphotungstic acid (Lillie, 1952). Oxalic acid intensifies the staining of aniline blue (Mallory, 1944).

When a tissue is stained with Mallory, aniline blue (acidic) stains connective tissue and cartilage, Orange G (acidic) stains blood cells, myelin and muscle, and the remaining tissue is stained by acid fuchsin in shades of pink to red. However, nuclei when stained fade in few years. Mallory staining may be modified by using basic fuchsin or carmine instead of acid fuchsin or azocarmine. Here the basic stains react readily with the nuclei than the acid stains.

Certain tissues react more specifically with certain elements, for example collagen to molybdenum, fibroglia, myoglia, neuroglia and epithelial fibres to tungsten.

Oxalic acid in Mallory's triple accelerates the aniline blue stain more rapidly and intensely (Mallory, 1944). According to Baker (1958), oxalic acid lowers the pH and accelerates the staining of orange G and aniline blue.

In Mallory's triple stain, aniline blue stains connective tissue and cartilage, orange G stains the blood cells, acid fuchsin stains the rest of the tissue (either pinkish or reddish).

HISTOCHEMICAL TECHNIQUES FOR CONNECTIVE TISSUE

Pantin Method (1946)

Fixation

Any general fixative

Reagents Required

Acid fuchsin
Phosphomolybdic acid
Aniline blue
Orange G

Preparation of Reagents

Mallory 1

Solution 1 (Acid fuchsin stain)
Acid fuchsin   1.0 g
Distilled water   100.0 ml

Solution 2 (Phosphomolybdic acid)
Phosphomolybdic acid 1.0 g
Distilled water   100.0 ml

Mallory II

Solution 3 (Aniline–orange G solution)
Aniline blue   500 mg
Orange G   2.0 g
Distilled water   100.0 ml

Procedure

1. Deparaffinize and hydrate slides to water.
2. Stain in solution 1 (Mallory I) for 15 seconds.
3. Rinse in distilled water.
4. Transfer to solution 2 (phosphomolybdic acid) for 5 minutes.
5. Rinse in distilled water.
6. Transfer to solution 3 for 2 minutes.
7. Rinse in distilled water.
8. Differentiate in 95 per cent alcohol.
9. Dehydrate, clear and mount.

Result

Muscle and cytoplasmic elements	Red to orange
Nuclei	Red
Collagen	Dark blue
Mucus, connective tissue, hyaline substance	Blue
Myelin, yellow blood cells	Yellow and orange
Dense cellular tissue	Pink with red nuclei
Bone matrix	Red

Remarks

For best nuclear results, before staining with Mallory I, it should be preceded by alum haematoxylin. Lendrum and McFarlane (1940) recommend addition of celestian blue. Krichesky (1921) proposed the following for Mallory II solution.

1)	Aniline blue	2.0 g
	Distilled water	100.0 ml
2)	Orange G	1.0 g
	Distilled water	100.0 ml
3)	Phosphomolybdic acid	1.0 g
	Distilled water	100.0 ml

Just before use, mix equal parts of each. If these solutions are kept in separate bottles and mixed before use, staining is brilliant.

Azan stain–Mallory–Heidenhain's (Modified after Koneff, 1938)

Fixation

Zenker formol. For better fixation, slides should be treated with 3 per cent potassium dichromate overnight.

Reagents Required

Azocarmine
Glacial acetic acid
Aniline oil
Phosphotungstic acid
Aniline blue
Orange G
Oxalic acid

Preparation of Reagents

Solution 1 (Azocarmine)

Azocarmine	0.2 g or 1 g
Distilled water	100.0 ml
Glacial acetic acid	1.0 ml

Boil azocarmine in water for 5 minutes, cool, filter and add acetic acid.

Solution 2 (Aniline alcohol)
    Aniline oil   1.0 ml
    90 per cent alcohol   100.0 ml

Solution 3 (Acetic alcohol)
    Glacial acetic acid   1.0 ml
    95 per cent alcohol   100.0 ml

Solution 4 (Phosphotungstic acid)
    Phosphotungstic acid   500 mg
    Distilled water   100.0 ml

Solution 5 (Aniline blue stain)
    Aniline blue   500 mg
    Orange G   2.0 g
    Oxalic acid   2.0 g
    Distilled water 100.0 ml

Solution 6 (Acidulated water)
    Glacial acetic acid   1.0 ml
    Distilled water   100.0 ml

Procedure

1. Deparaffinize and hydrate slides to water.
2. Stain in solution 1 at 56°C for 1–2 hours.
3. Rinse in distilled water.
4. Differentiate in solution 2 (aniline alcohol).
5. Differentiate in solution 3.
6. Place in solution 4 for 2 hours.
7. Rinse in distilled water.
8. Transfer to solution 5 for 1 hour.
9. Differentiate in solution 6.
10. Rinse in distilled water.
11. Rinse in 70 per cent alcohol.
12. Dehydrate, clear and mount.

Result

Nuclei	Brilliant red
Collagen and reticulin	Blue
Muscle	Red and yellow
Basophil cytoplasm	Light blue
Acidophil cytoplasm	Orange-red

Remarks

Azan has been considered an outstanding stain for connective tissue. Romeis (1948) recommends the use of alizarine blue instead of Azocarmine. It shortens the time.

Mallory-Heidenhain Rapid One Step Method (Cason, 1950)

Fixation

Any general fixative

Reagents Required

Phosphotungstic acid
Orange G
Aniline blue
Acid fuchsin

Preparation of Reagents

Phosphotungstic acid	1.0 g
Orange G	2.0 g
Aniline blue (water soluble)	1.0 g
Acid fuchsin	3.0 g
Distilled water	200.0 ml

Add in order to distilled water. This solution lasts for several months.

Procedure

1. Deparaffinize and hydrate slides to water.
2. Stain in staining solution for 5 minutes.
3. Wash in running water.
4. Dehydrate, clear and mount.

Result

Collagen	Blue
Matrix	Blue
Fibrin	Red
Elastin	Yellow

Masson's Trichrome Stain (1928) (Modified by Gurr, 1956)

Fixation

Any general fixative

Reagents Required

Ferric ammonium sulphate
Acid fuchsin
Glacial acetic acid
Haematoxylin
Ponceau de xylidine
Fast green
Phosphomolybdic acid
Picric acid

Preparation of Reagents

Solution 1 (Iron alum)
Ferric ammonium sulphate   4.0 g
Distilled water   100.0 ml

Solution 2 (Haematoxylin) (*see* Chapter 7)

Solution 3 (Acid fuchsin)
Acid fuchsin   1.0 g
Distilled water   100.0 ml
Glacial acetic acid   1.0 ml

Solution 4 (Ponceau de xylidine)
Ponceau de xylidine   1.0 ml
Distilled water   100.0 ml
Glacial acetic acid   1.0 ml

Solution 5 (Phosphomolybdic acid)
Phosphomolybdic acid   1.0 g
Distilled water   100.0 ml

Solution 6 (Acid water)
Glacial acetic acid   1.0 ml
Distilled water   100.0 ml

Procedure

1. Deparaffinize and hydrate slides to water.
2. Mordant in solution 1 for 30 minutes.
3. Wash in running water.
4. Transfer to solution 2 (Delafield's haematoxylin) for 30 minutes.
5. Wash in running water.
6. Differentiate in saturated aqueous picric acid.
7. Wash thoroughly in running water for 10 minutes.
8. Transfer to solution 3 (acid fuchsin) for 5 minutes.
9. Rinse in distilled water.
10. Place in solution 4 (ponceau de xylidine) for 5 minutes.
11. Rinse in tap water.
12. Differentiate in solution 5 (phosphomolybdic acid) for 5 minutes.
13. Place directly in fast green for 2 minutes
14. Differentiate in solution 6.
15. Dehydrate, clear and mount.

Result

Nuclei	Blue or black
Cytoplasmic elements	Varying shades of red
Muscle	Red
Collagen	Green

Remarks

Lillie (1940) suggested the following as alternatives: ponceau 2R, nitrizine yellow, Beibrich scarlet, azofuchsin 3B, G and 4G. Bordeaux red, chromotrope 2R, chrysoidin, eosin Y and orange G. These could replace ponceau de xylidine.

Pollak's Rapid Method

Fixation

Any general fixative

Reagents Required

Haematoxylin
Acid fuchsin
Ponceau 2R
Light green
Orange G
Phosphotungstic acid
Phosphomolybdic acid
Glacial acetic acid
Ethyl alcohol

Preparation of Reagents

Solution 1

Acid fuchsin	500 mg
Ponceau 2R	1.0 g
Light green yellowish	450 mg
Orange G	750 mg
Phosphotungstic acid	1.5 g
Phosphomolybdic acid	1.5 g
Glacial acetic acid	3.0 ml
50 per cent ethyl alcohol	300.0 ml

Acetic acid is added to alcohol. Divide this 300 ml liquid into 4 parts. To one part add acid fuchsin and ponceau 2R, add light green to second part, orange G and phosphotungstic acid to third part and phosphomolybdic acid to fourth part. Mix the four solutions and filter. It keeps for several days.

Solution 2 (Acidified water)

Glacial acetic acid	0.2 ml
Distilled water	100.0 ml

Procedure

1. Deparaffinize and hydrate slides to water.
2. Place in Mayer's haematoxylin for 10 minutes.
3. Wash in running water.
4. Transfer to solution 1 (Trichrome stain) for 7 minutes.
5. Wash in running water.
6. Differentiate in acid water (solution 2)

7. Rinse in 70 per cent alcohol.
8. Dehydrate, clear and mount.

Result

Nuclei	Dark blue
Muscle, elastin	Red
Fibrin, calcium	Pale blue
Hyaline	Pale blue
Collagen	Bluish green

Gomori's One Step Trichrome (Gomori, 1950b)

Fixation

Neutral buffered formalin

Reagents Required

Chromotrope 2R
Light green
Glacial acetic acid
Phosphotungstic acid
Acetic acid

Preparation of Reagents

Solution 1 (Bouin's fluid) (*see* Chapter 1)

Solution 2 (Weigert's haematoxylin) (*see* Chapter 7)

Solution 3 (Gomori's trichrome stain)

Chromotrope 2R	600 mg
Light green	300 mg
Glacial acetic acid	1.0 ml
Phosphotungstic acid	800 mg
Distilled water	100.0 ml

Procedure

1. Dewax and hydrate slides to water.
2. Keep in solution 1 overnight at room temperature or for 1 hour at 56°C.
3. Wash in running water for 3 to 5 minutes.
4. Treat with solution 2 for 10 minutes.
5. Rinse in tap water.
6. Treat with solution 3 for 20 minutes.
7. Rinse in 1 per cent acetic acid.
8. Dehydrate, clear and mount.

Results

Muscle fibres	Red
Collagen	Green
Nuclei	Blue to black

Gomori's Method (1950b) (Modified by Sweat *et al.*, 1968)

Fixation

Any general fixative

Reagents Required

Chromotrope 2R
Light green
Phosphomolybdic acid
Hydrochloric acid

Preparation of Reagents

Solution 1 (Bouin's fixative) (*see* Chapter 1)

Solution 2 (Trichrome solution)

Chromotrope 2R	600 mg
Light green	300 mg
Phosphomolybdic acid	1.0 g
Distilled water	100.0 ml
HCl	1.0 ml

Allow it to stand for 24 hours and keep in refrigerator. Use cold solution.

Solution 3 (Acid water)

Glacial acetic acid	0.2 ml
Distilled water	100.0 ml

Procedure

1.   Dewax and hydrate slides to water.
2.   Immerse in solution 1 for 1 hour at 56°C.
3.   Wash in running water until it turns colourless.
4.   Treat with solution 2 for 1 minute.
5.   Rinse in distilled water.
6.   Rinse in solution 3 for 20 seconds.
7.   Dehydrate, clear and mount.

Result

Muscle fibres are stained red, collagen green and nuclei blue to black. Sweat *et al.* (1968) method improves the colours of fine connective tissue.

Puchtler Sweat Method for Basement Membrane (Puchtler and Sweat, 1964))

Fixation

Carnoy's fixative

Absolute alcohol	60.0 ml
Chloroform	30.0 ml
Glacial acetic acid	10.0 ml

Reagents Required

Periodic acid
Sodium sulphite
Resorcin fuchsin
Fast red
Aluminium sulphate
Hydrochloric acid

Preparation of Reagents

Solution 1

Periodic acid	500 mg
Distilled water	100.0 ml

Solution 2

Sodium bisulphite	20.0 g
Distilled water	40.0 ml
Alcohol 100 per cent	10.0 ml

Solution 3 (Stock solution)

Resorcin fuchsin	2.0 g
Resorcinol	4.0 g
Distilled water	200.0 ml

Boil for 1 minute. Then add 25 ml of 29 per cent ferric chloride, cool and filter-dry.
Keep the precipitate in a porcelain dish, add 200 ml of 95 per cent alcohol. Then add
4 ml of concentrated hydrochloric acid.

Working Solution

Stock solution	10.0 ml
1 per cent acetic acid	100.0 ml

Solution 4 (Kernechtrot solution)

Fast red	100 mg
5 per cent aluminium sulphate	10.0 ml
(5 g/100 ml acid)	

Boil, cool and filter.

Procedure

1. Deparaffinize and bring slides to water.
2. Transfer to solution 1 for 5 minutes.
3. Rinse in distilled water.
4. Transfer to solution 2 for 15 hours.
5. Rinse in distilled water.
6. Transfer to solution 3 for 4 hours.
7. Wash in distilled water.
8. Counterstain in solution 4 for 5 minutes.
9. Rinse in distilled water.
10. Dehydrate, clear and mount in permount.

Result

Basement membrane	Black
Nuclei	Pink or red

ELASTIN STAINING

Verhoeff's Elastin Stain (Plate 18, Figures 1 and 2)

Fixation

Any general fixative

Reagents Required

Potassium iodide
Iodine
Ferric chloride
Ponceau S
Acetic acid

Preparation of Reagents

Verhoeff's stain

Dissolve 3 g of haematoxylin in 66 ml absolute ethyl alcohol. Cool, filter and add 24 ml of 10 per cent aqueous ferric chloride and 24 ml of Verhoeff's iodine solution. Solution lasts for 2 weeks.

Solution 1 (Verhoeff's iodine)

Potassium iodide	4.0 g
Distilled water	100.0 ml
Dissolve and add iodine	2.0 g

Solution 2 (Ferric chloride solution 10 per cent)

Ferric chloride	10.0 g
Distilled water	100.0 ml

Solution 3 (Ferric chloride solution 2 per cent)

10 per cent ferric chloride	20.0 ml
Distilled water	100.0 ml

Solution 4 (Picro-Ponceau solution)

Ponceau S, C.127195 (1 per cent aq.)	10.0 ml
Picric acid saturated	86.0 ml
Acetic acid 1 per cent aqueous	4.0 ml

Procedure

1. Deparaffinize and bring slides to 70 per cent alcohol.
2. Stain in solution 1 (Verhoeff's stain) for 15–30 minutes.
3. Rinse in distilled water.
4. Differentiate in solution 3 (2 per cent ferric chloride) for few minutes. If the section is destained, bring them back to solution 1.

5. Transfer to 5 per cent sodium thiosulphate for 1 minute.
6. Wash in running water.
7. Counterstain in solution 4 (Picro-Ponceau).
8. Differentiate in 95 per cent alcohol.
9. Dehydrate, clear and mount.

Results

Elastic fibres Brilliant black
Nuclei   Blue to brownish black
Collagen   Red

Remarks

Verhoeff's stain can be combined with Perl's staining to show both iron and elastin (Pickett and Klavins, 1961).

Iron Gallein Elastin Stain (Churukian and Schenk, 1976)

Fixation

Any general fixative

Reagents Required

Gallein
Ethylene
Ferric chloride
Concentrated HCl

Preparation of Reagents

Solution 1
Gallein   1.0 g
Ethylene glycol   20.0 ml

Add 80 ml absolute alcohol and mix.

Solution 2
Ferric chloride (20 per cent)   4.0 ml
Distilled water   95.0 ml
Conc. hydrochloric acid   1.0 ml

Procedure

1. Dewax and hydrate slides to water.
2. Place in solution 1 for 30 minutes.
3. Wash in running water.
4. Immerse in solution 2 for 2 minutes.
5. Wash in water.
6. Counterstain if necessary.

Result

Elastin is stained black.

Basic Fuchsin Stain (Horobin *et al.*, 1974)

Reagents Required

Basic fuchsin
Ferric chloride
Hydrochloric acid

Preparation of Reagents

Solution 1

Add 2.0 g of basic fuchsin to 200 ml of boiling water. Then add 25 ml of 30 per cent ferric chloride and again boil for 5 minutes. Filter and dry the precipitate. Dissolve the precipitate in 200 ml of 95 per cent alcohol. Store this solution in refrigerator. This solution lasts for several months.

Procedure

1.   Dewax and hydrate slides to water.
2.   Keep in solution 1 for 30 minutes.
3.   Differentiate in 70 per cent alcohol.
4.   Differentiate in 1 per cent HCl in 70 per cent alcohol (acid alcohol) for 5 minutes.
5.   Dehydrate, clear and mount.

Results

| Elastin | Deep purple |
| Basophilic substance | Purple |

Orcinol–New fuchsin (Fullmer and Lillie, 1956)

Fixation

Any general fixative

Reagents Required

New fuchsin
Orcinol
Ferric chloride

Preparation of Reagents

Solution 1

New fuchsin C. 142520	2.0 g
Orcinol	4.0 g
Distilled water	200.0 ml

Boil for 5 minutes and then add 25 ml of 29.1 per cent aqueous ferric chloride. Boil for 5 minutes, collect precipitate on a filter paper and dissolve in 100 ml of 95 per cent ethyl alcohol.

Procedure

1. Deparaffinize and transfer slides to absolute alcohol.
2. Stain in solution 1 at 37°C for 15 minutes.
3. Differentiate in 70 per cent alcohol giving 3 changes of 5 minutes each.
4. Dehydrate in 95 per cent absolute alcohol, clear and mount.

Result

Elastic fibres   Deep violet
Collagen   Unstained

If desired, counterstain with safranin.

Aldehyde fuchsin (Gomori, 1950a)
(Modified from Cameron and Steele Method, 1959)

Fixative

Any fixative without dichromate

Reagents Required

Potassium permanganate
Sulphuric acid
Sodium bisulphite
Basic fuchsin
Paraldehyde
Hydrochloric acid

Preparation of Reagents

Solution 1 (Potassium permanganate)

Potassium permanganate	300 mg
Distilled water	100.0 ml
H_2SO_4 conc.	0.3 ml

Solution 2 (Sodium bisulphite) (2.5 per cent)

Sodium bisulphite	2.5 g
Distilled water	100.0 ml

Solution 3 (Aldehyde fuchsin)

Add 1 g of basic fuchsin to 200 ml boiling water. Boil, cool and filter. Add 2 ml of conc. HCl and 2 ml paraldehyde. Leave stoppered at room temperature. When mixture has lost reddish fuchsin colour and is deep purple (3–4 days) filter it and discard filtrate. Dry precipitate on filter paper in an oven. Remove and store in a bottle. To make staining solution, dissolve 0.25 g in 50 ml of 70 per cent alcohol. It keeps for 6 months.

Procedure

1. Hydrate slides to water and remove $HgCl_2$.
2. Oxidize in solution 1 for 1 minute.

3. Rinse in distilled water.
4. Bleach in solution 2 until permanganate colour is removed.
5. Wash.
6. Transfer to 70 per cent alcohol.
7. Stain in solution 3 (aldehyde fuchsin) for 10 minutes.
8. Wipe off back of slide and rinse in 95 per cent alcohol.
9. Dehydrate, clear and mount.

Result

Elastin   Deep purple
Mast cells, etc.   Purple

This method can be combined with Pearl's reaction to show both iron and elastin (Pickett and Klavins, 1961).

Orcein Method (Plate 18, Figure 3)

Fixation

Any general fixative

Reagents Required

Orcein
Hydrochloric acid

Preparation of Reagents

Solution 1

Orcein   1.0 g
70 per cent ethyl alcohol   100.0 ml
Hydrochloric acid   1.0 ml
(Assay 37.38 per cent)

Procedure

1. Deparaffinize and hydrate slides to water.
2. Stain in solution 1 for 30–60 minutes.
3. Wash briefly in distilled water.
4. Dehydrate in 95 per cent alcohol for 2 minutes.
5. Differentiate in absolute alcohol.
6. Rinse in fresh absolute alcohol.
7. Clear and mount.

Result

Elastin is stained red.

Safranin–Orcein Method

Fixation

Any general fixative

Reagents Required

Orcein
Safranin
Hydrochloric acid
Aniline water

Preparation of Reagents

Solution 1

Orcein	1.0 g
90 per cent alcohol	100.0 ml
Conc. hydrochloric acid	1.0 ml

Solution 2 (Safranin O)

Safranin	1.0 g
Aniline water	48.0 ml
96 per cent alcohol	52.0 ml

As an alternative, use 1 per cent aqueous solution of the dye. To prepare aniline water, add 1 ml of aniline oil to 50 ml of distilled water and shake vigorously and filter.

Procedure

1. Deparaffinize.
2. Stain in solution 1 for 1 hour.
3. Rinse in acid alcohol till no more stain comes out of slide.
4. Wash.
5. Stain in solution 2 for 10 minutes.
6. Rinse in water.
7. Dip in 90 per cent alcohol.
8. Dehydrate, clear and mount.

Result

Elastic fibres	Dark brown or reddish brown
Cell nuclei	Bright red
Ground substance	Yellow

Orcein and Van Gieson

This combination is sometimes useful for determining both elastic and collagen fibres in one and the same specimen.

Fixative

Any general fixative

Reagents Required

Orcein
Hydrochloric acid
Picric acid
Acid fuchsin

Preparation of Reagents

Solution 1 (Orcein solution)

Orcein	1.0 g
95 per cent alcohol	100.0 ml
Conc. hydrochloric acid	1.0 ml

Solution 2 (Van Gieson)

Saturated picric acid	100.0 ml
1 per cent acid fuchsin	10.0 ml

Procedure

1. Deparaffinize and hydrate slides to water.
2. Stain in solution 1 for 30 minutes.
3. Rinse in acid alcohol.
4. Pass through 70 per cent alcohol.
5. Wash in water.
6. Stain in solution 2 (Van Gieson) for 3–5 minutes.
7. Rinse rapidly in water for a few seconds.
8. Dehydrate, clear and mount.

Result

Collagen fibres	Red
Elastic fibres	Brown
Epithelia and muscle	Yellow

Resorcinol Fuchsin Method

Fixative

Any general fixative

Reagents Required

Resorcinol
Basic fuchsin
Hydrochloric acid

Preparation of Reagents

Solution 1 (Resorcinol fuchsin solution)

Resorcinol	2.0 g
Basic fuchsin	500 mg
25 per cent hydrochloric acid	0.5 ml
95 per cent alcohol	100.0 ml

To this, add 15 ml 5 per cent potassium permanganate. Filter and use.

Procedure

1. Deparaffinize and bring sections to water.
2. Place the sections in solution 1 for 1 hour.

3. Rinse in tap water.
4. Stain nuclei with haematoxylin if desired.
5. Differentiate in acid alcohol.
6. Dehydrate, clear and mount.

Result

Elastin is stained brownish violet.

Weigert Resorcin Fuchsin for Elastin

Fixative

Any general fixative

Reagents Required

Basic fuchsin
Resorcin
Ferric chloride
Hydrochloric acid

Preparation of Reagents

Solution 1 (Basic fuchsin–resorcin solution)

Basic fuchsin	2.0 g
Resorcin	4.0 g
Distilled water	200.0 ml
29 per cent ferric chloride	25.0 ml
95 per cent alcohol	200.0 ml
Hydrochloric acid	4.0 ml

Boil fuchsin resorcin in distilled water. Add aqueous solution of ferric chloride and continue boiling for 2–5 minutes. Let it stand for precipitation to be completed. Cool and filter. Leave the precipitate to dry on the filter paper. In a fresh dish, add 200 ml of 95 per cent alcohol to the precipitate and heat while stirring. Remove the filter paper after the precipitate dissolves. Cool, filter and add 95 per cent alcohol to make up to 200 ml. Add hydrochloric acid. This solution keeps for a month.

Procedure

1. Deparaffinize and bring sections to water.
2. Stain in solution 1 for 20 minutes for 1 hour.
3. Wash off excess stain in 95 per cent alcohol.
4. Wash in tap water.
5. Stain in haematoxylin–eosin.
6. Dehydrate and mount.

Result

Elastic fibres are stained dark blue or black.

Aldehyde Fuchsin–Luxol Fast Yellow–Pontacyl Blue-black

Fixative

Any general fixative, preferably Bouin's.

Reagents Required

Pontacyl blue-black
Potassium dichromate
Fast yellow TN

Preparation of Reagents

Solution 1 (Aldehyde fuchsin as indicated (*see* Chapter 9)

Solution 2 (Pontacyl blue-black)
  Pontacyl blue-black   1.0 g
  Distilled water   100.0 ml

Solution 3 (Potassium dichromate)
  Potassium dichromate   2.0 g
  Distilled water   100.0 ml

Solution 4 (Working solution)
  Solution 2   30.0 ml
  Solution 3   10.0 ml

Solution 5
  Fast yellow TN   2.0 g
  90 per cent alcohol   100.0 ml

Procedure

1. Deparaffinize and bring sections to water.
2. Stain in solution 1 for 30 minutes.
3. Rinse in 70 per cent alcohol.
4. Rinse in distilled water.
5. Immerse in solution 4 (pontacyl blue-black) for 15 minutes.
6. Rinse in tap water.
7. Differentiate in 70 per cent alcohol.
8. Immerse in solution 5.
9. Wash, dehydrate, clear and mount.

Result

Elastin   Deep red
Collagen   Yellow
Mucin   Reddish purple

Elastin-trichrome Stain

Fixative

Any General Fixative

Reagents Required

Weigert's elastin stain
Hydrochloric acid
Ponceau acid fuchsin
Ponceau dixylidine
Glacial acetic acid
Phosphotungstic acid
Light green

Preparation of Reagents

Solution 1 (Weigert's elastic stain)

Weigert's elastin stain powder	1.0 g
Conc. hydrochloric acid	20.0 ml
Absolute alcohol	100.0 ml

The stain is dissolved and boiled for 2 minutes in a flask plugged tightly with cotton wool on a water bath. Cool and filter. Make up the volume to 100 ml with absolute alcohol. This solution is useless after a month.

Solution 2 (Ehrlichs' haematoxylin) (See Chapter 7)

Solution 3 (Staining solution)

Acid fuchsin	300 mg
Ponceau dixylidine	700 mg
Distilled water	100.0 ml
Glacial acetic acid	1.0 ml

Solution 4 (Phosphotungstic acid solution)

Phosphotungstic acid	3.0 g
Distilled water	100.0 ml

Solution 5 (Light green solution)

Light green	1.0 g
Distilled water	100.0 ml

Procedure

1. Deparaffinize and hydrate slides to water.
2. Stain in solution 1 for 10 minutes.
3. Differentiate in absolute alcohol.
4. Stain in solution 2 (Ponceau and fuchsin) for 5 minutes.
5. Wash in solution 3 (PTA) for 10 minutes.
6. Wash and transfer to solution 4 for 5 minutes.
7. Wash in distilled water and stain with solution 5.
8. Flood the preparation with 1 per cent acetic acid.
9. Dehydrate, clear and mount.

Result

Elastin	Blue-black
Collagen	Green

Orcein–aniline blue–orange G (Plate 18, Figure 6)

Fixative

Bouin's solution

Reagents Required

Orcein
Hydrochloric acid
Aniline blue
Orange G
Phosphomolybdic acid

Preparation of Reagents

Solution 1 (Orcein solution)

Orcein	1.0 g
70 per cent alcohol	100.0 ml
Conc. hydrochloric acid	0.6 ml

Solution 2 (Acid alcohol)

50 per cent alcohol	49.0 ml
Conc. hydrochloric acid	0.5 ml

Solution 3 (Aniline blue solution)

Aniline blue (aqueous)	500 mg
Orange G	2.0 g
1 per cent phosphomolybdic acid	100.0 ml

Solution 4 (Orange G solution)
Orange G, 6 per cent in absolute alcohol.

Procedure

1. Deparaffinize and bring slides to water.
2. Stain in solution 1 for 30 minutes.
3. Differentiate in solution 2.
4. Wash in tap water.
5. Immerse in solution 3 with an equal volume of distilled water for 1–2 minutes.
6. Rinse in 95 per cent alcohol.
7. Rinse in solution 4.
8. Wash, dehydrate and mount.

Result

Elastic fibres	Red
Collagen	Blue
Keratin	Bright yellow

Orcein–Aniline–Safranin Method

Fixative

Bouin's fixative

Reagents Required

Orcein
Hydrochloric acid
Safranin O
Aniline blue

Preparation of Reagents

Solution 1 (Orcein solution)
    Orcein ½ to 1 per cent in 80 per cent alcohol   100.0 ml
    Conc. Hydrochloric acid   1.0 ml

Solution 2 (Safranin solution)
    Safranin O (water-soluble)   1.0 g
    Aniline blue   38.0 ml
    Absolute alcohol   52.0 ml

Procedure

1. Deparaffinize and bring slides to 70 per cent alcohol.
2. Stain in solution 1 for 20–30 minutes.
3. Rinse in acid alcohol.
4. Rinse in 70 per cent alcohol.
5. Stain in solution 2 for 5 minutes.
6. Differentiate in 96 per cent alcohol.
7. Dehydrate, clear and mount.

Result

Elastic fibres   Reddish brown
Nuclei   Bright red

Hart's Modification of Weigert's Elastic Stain

Fixation

Any general fixative or 10 per cent formalin

Reagents Required

Basic fuchsin
Resorcin
Glacial acetic acid

Preparation of Reagents

Solution 1 (Weigert's stain)

Basic fuchsin	2.0 g
Resorcin	4.0 g
Distilled water	200 ml

The stain is diluted with 1 per cent acid alcohol.

Procedure

1. Deparaffinize and bring slides to water.
2. Keep in solution 1 at room temperature.
3. Differentiate in acid alcohol and wash in water.
4. Counterstain with haematoxylin and Van Geison or 1 per cent neutral red.
5. Dehydrate, clear and mount.

Result

Elastic tissue stains blue-black.

Sheridan's Resorcin–Crystal Violet Method for Elastic Fibres

Fixation

Any general fixative

Reagents Required

Crystal violet
Dextrin
Basic fuchsin
Oxalic acid

Preparation of Reagents

Solution 1 (Crystal violet solution) (Sheridan's stain)

Crystal violet	2.0 g
Dextrin	2.0 g
or	
Basic fuchsin	2.0 g

Solution 2 (Potassium permanganate solution)

Potassium permanganate	500 mg
Distilled water	100.0 ml

Solution 3

Oxalic acid	5.0 g
Distilled water	100.0 ml

Procedure

1. Deparaffinize and hydrate slides to water.
2. To oxidize, place in solution 2 (Potassium permanganate).
3. Rinse in tap water for 1–2 minutes.

4. Decolorize with solution 3 for 1 minute.
5. Wash in water.
6. Wash in distilled water.
7. Rinse in distilled water.
8. Transfer to solution 1 for 1 hour at 37°C or overnight at room temperature.
9. Rinse in absolute alcohol.
10. Rinse in distilled water.
11. Counterstain with 1 per cent neutral red.
12. Rinse in distilled water and later in absolute alcohol.
13. Flood slides with 0.5 per cent picric acid.
14. Rinse in water and blot.
15. Dehydrate rapidly, clear and mount.
16. Wash.
17. Dehydrate, clear and mount in DPX.

Result

Elastin fibres	Blue-green
Nuclei	Red

Fraenkel's Method for Elastic Tissue
(Lillie and Fullmer, 1976; Described by Schmorl, 1928)

Fixation

Any general fixative

Reagents Required

Orcein
Nitric acid
Indigo carmine
Picric acid

Preparation of Reagents

Solution 1 (Stock solution)

Orcein	1.5 g
95 per cent alcohol	125.0 ml
Distilled water	60.0 ml
Nitric acid	6.0 ml

To 3 per cent hydrochloric acid in 70 per cent alcohol add enough stock solution until it is brown.

Solution 2

Indigo carmine	250 mg
Saturated picric acid	100.0 ml

Procedure

1. Deparaffinize and hydrate slides to water.
2. Stain for nuclei in Orth's lithium carmine for 2 minutes (2.5 g in 100 ml saturated lithium carbonate).
3. Differentiate in acid alcohol (1 per cent alcoholic HCl).
4. Transfer to solution 1 for 24 hours.
5. Rinse in 80 per cent alcohol.
6. Transfer to solution 2 for 10–15 minutes.
7. Rinse in 3–5 per cent acetic acid.
8. Dehydrate, clear and mount in synthetic resin.

Result

Elastin	Dark brown
Collagen	Blue-green
Muscle	Greenish yellow
Nuclei	Red

Pinkus Acid–Orcein Giemsa Method (Pinker, 1944)

Fixation

10 per cent neutral buffered formalin.

Reagents Required

Orcein
Conc. HCl
Giemsa stain
Eosin

Preparation of Reagents

Solution 1 (Acid orcein)

Orcein	200 mg
70 per cent alcohol	100.0 ml
Hydrochloric acid	0.5 ml

Solution 2 (Giemsa solution)

Giemsa stock	1 drop
Distilled water	200.0 ml

Solution 3 (Alcoholic eosin)

1 per cent alcoholic eosin	2.0 ml
95 per cent alcohol	100.0 ml

Procedure

1. Deparaffinize and hydrate slides to water.
2. Transfer to solution 1 for 30 minutes.

3. Wash in running water for 15 minutes.
4. Transfer to solution 2 overnight.
5. Drain off excess.
6. Stain with solution 3.
7. Differentiate in 70 per cent alcohol.
8. Dehydrate and clear.
9. Mount in permount.

Result

Nuclei	Deep blue
Cytoplasm	Light blue
Elastic fibres	Dark brown
Collagen	Rose
Mast cells	Purple

Gomori's Aldehyde Fuchsin Method (Gomori, 1946)

Fixation

10 per cent neutral buffered formalin

Reagents Required

Basic fuchsin
Peraldehyde
Hydrochloric acid
Acid fuchsin
Picric acid
Metanil yellow
Glacial acetic acid

Preparation of Reagents

Solution 1 Aldehyde fuchsin (*see* Chapter 9)

Solution 2 Van Gieson (*see* page 571)

Solution 3 (Metanil yellow)

Metanil yellow	250 mg
Distilled water	100.0 ml
Glacial acetic acid	0.25 ml

Procedure

1. Deparaffinize and hydrate slides to water.
2. Transfer to solution 1 for 30 minutes.
3. Rinse in 70 per cent alcohol.
4. Counterstain either in solution 2 or 3.
5. Wash.
6. Dehydrate, clear and mount.

Result

Elastin fibres and mucin are stained deep purple.

Kornhauser Quad Method (Kornhauser, 1943)

Fixation

Zenker's or Helly's fixative

Reagents Required

Orcein
Nitric acid
Alizarin blue-2B
Aluminium sulphate
Glacial acetic acid

Preparation of Reagents

Solution 1 (Orcein solution)

Orcein	400 mg
Conc. nitric acid	0.4 ml
90 per cent alcohol	100.0 ml

Solution 2 (Alizarine blue solution)

Alizarine blue 2B	350 mg
10 per cent aluminium sulphate	100.0 ml
(10 g/100 ml water)	

Boil, cool and filter. The final pH should be 2.9 with addition of acetic acid (0.5 ml for 100 ml).

Solution 3 (Orange G solution)

Orange G	2.0 g
Fast green FCF	200 mg
Glacial acetic acid	2.0 ml
Distilled water	100.0 ml

Solution 4 (PTA & PMA solution)

Phosphotungstic acid	4.0 g
Phosphomolybdic acid	1.0 g
Distilled water	100.0 ml

Procedure

1. Deparaffinize and keep in 70 per cent alcohol.
2. Transfer to solution 1 for 1 hour.
3. Wash excess orcein in alcohol and then in water.
4. Transfer to solution 2 for 10 minutes.
5. Rinse in distilled water.
6. Mordant in solution 4 for 10–20 minutes.
7. Rinse in water.

8. Stain in solution 3 for 10 minutes.
9. Dehydrate, clear and mount in synthetic resin.

Result

Elastin	Red-brown
Collagen, reticulum, basement membrane	Green
Myelin sheath	Orange

COLLAGEN STAINING

Picroponceau with Haematoxylin
(Gurr, 1956) (Van Gieson Substitute (Non-fading))

Fixation

Any general fixative

Reagents Required

Ponceau S
Picric acid
Acetic acid

Preparation of Reagents

Solution 1 (Ponceau S solution)

Ponceau S (1 per cent)	10.0 ml
Picric acid saturated	86.0 ml
1 per cent acetic acid	4.0 ml

Procedure

1. Deparaffinize and hydrate slides to water.
2. Option-mordant with iron alum as done for haematoxylin.
3. Overstain in haematoxylin for 5–15 minutes.
4. Wash thoroughly in running water until slides are deep blue, for 10 minutes.
5. Transfer slides to solution 1 for 3–5 minutes or longer.
6. Rinse in distilled water.
7. Rinse in 70 per cent alcohol.
8. Dehydrate, clear and mount.

Result

Nuclei	Brown to brownish or bluish black
Collagen and reticular fibres	Red
Elastin fibres, erythrocytes, epithelia	Yellow

Van Gieson's Picrofuchsin Method (Plate 19, Figure 2)

Fixation

Any alcohol

Reagents Required

Harris haematoxylin (*see* Chapter 7)
Acid fuchsin
Picric acid

Preparation of Reagents

Solution 1
   Harris haematoxylin (see Chapter 7)

Solution 2 (Van Gieson No. 1)
   1 per cent aqueous acid fuchsin   5.0 ml
   Saturated aqueous picric acid   100.0 ml

Solution 3 (Saturated picric acid)
   Picric acid saturated in 95 per cent alcohol
   Terpineol or organum (for celloidin section)

Procedure

1.   Deparaffinize and hydrate slides to water.
2.   Stain nuclei deeply (for about 10 minutes) in solution 1 (Harris haematoxylin).
3.   Stain in solution 2 (Van Gieson's solution) for 1 minute.
4.   Wash quickly in water.
5.   Decolorize and differentiate in (solution 3) 95 per cent alcohol saturated with picric acid until sections are pale yellow.
6.   Dehydrate in alcohol, clear and mount.

Result

Collagen   Bright red
Nuclei   Blue or black
Other tissue elements Yellow

Biebrich Scarlet–Picro Aniline Blue

Fixation

Any general fixative

Reagents Required

Biebrich scarlet
Glacial acetic acid
Aniline blue
Picric acid

Preparation of Reagents

Solution 1 (Weigert's haematoxylin) (*see* Chapter 7)

Solution 2 (Biebrich scarlet solution)
   Biebrich scarlet   200 mg
   Glacial acetic acid   1.0 ml
   Distilled water   100.0 ml

Solution 3 (Aniline blue solution)
    Aniline blue   100 mg
    Saturated aq. picric acid   100.0 ml

Solution 4 (1 percent aqueous acetic acid)

Procedure

1. Deparaffinize and bring sections to water.
2. Stain in solution 1 (Weigert's haematoxylin) for 5 minutes.
3. Wash in water.
4. Stain in solution 2 (Biebrich scarlet) for 4 minutes.
5. Rinse in water.
6. Stain in solution 3 (aniline blue) for 5 minutes.
7. Wash in solution 4.
8. Dehydrate, clear and mount.

Result

Erythrocytes	Orange
Muscles	Pink
Cytoplasm	Pink
Nuclei	Black
Reticulum	Blue
Connective tissue	Blue

Mallory's Aniline Blue for Collagen (Plate 18, Figure 4; Plate 19, Figure 1)

Fixation

Zenker, mercury-containing fixative

Reagents Required

Acid fuchsin
Aniline blue
Orange G
Phosphotungstic acid

Preparation of Reagents

Solution 1 (Acid fuchsin solution)
    Acid fuchsin   500 mg
    Distilled water   100.0 ml

Solution 2 (Aniline blue–orange G solution)
    Aniline blue (water-soluble)   500 mg
    Orange G   2.0 g
    1 per cent phosphotungstic acid   100.0 ml

Procedure

1. Deparaffinize and hydrate slides to water.
2. Place in solution 1 (acid fuchsin solution) for 1–3 minutes.

3. Transfer directly to solution 2 (aniline blue–orange G solution) for 20–60 minutes.
4. Rinse in 95 per cent alcohol giving several changes.
5. Dehydrate, clear and mount.

Result

Collagen	Intense blue
Hyaline tissue	Blue
Erythrocytes and myelin	Yellow to orange

Haemalum–Eosin Method

Fixation

Any general fixative

Reagents Required

Haematoxylin (*see* Chapter 7)
Eosin bluish

Preparation of Reagents

Haemalum (Mayer's)
Eosin bluish (0.2 per cent in 20 per cent alcohol)

Procedure

1. Hydrate slides to water.
2. Stain in haemalum for 5–10 minutes.
3. Rinse in tap water.
4. Stain in eosin for 2–3 seconds.
5. Rinse in water.
6. Dehydrate and mount.

Result

Collagen	Deep pink
Smooth muscle	Pink
Cytoplasm	Pale pink
Nuclei	Blue

MUSCLE STAINING

Milligan's Trichrome Stain (1946)

Fixation

10 per cent formalin in normal saline

Reagents Required

Potassium dichromate
Conc. HCl

Acid fuchsin
PMA
Orange G
Fast green

Preparation of Reagents

Solution 1 (Mordant)
 Solution A
  Potassium dichromate   3.0 g
  Distilled water   100.0 ml

 Solution B
  Conc. HCl   10.0 ml
  95 per cent alcohol   100.0 ml

To 3 parts of A add 1 part of B. Use within 4 hours.

Solution 2 (Acid fuchsin)
 Acid fuchsin   100 mg
 Distilled water   100.0 ml

Solution 3 (Phosphomolybdic acid)
 Phosphomolybdic acid   2.0 g
 Distilled water   200.0 ml

Solution 4 (Orange G)
 Orange G   2.0 g
 71 per cent PMA   100.0 ml

Solution 5 (Fast green stock solution)
 Fast green   1.0 g
 2 per cent acetic acid   100.0 ml

Solution 6 (Working solution)
 Fast green stock solution   10.0 ml
 Distilled water   90.0 ml

Procedure

 1. Deparaffinize and keep in absolute alcohol.
 2. Mordant in solution 1 (potassium dichromate–HCl solution) for 7 minutes.
 3. Rinse in distilled water.
 4. Transfer to solution 2 (acid fuchsin) for 8 minutes.
 5. Rinse in distilled water.
 6. Transfer to solution 3 (PMA) for 5 minutes.
 7. Stain with solution 4 (Orange G) for 10 minutes.
 8. Rinse in distilled water.
 9. Treat with 1 per cent acetic acid.
 10. Transfer to solution 6 Fast green for 10 minutes.
 11. Treat with 1 per cent acetic acid.
 12. Rinse in 95 per cent alcohol.
 13. Dehydrate, clear and mount.

Result

Muscle	Magenta
Collagen	Green
Red blood cells	Orange

For further details on muscle staining, consult Poley and Forbes (1964), Puchtler *et al.* (1969), Churikian (1993b), Shapiro and Sohns (1994) and Gowali (1995).

Haematoxylin (Heidenhain)–Van Gieson Stain

Fixatives

Bouin's, Carnoy and Susa fixative, or 10 per cent formalin.

Reagents Required

Haematoxylin
Van Gieson

Preparation of Reagents

Solution 1 (Haematoxylin (Heidenhain's)) (*see* Chapter 7)
Solution 2 (Van Gieson stain (Picro fuchsin)) (*see* page 535)

Procedure

1. Hydrate slides to water.
2. Immerse in solution 1 for ½ to 1 hour.
3. Rinse in water.
4. Stain in solution 2 for ½ to 1 hour.
5. Rinse in water.
6. Differentiate in solution 1.
7. Wash.
8. Stain in Van Gieson for 3–5 minutes.
9. Wash, dehydrate and mount.

Result

Collagen fibres	Bright red
Nuclei	Black or brown
Erythrocytes	Yellow

Safranin–Water Blue Stain for Collagen

Fixatives

1 per cent picric acid or absolute alcohol.

Reagents Required

Safranin O
Water blue
Tannic acid

Preparation of Reagents

Solution 1 (Safranin solution)
    1 per cent safranin O   1 g/100 ml

Solution 2 (Water blue–tannic acid solution)
    Water blue 1 per cent aqueous   10.0 ml
    Tannic acid 33 per cent aqueous   10.0 ml
This solution must be freshly prepared.

Procedure

1. Hydrate slides to water.
2. Stain in solution 1 (safranin O solution) for 5–10 minutes.
3. Stain in solution 2 for 10–15 minutes.
4. Wash, dehydrate.
5. Clear in Bergamotoil and mount in DPX.

Result

Collagen fibres   Blue
Nuclei   Red

Hart's Method for Elastic Fibres

Fixation

10 per cent buffered neutral formalin

Reagents Required

Potassium permanganate
Oxalic acid
Resorcin fuchsin
Picric acid
Acid fuchsin
Conc. hydrochloric acid

Preparation of Reagents

Solution 1 (Permanganate solution)
    Potassium permanganate   250 mg
    Distilled water   100.0 ml

Solution 2 (Oxalic acid solution)
    Oxalic acid   5.0 g
    Distilled water   100.0 ml

Solution 3 (Resorcin fuchsin solution)
    Resorcin fuchsin solution (*see* page 524)

Solution 4 (Van Gieson solution) (*see* page 535)

Procedure

1. Deparaffinize and hydrate slides to water.
2. Keep in solution 1 for 5 minutes.
3. Rinse in distilled water.
4. Keep in solution 2 for 5 minutes.
5. Wash in tap water.
6. Transfer to solution 3 overnight.
7. Wash in tap water.
8. Rinse in distilled water.
9. Transfer to solution 4 for 1 minutes.
10. Dehydrate, clear and mount.

Result

Elastic fibres   Blue-black
Nuclei   Blue
Collagen   Pink or red

Van Gieson/Haematoxylin (Van Gieson, 1889; Mallory, 1938)

Fixation

Any general fixative

Reagents Required

Alum haematoxylin
Sodium iodate
Acid fuchsin
Picric acid

Preparation of Reagents

Solution 1 (Alum haematoxylin or iron haematoxylin) (*see* Chapter 7)

Solution 2 (Acid fuchsin solution)
Acid fuchsin   1.0 g
Distilled water   100.0 ml

Solution 3
1 per cent acid fuchsin   15.0 ml (solution 2)
Saturated picric acid   100.0 ml

Procedure

1. Deparaffinize and hydrate slides to water.
2. Place in solution 1 for 5 minutes.
3. Wash in water.
4. Transfer to solution 3 for 5 minutes.
5. Wash in water.
6. Dehydrate, clear and mount.

Result

Collagen	Red
Striated muscle	Yellow

Van Gieson/Trypan Blue (Clark, 1971)

Fixation

Formalin

Reagents Required

Naphthol yellow
Acid fuchsin
Trypan blue
Gallocyanin
Haematoxylin

Preparation of Reagents

Solution 1 (Naphthol yellow solution)
Naphthol yellow   300 mg
Distilled water   100.0 ml

Solution 2 (Acid fuchsin solution)
Acid fuchsin   200 mg
Distilled water   100.0 ml

Solution 3 (Trypan blue solution)
Trypan blue   200 mg
Distilled water   100.0 ml

Solution 4 (Working solution)
Solution 1   40.0 ml
Solution 2   25.0 ml
Solution 3   2.0 ml

Add 0.4 ml concentrated HCl and solution 1 (40.0 ml).

Procedure

1. Deparaffinize and hydrate slides to water.
2. Place in gallocyanin chrome alum (see Chapter 7) for 10 minutes.
3. Wash thoroughly.
4. Transfer to solution 4 for 10 minutes.
5. Rinse in water.
6. Dehydrate, clear and mount.

Result

Collagen large fibres	Red
Collagen small fibres	Blue
Nucleus	Blue-black

Allochrome Method (Lillie, 1951)

Fixation

Formalin

Reagents Required

Picromethyl blue
Methyl blue
Picric acid
Schiff's reagent (*see* Chapter 9)
Sodium metabisulphite

Preparation of Reagents

Solution 1 (Schiff's reagent) (*see* Chapter 9)

Solution 2 (Sodium metabisulphite)
Sodium metabisulphite   500 mg
Distilled water   100.0 ml

Solution 3 (Iron haematoxylin) (*see* Chapter 7)

Solution 4 (Methyl blue solution)
Methyl blue   400 mg
Saturated aqueous picric acid   100.0 ml

Alternative to solution 4 (Sirius blue solution)
Sirius supra blue   700 mg
Saturated aq. picric acid   100.0 ml

Procedure

1. Deparaffinize and hydrate slides to water.
2. Place in 1 per cent periodic acid for 10 minutes.
3. Wash in water.
4. Transfer to solution 1 (Schiff's reagent) for 10 minutes.
5. Transfer to solution 2 for 2 minutes.
6. Wash.
7. Stain in solution 3 for 10 minutes.
8. Rinse in running water.
9. Transfer to either solution 4 (picromethyl blue) or sirius blue solution for 5 minutes.
10. Rinse in alcohol, dehydrate, clear and mount.

Result

Collagen and reticulin   Blue
Basement membrane   Red
Silver impregnation for reticulin
(Subbed slides are good. No loss of sections.)

Del Rio-Hortega Method (Mallory, 1938)

Fixation

Any good general fixative

Reagents Required

Silver nitrate
Lithium carbonate

Preparation of Reagents

Solution 1 (Ammoniacal silver carbonate)
   10 per cent silver nitrate   10.0 ml
   Saturated lithium carbonate   10.0 ml

A precipitate occurs. Decant and wash precipitate 5 times. Add 25 ml of distilled water. Add ammonia (28 per cent) drop by drop till precipitate dissolves. Add 100 ml 95 per cent alcohol. Filter. Warm at 50°C for 30 minutes. Filtering should be repeated every time before use.

Solution 2 (Formalin)
   Formalin   10.0 ml
   Distilled water   50.0 ml

Solution 3 (Gold chloride solution)
   Gold chloride 1 per cent   12.5 ml
   (1 g + 100 ml water)
   Distilled water   50.0 ml

Procedure

1. Deparaffinize and hydrate slides to water.
2. Treat with 0.2 per cent potassium permanganate (200 mg + 100 ml water) for 2 minutes.
3. Wash in distilled water.
4. Bleach in 5 per cent oxalic acid (5 g + 100 ml water) for 3 minutes.
5. Wash in distilled water for 10 minutes.
6. Impregnate with solution 1 (silver solution) at 37°C for 30 minutes (do not expose to light).
7. Rinse quickly in distilled water.
8. Reduce in solution 2 (formalin solution) for 3 minutes.
9. Wash in distilled water for 3 minutes.
10. Tone with solution 3 until yellow colour turns purple grey.
11. Rinse in distilled water.
12. Fix in sodium thiosulphate (5 g + 100 ml water) for 3 minutes.
13. Wash in running water.
14. Counterstain (if desired) in haematoxylin.
15. Dehydrate, clear and mount.

Result

Reticulin	Black
Collagen	Red to rose
Nuclei	Black or blue
Cytoplasm	Greenish yellow
Muscle fibres, elastin	Light yellow

Gridley's Method (1951)

Fixation

Any good general fixative.

Reagents Required

Silver nitrate
Sodium hydroxide
Periodic acid

Preparation of Reagents

Solution 1 (Ammoniacal silver oxide)

5 per cent silver nitrate	20.0 ml
(5 g + 100 ml water)	
10 per cent sodium hydroxide	20 drops
(10 g + 100 ml water)	

Add 28 per cent ammonia dropwise to the above mixture until precipitate dissolves.
Add distilled water up to 50 ml.

Procedure

1. Deparaffinize and hydrate slides to water.
2. Treat with 0.5 per cent periodic acid (500 mg + 100 ml water) for 15 minutes.
3. Rinse in distilled water.
4. Treat with 2 per cent silver nitrate (2 g + 100 ml water) for 30 minutes at room temperature.
5. Rinse in 2 changes of distilled water.
6. Impregnate in solution 1 (ammoniacal silver solution) for 15 minutes at room temperature.
7. Rinse in distilled water.
8. Reduce in 30 per cent formalin (30 ml + 70 ml water) for 3 minutes.
9. Rinse in distilled water giving 4 changes.
10. Tone in gold chloride (10 ml 1 per cent stock solution/40 ml water) until yellow brown colour has changed to lavender grey.
11. Rinse in distilled water.
12. Fix in 5 per cent sodium thiosulphate (95 g + 100 ml water) for 3 minutes.
13. Wash in running water.
14. Counterstain if desired.
15. Dehydrate, clear and mount.

Result

Reticulin fibres	Black
Other tissue elements	Depends on counterstain used

Gomori's Method (Mallory, 1944)

Fixation

10 per cent neutral formalin

Reagents Required

Silver nitrate
Potassium hydroxide
Potassium metabisulphite
Ferric ammonium sulphate
Gold chloride
Sodium thiosulphate

Preparation of Reagents

Solution 1a (Silver nitrate solution)
Silver nitrate   10.0 g
Distilled water   100.0 ml

Solution 1b (Potassium hydroxide solution)
Potassium hydroxide   10.0 g
Distilled water   100.0 ml

Solution 2 (Working solution)
To 200 ml of silver nitrate, add 4–5 ml of potassium hydroxide, and also 28 per cent ammonia water, drop by drop. Shake the flask continuously until the precipitate formed dissolves. Make up the solution with distilled water to twice its volume.

Solution 3 (Potassium permanganate solution)
Potassium permanganate   500 mg
Distilled water   100.0 ml

Solution 4 (Potassium metabisulphite solution)
Potassium metabisulphite   2.0 g
Distilled water   100.0 ml

Solution 5 (Ferric ammonium sulphate solution)
Ferric ammonium sulphate   2.0 g
Distilled water   100.0 ml

Solution 6 (Formaldehyde)
Formaldehyde   20.0 ml
Distilled water   80.0 ml

Solution 7 (Gold chloride solution)
Gold chloride stock solution   20.0 ml
Distilled water   80.0 ml

Solution 8 (Sodium thiosulphate solution)
    Sodium thiosulphate   2.0 g
    Distilled water   100.0 ml

Procedure

1. Deparaffinize and hydrate slides to water.
2. Oxidize in solution 3.
3. Wash in tap water.
4. Bleach in solution 4 (potassium metabisulphite) for 1 minute.
5. Wash in tap water.
6. Sensitize in solution 5 (ferric ammonium sulphate) for 1 minute.
7. Wash in tap water for 2 minutes.
8. Impregnate in solution 2 (silver solution) for 1 minute.
9. Rinse in distilled water for 20 seconds.
10. Reduce in solution 6 (formalin solution) for 3 minutes.
11. Wash in tap water.
12. Tone in solution 7 (gold chloride) for 10 minutes Sections become purplish grey.
13. Rinse in distilled water.
14. Reduce in solution 4 (potassium metabisulphite) for 1 minute.
15. Fix in solution 8 (sodium thiosulphate) for 1 minute.
16. Wash in tap water for 2 minutes.
17. Counterstain if desired.
18. Dehydrate, clear and mount.

Result

Reticulin fibres stain black.

Laidlaw's Method (1929)

Fixation

Preferably Bouin's

Reagents Required

Silver nitrate
Lithium carbonate
Ammonia
Gold chloride

Preparation of Reagents

Solution 1

    Dissolve 6 g of silver nitrate in 10 ml of distilled water. To this, add 115 ml of saturated lithium carbonate in distilled water. Shake well. Settle precipitate and pour off solution to leave about 35 ml of precipitate. Wash well with distilled water four to five times. Allow to settle, pour off water, and gradually add ammonia, shake well until

the fluid is clear. A few grains of precipitate should remain. Add distilled water to a total of 60 ml. Shake and filter into brown stock bottle. Keeps for several months.
Solution 2 (Gold chloride solution)
    Gold chloride stock solution   10.0 ml
    (1 g/100 ml water)
    Distilled water   40.0 ml

Procedure

1. Deparaffinize and hydrate slides to water.
2. Oxidize in potassium permanganate for 3 minutes.
3. Wash.
4. Bleach in oxalic acid (1 g/10 ml water) for 3 minutes.
5. Wash in running water for 10 minutes.
6. Wash in distilled water giving 3 changes of 5–10 minutes each.
7. Impregnate in solution 1 (ammoniacal silver carbonate) at 50°C for 5 minutes.
8. Rinse in distilled water.
9. Reduce in 1 per cent formalin (1 ml + 99 ml water) for 3 minutes.
10. Rinse in distilled water.
11. Tone in solution 2 (gold chloride) until sections turn purplish grey.
12. Rinse in distilled water.
13. Fix in 5 per cent sodium thiosulphate (5 g + 100 ml water) for 3 minutes.
14. Wash in running water for 5–10 minutes.
15. Dehydrate, clear and mount.

Result

Reticulin   Black
Collagen   Reddish

Nassar and Shanklin's Method (1961)

Fixation

Any general fixative

Reagents Required

Potassium permanganate
Sulphuric acid

Preparation of Reagents

Solution 1 (Potassium permanganate solution)
    Potassium permanganate   0.5 g
    Distilled water   100.0 ml

Solution 2 (Conc. sulphuric acid)
    Conc. sulphuric acid   0.5 ml
    Distilled water   100.0 ml
Mix equal parts of 1 and 2. Use freshly mixed solutions.

Solution 3 (Oxalic acid solution)
    Oxalic acid   2.0 g
    Distilled water   100.0 ml

Solution 4 (Silver nitrate solution)
    Silver nitrate   1.0 g
    Distilled water   50.0 ml
    Pyridine   15 drops

Solution 5 (Silver diamine nitrate)

To 1 ml of 28 per cent ammonia, add 7 ml of 10 per cent silver nitrate (2.5 g/25 ml water). Continue to add 10 per cent $AgNO_3$ drop by drop while shaking the ammonia solution until a faint permanent turbidity remains. Dilute with equal amount of distilled water.

Solution 6 (Reducing solution)
    2 per cent neutral formalin   50.0 ml
        (1 ml + 49 ml water)
    Absolute ethyl alcohol   50.0 ml

Solution 7 (Gold chloride)
    1 per cent gold chloride stock solution   10.0 ml
        (1 g/100 ml water)
    Distilled water   40.0 ml

Procedure

1. Deparaffinize and hydrate slides to water.
2. Oxidize in solution 2 (permanganate solution) for 1–2 minutes.
3. Wash in distilled water.
4. Bleach in solution 3 (oxalic acid) for 2 minutes.
5. Wash in running water for 5 minutes.
6. Transfer to 70, 90, 95 per cent alcohol.
7. Immerse in solution 4 (silver nitrate solution) at 50°C for ½ to 1 hour and at room temperature for 1½ to 2 hours.
8. Rinse in 95 per cent alcohol.
9. Impregnate in solution 5 (silver diamine solution) at 50°C for 5 minutes.
10. Rinse in 95 per cent alcohol.
11. Reduce in solution 6 (formalin alcohol) for 2 minutes.
12. Wash in distilled water.
13. Tone in solution 7 (gold chloride) until colour turns purplish grey.
14. Rinse in distilled water.
15. Fix in 5 per cent sodium thiosulphate (5 g/100 ml water) for 3–5 minutes.
16. Wash in running water.
17. Counterstain with haematoxylin.
18. Dehydrate, clear and mount.

Result

Reticulin fibres are stained black.
[See also Staples and Clark (1990).]

Wilder's Method (1935)

Fixation

Any general fixative

Reagents Required

Phosphomolybdic acid
Uranium nitrate
Silver nitrate
Ammonia

Preparation of Reagents

Solution 1 (Phosphomolybdic acid)
    Phosphomolybdic acid   10.0 g
    Distilled water   100.0 ml

Solution 2 (Uranium nitrate solution)
    Uranium nitrate   1.0 g
    Distilled water   100.0 ml

Solution 3 Ammoniacal silver nitrate
    Add ammonia (28 per cent) to 5 ml of 10 per cent silver nitrate drop by drop. Add
5 ml of 3.1 per cent sodium hydroxide until the precipitate formed dissolves. Dissolve
the precipitate with addition of ammonia. Make the solution up to 500 ml with
distilled water. Use immediately.

Solution 4 (Reducing solution)
    Distilled water   50.0 ml
    Formalin   0.5 ml
    Uranium nitrate (1 per cent)   1.5 ml

Make fresh solution each time.

Solution 5 (Gold chloride solution)
    Gold chloride stock solution (1 g/100 ml)   10.0 ml
    Distilled water   40–80 ml

Procedure

1. Deparaffinize and hydrate slides to water.
2. Wash.
3. Treat with solution 1 phosphomolybdic acid for 1 minute.
4. Wash in running water for 5 minutes.
5. Treat with solution 2 (uranium nitrate) for 5 seconds.
6. Rinse in distilled water.

7. Impregnate with solution 3 (ammoniacal silver nitrate) for 1 minute.
8. Dip quickly in 95 per cent alcohol, then into solution 4 (reducing solution) for 1 minute.
9. Wash in distilled water.
10. Tone in solution 5 (gold chloride) until yellow colour turns purplish grey.
11. Rinse in distilled water.
12. Fix in sodium 5 per cent thiosulphate (5 g/100 ml water) for 3–5 minutes.
13. Wash in running water.
14. Counterstain if desired.
15. Dehydrate, clear and mount.

Result

Reticulin fibres are stained black.

Comments

Phosphomolybdic acid can be used as an alternative to potassium permanganate. Sensitization with uranium nitrate reduces the time and eliminates the heat required by reticulum methods. Uranium nitrate is not a sensitizer but it is an oxidizer (Lillie, 1946). Staple and Grizzle (1986a, 1986b) recommend methods based on the argyrophil reaction for reticulum and argentaffin granules.

Foot's Silver Impregnation Method for Reticulin

Fixation

10 percent formalin

Reagents Required

Silver nitrate
Sodium carbonate
Ammonia
Gold chloride
Sodium thiosulphate
Safranin
Potassium permanganate
Oxalic acid

Preparation of Reagents

10 per cent silver nitrate   10.0 ml
(10 g/100 ml water)
5 per cent sodium carbonate   40.0 ml
(5 g/100 ml water)

Allow the precipitate formed to settle. Pour out the supernatant and wash the deposit several times with distilled water. Add strong ammonia drop by drop to dissolve precipitate completely. Make up to 100 ml with distilled water. This solution keeps for a few weeks only. Used solution should be discarded.

Procedure

1. Deparaffinize and hydrate slides to water.
2. Oxidize in 0.25 per cent permanganate for 3 minutes.
3. Rinse in tap water.
4. Bleach in 5 per cent oxalic acid (5 g/100 ml water) for 10 minutes.
5. Wash.
6. Rinse in distilled water.
7. Flood slide with 1 per cent formalin.
8. Wash.
9. Tone in 0.2 per cent gold chloride for 15 minutes.
10. Rinse in distilled water.
11. Fix in 5 per cent sodium thiosulphate for 2 minutes.
12. Wash.
13. Counterstain in 1 per cent safranin.
14. Dehydrate, clear and mount.

Result

Reticulin fibres	Black
Nuclei	Red

Naoumenko and Feigin Method (1974)

Fixation

Any general fixative

Reagents Required

Silver nitrate
Ammonium nitrate
Potassium permanganate
Oxalic acid
Formaldehyde
Gold chloride

Preparation of Reagents

Solution 1 (Silver solution)

Distilled water	35.0 ml
8 per cent ammonium nitrate	7.0 ml
4 per cent sodium hydroxide	8.0 ml
10 per cent aq. silver nitrate	3.8 ml

Solution 2 (Potassium permanganate)

0.25 per cent potassium permanganate	45.0 ml
0.67 per cent acetic acid (aqueous)	0.5 ml

Mix fresh.

Solution 3 (Oxalic acid)
    Oxalic acid   1.0 g
    Distilled water   100.0 ml

Solution 4
    Formaldehyde 0.2 ml
    Distilled water   100.0 ml

Solution 5 (Gold chloride)
    Gold chloride   1.0 g
    Distilled water   100.0 ml

Procedure

1. Dewax and bring slides to water.
2. Place in solution 2 for 2 minutes.
3. Wash in distilled water for 15 seconds.
4. Bleach in solution 3 for 2 minutes.
5. Wash in distilled water.
6. Impregnate with solution 1 for 6 minutes.
7. Dip in 70 per cent alcohol.
8. Reduce in solution 4 giving 2 changes of 2 minutes each.
9. Wash in distilled water.
10. Tone in solution 5 for 1 minute.
11. Rinse in distilled water for 1 minute.
12. Fix in 5 per cent sodium thiosulphate for 1 minute.
13. Rinse in distilled water for 1 minute.
14. Counterstain.
15. Dehydrate, clear and mount.

Process the slides individually.

Result

Reticulin is stained black.

Silver Impregnation for Reticulin (Gordon and Sweet, 1936)

Fixation

Any general fixative

Reagents Required

Potassium permanganate
Sulphuric acid
Silver nitrate
Ammonia
Sodium hydroxide
Gold chloride
Iron alum

Preparation of Reagents

Solution 1 (Potassium permanganate solution)
Potassium permanganate   47.5 ml
3 per cent sulphuric acid   2.5 ml

Solution 2 (Oxalic acid solution)
Oxalic acid   1.0 g
Distilled water   100.0 ml

Solution 3 (Silver nitrate solution)
10 per cent silver nitrate   5.0 ml
25 per cent ammonia   Dropwise
3 per cent sodium hydroxide   5.0 ml

Solution 4 (Alum)
Iron alum   2.0 g
Distilled water   100.0 ml

Procedure

1. Hydrate slides to water.
2. Oxidize in solution 1 for 1 minute.
3. Bleach in solution 2 (1 per cent oxalic acid) for 1 minute.
4. Wash in distilled water.
5. Mordant in solution 4 iron alum for 15 minutes.
6. Wash in distilled water.
7. Impregnate in solution 3 for 5–7 seconds, add ammonia drop by drop until a brown precipitate is formed. Add 5 ml of 3 per cent NaOH. Again add ammonia drop by drop until the solution is clear.
8. Wash briefly.
9. Reduce with 10 per cent neutral formalin.
10. Wash in water.
11. Tone in 2 per cent gold chloride for 3 minutes.
12. Wash.
13. Fix in 5 per cent sodium thiosulphate for 5 minutes.
14. Wash.

Result

Reticulin   Black
Collagen   Yellow

Menzies Method for Reticulin (Menzies, 1961)

Fixation

10 per cent buffered formalin

Reagents Required

Uranium nitrate
Silver nitrate

Gold chloride
Ammonium hydroxide
Sodium thiosulphate
Nuclear fast red

Preparation of Reagents

Solution 1 (Uranium nitrate)
Uranium nitrate   1.0 g
Distilled water   100.0 ml
Solution 2
Dissolve 100.0 g of silver nitrate in 1000.0 ml of distilled water. Take 930.0 ml of $AgNO_3$ solution and add 60.0 ml of 28 per cent ammonium hydroxide. Shake vigorously until clear.

Solution 3 (1 per cent formalin)

Solution 4 (1 per cent gold chloride solution (1 g in 100 ml water))

Solution 5 (Sodium thiosulphate solution)
Sodium thiosulphate   5.0 g
Distilled water   100.0 ml

Solution 6 (Nuclear fast red solution) (*see* Chapter 15)

Procedure

1. Dewax and hydrate slides to water.
2. Sensitize in solution 1 for 2 minutes.
3. Rinse in water.
4. Impregnate in solution 2 for 1 minute.
5. Wash in running water.
6. Develop in solution 3 for 1 minute.
7. Wash in running water.
8. Tone in solution 4 for 1 minute.
9. Wash in running water.
10. Reduce in solution 5 for 1 minute.
11. Wash in running water.
12. Counterstain in solution 6 for 5 minutes.
13. Wash, dehydrate and mount.

Result

Reticulin   Blue
Nuclei   Red

Snook's Method for Reticulin (Snook, 1944)

Fixation

10 per cent buffered formalin

Reagents Required

1 per cent potassium permanganate
Oxalic acid
Uranium nitrate
Formaldehyde
Gold chloride
Sodium thiosulphate
Nuclear fast red
Silver nitrate
Sodium hydroxide
Ammonium hydroxide

Preparation of Reagents

Solution 1 ($KMnO_4$, 0.25 per cent solution) (*see* Chapter 13)

Solution 2 (Oxalic acid solution) (*see* Chapter 13)

Solution 3 (Uranium nitrate solution) (*see* Chapter 13)

Solution 4
    Formaldehyde (40 per cent)   1.0 ml
    Distilled water   100.0 ml

Solution 5 (Gold chloride)
    Gold chloride   1.0 g
    Distilled water   100.0 ml

Solution 6 (Sodium thiosulphate 5 per cent (5 g in 100 ml water))

Solution 7 (Nuclear Fast red (*see* Chapter 7))

Solution 8 (Silver nitrate solution)
    Silver nitrate   5.0 g
    Distilled water   100.0 ml

Solution 9 (Sodium hydroxide solution)
    Sodium hydroxide   20.0 ml
    Distilled water   100.0 ml

Solution 10 (Ammoniacal silver nitrate solution)

    Solution 8   20.0 ml
    Solution 9   10 drops

Then add ammonium hydroxide drop by drop until only a few granules of the precipitate remain. Make up to 60 ml with distilled water.

Procedure

1. Dewax and hydrate slides to water.
2. Oxidize in solution 1 for 5 minutes.
3. Wash.
4. Bleach in solution 2 for 5 minutes.
5. Wash in tap water.

6. Mordant in solution 3 for 5 seconds.
7. Wash in running water.
8. Transfer to solution 10 for 1 minute.
9. Rinse in running water.
10. Transfer to solution 4 for 1 minute.
11. Wash in tap water.
12. Tone in solution 5 for 1 minute.
13. Wash in running water.
14. Transfer to solution 6 for 1 minute.
15. Wash in running water.
16. Counterstain in solution 7 for 5 minutes.
17. Rinse in distilled water.
18. Dehydrate, clear and mount.

Result

Reticulin fibres	Grey to black
Background	Pink

Mallory PTAH Method for Fibrin

Fixation

Any general fixative

Reagents Required

Potassium dichromate
Potassium permanganate
Oxalic acid
Haematoxylin
Phosphotungstic acid

Preparation of Reagents

Solution 1 (Potassium dichromate solution)
  3 per cent potassium dichromate   3 parts
  10 per cent hydrochloric acid   1 part

Solution 2 (Potassium permanganate solution)
  0.5 per cent permanganate   47.5 ml
  3 pr cent sulphuric acid   2.5 ml

Solution 3 (Oxalic acid solution)
  Oxalic acid   1.0 g
  Distilled water   100.0 ml

Solution 4 (PTAH)
  Haematoxylin   100 mg
  Phosphotungstic acid   2.0 g
  Distilled water   100.0 ml
Allow to ripen for 6 months.

Procedure

1. Deparaffinize and hydrate slides to water.
2. Postchrome in solution 1 for 30 minutes.
3. Wash in water.
4. Oxidize in solution 2 for 1 minute.
5. Wash in water.
6. Bleach in solution 3.
7. Rinse in distilled water.
8. Transfer to solution 4 for 12–24 hours.
9. Dehydrate, clear and mount.

Result

Fibrin	Dark blue
Nuclei	Light blue
Collagen	Rose red

Picro-Mallory V for Fibrin (Lendrum *et al.*, 1962)

Fixation

Any general fixative

Reagents Required

Picric acid
Orange G
Lissamine fast yellow
Acid fuchsin
Acetic acid
Phosphotungstic acid
Acid blue
Light green

Preparation of Reagents

Solution 1

Saturated picric acid in 200 ml ethanol. Then add 400 mg orange G, 400 mg Lissamine yellow 26 (yellow mordant).

Solution 2 (Acetic acid fuchsin)

Acid fuchsin	1.0 g
Acetic acid 1 per cent	100.0 ml

Solution 3 (PTA solution)

Phosphotungstic acid	1.0 g
Distilled water	100.0 ml

Solution 4 (Light green solution)

Light green	2.0 g
Distilled water	100.0 ml

Solution 5 (Mordant)
    Yellow mordant (solution 1)   30.0 ml
    Ethanol (80 per cent)   70.0 ml

Procedure

1. Hydrate slides to water.
2. Immerse in solution 5 for 5 minutes.
3. Wash.
4. Stain in solution 2.
5. Rinse in tap water.
6. Transfer to solution 3.
7. Rinse in tap water.
8. Stain in solution 4 for 2 minutes.
9. Rinse in tap water.
10. Dehydrate, clear and mount.

Result

Nuclei	Blue-black
Collagen	Bluish grey
Fibrin	Red

Maritius–Scarlet–blue Method for Fibrin (Lendrum *et al.*, 1962)

Fixation

Any general fixative

Reagents Required

Maritius yellow
Phosphotungstic acid
Brilliant crystal scarlet 6R
Acetic acid

Preparation of Reagents

Solution 1 (Maritius yellow solution)
    Maritius yellow   500 mg
    Ethanol 95 per cent   100.0 ml
    Phosphotungstic acid 2.0 g
Solution 2 (Crystal scarlet solution)
    Brilliant crystal scarlet   1.0 g
    2.5 per cent acetic acid   100.0 ml
Solution 3 (PTA solution)
    Phosphotungstic acid 1.0 g
    Distilled water   100.0 ml
Solution 4 (Soluble blue)
    Soluble blue   1.0 g
    1 per cent acetic acid   100.0 ml

Procedure

1. Hydrate slides to water.
2. Stain with solution 1 for 2 minutes.
3. Wash in water.
4. Stain in solution 2 for 10 minutes.
5. Wash.
6. Treat with solution 3 for 5 minutes.
7. Wash.
8. Treat with solution 4 for 10 minutes.
9. Wash.
10. Dehydrate, clear and mount.

Result

Fibrin	Red
Nuclei	Blue-black
Red cells	Yellow

Masson 44/41 Method for Fibrin (Lendrum *et al.*, 1962)

Fixation

Any general fixative

Reagents Required

Phosphotungstic acid
Picric acid
Acid red 44 LB
Mercuric chloride
Celestian blue
Brilliant crystal scarlet 6R
Naphthalene blue-black Cs

Procedure

1. Remove wax with xylene. Rinse with trichloroethylene and immerse in a closed jar of trichloroethylene for 48 hours.
2. Rinse in absolute alcohol.
3. Place in a closed jar containing absolute ethanol saturated with picric acid containing 3 per cent mercuric chloride.
4. Wash and remove mercury by treating with Lugol's iodine followed by thiosulphate.
5. Wash till yellow colour disappears.
6. Stain nuclei with celestian blue haemalum sequence. Rinse in tap water.
7. Differentiate with 0.25 per cent HCl in 70 per cent ethanol.
8. Wash in running water.
9. Stain with brilliant crystal scarlet 6R in 1 per cent acetic acid for 5 minutes.
10. Rinse in water and treat with 1 per cent PTA for 5 minutes.

11. Rinse in water, treat with 1 per cent naphthalene blue black CS in 1 per cent aqueous acetic acid for 30 minutes.

12. Wash, dehydrate, clear and mount in DPX.

Result

Nuclei	Black
Red cells	Red
Fibrin	Deep black

Fraser-Lendrum Method for Fibrin

Fixation

Zenker's fixative

Reagents Required

Celestian blue
Ferric ammonium sulphate
Haematoxylin
Orange G
Picric acid
Acid fuchsin
Acetic acid
Phosphotungstic acid
Light green

Preparation of Reagents

Solution 1 (Celestian blue)

Ferric ammonium sulphate	2.5 g
Distilled water	50.0 ml

Dissolve for 12 hours at room temperature and add

Celestian blue   250 mg

Boil for 3 min and filter. Add 7 ml glycerine.

Solution 2 (Mayer's haematoxylin (*see* Chapter 7))

Solution 3 (Orange G picric acid solution)

Saturated picric acid 80 per cent	200.0 ml
Orange G	400 mg

Solution 4 (Acid fuchsin solution)

Acid fuchsin	1.0 g
Distilled water	99.0 ml
Glacial acetic acid	1.0 ml

Solution 5 (MacFarlanes solution stock)

Phosphotungstic acid	25.0 g
Picric acid	2.5 g
95 per cent alcohol	100.0 ml

Solution 6 (Working solution)
Solution 5	40.0 ml
95 per cent alcohol	40.0 ml
Distilled water	20.0 ml

Solution 7
Light green	2.0 g
Distilled water	98.0 ml
Glacial acetic acid	1.0 ml

Solution 8
Solution 3	30.0 ml
80 per cent alcohol	70.0 ml

Procedure

1. Deparaffinize and hydrate slides to water.
2. Transfer to solution 1 for 5 minutes.
3. Wash in tap water.
4. Transfer to solution 2 for 5 minutes.
5. Wash in tap water.
6. Transfer to solution 3 for 5 minutes.
7. Wash in tap water.
8. Keep slides in solution 4 for 5 minutes.
9. Wash in tap water.
10. Transfer to solution 8 for 15 seconds.
11. Wash in tap water.
12. Transfer to solution 6 for 5 minutes.
13. Wash in tap water.
14. Counterstain in solution 7 for 1 minute.
15. Dehydrate, clean and mount.

Result

Fibrin, keratin	Red
Collagen	Green
Erythrocytes	Orange

Ayoub Shiklar Method for Keratin and Pre-keratin (Ayoub and Shiklar, 1963)

Fixation

10 per cent buffered neutral formalin

Reagents Required

Fuchsin
Aniline blue
Orange G

Preparation of Reagents

Solution 1 (Acid fuchsin solution)
    Acid fuchsin   5.0 g
    Distilled water   100.0 ml
Solution 2 (Aniline blue–Orange G solution) (*see* page 528)

Procedure

1. Deparaffinize and hydrate slides to water.
2. Transfer to solution 1 for 5 minutes.
3. Transfer directly to solution 2 for 45 minutes.
4. Transfer direct to 95 per cent alcohol.
5. Dehydrate, clear and mount.

Result

Keratin   Brilliant red
Connective tissue   Deep blue

Dane's Method for Pre-keratin, Keratin and Mucin (Dane and Herman, 1963)

Fixation

10 per cent buffered formalin

Reagents Required

Haematoxylin
Phloxine
Alcian blue
Orange G
Glacial acetic acid
Phosphotungstic acid

Preparation of Reagents

Solution 1 (Mayer's haematoxylin (*see* Chapter 7))
Solution 2
    Phloxine   1.0 g
    Distilled water   100.0 ml

Solution 3 (Alcian blue solution)
    Alcian blue   50.0 ml
    Glacial acetic acid   50.0 ml

Solution 4
    Orange G   0.5 g
    Phosphotungstic acid 2.0 g
    Distilled water   100.0 ml

Procedure

1. Deparaffinize and hydrate slides to water.
2. Transfer to solution 1 for 10 minutes.
3. Wash in running water.
4. Transfer to solution 2 for 5 minutes.
5. Wash in running water.
6. Rinse in distilled water.
7. Transfer to solution 3 for 5 minutes.
8. Wash in tap water.
9. Transfer to solution 4 for 15 minutes.
10. Transfer to 95 per cent alcohol.
11. Dehydrate, clear and mount.

Result

Acid mucopolysaccharides	Torquoise blue
Pre-keratin and keratin	Orange-red, orange
Nuclei	Orange to brown

Rhodamine B Method for Keratin (Plate 18, Figure 5; Plate 20, Figure 1)

Fixation

Any fixative or 10 per cent formalin

Reagents Required

Rhodamine B
McIlvaine buffer (*see* Chapter 5)
Toluidine blue

Preparation of Reagents

Solution 1
   Toluidine blue   100 mg
   Distilled water   100 ml

Solution 2
   Rhodamine B   100 mg
   McIlvaine's buffer (pH 3.6)   100 ml

Procedure

1. Deparaffinize and hydrate slides to water.
2. Transfer to solution 1 for 10 minutes.
3. Wash in distilled water.
4. Transfer to solution 2 for 10 minutes.

5. Wash in distilled water.

6. Dehydrate, clear and mount.

Result

Keratin is stained red.

SUGGESTED READINGS

Poley and Forbes (1964), Puchtler *et al.* (1969), Laurie *et al.* (1982), Laurie and Leblond (1982a, 1982b), Warburton *et al.* (1982), Horton (1984), Horton *et al.* (1984), Rojkund *et al.* (1984), Sternberg *et al.* (1985), Churukian (1993), Shaprio and Solins (1994), Gowali (1995), and Hanumantha Rao and Shyamasundari (2003).

Table 19.1 Histochemical techniques applied for the demonstsration of connective tissue

Technique	Fixative	Muscle	Basement membrane	Elastin	Collagen	Reticulin	Fibrin
Pantin method	Any general fixative	Red to orange			Dark blue		
Azan stain	Zenker formol	Red and yellow			Blue	Blue	
Mallory Rapid method	Any general fixative			Yellow	Blue		Red
Masson's Trichrome stain	Any general fixative	Red			Green		
Pollak's Rapid method	Any general fixative	Red		Red	Bluish green		Pale blue
Putchler and Sweat method	Carnoy		Black				
Verhoeff's Elastin stain	Any general fixative			Brilliant black	Red		
Iron Gallein Elastin stain	Any general fixative			Black			
Orcinol–new fuchsin	Any general fixative			Deep violet	Unstained		
Aldehyde fuchsin	Any general fixative without dichromate			Deep purple			
Orcein method	Any general fixative			Red			

Table 19.1   (Continued)

Technique	Fixative	Muscle	Basement membrane	Elastin	Collagen	Reticulin	Fibrin
Safranin–Orcein method	Any general fixative			Reddish brown			
Orcein and Van Gieson	Any general fixative	Yellow		Brown	Red		
Resorcinol Fuchsin method	Any general fixative			Brownish violet			
Weigert Resorcin Fuchsin	Any general fixative			Dark blue or black			
Aldehyde Fuchsin–Luxol Fast Yellow–Pontacyl blue-black	Bouin's			Deep red	Yellow		

Table 19.2 Histochemical techniques applied for the demonstration of connective tissue

Technique	Fixative	Elastin	Collagen	Reticulin	Fibrin	Pre-keratin	Keratin	Muscle
Elastin-Trichrome stain	Any general fixative	Blue-black	Green					
Orcein–aniline blue–Orange G	Bouin's	Red	Blue				Bright yellow	
Orcein–aniline blue-safranin	Bouin's	Reddish brown						
Sheridan's Resorcin–Crystal Violet method	Any general	Blue-green						
Hart's modification of Weigert	Any general	Blue-black						
Fraenkel's method	Any general	Dark brown	Blue-green					
Pinkus acid-Orcein-Giemsa	10% neutral buffered formalin	Dark brown	Rose					
Gomori's Aldehyde Fuchsin	10% neutral buffered formalin	Deep purple						

(Contd.)

Table 19.2 (Continued)

Technique	Fixative	Elastin	Collagen	Reticulin	Fibrin	Pre-keratin	Keratin	Muscle
Kornhauser	Zenker	Red-brown	Green					
Picroponceau with haematoxylin	Any general	Yellow	Red					
Van Gieson's Picrofuchsin	Alcohol		Bright red					
Biebrich Scarlet–Picroaniline Blue	Any general							Pink
Mallory's Aniline Blue	Zenker, mercury-containing fixative		Intense blue					
Haemalum–Eosin method	Any general		Deep pink					
Haematoxylin–Van Gieson	Bouin's, Carnoy		Bright red					
Safranin–Water Blue	1% picric acid or alcohol		Blue					
Hart's method	10% neutral buffered formalin	Blue-black	Pink or red					

Van Gieson/ haematoxylin	Any general	Red	
Van Gieson/Trypan blue	Formalin	Red	
Allochrome method	Formalin	Blue	Blue
Del Rio-Hortega method	Any general fixative	Red to rose	Black
Gridley's method	Any general fixative		Black
Gomori's method	10% neutral buffered formalin		Black
Laidlow's method	Bouin's	Red	Black
Nassar and Shanklin's method	Any general fixative		Black
Wilder's method	Any general fixative		Black
Foot's Silver impregnation	10% formalin		Black

Table 19.2   (Continued)

Technique	Fixative	Elastin	Collagen	Reticulin	Fibrin	Pre-keratin	Keratin	Muscle
Naoumenko's and Feigin's method	Any general 10% formalin			Black				
Silver impregnation	10% formalin		Yellow	Black				
Menzies method	10% buffered formalin			Blue				
Snook's method	10% buffered formalin			Grey to black				
Mallory's PTAH	Any general fixative		Rose-red		Dark blue			
Picro Mallory	Any general fixative		Bluish grey		Red			
Mauritius–Scarlet–Blue	Any general fixative				Red			
Masson's 44/41 method	Any general fixative				Deep black			
Fraser-Lendrum method	Zenkers		Green		Red		Red	
Ayoub Shiklar method	10% buffered neutral formalin						Brillia nt red	
Dane's method	10% buffered formalin					Orange-red	Orange	

NEUROLOGICAL STUDIES

20

When nervous tissue is processed through alcohols and xylene, there is every chance of lipids being removed which results in no impregnation of oligodendroglia and microglia. So frozen sections are recommended for nervous tissue. Action of microglia and oligodendroglia is weakened by oxidation with periodic acid and chromic acid. But these methods are suggested for connective tissue.

Myelin has a high content of cholesterol, cerebroside and phospholipids. When treated with alcohols for dehydration and xylene for clearing, most lipid material is removed. However, with osmium tetroxide (though it colours it black) fixation is effective and normal dehydration, clearing and embedding can be carried out. Myelin can be preserved when tissue is post-chromated in 3 per cent potassium dichromate for 12–15 hours.

Nervous tissue is of soft consistency due to the absence of supporting tissue commonly seen in ordinary connective tissue with tough substances like collagen and elastin. Only neurological cells and processes support the nerve tissue by lying between and binding together the nerve fibres and blood vessels. Because of this factor, the tissue has to be fixed for sufficient time and hardened so that the sections will adhere to the slides while staining.

Thus when nervous tissue has to be paraffin-embedded, long periods of fixation, dehydration, clearing and embedding may be necessary. Many technicians avoid highly specialized neurological technique where duration of fixation, composition of fixative, washing time, etc., have to be carefully planned.

Glass distilled water is recommended for preparing silver solution. Reagent-grade chemicals have to be used. Metal containers and rubber corks have to be avoided. Fresh solutions are necessary especially gold chloride which leaves an yellowish brown hue with ageing. Embedding and sectioning depend upon the staining technique. As stated earlier, frozen sections are the best, though paraffin and nitrocellulose methods may be followed. Subbed slides are necessary.

Some neurological elements are

1. Nervous tissue proper consisting of
   - Nerve cells
   - Nerve cell bodies
   - Nerve fibres
   - Nerve endings
2. Neuroglia which include
   - Macroglia
   - Fibrous astrocytes
   - Oligodendroglia
   - Microglia

Nerve Tissue Proper

This constitutes the nerve tissue neurons (nerve cells) and their processes are called nerve fibres. The fibres called axons are long, and conduct impulses from the cells and dendrites. These axons extend and end by branching extensively into fine branches which come in contact with the next neuron where they meet the dendrites of another cell. This is a synapse.

Nerve cells   Neurons vary in size (5–150 m in diameter) and are categorized based on their processes as unipolar, bipolar and multipolar. Multipolar neurons are common, occurring in most part of the nervous system. Unipolar neurons are rare. Nucleus is large and takes a pale shade when stained and has a prominent nucleolus.

Within the cytoplasm of the nerve cells there are neurofibrils, Nissl substance, granules, mitochondria and Golgi complex. Neurofibrils and fine fibres spread through the cytoplasm. Nissl substance is distributed throughout the cytoplasm except at the origin of the axon sometimes extending into dendrites. It takes a deep stain with all basic dyes and when a nerve fibre is damaged, there may be loss of nissl substance. Mitochondria and Golgi complex can be demonstrated well in the nerve cell.

Ageing nerve cells display yellow lipofuchsin granules.

Nerve fibres   These contain axis cylinders (naked axons), axons with sheath and myelinated nerve fibres.

Axis cylinders are naked without any type of covering and are distributed in the grey matter and spinal cord.

Axons have a cellular sheath surrounding them and are distributed in the sympathetic nervous system. Nucleus stains intensely.

Myelinated nerve fibres are surrounded by a lipid sheath which in turn is surrounded by the neurilemmal sheath.

Nerve endings   Two types of nerve endings occur:

- Sensory endings transmitting impulses from the peripheral organs to the nervous system. These are dendrites of a nerve cell and they may end either as a complex branching organ or a simple branching organ.

- Motor endings transmit impulses from the nervous system to the peripheral organs. These are terminations of nerve axons.

Neuroglia

This is a type of connective (special) tissue composed of neuroglial fibres and cells. Neuroglial cells are of two types—macroglia and microglia.

1. Macroglia is ectodermal in origin and may be protoplasmic astrocytes, fibrous astrocytes or oligodendroglia.

   - Protoplasmic astrocytes have long branching processes which are protoplasmic in origin having a large nucleus and granular cytoplasm, mostly in grey matter. These are without fibres.

   - Fibrous astrocytes have thick fibres. They are attached to the capillaries and their processes are very long. They are more abundant in white matter.

   - Oligodendroglia are small cells without fibres or they are attached to the vascular system. Processes are very fine and the cells stain deeply.

2. Microglia are small oval cells with a deeply staining nucleus. A thick process arises from each cell, branches extensively and gives rise to small processes.

HISTOCHEMICAL TESTS FOR NERVOUS SYSTEM

Biels Chowsky's Method for Nerve Cells (Davenport *et al.*, 1934)

Fixation

10 per cent formalin for 48 hours

Reagents Required

Silver nitrate
Sodium hydroxide
Ammonia
Sodium thiosulphate

Preparation of Reagent

Take 5 ml of 20 per cent silver nitrate and to this add 6 drops of 40 per cent sodium hydroxide. A precipitate is formed. To this add strong ammonia (0.880) drop by drop until the precipitate dissolves. Now add distilled water to make it 25 ml.

Procedure

1. After fixation wash thoroughly.
2. Cut frozen sections.
3. Wash sections in distilled water for 2–4 hours.
4. Impregnate with aqueous silver nitrate for 3 days at 37°C.
5. Wash.
6. Treat with fresh ammoniacal silver nitrate for 20 minutes until a tobacco brown colour develops.
7. Rinse in distilled water.
8. Place in 10 per cent formalin in water for 10 minutes.
9. Wash in distilled water.
10. Tone in 0.2 per cent gold chloride for 15 minutes.
11. Rinse in distilled water.
12. Place in 5 per cent sodium thiosulphate for 5 minutes.
13. Wash in distilled water.
14. Float on clean slides, dehydrate, clear and mount in Canada balsam.

Result

Nerve cells are stained brown to black.

Ramony's Cajal's Gold Chloride Sublimate (Mallory, 1938)

Fixation

Formalin   15.0 ml
Ammonium bromide   2.0 g
Distilled water   85.0 ml

Fix for 24 hours at 37°C (Frozen sections).

Reagents Required

Mercuric chloride
Gold chloride
Creosote
Phenol
Xylene

Preparation of Reagents

Solution 1 (Mercuric gold chloride solution)
  Mercuric chloride   500 mg
  Gold chloride (1 per cent aqueous)   6.0 ml
  Distilled water   35.0 ml

Powder mercuric chloride and add distilled water to it. Heat gently and add gold chloride and filter.

Solution 2 (Del Rio Hortega's Carbol–Xylol Creosote)
    Creosote   10.0 ml
    Phenol   10.0 ml
    Xylene   80.0 ml

Procedure

1. Keep sections in 1 per cent formalin.
2. Wash.
3. Transfer the slides to freshly prepared gold chloride (Solution 1) and keep in dark at room temperature for 6 hours. A purplish colour develops.
4. Wash in distilled water.
5. Fix in 5 per cent sodium thiosulphate.
6. Wash in distilled water.
7. Dehydrate with 95 per cent alcohol.
8. Clear in del Rio Hortega's fluid (Solution 2).
9. Blot and mount.

Result

Nerve cells	Pale red
Astrocytes	Black

Nerve fibres are unstained.

Golgi Rapid Method

Fixation

Potassium dichromate   40.0 ml
(2.5 g in 100 ml water)
Osmic acid   10.0 ml
   (1 per cent aqueous)

Fix for a week. Renew fixative once in 2 days.

Reagents Required

Silver nitrate
Celloidin

Preparation of Stain

Solution 1 (Silver nitrate)
    Silver nitrate   750 mg
    Distilled water   100.0 ml

Procedure

1. Blot pieces of tissue on a filter paper and transfer to solution 1 for 24–48 hours.
2. Dehydrate in 40 per cent alcohol for 2 hours.
3. Dehydrate in 80 per cent alcohol for 1 hour.
4. Dehydrate in 80 per cent absolute alcohol for 12 hours.

5. Infiltrate with 4 per cent nitrocellulose for 24 to 48 hours.
6. Embed and cut 20–25 µm sections.
7. Wash sections in 80 per cent alcohol and absolute alcohol.
8. Clear in clove oil.
9. Mount.

Result

Nerve cells   Deep black
Back ground   Light yellow

Ramony's Cajal's Pyridine Silver Method (Modification by Davenport *et al.*, 1934)

Fixation

Absolute alcohol   98.0 ml
Conc. ammonia   2.0 ml

Fixation time is 1 week

Reagents Required

Silver nitrate
Pyrogallol

Procedure

1. Keep tissue blocks in 5 per cent aqueous pyridine for 24 hours.
2. Wash in distilled water for 8 hours.
3. Impregnate in 2 per cent silver nitrate (2 g/100 ml water) at 37°C for 72 hours. The longer the tissue is impregnated, the greater is the tendency for the tissue to stain.
4. Wash in distilled water for 1 hour.
5. Reduce in 4 per cent pyrogallol for 4 hours.
6. Dehydrate, clear, embed and section.
7. Mounted slides are deparaffinized, cleared and mounted.

Result

Nerve cells	Yellow
Neurofibrils	Brown to black
Axis cylinders of myelinated fibres	Yellow to brown
Axis cylinders of non-myelinated fibres	Black

Gros–Schultze Method for Fibres and Nerve Endings

Fixation

Frozen sections

Reagents Required

Silver nitrate
Ammonia

Preparation of Reagents

Same as for Biel Chowsky's method.

Procedure

1. Wash sections in distilled water.
2. Keep the slides in 20 per cent silver nitrate and place in a dark chamber for 5 minutes.
3. Place in 20 per cent formalin.
4. Impregnate in ammoniacal silver nitrate.
5. Place in 5 per cent ammonia.
6. Rinse in 1 per cent acetic acid.
7. Rinse in distilled water.
8. Tone in 0.02 per cent gold chloride for 10 minutes.
9. Wash in distilled water.
10. Transfer to 5 per cent sodium thiosulphate for 5 minutes.
11. Wash in tap water.
12. Dehydrate, clear and mount.

Result

Nerve cells and neurofibrils	Brown to black
Axons	Black

Nauta and Gygax Method (Modified by Powell and Brown, 1975)

Fixation

Perfusion method
1. Perfuse with 0.9 per cent sodium chloride.
2. Perfuse with 500 ml of 5 per cent aqueous potassium dichromate and 2.5 per cent aqueous potassium chlorate.
3. Fix in 10 per cent formalin neutralized with carbonate for 1 week (Nauta and Gygax recommend 1–3 months).

Reagents Required

Silver nitrate
Uranyl nitrate
Hydroquinone
Oxalic acid
Ammonium hydroxide
Sodium hydroxide
Citric acid
Formalin

Preparation of Reagents

Solution 1 (Bleaching solution)
1 per cent hydroquinone (fresh)	1 part
1 per cent oxalic acid	1 part

Solution 2 (Silver solution 1)
    0.5 per cent aqueous uranyl nitrate   30.0 ml
    2.5 per cent aqueous silver nitrate   36.0 ml
    Distilled water   56.0 ml

Solution 3 (Silver solution 2)
    0.5 per cent uranyl nitrate   40.0 ml
    2.5 per cent silver nitrate   60.0 ml

Solution 4 (Silver solution 3)
    2.5 per cent silver nitrate   60.0 ml
    Concentrated ammonium hydroxide   5.0 ml
    2.5 per cent aqueous sodium hydroxide   3.0 ml

Mix all solutions (2,3,4) just before use.

Solution 5 (Reducing solution)
    Distilled water   405.0 ml
    95 per cent alcohol   45.0 ml
    1 per cent acetic acid   13.5 ml
    Aqueous 10 per cent formalin   13.5 ml

Solution 6 (Gelatin alcohol)
    Dissolve 6 g of gelatin is dissolved in 400.0 ml of hot distilled water. Cool and add 400 ml of 80 per cent alcohol.

Procedure

1. Rinse sections in distilled water giving 4 changes of 5 minutes each.
2. Keep in 0.05 per cent potassium permanganate for 26 minutes.
3. Rinse in distilled water.
4. Keep in mixed silver solution for 30 seconds.
5. Rinse in distilled water over night.
6. Treat with fresh solution 2 for 30 minutes.
7. Transfer to fresh solution 3 for 30 minutes.
8. Rinse in distilled water.
9. Transfer to fresh solution 4 for 2 minutes.
10. Transfer to solution 5 for 2 minutes.
11. Fix in 0.5 per cent sodium sulphate for 1 minute.
12. Rinse in distilled water.
13. Place in solution 6 overnight.
14. Mount on slide.
15. Dehydrate and clear in 2 changes xylene and mount.

Result

Degenerating axon   Black
Normal axon   Brown

Holmes Method for Nerve Cells and Fibres (Holmes, 1943)

Fixation

10 per cent neutral buffered formalin

Reagents Required

Silver nitrate
Borax
Boric acid
Pyridine
Hydroquinone
Sodium sulphate
Gold chloride

Preparation of Reagents

Solution 1 (Silver nitrate)
Silver nitrate   22.0 g
Distilled water   100.0 ml

Solution 2 (Boric acid buffer)
Boric acid   12.4 g
Distilled water   1000.0 ml

Solution 3 (Borax buffer)
Sodium borate   19.0 ml
Distilled water   1000.0 ml

Solution 4 (1 per cent silver nitrate)
Silver nitrate   1.0 g
Distilled water   100.0 ml

Solution 5 (Pyridine solution)
Pyridine   10.0 ml
Distilled water   100.0 ml

Solution 6 (Impregnating solution)
Solution 2   55.0 ml
Solution 3   45.0 ml
Distilled water   394.4 ml
Solution 4   1.0 ml
Solution 5   5.0 ml

Solution 7 (Reducing solution)
Hydroquinone   1.0 g
Sodium sulphate   10.0 g
Distilled water   100.0 ml
This solution lasts for few days.

Solution 8 (Gold chloride)
1 per cent gold chloride   2.0 ml
Distilled water   100.0 ml

Solution 9 (Oxalic acid)

Procedure

1. Dewax and bring slides to water.
2. Transfer to solution 1 for 2 hours.
3. Rinse in distilled water giving 3 changes.
4. Treat with solution 6 at 37°C overnight.
5. Drain off excess solution.
6. Wash in running water.
7. Rinse in solution 7 for 2 minutes.
8. Wash.
9. Transfer to solution 8 for 5 minutes.
10. Rinse in distilled water.
11. Place in solution 9 for 5 minutes.
12. Wash.
13. Transfer to reducing solution for 5 minutes.
14. Wash, dehydrate, clear and mount.

Result

Axis cylinders	Blue to black
Nerve and nerve endings	Black
Background	Grey to rose

Hirano–Zimmerman Method for Nerve Cells and Fibres (1962)

Fixation

10 per cent neutral buffered formalin

Reagents Required

Silver nitrate
Ammonium hydroxide
Formalin
Gold chloride
Sodium thiosulphate

Preparation of Reagents

Solution 1 (Silver nitrate)
Silver nitrate	10.0 g
Distilled water	100.0 ml

Solution 2 (Ammonia water solution)
20 per cent ammonium hydroxide	2 drops
Distilled water	100.0 ml

Solution 3 (Formalin)
Formalin 40 per cent	50.0 ml
Tap water	50.0 ml

Solution 4 (Gold chloride solution)
     1 per cent gold chloride   10.0 ml
          (1 g/100 ml water)
     Distilled water   200.0 ml

Solution 5 (5 per cent sodium thiosulphate) (5 g/100 ml water)

Procedure

1. Dewax and bring down slides to water.
2. Transfer to solution 1 for 2 hours.
3. Transfer directly to solution 2 for 3 minutes.
4. Place sections directly in solution 3 for 3 minutes.
5. Wash in distilled water.
6. Transfer again to solution 1 for 5 minutes.
7. Tone in solution 4 for 20 minutes.
8. Transfer sections directly to solution 5.
9. Rinse in distilled water.
10. Dehydrate, clear and mount.

Result

Neurofibrils, dendrites and axis cylinders   Black
Cytoplasm of astrocytes   Grey

PTAH Method for Central Nervous System

(*see* Chapter 7)

Silver Impregnation for Nerve Tissue
(Garvey *et al.*, 1987; Staples, 1991)

Fixation

10 per cent neutral buffered formalin

Reagents Required

Silver nitrate
Gum mastic
Hydroquinone
Gold chloride
Sodium thiosulphate

Preparation of Reagents

Solution 1 (1 per cent silver nitrate)
     Silver nitrate   1.0 g
     Distilled water   100.0 ml

Solution 2 (3 per cent gum mastic)
     Gum mastic   3.0 g
     Absolute alcohol   100.0 ml

Solution 3 (2 per cent hydroquinone)
    Hydroquinone   0.25 g
    Distilled water   25.0 ml

Prepare it just before use.

Solution 4 (Developing solution)
    Solution 3   25.0 ml
    Solution 2   15.0 ml
    Solution 1   1.2 ml

First mix solution 3 and 2 and filter it. Add solution 1 just before use.

Solution 5 (0.25 per cent gold chloride)
    10 per cent gold chloride   25.0 ml
    Distilled water   75.0 ml

Solution 6 (1 per cent sodium thiosulphate)
    Sodium thiosulphate   1.0 g
    Distilled water   100.0 ml

Procedure

1. Dewax and hydrate slides to water.
2. Keep in solution 1 at 4°C for 30 minutes.
3. Rinse in distilled water giving 3 changes.
4. Rinse in 95 per cent alcohol giving 2 changes.
5. Rinse in absolute alcohol giving 3 changes.
6. Place slides in solution 2 for 5 minutes.
7. Rinse in absolute alcohol giving 3 changes.
8. Drain off and place slides in solution 4 at 60°C for 2 minutes.
9. Rinse in distilled water giving 3 changes.
10. Treat with solution 6 for 5 minutes.
11. Rinse in distilled water.
12. Keep in solution 5 for 2 minutes.
13. Rinse in water.
14. Dehydrate, clear and mount.

Result

Axons, dendrites, neurofibrils   Black
Background   Grey

Cresyl Violet Method (Powers and Clark, 1955)

Fixation

Bouin's fixative

Reagents

Cresyl violet
Acetic acid
Sodium acetate

Preparation

Solution 1 (Cresyl violet solution)
Cresyl violet acetate 200 mg
Distilled water 100.0 ml

Solution 2 (Buffer)
0.1 M acetic acid (6 ml glacial acetic acid/1000 ml water) 91.0 ml
0.1 M sodium acetate (13.6 g of 6.0 ml
sodium acetate/1000 ml water)

Solution 3 (Working solution)
Solution 1 6–12 ml
Solution 2 100.0 ml

Procedure

1. Dewax and hydrate slides to water.
2. Transfer to solution 3 for 20 minutes.
3. Rinse in 70 per cent alcohol and then 95 per cent alcohol.
4. Dehydrate in isopropyl alcohol.
5. Clear and mount.

Result

Nissl substance is stained purple.

Thionin Method (Clark and Sperry, 1945)

Fixation

Bouin's fixative

Reagents Required

Lithium carbonate
Thionin
Butyl alcohol

Preparation of Reagents

Solution 1 (Lithium carbonate)
Lithium carbonate 5.5 g
Distilled water 1000.0 ml

Solution 2 (Thionin solution)
Thionin 250 mg
0.55 per cent lithium carbonate 100.0 ml

Procedure

1. Hydrate slides to water.
2. Transfer to solution 1 for 5 minutes.
3. Overstain in solution 2 for 10 minutes.

4. Rinse in distilled water.
5. Dehydrate in butyl alcohol, clear and mount.

Result

Nissl susbtance is stained bright blue.

Gallocyanin Method

Fixation

Any general fixative

Reagents Required

Gallocyanin
Chrome alum

Preparation of Reagent

Solution 1 (Gallocyanin–chrome alum)
    Gallocyanin   150 mg
    Chrome alum   100.0 ml
    (15 per cent aqueous)

Procedure

1. Deparaffinize and hydrate slides to water.
2. Transfer to solution 1 for 3 hours at 56°C overnight at room temperature.
3. Wash.
4. Dehydrate, clear and mount.

Result

Nissl substance is stained blue.

Thionin and Toluidine Blue Method

Fixation

Alcohol or formol saline

Reagents Required

Creosote
Cajuput oil
Xylol
Absolute alcohol

Preparation

Solution 1 (Gothard's differentiator)
    Creosote   50.0 ml
    Cajuput oil   40.0 ml

        Xylol   50.0 ml
        Absolute alcohol   150.0 ml

Procedure

1. Deparaffinize and hydrate slides to water.
2. Transfer to 1 per cent toluidine blue in 1 per cent borax at 50°C for 30 minutes.
3. Rinse in water.
4. Treat with 90 per cent alcohol for 30 seconds.
5. Differentiate in solution 1.
6. Rinse in absolute alcohol.
7. Clear and mount.

Result

Nissl substance   Deep blue
Nucleoli   Blue

Einarson's Gallocyanin Method (1932)

Fixation

Formol saline. Carnoy's fixative is preferable.

Reagents Required

Gallocyanin
Chrome alum

Preparation

Solution 1 (Gallocyanin–Chrome alum)
        Gallocyanin   300 mg
        Chrome alum   10.0 g
        Distilled water   100.0 ml
Heat chrome alum in distilled water. Then add gallocyanin. Cool it and add distilled water and filter.

Procedure

1. Deparaffinize and hydrate slides to water.
2. Stain with solution 1 (gallocyanin) overnight.
3. Rinse in water.
4. Dehydrate, clear and mount.

Result

Nissl substance   Deep blue
Nuclei   Deep blue

Weigert–Pal Technique

Fixation

Formol saline

Reagents Required

Potassium dichromate
Fluorochrome
Copper acetate
Haematoxylin
Glacial acetic acid
Oxalic acid
Sodium sulphate

Preparation

Solution 1 (Potassium dichromate)

Potassium dichromate	5.0 g
Fluorochrome (Chromium fluoride)	2.5 g
Distilled water	100.0 ml

Boil water, add dichromate and then add fluorochrome. Cool and filter.

Solution 2 (Gliabeize)

Fluorochrome	2.5 g
Copper acetate	5.0 g
Distilled water	100.0 ml

Boil water and add fluorochrome and copper acetate. When boiling, add 3 ml of acetic acid, cool and filter.

Solution 3 (Kultschitzsky's haematoxylin)

Ripened 10 per cent haematoxylin in absolute alcohol	10.0 ml
Glacial acetic acid	2.0 ml
Distilled water	90.0 ml

Solution 4 (Pal's solution)

Oxalic acid	1.0 g
Sodium sulphate	1.0 g
Distilled water	100.0 ml

Procedure

1. Hydrate slides to water.
2. Place them in solution 1 and mordant for at least 10 days.
3. Wash overnight.
4. Dehydrate and embed in celloidin.
5. Cut 12–20 μ sections.
6. Keep the slides in solution 2 for 30 minutes.
7. Wash in water.
8. Transfer to solution 3 for 12–24 hours.
9. Rinse in distilled water.
10. Transfer to 0.25 per cent potassium permanganate for 1 minute.
11. Rinse in water.
12. Transfer to solution 4 for 2 minutes.

13. Wash.
14. Dehydrate, clear and mount.

Results

Myelin sheaths   Deep blue
Others   Colourless

Luxol Fast Blue–Cresyl Violet for Myelin (Klüver and Barrera, 1953)

Fixation

10 per cent formalin

Reagents Required

Luxol fast blue
Acetic acid
Isopropanol
Cresyl violet
Lithium carbonate
Chloroform

Preparation of Reagents

Solution 1 (Luxol fast blue)
Luxol fast blue   1.0 g
95 per cent alcohol   1000.0 ml
10 per cent acetic acid   5.0 ml

Solution 2 (Alternative for Luxol fast blue)
Isopropanol   100.0 ml
Luxol fast blue   100 mg

Solution 3 (Cresyl violet)
Cresyl violet   100 mg
Distilled water   100.0 ml

Solution 4
Lithium carbonate   5 mg
Distilled water   1000.0 ml

Solution 5 (Cresyl violet differentiator)
90 per cent alcohol   90.0 ml
Chloroform   10.0 ml
Glacial acetic acid   3 drops

Procedure

1. Deparaffinize and bring slides to 95 per cent alcohol.
2. Transfer to solution 1 or 2 for overnight at 37°C.
3. Differentiate in 95 per cent alcohol.
4. Wash in water.

5. Immerse slides in solution 4 (lithium carbonate) to differentiate for 20 seconds.
6. Differentiate in 70 per cent alcohol.
7. Rinse in distilled water.
8. Transfer to solution 3 for 10 minutes
9. Wash in distilled water.
10. Differentiate in solution 5 for 2 seconds.
11. Rinse in 95 per cent alcohol.
12. Rinse in absolute alcohol.
13. Clear in xylol and mount.

Results

Myelin   Blue
Nuclei   Purple

Luxol Fast Blue–PAS

Fixation

10 per cent formalin
10 per cent formalin or Formol calcium
Calcium chloride   10.0 g
Distilled water   900.0 ml

After calcium chloride dissolves, add 5.0 g of cetyl trimethyl ammonium bromide. Then add 100.0 ml of concentrated formalin (40 per cent).

Reagents Required

Luxol Fast blue
Acetic acid
Isopropyl alcohol
Periodic acid
Lithium carbonate
Schiff's reagent
Hydrochloric acid
Mayer's haematoxylin

Preparation of Reagents

Proceed as in previous case.

Procedure

Proceed as in the previous case. After 8th step, apply PAS technique (*see* Chapter 9)

Luxol Fast Blue–Phosphotungstic Acid–Haematoxylin Method

Reagents Required

As in previous experiment

Procedure

1. Deparaffinize and hydrate slides to water.
2. Keep the slides in 0.25 per cent potassium permanganate for 5 minutes.
3. Bleach in 5 per cent oxalic acid for 5 minutes
4. Wash in distilled water.
5. Then proceed as in the previous case (steps 1–7).
6. Proceed for PTA haematoxylin technique (*see* Chapter 19).

Results

Myelin   Blue
Nuclei   Purple

Luxol Fast Blue–Holmes Silver Nitrate (Margolie and Pickett, 1956)

Fixation

10 per cent formalin

Reagents

Silver nitrate
Boric acid
Borax
Hydroquinone
Sodium sulphite

Preparation

Solution 1 (Silver nitrate)
    Silver nitrate   20.0 g
    Distilled water   100.0 ml

Solution 2 (Boric acid)
    Boric acid   12.4 g
    Distilled water   1000.0 ml

Solution 3
    Borax   19.0 g
    Distilled water   1000.0 ml

Solution 4 (Impregnating fluid)
    Solution 2   55.0 ml
    Solution 3   45 ml
    10 per cent silver nitrate   1.0 ml
    10 per cent pyridine   5.0 ml

Solution 5 (Reducer)
    Hydroquinone   1.0 g
    Sodium sulphite   10.0 g
    Distilled water   100.0 ml

Procedure

1. Deparaffinize and bring slides to water.
2. Place in solution 1 for 1 hour in a dark chamber.
3. Wash in distilled water.
4. Transfer to solution 4 at 37°C overnight.
5. Treat with solution 5 for 2 minutes.
6. Wash.
7. Tone in gold chloride (0.2 per cent) for 3 minutes
8. Rinse.
9. Transfer to 2 per cent oxalic acid for 3 minutes
10. Wash.
11. Fix in sodium thiosulphate.
12. Wash.
13. Stain in Luxol fast blue overnight.
14. Rinse in 95 per cent alcohol.
15. Transfer to lithium carbonate (as in previous).
16. Differentiate in 70 per cent and 90 per cent alcohol.
17. Dehydrate, clear and mount.

Result

Myelin	Greenish blue
Axis cylinders	Black

Weil Stain (Berube *et al.*, 1965)

Fixation

10 per cent formalin

Reagents Required

Haematoxylin
Ferric ammonium sulphate
Potassium ferricyanide
Darrow red
Borax

Preparation of Reagents

Solution 1 (Haematoxylin)
Haematoxylin   1.0 g
Ethyl alcohol   100.0 ml

Solution 2 (Ferric ammonium sulphate)
Ferric ammonium sulphate   4.0 g
Distilled water   100.0 ml

Solution 3 (Potassium ferricyanide)

Potassium ferricyanide	12.5 g
Borax	2.5 g
Distilled water	1000.0 ml

Solution 4 (Staining solution)

Solution 1	1 part
Solution 2	1 part
Distilled water	3 parts

Procedure

1. Deparaffinize and bring slides to water.
2. Wash in distilled water.
3. Place in solution 4. After 20 minutes a precipitate develops.
4. Wash in water.
5. Differentiate in solution 2.
6. Wash in distilled water.
7. Differentiate in solution 3.
8. Wash in several changes of distilled water.
9. Counterstain if desired in darrow red.
10. Dehydrate, clear and mount.

Result

Myelin sheaths are stained dark blue or black.

Lapham's Method for Myelin and Glial Fibres (Lapham *et al.*, 1964)

Fixation

10 per cent buffered formalin

Reagents Required

Gallocyanin
Chrome alum
Phloxine B
Phosphotungstic acid
Glacial acetic acid
Fast green

Preparation of Reagents

Solution 1 (Chrome alum solution)

Chrome alum	10.0 g
Distilled water	200.0 ml
Gallocyanin	300 mg

Heat slowly to boiling for 10–35 minutes with stirring. Cool and filter.

Solution 2 (Phloxine)
(0.5 per cent phloxine (500 mg in 100 ml of water))

Solution 3 (5 per cent Phosphotungstic acid (5 g in 100 ml of water))

Solution 4 (1 per cent Glacial acetic acid (1 g in 100 ml of water))

Solution 5 (0.05 per cent Fast green)
    Fast green FCF   50 mg
    Distilled water   100.0 ml

Procedure

1. Dewax and hydrate slides to water.
2. Transfer to solution 1 and keep overnight.
3. Wash in running water for 5 minutes.
4. Transfer to solution 2 for 5 minutes.
5. Wash in running water.
6. Place in solution 3 for 1 minute.
7. Wash in running water.
8. Treat with solution 4 for 2 minutes.
9. Differentiate in 80 per cent alcohol.
10. Again put it back in solution 4 for 2 minutes.
11. Directly transfer to solution 5 for 1 minute.
12. Place it back in solution 4 for 1 minute.
13. Dehydrate, clear and mount.

Results

Myelin	Magenta
Glial fibres	Bright green
Nuclei	Blue-black
Collagen	Blue-green
Erythrocytes	Orange to red

Woelcke's Method for Myelin Sheath (Clark and Ward, 1934)

Fixation

10 per cent buffered formalin

Reagents Required

Ferric alum
Haematoxylin
Lithium carbonate

Preparation of Reagents

Solution 1 (2.5 per cent ferric alum)
    Ferric alum   2 g
    Distilled water   100.0 ml

Solution 2 (10 per cent alcoholic haematoxylin stock)
    Haematoxylin   10.0 g
    100 per cent alcohol   100.0 ml (ripen for 6 months)

Solution 3 (Saturated lithium carbonate solution)

Solution 4 (Haematoxylin working solution)
Distilled water   45.0 ml
Solution 2 (Haematoxylin stock) 10.0 ml
Solution 3 (Lithium carbonate)   7.0 ml

Procedure

1. Dewax and hydrate slides to water.
2. Transfer slides to solution 1 and keep overnight.
3. Rinse in distilled water.
4. Transfer to solution 4 for 2 hours.
5. Rinse in distilled water.
6. Rinse in 80 per cent alcohol.
7. Dehydrate, clear and mount.

Result

Myelin sheath   Blue
Glial cells   Black
Nucleoli of nerve cells   Black

Chromic Acid Haematoxylin Method (Lillie and Henderson, 1968)

Fixation

10 per cent formalin

Reagents Required

Chromic acid
Haematoxylin
Glacial acetic acid
Borax
Potassium ferricyanide

Preparation of Reagents

Solution 1 (Chromic acid)
Chromic acid   200 mg
Distilled water 100.0 ml

Solution 2 (Acetic haematoxylin)
1 per cent stock alcoholic haematoxylin 10.0 ml
Distilled water   89.0 ml
Glacial acetic acid   1.0 ml
        or
1 per cent acetic acid   100.0 ml
Haematoxylin   100.0 mg

Solution 3 (Borax–ferricyanide)
Borax   1.0 g

Potassium ferricyanide   1.0 g
Distilled water   100.0 ml

Procedure

1. Cut 10–15 μm frozen sections and place them on subbed slide.
2. Immerse in 1 per cent aqueous gelatin for 2 minutes.
3. Drain and later fix gelatin in formalin vapour for 30 minutes.
4. Place in 10 per cent formalin.
5. Wash in running water for 10 minutes.
6. Place in solution 1 (for mordanting) for 4 hours.
7. Wash in running water.
8. Treat with solution 2 for 2 hours.
9. Rinse in distilled water.
10. Differentiate in solution 3 for 1 hour.
11. Dehydrate, clear and mount.

Result

Myelin is stained blue-black.

DEGENERATING MYELIN

Marchis Method (Swank and Davenport, 1935)

Fixation

Formol calcium–post-chromed for 4–8 days in 3 per cent $K_2Cr_2O_7$.

Reagents Required

Potassium dichromate
Osmium tetroxide

Preparation of Reagents

Solution 1 (Marchis fluid)
3 per cent potassium dichromate   40.0 ml
1 per cent osmium tetroxide   20 ml

Procedure

1. Fix tissue in formalin and post-chrome for 8 days in 3 per cent potassium dichromate.
2. Transfer to solution 1 for 12 days.
3. Wash in running water.
4. Dehydrate and embed in celloidin or paraffin.
5. Cut 30-μm sections and mount them in Canada balsam.

Result

Degenerating myelin stains black.

Swank Davenport Method (1935)

Fixation

Formol saline

Reagents Required

Potassium chlorate
Osmium tetroxide
Concentrated formalin

Preparation of Reagents

Solution 1

1 per cent potassium chlorate	60 ml
(1 g + 100 ml water)	
1 per cent osmium tetroxide	20 ml
(1 g + 100 ml water)	
Formalin	10.0 ml
Glacial acetic acid	1.0 ml

Procedure

1. After fixation, transfer to solution 1 for one week.
2. Wash in running water.
3. Embed in celloidin.
4. Cut 30-μ sections and mount in Gurr's neutral mountant.

Results

Degenerating myelin	Black
Neutral fats	Black

Luxol Fast Blue–Oil red O technique

Reagents required

Luxol fast Blue
Lithium Carbonate
Oil redO

Preparation of Reagents

Solution 1
    Luxol fast blue   1.0 g
    Isopropanol   100 ml
Solution 2
    Lithium carbonate   5–0.005 mg
    Distilled water   100 ml
Solution 3
    Oil red O

Procedure

1. Cut frozen sections of formalin-fixed tissue about 15–20 µm thick.
2. Rinse in 70 per cent alcohol.
3. Transfer to luxol fast blue solution for 3 hours at room temperature.
4. Rinse in 70 per cent alcohol.
5. Rinse in distilled water.
6. Differentiate in lithium carbonate for 20 seconds.
7. Rinse in 70 per cent alcohol.
8. Wash in distilled water.
9. Stain in Lillie and Ashburn's Oil Red O from stages 3–10.

Result

Normal myelin   Blue-green
Degenerating myelin   Red

Guillery's Method for Degenerating Nerve Fibres (Guillery *et al.*, 1961)

Fixation

10 per cent formol saline by perfusion

Reagents Required

Ammonium hydroxide
Silver nitrate
Sodium hydroxide
Pyridine
Citric acid
Formalin

Preparation of Reagents

Solution 1 (Ammonium hydroxide)
    28 per cent ammonium hydroxide   1.0 ml
    50 per cent alcohol   99.0 ml

Solution 2 (Silver nitrate)
    1.5 per cent silver nitrate

Solution 3 (Silver nitrate–pyridine)
    Pyridine   5.0 ml
    Silver nitrate solution   95.0 ml

Solution 4 (Silver nitrate stock)
    Silver nitrate   4.5 g
    Distilled water   100.0 ml

Solution 5 (Sodium hydroxide stock)
    Sodium hydroxide 2.5 g
    Distilled water   100.0 ml

Solution 6 (Ammoniacal silver solution)

Solution 4	20.0 ml
100 per cent alcohol	10.0 ml
Solution 1	1.5 ml
Solution 5	1.5 ml

This solution should be placed in a warm dish on a hot plate.

Solution 7 (Reducing solution)

Citric acid	1.0 g
Distilled water	100.0 ml
Formalin 10 per cent	100.0 ml

Mix equal parts of these two solutions and add 450.0 ml of 10 per cent alcohol.

Solution 8 (Hypo (see Chapter 5))

Procedure

1. Dewax and hydrate slides to water.
2. Transfer to solution 1 for 6 hours.
3. Rinse in distilled water.
4. Transfer to solution 3 for 24 hours in the dark.
5. Treat with pre-heated solution 6.
6. Rinse in distilled water.
7. Reduce in solution 7.
8. Treat with solution 8 for 5 minutes.
9. Wash.
10. Dehydrate, clear and mount.

Results

Degenerating fibres	Black
Normal fibres	Light black

Anderson's Victoria Blue Method

Fixation

Frozen or formalin-fixed

Reagents Required

Sodium sulphite
Oxalic acid
Potassium iodide
Iodine
Acetic acid
Potassium permanganate
Victoria blue
Lugol's iodine
Aniline oil

Preparation of Reagents

Solution 1 (Neuroglia mordant)

Sodium sulphite	5.0 g
Oxalic acid	2.5 g
Potassium iodide	5.0 g
Iodine	2.5 g
Distilled water	100.0 ml

Mix them in order and add 5 ml of acetic acid. The solution tends to turn brown, then add two crystals of sodium sulphite.

Procedure

1. Frozen sections are washed, paraffin sections are deparaffinized and hydrated to water.
2. Place slides in solution 1 with addition of 5 per cent ferric chloride for 15 minutes.
3. Wash.
4. Oxidize in 0.25 per cent potassium permanganate.
5. Place them in Pal's solution (Oxalic acid 1 g + sodium sulphite 1 g + distilled water 100 ml) for 5 minutes till the sections become colourless.
6. Rinse in distilled water.
7. Float frozen sections onto albuminized slides.
8. Transfer the slides to a boiled 1.5 per cent Victoria blue or pour the solution onto the slides for 5 minutes.
9. Drain excess stain and blot.
10. If the sections are not clear, flood with xylol.
11. Differentiate sections with a mixture of equal parts of aniline oil and xylol. Examine under microscope. Astrocytes and neuroglial fibres should be blue.
12. Wash with xylol and remove aniline oil.
13. Mount in Canada balsam.

Results

Neuroglial fibres and astrocytes	Blue
Cell nuclei	Blue

Cajal's Gold Sublimate for Astrocytes

Fixation

Formol ammonium bromide

Reagents Required

Formalin
Ammonium bromide
Mercuric chloride
Gold chloride
Sodium thiosulphate

Preparation of Reagents

Solution 1 (Formol ammonium bromide FAB)
Formalin	15.0 ml
Ammonium bromide	2.0 g
Distilled water	85.0 ml

Solution 2 (Gold sublimate)
Mercuric chloride	400 mg
1 per cent gold chloride	10.0 ml
Distilled water	60.0 ml

Heat distilled water and add mercuric chloride, heating till it dissolves. Cool and then add gold chloride. This solution must be prepared afresh.

Procedure

1. Cut frozen sections 20 μ thick.
2. Sections can be stored in solution 1 till required.
3. Rinse in distilled water.
4. Transfer to solution 2. Keep the slides in a petri dish and pour solution 2. Keep them at 22°C for 48 hours until they are deep purple.
5. Rinse (rapidly) in distilled water.
6. Transfer to sodium thiosulphate (5 per cent).
7. Wash in distilled water.
8. Dehydrate, clear and mount.

Results

Astrocytes are stained reddish black.

Scharenberg's Triple Impregnation Method for Glial Cells

Fixation

10 per cent formol saline

Reagents Required

Silver nitrate
Ammonia
Pyridine
Sodium carbonate

Preparation of Reagents

Solution 1 (Silver carbonate solution)
10 per cent silver nitrate	100.0 ml
(10 g/100 ml water)	
5 per cent sodium carbonate	300.0 ml
(5 g/100 ml water)	

Add ammonia drop by drop till the precipitate formed dissolves. Make up the solution to 700 ml. It lasts for 3 months. Keep in a light dark bottle.

Solution 2

Take 2 g of silver nitrate and add 100.0 ml water. To this add ammonia (strong) drop by drop until the precipitate formed dissolves. This solution keeps indefinitely if stored in the dark.

Procedure

1. Cut 20 μ thick frozen sections.
2. Wash sections in 1 per cent ammonia overnight.
3. Place slides in 5 per cent hydrobromic acid for 2–3 hours at 37°C.
4. Sensitize sections for 15 minutes at 60°C in 50 ml of 2 per cent silver nitrate to which 20 drops of pyridine are added.
5. Transfer to solution 1 for 15 minutes at 60°C. Drops of pyridine are added to this solution.
6. Transfer to solution 2 at room temperature for 5 minutes.
7. Place in 1 per cent formaldehyde for 2–3 minutes.
8. Wash.
9. Store in 0.2 per cent gold chloride for 5 minutes.
10. Wash and mount on albuminized slides.
11. Carefully blot and dehydrate.
12. Clear and mount.

Results

Astrocytes and neuroglial fibres stain jet black

Nassar and Shanklin's Silver Impregnation Method (1951)

Fixation

In a solution containing 70.0 ml formalin, add 14.0 g ammonium bromide and 680.0 ml distilled water. After fixation for 4 days, wash in distilled water with strong ammonia (50 drops of ammonia in 100.0 ml water) for 10 hours . Wash in distilled water for 2 hours, dehydrate, clear and mount.

Reagents Required

Ammonia
Silver nitrate
Sodium sulphite
Gold chloride
Sodium thiosulphate

Preparation of Reagents

Solution 1

Concentrated ammonia   1.0 ml

Solution 2

Silver nitrate   10.0 g
Distilled water   100.0 ml

To 1 ml of ammonia, add 7–8 ml of solution 2 (silver nitrate) drop by drop until a faint turbid liquid remains. Agitate, and dilute with an equal amount of distilled water.

Procedure

1. Deparaffinize and hydrate slides to water.
2. Sensitize in 5 per cent sodium sulphite (5 g/100 ml water) for 2 hours.
3. Place in distilled water.
4. Impregnate in ammoniated silver solution (solution 2) for 2–5 minutes.
5. Rinse in distilled water.
6. Reduce in 2 per cent formalin (2 ml/98 ml water) for 1 minute.
7. Wash in distilled water.
8. Tone in gold chloride (1 g/50 ml water) for 2 minutes.
9. Fix in 5 per cent sodium thiosulphate (5 g/100 ml water) for 2 minutes.
10. Wash, dehydrate and mount.

Result

Neuroglia are stained dark grey to black.

HISTOCHEMICAL TECHNIQUES FOR OLIGODENDROGLIA AND MICROGLIA

Penfields Modification of Hortega's Technique

Fixation

Cold formalin, frozen

Reagents Required

Silver nitrate
Silver carbonate
Ammonia
Hydrobromic acid

Preparation of Reagents

Solution 1 (Silver carbonate solution)
20 per cent silver nitrate	10.0 ml
5 per cent sodium carbonate	20.0 ml

To this solution, add strong ammonia drop by drop until the precipitate dissolves. Make up the volume to 75 ml with distilled water.

Procedure

1. Cut 15 μ sections and wash in 1 per cent ammonia overnight.
2. Place the slides in 5 per cent hydrobromic acid.
3. Wash in distilled water.
4. Place the slides in solution 1 (5 minutes) until sections turn light brown.
5. Transfer to 1 per cent formalin for 3 minutes.

6. Wash in distilled water.
7. Tone in 0.2 per cent gold chloride for 15 minutes.
8. Rinse in distilled water.
9. Fix in 5 per cent sodium thiosulphate.
10. Wash, dehydrate, clear and mount.

Results

Oligodendroglia and microglia are stained black.

Weil Davenport's Technique (1930)

Fixation

Frozen sections

Reagents Required

Ammonia
Silver nitrate
Formalin
Gold chloride
Sodium thiosulphate

Preparation of Reagents

Add 2.3 ml of strong ammonia to 10 per cent silver nitrate drop by drop.

Procedure

1. Cut 12–15 μ thick sections.
2. Wash in distilled water.
3. Place in solution 1 for 20 seconds.
4. Place them in 15 per cent formalin until sections are coffee brown in colour.
5. Rinse in distilled water.
6. Tone in 0.2 per cent gold chloride for 15 minutes.
7. Rinse in distilled water.
8. Fix in 5 per cent sodium thiosulphate for 5 minutes.
9. Wash.
10. Float on clean slide, dehydrate, clear and mount.

Results

Oligodendroglia and microglia stain black.

Bodian Method (Russell modification, 1973)

Fixation

Formalin	40.0 ml
Glacial acetic acid	10.0 ml

80 per cent ethyl alcohol   100.0 ml
Picric acid   2.0 g

Reagents Required

Protargol
Hydroquinone
Sodium sulphite
Gold chloride
Glacial acetic acid

Preparation of Reagents

Solution 1 (Protargol solution)
   Protargol   1.0 g
   Distilled water 100.0 ml

Protargol dissolves in water at 37°C. Keep 100 ml of water on a hot plate and sprinkle protargol.

Solution 2 (Hydroquinone–sodium sulphite)
   Hydroquinone   1.0 g
   Sodium sulphite 5.0 g
   Distilled water   100.0 ml
Solution 3 (Gold chloride solution)
   Gold chloride   1.0 g
   Distilled water 100.0 ml
Solution 4 (Acetic water)
   Glacial acetic acid   500 mg
   Distilled water   100.0 ml

Procedure

1. Hydrate slides to water.
2. Impregnate in solution 1.
3. Wash in distilled water.
4. Place in solution 2 for 10 minutes.
5. Rinse in distilled water.
6. Tone in solution 3 for 5 minutes..
7. Rinse.
8. Develop in 2 per cent oxalic acid, examine under microscope until background is grey.
9. Wash.
10. Fix in 5 per cent sodium thiosulphate for 5 minutes..
11. Wash.
12. Treat with solution 4.
13. Dehydrate, clear and mount.

Result

Nerve fibres are stained black.

Chlorol hydrate Silver Method (Nonidez, 1939)

Fixation

Chlorol hydrate   25.0 g
50 per cent alcohol   100.0 ml

Reagents Required

Ammonia
Silver nitrate
Pyrogallol

Preparation of Reagents

Solution 1 (Alcoholic ammonia)
    Ammonia (conc.)   4 drops
    95 per cent alcohol   60.0 ml

Solution 2 (Silver nitrate solution)
    Silver nitrate   2 g
    Distilled water   100.0 ml

Solution 3 (Pyrogallol solution)
    Pyrogallol   2.0 g
    Formalin   8.0 ml
    Distilled water   100.0 ml

Procedure

1. After fixation, keep block of tissue in solution 1 for 24 hours.
2. Rinse in distilled water.
3. Transfer to solution 2 at 37°C in dark (Replace in fresh solution every 2 days).
4. Rinse in distilled water.
5. Reduce in solution 3 for 24 hours.
6. Wash.
7. Dehydrate, clear and mount.
8. Section, deparaffinize, clear and cover.

Results

Nerve fibres are stained brownish black.

Nauta and Gygax method (1951)

Fixation

Frozen sections 15–20 μ thick.

Reagents Required

Silver nitrate
Pyridine

Sodium hydroxide
Citric acid
Formalin
Sodium thiosulphate

Preparation of Reagents

Solution 1

Silver nitrate	1.5 g
Distilled water	100.0 ml
Pyridine	5.0 ml

Solution 2

Dissolve 450 mg of silver nitrate in 20.0 ml distilled water. Add 10.0 ml of 95 per cent ethyl alcohol. Add 2.0 ml conc. ammonia and 2.2 ml of 2.5 per cent sodium hydroxide and mix well.

Solution 3 (Reducing solution)

10 per cent ethyl alcohol	45.0 ml
10 per cent formalin	2.0 ml
1 per cent citric acid	1.5 ml

Procedure

1. Demyelinate sections in 50 per cent ethyl alcohol plus 1.0 ml ammonia per 100.0 ml for 6–12 hours.
2. Wash in distilled water.
3. Impregnate in silver solution (solution 1).
4. Place in solution 2 for 5 minutes.
5. Transfer directly to reducing solution (solution 3) till sections turn golden brown in colour.
6. Treat with 2.5 sodium thiosulphate (2.5 g/100 ml water) for 2 minutes.
7. Wash in distilled water.
8. Dehydrate, clear and mount.

Result

Nerve fibres and endings	Black
Cells	Pale yellow

Steins Method for Microglia and Oligodendroglia (Modified by Weil and Davenport, 1935)

Fixation

Formalin (1 : 10)

Reagents Required

Ammonium hydroxide
Silver nitrate

Preparation of Reagents

Solution 1 (Ammoniacal silver nitrate)
>   Conc. ammonium hydroxide   2.0 ml
>   10 per cent silver nitrate   20.0 ml
>   (10 g/100 ml water)

Procedure

1. Cut celloidin sections 15 µ in thickness.
2. Stain in solution 1 for 20 seconds.
3. Place in formalin 3:17, section should take a coffee brown colour.
4. Wash in tap water.
5. Dehydrate up to 95 per cent alcohol.
6. Transfer to n-butyl alcohol.
7. Clear and mount.

Result

Microglia	Black
Background	Yellow or brownish yellow

Heidelberger Victoria Blue (Modified by Proesches, 1934)

Fixation

10 per cent formalin

Reagents Required

Victoria blue
N/20 iodine solution

Preparation of Reagents

Solution 1
>   Saturated solution of Victoria blue

Procedure

1. Cut frozen sections, 10–15 µ thick.
2. Store in 10 per cent formalin for 24 hours.
3. Transfer to solution 1 (Victoria blue solution) for 12–24 hours.
4. Wash rapidly.
5. Mount sections on albuminized slides.
6. Expose to ultraviolet light.
7. Dip in N/20 iodine solution for few seconds.
8. Blot and dry.
9. Differentiate in xylene–aniline (1:1).
10. Clear in xylene and mount.

Result

Glial cell bodies and fibrils stain deep blue.

Holczer Method for Glial Fibres (Holczer, 1921)

Fixation

Formol alcohol or 10 per cent neutral buffered formalin.

Reagents Required

Phosphomolybdic acid
Chloroform
Crystal violet
Potassium bromide

Preparation of Reagents

Solution 1

0.5 per cent phosphomolybdic acid	50.0 ml
95 per cent alcohol	100.0 ml

Solution 2 (alcohol–chloroform)

100 per cent alcohol	40.0 ml
Chloroform	160.0 ml

Solution 3 (Crystal violet solution)

Crystal violet	5.0 g
100 per cent alcohol	20.0 ml
Chloroform	80.0 ml

Solution 4 (Potassium bromide)

Potassium bromide	100.0 g
Distilled water	1000.0 ml

Solution 5 (Differentiating solution)

Aniline	100.0 ml
Chloroform	180.0 ml
28 per cent ammonium hydroxide	20 drops

Procedure

1. Dewax and hydrate slides to water.
2. Transfer to solution 1 for 3 minutes.
3. Place slides in solution 2 until sections are clear.
4. Pour solution 3 on slides for 30 minutes and blot dry.
5. Again flood slides with solution 4, blot and dry.
6. Differentiate in solution 5 for 30 seconds.
7. Rinse in xylene and clear.
8. Mount.

Result

| Glial fibres | Deep violet |
| Background | Pale violet |

Bosch's Method for Negri Bodies (Bosch, 1966)

Fixation

Bouin's or 10 per cent neutral buffered formalin.

Reagents Required

Solution 1 (Mayer's haematoxylin solution (*see* Chapter 7))

Solution 2 (Phenol–phloxine B solution)

| Phloxine B | 1.5 g |
| Distilled water | 5.0 ml |

Add while shaking,

5 per cent phenol   100.0 ml

Add glacial acetic acid drop by drop until solution becomes cloudy and heat at 56°C for 24 hours. Filter and use supernatant as working solution.

Solution 3

| Lithium carbonate | 200 mg |
| Distilled water | 100.0 ml |

Procedure

1. Dewax and hydrate slides to water.
2. Keep slides in Bouin's solution for mordanting overnight.
3. Wash first in running water and then in distilled water.
4. Place in solution 2 for 5 minutes.
5. Rinse in distilled water for 1 minute.
6. Treat with solution 1 for 5 minutes.
7. Wash in distilled water.
8. Differentiate in solution 3 for few seconds until the sections are pale pink.
9. Wash in distilled water.
10. Dehydrate, clear and mount.

Result

Negri bodies	Deep red
Nuclei	Blue
Background	Pink
Erythrocytes	Bright red

Massignani–Malferrari Method for Negri Bodies (Massignani and Malferrari, 1961)

Fixation

10 per cent neutral buffered formalin.

Reagents Required

Haematoxylin
Hydrochloric acid
Lithium carbonate
Eosin
Phosphotungstic acid

Preparation of Reagents

Solution 1 (Harris haematoxylin (*see* Chapter 7)

Solution 2 (0.5 per cent hydrochloric acid (0.5 ml in 100 ml distilled water))

Solution 3 (Lithium carbonate)
Saturated lithium carbonate   1.0 ml
Distilled water   200.0 ml

Solution 4 (Eosin solution)
Eosin   1.0 g
Phosphotungstic acid   700 mg

Grind the two solutions. Add 10 ml distilled water and then make up to 200.0 ml with absolute alcohol. Add 2 drops of solution 3, stir and filter.

Procedure

1.   Dewax and hydrate slides to water.
2.   Place slides in solution 1 for 2 minutes.
3.   Wash in distilled water.
4.   Differentiate in solution 2.
5.   Transfer to solution 3 for 1 minute.
6.   Wash in running water for 10 minutes.
7.   Dehydrate and transfer to solution 4 for 8 minutes.
8.   Blot and dry.
9.   Dehydrate, clear and mount.

Result

Negri bodies   Deep red
Nuclei   Light blue

Schleifstein's Method for Negri Bodies (Schleifstein, 1937)

Fixation

Zenker's fixative

Reagents Required

Basic fuchsin
Methylene blue
Glycerine
Methyl alcohol
Potassium hydroxide

Preparation of Reagents

Solution 1 (Schleifstein's stock solution)
    Basic fuchsin   1.8 g
    Methylene blue 1.0 g
    Glycerine   100.0 ml
    Methyl alcohol 100.0 ml

Solution 2 (Potassium hydroxide solution)
    Potassium hydroxide   10 mg
    Tap water   4000.0 ml

Solution 3 (Working solution)
    Solution 1   10 drops
    Solution 2   20.0 ml
This should be prepared just before use.

Procedure

1. Dewax and bring slides to water.
2. Flood slides with solution 3. Gently heat until vapour is produced and later cool.
3. Wash quickly.
4. Differentiate in 90 per cent alcohol.
5. Dehydrate, clear and mount.

Result

Negri bodies   Deep magenta
Cytoplasm   Bluish violet

Aldehyde–Thionin/PAS Method for Central Nervous System

Fixation

10 per cent neutral buffered formalin.

Reagents Required

Sulphuric acid
Potassium permanganate
Paraldehyde
Thionin
Periodic acid
Basic fuchsin
Potassium metabisulphite
Hydrochloric acid
Orange G
Phosphotungstic acid

Preparation of Reagents

Solution 1 (Sulphuric acid)
    Conc. sulphuric acid   0.5 ml
    Distilled water   99.5 ml

Solution 2 (0.5 per cent potassium permanganate (500 mg in 100 ml of distilled water))

Solution 3 (Potassium metabisulphate)
    Potassium metabisulphite   2.0 g
    Distilled water   100.0 ml

Solution 4 (Aldehyde–thionin solution)
    Thionine   0.5 g
    70 per cent alcohol   91.5 ml
    Paraldehyde   7.5 ml
    Conc. hydrochloric acid   1.0 ml

Solution 5 (Periodic acid solution (*see* Chapter 9))

Solution 6 (Schiff's reagent (*see* Chapter 19))

Solution 7 (Orange G solution (*see* Chapter 19))

Solution 8 (Glacial acetic acid solution (1 ml + 100 ml distilled water))

Solution 9 (Phosphotungstic acid solution) (*see* Chapter 19)

Procedure

1. Dewax and hydrate slides to water.
2. Place slides in a solution containing equal parts of solution 1 and solution 2 for 2 minutes.
3. Bleach in solution 3 for 1 minute.
4. Rinse in distilled water.
5. Transfer slides to solution 4 for 1 hour.
6. Rinse in distilled water.
7. Place slides in solution 5 for 5 minutes.
8. Rinse in distilled water.
9. Transfer slides to solution 6 for 15 minutes.
10. Wash in running water.
11. Transfer to solution 7 for 3–5 minutes.
12. Treat slides with solution 9 for 1 minute.
13. Differentiate in solution 8.
14. Dehydrate, clear and mount.

Results

Nissl substance is stained blue-black.

Mann's Method for Staining Negri Bodies (Mann, 1894)

Fixation

Zenker's fixative

Reagents Required

Eosin
Methyl blue
Sodium hydroxide

Preparation of Reagents

Solution 1 (Stock eosin solution)
Eosin Y   1.0 g
Distilled water   100.0 ml

Solution 2 (Stock methyl blue solution)
Methyl blue   1.0 g
Distilled water   100.0 ml

Solution 3 (Alkaline ethanol)
0.1N sodium hydroxide   0.4 ml
Absolute alcohol   100.0 ml

Solution 4 (Working solution)
Solution 2 (Methyl blue stock)   18.0 ml
Solution 1 (Eosin Y stock)   23.0 ml
Distilled water   49.0 ml
Prepare this solution just before use.

Procedure

1.   Deparaffinize and bring down slides to water.
2.   Immerse in solution 4 for 24 hours (Fresh).
3.   Differentiate in solution 3 for 10 minutes.
4.   Wash in water.
5.   Dehydrate rapidly, clear and mount.

Results

Negri bodies	Dark red
Erythrocytes	Pink
Nucleoli	Violet
Cytoplasm	Blue

Stovall–Black Method for Negri Bodies (Stovall and Black, 1940)

Fixation

Acetone

Reagents Required

Ethyl eosin
Hydrochloric acid
Methylene blue
Acetate buffer
Acetic acid

Preparation of Reagents

Solution 1 (Alcoholic eosin solution)
Sodium ethyl eosinate   1.0 g
Absolute alcohol   100.0 ml
pH of the solution should be 3, adjusted by 0.1N hydrochloric acid.

Solution 2 (Methylene blue stock)

Methylene blue	1.0 g
95 per cent alcohol	100.0 ml

Solution 3 (Working solution)

Solution 2	10.0 ml
0.2 M acetate buffer	10.0 ml
Distilled water	20.0 ml

Procedure

1. Deparaffinize and hydrate slides to water.
2. Place in solution 1 for 2 minutes.
3. Rinse in distilled water.
4. Transfer to solution 3 for 30 seconds.
5. Differentiate in 0.38 per cent acetic acid in water until sections are brownish red.
6. Wash, dehydrate and mount.

Result

Negri bodies	Brownish red
Nuclei	Pale blue

Zlotnik's Method for Negri Bodies (1953)

Fixation

Acetone

Reagents Required

Haematoxylin
Picric acid
Orange G
Acid fuchsin
Phosphotungstic acid
Acetic acid
Aniline blue

Preparation of Reagents

Solution 1 (Ehrlich haematoxylin) (*see* Chapter 7)

Solution 2 (Saturated picric acid)

Picric acid saturated	100.0 ml
Orange G	500 mg

Solution 3 (Acid fuchsin–PTA)

Acid fuchsin	500 mg
Phosphotungstic acid	500 mg
1 per cent acetic acid	100.0 ml

Solution 4

Phosphotungstic acid	2.0 g
Phosphomolybdic acid	2.0 g
100 per cent alcohol	30.0 ml
Saturated picric acid	70.0 ml

Solution 5 (1 per cent aniline blue)

Aniline blue	1.0 g
2 per cent acetic acid	100.0 ml

Procedure

1. Deparaffinize and hydrate slides to water.
2. Place slides in solution 1 for 5 minutes.
3. Wash in water.
4. Counterstain in solution 2.
5. Wash in water.
6. Place slides in solution 3 for 10 minutes.
7. Rinse in distilled water.
8. Transfer to solution 4 for 5 minutes.
9. Rinse in 1 per cent acetic acid.
10. Place slides in solution 5 for 15 minutes.
11. Rinse in acetic acid (1 per cent).
12. Dehydrate, clear and mount.

Result

Negri bodies	Purplish blue
Nucleoli	Dark purple
Erythrocytes	Yellow

Penfield Modification of Del Rio-Hortega Silver Carbonate Method (McClung, 1950)

Fixation

10 per cent neutral buffered formalin	1 week
Formaldehyde	14.0 ml
Ammonium bromide	2.0 g
Distilled water	86.0 ml

Reagents Required

Hydrobromic acid
Silver nitrate
Sodium thiosulphate
Sodium carbonate
Gold chloride

Preparation of Reagents

Solution 1

40 per cent hydrobromic acid   5 ml
Distilled water   95 ml

Solution 2

Silver carbonate solution
10 per cent silver nitrate   5.0 ml
Sodium carbonate   20.0 ml

Solution 3

Gold chloride   1.0 g
Distilled water   500.0 ml

To 5 ml of 10 per cent silver nitrate, add 20 ml of sodium carbonate. Add ammonium hydroxide dropwise. When the precipitate dissolves, add 75 ml of distilled water. Filtered solution will keep long if kept in a dark bottle.

Procedure

1. Keep frozen sections in 1 per cent distilled water.
2. Place the slides in a combination of 1 per cent ammonia and distilled water overnight.
3. Transfer to solution 1 at 38°C for an hour.
4. Wash in distilled water.
5. Transfer slides to a mordant (5 per cent sodium carbonate) and leave them for 1 hour.
6. Impregnate in solution 2 for 3 minutes till sections turn grey.
7. Reduce in 1 per cent formalin (agitate).
8. Wash in distilled water.
9. Tone in solution 3 until the section turns bluish grey.
10. Place in 5 per cent sodium thiosulphate for 3 minutes.
11. Wash in distilled water.
12. Dehydrate, clear and mount.

Results

Oligodendroglia and microglia dark grey.

Phosphotungstic Acid Haematoxylin (Mitchell, 1975)

Fixation

10 per cent neutral buffered formalin.

Reagents Required

Potassium permanganate
Oxalic acid
Phosphotungstic acid haematoxylin

Preparation of Reagents

Solution 1 (Potassium permanganate)
Potassium permanganate   250 mg
Distilled water   100.0 ml

Solution 2 (Oxalic acid)
Oxalic acid   1.0 g
Distilled water   100.0 ml

Solution 3 PTAH (Phosphotungstic acid haematoxylin) (*see* Chapter 7)

Procedure

1.   Dewax and hydrate slides to water.
2.   Mordant in Susa fixative (*see* Chapter 1).
3.   Wash in distilled water.
4.   Oxidize in solution 1 for 5 minutes.
5.   Bleach in solution 2 for 5 minutes.
6.   Wash in distilled water.
7.   Transfer to solution 3 for 4 hours.
8.   Wash.
9.   Dehydrate, clear and mount.

Result

Astrocytes   Blue
Nuclei   Blue
Collagen   Red
Myelin   Blue

Silver Impregnation for Nerve Tissue
(Garvey *et al.*, 1987; Staples, 1991)

Fixation

10 per cent neutral buffered formalin.

Reagents Required

Silver nitrate
Gum mastic
Hydroquinone
Gold chloride
Sodium thiosulphate

Preparation of Reagents

Solution 1 (1 per cent silver nitrate)
Silver nitrate   1.0 g
Distilled water   100.0 ml

Solution 2 (3 per cent gum mastic)
    Gum mastic   3.0 g
    Distilled water   25.00 ml

Solution 3 (Hydroquinone)
    Hydroquinone crystals   0.25 g
    Distilled water   100.0 ml

Solution 4 (Developing solution)
    Solution 3   25.0 ml
    Solution 2   15.0 ml
    Solution 1   1.2 ml

Solution 5 (0.25 per cent gold chloride)
    10 per cent gold chloride solution   25.0 ml
    Distilled water   100.0 ml

Solution 6 (1 per cent sodium thiosulphate)
    Sodium thiosulphate   1.0 g
    Distilled water   100.0 ml

Procedure

1. Dewax and hydrate slides to water.
2. Place in solution 1 for 30 minutes at 4°C.
3. Rinse in distilled water giving 3 changes.
4. Rinse in 95 per cent alcohol giving 2 changes.
5. Rinse in absolute alcohol giving 2 changes.
6. Transfer to solution 2 for 5 minutes
7. Rinse in absolute alcohol giving 3 changes.
8. Place slides in solution 4 at 60°C.
9. Rinse in distilled water giving 3 changes.
10. Transfer to solution 5 for 5 minutes.
11. Rinse in distilled water.
12. Transfer to solution 6 for 2 minutes.
13. Rinse in running water.
14. Dehydrate, clear and mount.

Result

Axons, dendrites and neurofibrils   Black
Background   Grey

Glees Method (Novotney Modification, 1974, 1977)

Fixation

Perfuse with neutral buffered formalin and store for one week, embed and use subbed slides.

Reagents Required

Silver nitrate
Formalin
Acetic acid
Luxol fast blue
Cresyl violet

Preparation of Reagents

Solution 1 (Silver nitrate)
 Silver nitrate 20.0 g
 Distilled water 100.0 ml

Solution 2 (Reducing solution 1)
 10 per cent formalin 13.5 ml
 1 per cent acetic acid 13.5 ml
 95 per cent alcohol 45.0 ml
 Distilled water 400.0 ml

Solution 3 (Ammoniacal silver solution)
 Silver nitrate 5.0 g
 80 per cent alcohol 100.0 ml

Add 25 per cent NH_4OH dropwise. A precipitate is formed and it dissolves. When a few grains remain, add 3 more drops. This solution must be freshly prepared. Avoid metal container.

Solution 4 (Reducing solution 2)
 10 per cent formalin 400.0 ml
 95 per cent alcohol 50.0 ml
 1 per cent acetic acid 20.0 ml

Solution 5 (Luxol fast blue) (*see* Chapter 19)

Solution 6 (Crystal violet) (*see* Chapter 11)

Procedure

1. Dewax and hydrate slides to water.
2. Keep in solution 1 for 2 hours at room temperature till it turns brown.
3. Transfer to solution 2 for 10 minutes.
4. Directly transfer to solution 3 for 15 minutes. Sections should be orange-brown.
5. Rinse in absolute alcohol.
6. Transfer to solution 4 for 10 minutes.
7. Wash in running water.
8. Fix in 5 per cent sodium thiosulphate for 3 minutes
9. Wash in running water.
10. Dehydrate in 95 per cent alcohol.
11. Transfer to solution 5 and keep overnight at 56°C.
12. Rinse in absolute alcohol.
13. Rinse in distilled water.

14. Differentiate in 70 per cent alcohol until neutrophils are yellow, nerve cells are orange-brown and myelin is blue.
15. Rinse in 1 per cent acetic acid.
16. Transfer to solution 6.
17. Differentiate in 95 per cent alcohol.
18. Clear and mount.

Result

Axons	Black or brown
Myelin	Blue
Nissl bodies	Violet
Glial cells	Violet
Neutrophil	Yellow

For further information on neurological staining also refer the following. Culling (1957), Luna (1960, 1964, 1968), Lillie (1954), Mettler (1932), Mettler and Hanada (1942), Swank and Davenport (1934a, 1934b, 1935a, 1935b), Garvey *et al.* (1987, 1990, 1991), Koski and Reyes (1986), Lloyd *et al.* (1985), Murdock and Frabsesi (1988) and Vacc (1985).

Marsland, Glees and Erikson's Method for Axons (Marsland *et al.*, 1954)

Fixation

10 per cent formol saline.

Preparation of Reagents

Solution 1

Absolute alcohol	20.0 ml
20 per cent silver nitrate	30.0 ml

Add strong ammonia dropwise till the precipitate dissolves. Add some more (5 drops) ammonia.

Procedure

1. Dewax and rinse in absolute alcohol.
2. Dip in 0.5–1.0 per cent celloidin. Remove celloidin from back of slide and wash in 70 per cent alcohol for 5 minutes.
3. Wash in distilled water.
4. Transfer to solution 1 for 15–60 minutes at 37°C.
5. Wash in 10 per cent formalin for 10–15 seconds . Sections should be pale yellow-brown.
6. While washing again flood slides with solution 1 for 30 seconds.
7. Drain off excess solution and flood with 10 per cent formalin for 1–2 minutes. Examine under the microscope.
8. Wash in distilled water.
9. Fix in 5 per cent sodium thiosulphate for 1–5 minutes.
10. Wash, dehydrate and remove celloidin in absolute alcohol.

Results

Nerve fibres	Dark brown to black
Background	Light brown

Palmgren's Method for Nerve Fibres (Palmgren, 1948)

Fixation

Formol saline or Bouin's

Preparation of Reagents

Solution 1 (Acid formalin)

40 per cent formaldehyde	25.0 ml
Distilled water	75.0 ml
1 per cent nitric acid	0.2 ml

Solution 2 (Silver solution)

Silver nitrate	15.0 g
Potassium nitrate	10.0 g
Distilled water	100.0 ml
5 per cent acetic acid	1.0 ml

Solution 3 (Reducer)

Pyrogallol	10.0 g
Distilled water	450.0 ml
Absolute ethanol	550.0 ml
1 per cent nitric acid	2.0 ml

It can be used after 24 hours.

Solution 4 (Toning solution)

Gold chloride	1.0 g
Distilled water	200.0 ml
Glacial acetic acid	0.2 ml

Solution 5 (Aniline–alcohol)

50 per cent ethyl alcohol	100.0 ml
Aniline oil	2 drops

Solution 6

Sodium thiosulphate	5.0 g
Distilled water	100.0 ml

Procedure

1. Dewax and bring sections to water.
2. Immerse in solution 1 for 5 minutes.
3. Wash in distilled water for 5 minutes giving 3 changes.
4. Treat with solution 2–15 minutes at 20–25°C.
5. Treat with solution 3 (pre-heated to 40–45°C) leave for 1 minute.
6. Rinse in 50 per cent alcohol for 5–10 seconds.

7. Wash in distilled water.
8. Tone in solution 4 until yellow colour fades.
9. Treat with solution 5 for 15 seconds.
10. Wash in tap water.
11. Transfer to solution 6 for few seconds.
12. Wash in water.
13. Dehydrate, clear and mount.

Result

Nerve fibres are stained brown or black.

Eager's Method for Degenerating Axons (1972)

Fixation

Formol saline

Preparation of Reagents

Solution 1 (Ammoniacal silver solution)

1.5 per cent silver nitrate	40.0 ml
95 per cent ethanol	24.0 ml
Ammonia	4.0 ml
2.5 per cent sodium hydroxide	3.6 ml

Solution 2 (Reducer)

Absolute alcohol	90.0 ml
Distilled water	810.0 ml
1 per cent nitric acid	27.0 ml
10 per cent formalin	37.0 ml

Procedure

1. Keep frozen sections in 2 per cent formalin.
2. Rinse in distilled water.
3. Treat with 2.5 per cent uranyl nitrate for 5 minutes.
4. Rinse in distilled water and transfer to solution 1 till sections are brown for 3–15 minutes.
5. Treat directly with solution 2.
6. Rinse in distilled water.
7. Fix in 0.5 per cent sodium thiosulphate.
8. Wash, dehydrate, clear and mount.

Result

Degenerating fibres	Brown to black
Normal fibres	Pale yellow

Heidenhain's Myelin Stain

Fixation

Formol saline, formol calcium

Preparation of Reagents

10 per cent alcoholic haematoxylin	10.0 ml
Distilled water	83.0 ml
Saturated lithium carbonate	7.0 ml

Procedure

1. Dewax and bring sections to water.
2. Keep sections in 4 per cent alum for 3–24 hours to mordant.
3. Rinse in distilled water; if not properly stained leave them again in stain.
4. Wash well.
5. Dehydrate, clear and mount.

Results

Myelin	Blue black
Background	Greyish

Solochrome Cyanine Technique for Myelin (Page, 1965)

Fixation

Formol saline or formol calcium

Preparation of Reagent

Solochrome cyanine RS	200 mg
Distilled water	96.0 ml
10 per cent iron alum	4.0 ml
Concentrated sulphuric acid	0.5 ml

Procedure

1. Dewax and bring sections to water.
2. Keep sections in staining solution for 10–20 minutes at room temperature.
3. Wash in running water.
4. Differentiate in 5 per cent iron alum.
5. Wash in running water.
6. Counterstain if necessary.
7. Dehydrate and mount.

Result

Myelin sheaths are stained blue.

Steart's Modification of Holczer's Method for Astrocytic Processes

Fixation

Formalin, Helly or Bouin's

Preparation of Reagents

Solution 1 (Mordant)

1 per cent phosphomolybdic acid	10.0 ml
Absolute alcohol	40.0 ml

Solution 2 (Chloroform-Alcohol mixture)

Chloroform	160.0 ml
Absolute alcohol	40.0 ml

Solution 3 (Cresyl violet stain)

Cresyl violet	2.0 g
Absolute alcohol	20.0 ml
Chloroform	80.0 ml

Solution 4 (Differentiating solution)

Aniline oil	80.0 ml
Chloroform	120.0 ml
Concentrated ammonia	10 drops

Procedure

1. Dewax and place the slides in absolute alcohol.
2. Flood slides with solution 1 for 5–10 minutes.
3. Pour off stain and wash in absolute alcohol.
4. Flood slides with solution 2.
5. Drain off and flood with solution 3 for 30 seconds.
6. Flood with 10 per cent potassium bromide solution until green colouration disappears.
7. Drain, blot and dry.
8. Differentiate in solution 4 until the background is coloured. Examine under microscope.
9. Rinse in xylene.
10. Mount in synthetic mountant.

Result

Glial fibrils	Blue
Nuclei	Pale blue
Background	Colourless

Supravital Staining for Nerve Fibres and Endings

Preparation of Reagents

Solution 1 (Methylene blue solution)
Methylene blue (Zinc-free)	50 mg
Distilled water	100.0 ml
Sodium chloride	850 mg

Dissolve in order.

Solution 2 (Ammonium molybdate)
Ammonium molybdate	8.0 g
Distilled water	100.0 ml

Procedure

1. Inject solution 1 into the tissue and leave for 5–10 minutes.
2. Cut the tissue longitudinally into 3-mm thick strips.
3. Place the tissue on Kleenex tissue paper soaked in saline in a petri dish. A funnel is inverted over the tissue and oxygen is passed at the rate of 1–4 litres per minute for 1 hour. Turn the tissue in order to expose all surfaces.
4. Transfer tissue to cold ammonium molybdate and leave overnight at 4–6°C.
5. Wash several times in distilled water for half an hour.
6. Fix tissue in 10 per cent formalin for 24 hours.
7. Cut frozen sections at 50–100 μm.
8. Sections are dehydrated in 100% alcohol and cleared in xylene.

Result

Nerve fibre and endings are stained blue.

Table 20.1 Histochemical techniques applied for the demonstration of neurological elements

Technique	Fixative	Nerve cells	Neuro fibrils	NISSL substance	Myelin	Nerve fibres	Degenerating axons
Biels Chowsky's method	10% formalin	Brown to black					
Ramony's Cajal's Gold chloride method	15 ml formalin + 2 g ammonium bromide + 85.0 ml distilled water	Pale red				Unstained	
Golgi rapid method	40 ml $K_2Cr_2O_7$ + 10 ml osmic acid	Deep black	Brown to black				
Ramony's Cajal's Pyridine Silver method	Absolute alcohol 98 ml + conc. ammonia – 2 ml (Fixation time – 1 week)	Yellow				Brown black	
Gross-Schultze method	Frozen	Brown to black	Brown to black				
Nauta–Gygax method	Perfusion method						Black
Holme's method	10% neutral buffered formalin	Black					
Hirano–Zimmerman method	10% neutral buffered formalin		Black				
Silver impregnation method	10% neutral buffered formalin		Black				
Cresyl Violet method	Bouin's			Purple			
Gallocyanin method	Any general fixative			Blue			
Thionin method	Bouin's			Bright blue			

(*Contd.*)

Table 20.1 (Continued)

Technique	Fixative	Nerve cells	Neuro fibrils	Nissl substance	Myelin	Nerve fibres	Degenerating axons
Thionin and Toluidine blue	Alcohol or formol saline			Deep blue			
Einarson's Gallocyanin method	Carnoy or formol saline			Deep blue			
Weigert-Pal technique	Formol saline				Deep blue		
Luxol Fast Blue/Cresyl violet	10% formalin				Blue		
Luxol fast blue-PAS	10% formalin, formol calcium				Blue		
Heidenhain's method	Formol saline, formol calcium				Blue- black		
Luxol fast blue/Holmes silver nitrate	10% formalin				Greenish blue		

Table 20.2   Histochemical techniques applied for the demonstration of neurological elements

Technique	Fixative	Myelin	Degenerating myelin	Neuroglia	Astrocytes	Oligoden-droglia	Nerve fibres	Glial fibres
Weil stain method	10% formalin	Dark blue						
Lapham's method	10% buffered formalin	Magenta						Bright green
Woelcke's method	10% buffered formalin	Blue						
Solochrome Cyamine method	Formol saline, formol calcium	Blue						Black
Chromic acid haematoxylin method	10% formalin	Blue-black						
Marchi's method	Formol calcium		Black					
Swank Davenport method	Formol saline		Black					
Luxol Fast blue– Oil red O	10% formalin	Blue-green	Red					
Guillery's method	10% formol saline by perfusion		Black					

(*Contd.*)

Table 20.2   (Continued)

Technique	Fixative	Myelin	Degenerating myelin	Neuroglia	Astrocytes	Oligoden-droglia	Nerve fibres	Glial fibres
Anderson's Victoria blue	Frozen or formalin-fixed			Blue	Blue			
Stearts modification of Holczer's	Formalin, Helly, Bouins							Blue
Cajol's Gold Sublimate	Formol ammonium bromide for 3–8 days				Reddish black			
Scharenberg's triple impregnation method	10% formol saline			Jet black	Jet black			
Nassar and Shanklin Silver impregnation method	*see* page 658			Deep grey to black				
Penfield's modification of Hortega's technique	Frozen sections, cold formalin					Black		
Weil Davenport's technique	Frozen section					Black		

Method	Fixative		Result
Bodian method	40 ml formalin + 10 ml glacial acetic acid + 100 ml 80% ethyl alcohol + 2g picric acid		Black
Palmgren's method	Formol saline, Bouin's		Brown or black
Chlorol hydrate silver method	Chlorol hydrate + 50% alcohol		Brownish black
Nauta and Gygax method	Frozen sections		Black
Heidelberger's Victoria blue	10% formalin		Deep blue
Holczer method	Formol alchol or 10% neutral buffered formalin		Deep violet
Glees method	Perfuse with neutral buffered formalin	Blue	Violet

Table 20.3   Histochemical techniques applied for the demonstration of neurological elements

Technique	Fixative	Microglia	Myelin	Nerve fibres	Astrocytes	Negri bodies
Stein's method	Formalin	Black				
Bosch's method	Bouin, 10% neutral buffered formalin					Deep red
Massignani and Malferrari method	10% neutral buffered formalin					Deep red
Schleifstein's method	Zenker					Deep magenta
Mann's method	Zenker					Dark red
Stoval black method	Acetone					Brownish red
Zlotnik's method	10% neutral buffered formalin					Purplish blue
PTAH method	10% neutral buffered formalin		Blue		Blue	
Glees method	Perfuse with neutral buffered formalin		Blue			
Silver impregnation	10% neutral buffered formalin			Black		

ENDOCRINE GLANDS

21

PANCREAS

It is a well established fact that in all vertebrates except cyclostomes the pancreatic islet cells or islets of Langerhans are four major types, A(α or A_2 cells), B(β cells), D(γ or A_1 cells) and F cells. "A" cells produce glucagon, "B" cells produce insulin, "D" cells produce somatostatin and "F" cells produce pancreatic polypeptides. In addition to these hormones, these cells produce 12 peptides (Epple and Brinn, 1986).

Any general routine fixative gives good results except acetic acid which has a deleterious effect on granules. Of all, Zenker formol or formalin is the choice fixative. Tissue should be fixed as quickly as possible to avoid rapid autolysis.

Different techniques are used to demonstrate and identify these cells. The mostly routinely used stain is haematoxylin/eosin. But to demonstrate alpha (α), beta (β), D and F cells, special techniques are to be adopted. With Mallory, α-cells take a red shade and β-cells a blue shade. Azan method is better. Besides the Langerhans, zymogen granules (which are linked with pancreatic enzymes) require special attention.

	α-cells	β-cells	δ-cells
Masson's trichrome stain	Red	Orange	—
Periodic acid/Schiff (AS)	—	—	—
Aldehyde fuchsin	Purple	Purple	Purple
Aniline blue	—	—	Blue
Light green	—	—	Green
Chrome haematoxylin/phloxine	Red	Blue	Red
Mallory–Azan	Red	Orange-brown	Deep blue
Mallory–Azan after permanganate oxidation	Bright red	—	—
Phosphotungstic acid/ Haematoxylin (PTAH)	Deep blue	—	—

HISTOCHEMICAL METHODS

Histochemical demonstration of individual cells in endocrine glands in general depends upon the cell inclusions and functional state of the cell. So specific staining of specific cell types depends upon the cell inclusions. Though many methods are available, only a few techniques have been given here. Only a few histochemical techniques which demonstrate chemical groups which are associated with hormones (specific) elaborated by these cells are given. Identification of these cell types depends upon the granule size and morphology.

In most of the vertebrates, three types of granulated cells have been recognized (there are some additional types which may or may not be granular). The three recognized cells are α, β and δ cells.

α-cell (A_2) produces glucagon and is silver-negative.

β cell is an insulin-producing cell.

δ-cell (A_1) is a carboxyl-rich, silver-positive, pancreatic gastrin cell.

Mallory Heidenhain's Azan (Gomori, 1939)

Fixation

Bouin's or Helly

Reagents Required

Azocarmine G
Glacial acetic acid
Aniline oil
Iron alum
Aniline blue
Orange G

Preparation of Reagents

Solution 1 (Azocarmine solution)
    Azocarmine G   100 mg
    Distilled water   100.0 ml

Add azocarmine to distilled water and boil for 5 minutes. Cool and make up the volume to 100.0 ml. Add 2 ml acetic acid before use and store at room temperature.

Solution 2 (Aniline alcohol)
    Aniline   1.0 ml
    90 per cent alcohol   99.0 ml

Solution 3 (Mordant)
    Iron alum   5.0 g
    Distilled water   100.0 ml

Solution 4 (Aniline blue–Orange G)
    Aniline blue   500 mg

    Orange G   2.0 g
    Distilled water   100.0 ml

Procedure

1. Deparaffinize and hydrate slides to water.
2. Post-treat for picric acid.
3. Transfer to solution 1 kept at 56°C for 1 hour.
4. Rinse in distilled water.
5. Differentiate in solution 2 until excess of stain oozes out.
6. Rinse in distilled water.
7. Place in solution 3 for 5 minutes.
8. Rinse again in distilled water.
9. Transfer to solution 4 (Solution 4 is diluted at the ratio 1 : 3) for 15 minutes.
10. Wash and dehydrate quickly.
11. Clear in xylene and mount in Canada balsam.

Result

α-granules   Bright red
β-cells   Orange-brown
δ-cells   Dark blue

These results vary with the species as well as fixative.

Mallory Heidenhain Azan after Permanganate Oxidation (Gomori, 1939)

Fixation

Any fixative other than alcohol

Reagents Required

Azocarmine G
Glacial acetic acid
Aniline oil
Aniline blue
Orange G
Iron alum
Potassium permanganate
Concentrated sulphuric acid
Oxalic acid

Preparation of Reagents

All solutions from 1–4 are prepared in the same manner as in the previous procedure.
Solution 5 (Permanganate solution)

    Potassium permanganate   250 mg
    Conc. sulphuric acid   0.5 ml
    Distilled water   100.0 ml
Add sulphuric acid just before use.

Solution 6 (Bleaching solution)
    Oxalic acid   5.0 g
    Distilled water   100.0 ml

Procedure

1. Deparaffinize and hydrate slides to water.
2. Transfer to solution 5 for 30 seconds.
3. Wash in distilled water.
4. Bleach in solution 6 until sections are colourless.
5. Wash in distilled water.
6. Stain in solution 1 at 56°C for 1 hour.
7. Wash in solution 2 until excess stain comes out.
8. Remaining steps are same as in previous procedure.

Results

α-cells   Bright red
β and δ-cells cannot be distinguished.

Phosphotungstic Acid Haematoxylin (PTAH) (Levene and Feng, 1964)

Fixation

Bouin's or formalin

Reagents Required

Potassium permanganate
Sulphuric acid
Potassium metabisulphite
Iron alum
Haematoxylin
Phosphotungstic acid

Preparation of Solutions

Solution 1 (Potassium permanganate)
    Potassium permanganate   300 mg
    Distilled water   100.0 ml
    Conc. sulphuric acid   32 ml

Solution 2 (Bleaching solution)
    Potassium metabisulphite 4.0 g
    Distilled water   100.0 ml

Solution 3 (Mordant)
    Iron alum   4.0 g
    Distilled water   100.0 ml

Solution 4 (PTAH solution)
    Haematoxylin   100 mg

Warm distilled water   100.0 ml
Phosphotungstic acid   2.0 g

Prepare the solution with 2.5 ml of 1 per cent potassium permanganate. Allow it to stand for 48 hours.

Procedure

1. Deparaffinize and hydrate slides to water.
2. Transfer to solution 1 (Oxidation) for 50 seconds.
3. Rinse in distilled water.
4. Bleach in solution 2 for 5–10 seconds.
5. Wash in running water.
6. Transfer to solution 3 for 30 minutes to 2 hours.
7. Rinse in distilled water.
8. Transfer to solution 4 for 16–48 hours.
9. Pass rapidly through graded series of alcohol.
10. Clear and mount.

Results

α-cell granules are stained deep blue.

Post-coupled *p*-Dimethylaminobenzaldehyde (*p*-DMAB) Reaction (Barka and Anderson, 1963)

Alpha cells which contain lot of tryptophan are stained.

Fixation

Neutral buffered formalin

Reagents Required

p-dimethylaminobenzaldehyde
Concentrated hydrochloric acid
Glacial acetic acid
Diazotate of S-acid (8-amino 1-naphthol 5-sulphonic acid)
Sodium nitrite

Preparation of Reagents

Solution 1 (*p*-dimethylaminobenzaldehyde solution)
    p-dimethylaminobenzaldehyde   1.0 g
    Concentrated hydrochloric acid   10.0 ml
    Glacial acetic acid   30.0 ml

Solution 2 (S-acid solution)
    S-acid   240 mg
    1N HCl   3.0 ml
    Distilled water   6.0 ml
Cool it in an ice bath and then add
    1N $NaNO_3$ 1.0 ml
Cool it at 4°C.

Procedure

1. Deparaffinize and bring sections to absolute alcohol.
2. Transfer sections to solution 1 for 5 minutes.
3. Rinse in glacial acetic acid for 30 seconds going 2 changes.
4. Transfer sections to a solution containing 40 ml of glacial acetic acid and 1 ml of solution 2 for 5 minutes.
5. Wash in glacial acetic acid giving 2 changes.
6. Transfer to a mixture of (1:1) glacial acetic acid and xylene.
7. Clear in xylene.
8. Mount in synthetic resin.

Result

Alpha cells are stained intense blue.

Aldehyde Fuchsin (Gomori, 1950a)

Fixation

Formalin, Bouin's, and dichromate fixatives are avoided.

Reagents Required

Iodine
Potassium iodide
Sodium thiosulphate
Concentrated hydrochloric acid
Paraldehyde
Basic fuchsin
70 per cent ethanol

Preparation of Reagents

Solution 1 (Lugol's iodine)

Iodine	2.0 g
Potassium iodide	4.0 g
Distilled water	100.0 ml

Solution 2 (Bleaching solution)

Sodium thiosulphate	5.0 g
Distilled water	100.0 ml

Solution 3 (Aldehyde fuchsin)

Basic fuchsin	500 mg
70 per cent alcohol	100.0 ml
Conc. HCl	1.0 ml
Paraldehyde	1.0 ml

Allow it to stand for 48 hours. Keep it in refrigerator.

Procedure

1. Deparaffinize and hydrate slides to water.
2. Transfer to solution 1 to remove mercury and keep for 10 minutes.
3. Remove iodine with solution 2.
4. Wash in running water.
5. Transfer slides to 70 per cent alcohol for 1 minute.
6. Place slides in solution 3 for 5 minutes to 2 hours.
7. Differentiate in 70 per cent alcohol.
8. Dehydrate rapidly, clear in xylene and mount in Canada balsam.

Result

Beta cells are stained deep purple.

Alcoholic Silver Nitrate (Hellman and Hellerstorm, 1960)

Fixation

Bouin's, Romeis or 10 per cent formalin.

Reagents Required

Picric acid
Acetic acid
Formalin
Silver nitrate
Nitric acid
Pyrogallic acid

Preparation of Reagents

Solution 1 (Bouin's fluid)

Saturated picric acid	75.0 ml
Formalin	25.0 ml
Glacial acetic acid	5.0 ml

Solution 2 (Alcoholic silver nitrate)

Silver nitrate	10.0 g
Distilled water	10.0 ml
95 per cent alcohol	90.0 ml
IN Nitric acid	0.1 ml

Dilute the solution in the ratio 1 : 6 with distilled water and use.

Solution 3 (Developing solution)

Pyrogallic acid	5.0 g
95 per cent alcohol	100.0 ml
Formalin	5.0 ml

Procedure

1. Deparaffinize and place the slides in solution 1 at 37°C for 2 hours.
2. Wash till yellow colour disappears for 1 hour.

3. Place the slides in 100 and 95 per cent alcohol.
4. Transfer to solution 2 for 12–18 hours at 37°C.
5. Wash in 95 per cent alcohol for 10 seconds (agitate while washing).
6. Develop in solution 3 for 60 seconds
7. Rinse in 95 per cent alcohol.
8. Dehydrate, clear and mount in Canada balsam.

Result

Argyrophilic alphal cells (A) are impregnated with silver.

Maldonado's Method for Pancreatic Cells

Fixation

Bouin's without acetic acid

Reagents Required

Phloxine
Phosphotungstic acid
Azure II
Haematoxylin

Preparation of Reagents

Solution 1 (1 per cent phloxine solution)
Phloxine B   1.0 g
Distilled water   100.0 ml

Solution 2 (3 per cent phosphotungstic solution)
Phosphotungstic acid   3.0 g
Distilled water   100.0 ml

Solution 3 (0.05 per cent Azure II solution)
Azure II   50 mg
Distilled water   100.0 ml

Procedure

1. Deparaffinize and hydrate slides to water.
2. Transfer to solution 1 for 10 minutes.
3. Rinse in distilled water.
4. Transfer to solution 2 for 1 minute.
5. Rinse in distilled water.
6. Place slides in solution 3 for 30 seconds.
7. Rinse in distilled water.
8. Counterstain with Weigert's haematoxylin (*see* Chapter 7).
9. Wash, dehydrate and mount.

Result

A cells   Purple
B cells   Violet blue

| D cells | Light blue |
| Exocrine cells | Greyish blue |

Gomori's Method for Pancreatic Cells (Gomori, 1941b)

Fixation

Bouin's or 10 per cent neutral buffered formalin

Reagents Required

Potassium permanganate
Sodium bisulphite
Phloxine B
Phosphotungstic acid

Preparation of Reagents

Solution 1 (Potassium permanganate solution)
Potassium permanganate   300 mg
Distilled water   100.0 ml
Conc. sulphuric acid   0.5 ml

Solution 2 (5 per cent sodium bisulphite solution)
Sodium bisulphite   5.0 g
Distilled water   100.0 ml

Solution 3 (Chrome haematoxylin)
1 per cent haematoxylin (aqueous)   50.0 ml
3 per cent chrome alum   50.0 ml
   (3 g/100 ml water)

Solution 4 (1 per cent acid alcohol solution) (1 ml HCl + 100 ml water)

Solution 5 (0.5 per cent phloxine B solution)
Phloxine B   500 mg
Distilled water   100.0 ml

Procedure

1. Dewax and dehydrate slides to water.
2. Transfer to Bouin's solution for 18 hours.
3. Wash in tap water till sections are colourless.
4. Keep slides in solution 1 for 1 minute.
5. Bleach in solution 2 for 1 minute.
6. Wash in tap water.
7. Transfer to solution 3 for 10 minutes.
8. Differentiate in solution 4.
9. Wash in distilled water.
10. Transfer to solution 5 for 1 minute.
11. Wash in distilled water.
12. Dehydrate, clear and mount.

Result

Alpha cells	Red
Beta cells	Blue
D cells	Pink to red

PANCREATIC CELLS

Different techniques are being used to demonstrate and identify these cells. With Mallory 'A' cells take a red shade and 'B' cells a blue shade. Azan method is better. Gomori's chrome haematoxylin is much used.

Comments

Herlent stain is also good for pancreas. Tannic acid, haematoxylin, alcian blue and basic fuchsin used by Monroe and Spector (1963).

Kallman (1971) used the following method.

1. Material was fixed in Bouin's, Zenker, formalin, Helly, or Susa.
2. A and B cells for 1 hour in aldehyde fuchsin.
3. It was then rinsed in distilled water.
4. Again the section was stained in 0.05 per cent toluidine blue O in 0.2 M McIlvaine buffer (pH 5.0).
5. A cells take a light blue shade.
6. B cells take a violet-red shade.

For further details especially for 'D' cells consult Epple (1967), Solcia *et al.* (1968, 1969) and for staining 'B' cells, Bussoloti and Bassa (1974).

THYROID GLAND CELLS

In a simple way it can be stated that thyroid consists of secretory epithelial cells, and within the lumen the colloid which is an extracellular amorphous substance. As colloid contains basic types of protein like arginine, it can be demonstrated by Sakaguchi reaction, and S–H groups which could be demonstrated with ferric ferricyanide reduction test (Lillie, 1965). Not only proteins but also the carbohydrate content seems to be evident. Colloid is strongly PAS-positive and a positive dialysed iron reaction indicates the presence of mucopolysaccharides. The colloid is basophilic. Isotope tracer technique has been used in thyroid studies since thyroid uses iodine to make up its hormone, and since radioactive iodine was one of the first isotopes prepared. According to Balanger and Bois (1964), AFA fixation gives good results.

AFA Fixative

Acetic acid	1 part
Formalin	5 parts
Absolute alcohol	15 parts

Fixation time is 24 hours. This fixation gives details with any stain.

Cold fixation in calcium formalin (4–10°C) is also equally good (Pressnell and Schriebman, 1997). Maconail's lead haematoxylin (Solcea *et al.*, 1969) gives a good picture of "F" cells of thyroid which are engaged in the production of polypeptide hormones. Ljunberg (1970) cresyl fast violet is also good.

ADRENAL GLANDS

Adrenal glands in higher vertebrates are composite organs made of cortex and medulla. Both are different structurally and functionally. Whereas in lower verterbates cortical and medullary structures are termed as interrenal and chromaffin tissues respectively.

ADRENAL CORTEX OR INTERRENAL STRUCTURE

In all placental mammals, the interrenal tissue displays 3 layers. Cortical cells produce steroid hormones and almost all are morphologically similar. As a result, it is difficult to distinguish one from the other. They are distinguished from their location in the particular zone.

Two regions could be distinguished in the adrenals, the outer cortex and the central medulla which in turn is ensheathed by thin fibrous capsule. The outer cortex consists of three layers, the outer zona glomerulosa, middle zona fasciculata and inner zona reticulata.

In the zona glomerulosa, cells are columnar and arranged in single vertical rows. The cells are uninucleated and binucleated occasionally.

Zona fasciculata is the major portion of the cortex with cuboidal cells and some columnar cells. The cells are polygonal in shape and arranged in radiating columns. The cells and nuclei are larger than those of zona glomerulosa.

Zona reticulata is interspersed with sinusoids of various sizes giving the appearance of a broken net. The cells display picnotic nuclei with extra lipid vacuoles.

Medullary cells are arranged in irregular rows and are smaller than those of cortical cells. The cells are arranged in groups of 2 to 10 and mainly surrounded by thick strands of interlocking connective tissue.

The cortical cell types and the steroid hormones they produce are as follows.

- Glomerulosa cells produce aldosterone.
- Fasciculata cells produce cortisol.
- Reticularis cells produce sex hormones (androgens) and glucocorticoids.
- Cortical cells produce steroid hormones

HISTOCHEMICAL METHODS FOR ADRENAL CORTEX

Oil Red O Method

Same as described in chapter on lipids

Sudan Black B

Same as described in chapter on lipids

Result

With oil red O droplets, glomerulosa and reticularis cells stain bright red and nuclei stain blue. With Sudan black B, lipid droplets in cortical cells of adrenal cortex stain greenish black and nuclei pink to red.

Perchloric Acid Naphthoquinone Method for Cholesterol

Same as described in chapter on lipids.

Result

The cholesterol-rich lipid droplets of the adrenal cortex are stained dark blue.

ADRENAL MEDULLA OR CHROMAFFIN TISSUE

Epinephrine and norepinephrine are the two catecholamines produced by the cells of the interrenal medulla. According to their function, the two types of cells are named as epinephrine cells and norepinephrine cells. These cells react for chromaffin tests.

HISTOCHEMICAL METHODS FOR ADRENAL MEDULLA

Chromaffin Reaction (Lillie and Fullmer, 1976)

Fixation

Muller's or Orth's fixative

Reagents Required

Potassium dichromate
Sodium sulphate

Preparation of Reagents

Muller's Fluid

Potassium dichromate	2.5 g
Distilled water	100.0 ml
Sodium sulphate	1.0 g
Orths Fluid	

Just before use, add 10 ml formalin to Muller's (100 ml).

Procedure

1. Fix tissue in Muller's or Orth's fluid for 72 hours. Every 24 hours, change the fluid.
2. Wash in running water.
3. Proceed for embedding as usual.
4. Deparaffinize and mount in synthetic resin.
5. Paraffin sections may be stained with haematoxylin/eosin.
6. Frozen sections may be stained with Oil Red O or Sudan Black B.

Result

Chromaffin cells of adrenals	Brown
Epinephrine cells	Dark brown
Norepinephrine cells	Light brown to golden yellow

Ferric Ferricyanide Reaction Test

Procedure is same as described in chapter on proteins.

Result

Reducing sites in the chromaffin cells of the adrenal are stained dark blue.

ENTEROCHROMAFFIN CELL SYSTEM

Enterochromaffin cell system is a broad spectrum consisting of a wide variety of gut-associated epithelial cells. These cells though morphologically identical, are functionally and biochemically different. By using a wide range of techniques, these different cells could be differentiated.

The first person who could detect and term these as enterochromaffin cells was Ciaccio (1906) although Heidenhain in 1870 described this gastrointestinal epithelial cell to take a yellow stain with potassium dichromate. It was Kuttschizky (1897) who pointed out that these cells possess basal granules.

Enterochromaffin Reaction (Lillie and Fullmer, 1976)

Fixation

The tissue is fixed using Muller's solution for 4 weeks, changing the solution every alternate day. After that wash thoroughly for 24 hours.

Procedure

1. Proceed for routine paraffin embedding.
2. Take 5–7 µm sections.
3. Stain section with haematoxylin–eosin.
   Dehydrate, clear and mount.

Result

Enterochromaffin cells of gut and its associated structures take a deep yellow colour.

Grimelius Argyrophil Reaction (Grimelius, 1968)

Fixation

Bouin's, formalin, glutaraldehyde–picric acid mixture

Reagents Required

Silver nitrate
Sodium sulphate

Hydroquinone
Nuclear fast red
Aluminium sulphate

Preparation of Reagents

Solution 1 (Buffered silver nitrate)

0.2 M acetate buffer	10.0 ml
Distilled water	87.0 ml
Aq. 1 per cent silver nitrate	3.0 ml
(1 g/100 ml water)	

Solution 2 (Reducing solution)

Sodium sulphite	5.0 g
Distilled water	100.0 ml
Hydroquinone	1.0 g

Prepare it before use. Warm it at 45°C.

Solution 3 (Nuclear fast red)

Aluminium sulphate	5.0 g
Distilled water	100.0 ml
Nuclear fast red	100 mg

Add a pinch of thymol crystal to avoid formation of moulds.

Procedure

1. Deparaffinize and hydrate slides to water.
2. Transfer sections to a coplin jar containing solution 1 and incubate for 24 hours at 37°C or 3 hours at 60°C.
3. Drain solution and rinse in distilled water.
4. Transfer sections to solution 2 for 1 minute at 45°C.
5. Wash in distilled water.
6. Counterstain nuclei, if desired, in solution 3 for 1–10 minutes.
7. Wash in distilled water.
8. Dehydrate through graded series of alcohol. Clear in xylene and mount.

Results

Granules of argyrophil cells are stained dark brown or black.

Methenamine Silver Argentaffin Reaction (Burtner and Lillie, 1949)

Fixation

Formalin-containing fixatives, and Bouin's fixative. Avoid chromate and mercury-containing fixative.

Reagents Required

Methenamine
Silver nitrate
Holme's borate buffer (Boric acid, sodium tetraborate)

Potassium iodide
Iodine crystals
Gold chloride
Sodium thiosulphate

Preparation of Reagents

Solution 1 (Stock methenamine silver solution)

Methenamine	3.0 g
Distilled water	100.0 ml
5 per cent silver nitrate	5.0 ml
(5 g/100 ml water)	

A precipitate is formed. Keep the solution in a refrigerator.

Solution 2 (Holme's Borate buffer)

a) Boric acid   1.23 g
    Distilled water   100.0 ml

b) Sodium tetraborate   1.9 g
    Distilled water   100.0 ml
    pH of borate buffer should be 7.8. To obtain this solution C is prepared with a and b.

c) Solution a   16.0 ml
    Solution b   4.0 ml

Solution 3 (Silver solution)

Solution 1	30 ml
Solution 2c (Holme's borate buffer)	8.0 ml

Solution 4 (Weigert's iodine solution)

Potassium iodide	2.0 g
Distilled water	5.0 ml
Iodine crystals	1.0 g

Make up the volume to 100.0 ml.

Solution 5 (Toning solution)

Gold chloride	1.0 g
Distilled water	100.0 ml

Solution 6 (Bleaching solution)

Sodium thiosulphate	5.0 g
Distilled water	100.0 ml

Solution 7 (Nuclear stain)

Safranin O	100 mg
Distilled water	100.0 ml

Procedure

1. Deparaffinize and hydrate slides to water.
2. Place in solution 4 for 10 minutes.
3. Transfer to solution 6 for 2 minutes.
4. Wash in running water for 10 minutes.

5. Transfer slides to pre-heated solution 3 at 60°C for 3 hours.
6. Rinse in distilled water.
7. Tone in solution 5 for 10 minutes.
8. Rinse in distilled water.
9. Fix in solution 6 for 2 minutes.
10. Wash in running water for 5 minutes.
11. Counterstain in solution 7 for 1–5 minutes.
12. Dehydrate rapidly, clear in xylene, and mount.

Result

Argentaffin cells are stained black.

Alkaline Diazonium Method for Enterochromaffin Cells (Lillie *et al.*, 1961)

Fixation

Bouin's, formalin. Avoid mercury-containing fixatives.

Reagents Required

Sodium acetate
Sodium diethyl barbiturate
Hydrochloric acid
Fast garnet GBC or fast red salt B
Ammonium or potassium alum
Haematoxylin
Sodium iodate
Citric acid
Chlorol hydrate

Preparation of Reagents

Solution 1 (Veronol acetate buffer)
a)	Sodium acetate	1.17 g
	Veronol	2.94 g
	Distilled water	100.0 ml
b)	0.1 N HCl	8.5 ml
	(0.85 ml HCl/99 ml distilled water)	

To get pH of 8.0, mix the following
Solution 1a	25.0 ml
Solution 1b	8.5 ml
Distilled water	100.0 ml

Solution 2
Fast red salt or Fast garnet GBC	100 mg
Buffer (Solution 1)	100.0 ml

Solution 3 (Meyer's haemalum)
Ammonium of potash alum	5.0 g
Distilled water	100.0 ml

| Haematoxylin | 100 mg |
| Chlorol hydrate | 5.0 g |

Procedure

1. Deparaffinize and hydrate slides to water.
2. Keep slides in a coplin jar containing 40 ml of buffer solution (pH 8.0) (Solution 2) and add 40 mg of stable diazotate. Shake for 1–2 minutes.
3. Wash in running water.
4. Stain nuclei with solution 3.
5. Wash in running water.
6. Dehydrate, clear and mount.

Result

| Granules of enterochromaffin cells | Orange-red |
| Nuclei | Blue |

Clara's Haematoxylin (Clara, 1935)

Fixation

10 per cent neutral buffered formalin, Bouin's fixative

Reagents Required

Haematoxylin
Monobasic sodium phosphate
Dibasic sodium phosphate

Preparation of Reagents

Solution 1 (Stock haematoxylin)
Haematoxylin   1.0 g
Absolute alcohol   100.0 ml

Solution 2 (Phosphate buffer)
a)   Monobasic sodium phosphate   138 mg
Distilled water   100.0 ml
b)   Dibasic sodium phosphate   142 mg
Distilled water   100.0 ml

Solution 3
Solution 2a   36.0 ml
Solution 2b   14.0 ml

pH 6.5

Solution 4 (Dilute haematoxylin)
0.01 M Phosphate buffer (Solution 3)   50.0 ml
Alcoholic haematoxylin (Solution 1) (pH 6.5)   0.5 ml

Once prepared, it can be used only once.

Procedure

1. Deparaffinize and bring slides down to water.
2. Transfer slides to solution 4 for 24–36 hours. Now and then, check the slides under microscope.
3. Dehydrate rapidly, clear in xylene and mount.

Result

Enterochromaffin cells are stained blue to black.

Acid Hydrolysis Basophilia

Fixation

Bouin's fixative with 1 per cent acetic acid (w/v) or picro-glutaraldehyde (1 part 25 per cent glutaraldehyde, 3 parts saturated picric acid, 1 part of 1 per cent sodium acetate)

Reagents Required

Hydrochloric acid
Sodium acetate
Glacial acetic acid
Toluidine blue 'O'

Preparation of Reagents

Solution 1 (Hydrolysing solution)
 Conc. hydrochloric acid 1.4 ml
 Distilled water 98.3 ml

This is 0.2N HCl.

Solution 2 Acetate buffer pH 5.0
 a) Sodium acetate 1.64 g
  Distilled water 100.0 ml

 b) Glacial acetic acid 1.2 ml
  Distilled water 98.8 ml

 c) For buffer pH (5.0) 5.0
  2a solution 35.2 ml
  2b solution 14.8 ml

Solution 3 Staining solution
 Toluidine blue 10 mg
 Solution 2c (acetate buffer) 100.0 ml

Procedure

1. Deparaffinize and hydrate slides to water.
2. Wash in running water.
3. Keep the slides for hydrolysis in solution 1 at 60°C for 3–4 hours. Hydrolysis time varies with different fixatives.

4.   Wash in distilled water.
5.   Transfer to solution 3 for 20 minutes.
6.   Rinse in distilled water.
7.   Dehydrate rapidly, clear and mount.

Result

Pancreatic D cells
Enterochromaffin $\Big\}$ Metachromatic staining

Pancreatic A_2 cells
Adrenalin cells $\Big\}$ Orthochromatic staining

PITUITARY

The hypophysis or pituitary is composed of anterior and posterior lobes. In some vertebrates the two lobes are separate while in others they are partially fused as pars anterior and pars nervosa by a small pars intermedia. Pars anterior develops from oral ectoderm whereas pars nervosa is purely of nervous tissue and this can be demonstrated histochemically by suitable methods.

The adenohypophysis in all vertebrates is composed of cells that occur either in masses as in bony fishes or as randomized mixtures of hormone-producing cells in birds and mammals (Schreibman, 1986). According to him there are 5 to 6 types of cells which almost correlate with the number of hormones originating from the adenohypophysis.

Previously, staining techniques were applied singly or in combination and recognized them on the basis of their tinctorial affinities. It is very essential to analyse the pituitary structure and function to be able to identify the cell types based on their function.

In recent years, great progress has been made in the techniques of staining, from the routine haematoxylin/eosin to the trichrome methods. This enables to localize the specific chemical component of the hormones they produce by application for example of PAS, alcian blue, aldehyde fuchsin, lead haematoxylin and so on. Based on electron microscopical investigations, it is possible to identify the cells based on the granule size. These investigations resulted in the recognition of five to six different types of cells and the cell types correlated with the hormones suspected to originate from the adenohypophysis. If these investigations are coupled with autoradiography and physiological experiments, it is easy to pinpoint the cell according to its functional role, i.e., prolactin cell, somatotrope and so on. Added to these, immunohistochemical techniques were best used to confirm the functional role of the endocrine cells. More recently, the development of *in situ* hybridization has led us to further progress in functional cytology by allowing to distinguish the cells which actively produce, from the cells which store or incorporate it—Trichrome (Mallory) or tetrachrome (e.g. Herlant) PAS, aldehyde fuchsin and so on.

In the anterior part of the pituitary (pars anterior) the cells are classified as (1) Chromophobe cells which do not contain stainable granules (2) Chromophil cells which contain stainable granules. Chromophobes constitute about 50 per cent of the cells. Chromophils are subdivided into two further groups based on their staining reactions as (1) alpha (α) or acidophil cells and (2) Beta (β) cells. Alpha cells constitute 40 per cent of cells and beta cells the remaining 10 percent. α cells take a red shade with the trichrome stain and are PAS-negative. β cells take a blue or green shade with trichrome stain and are PAS-positive.

Neurosecretory substance (NSS) is a conglomeration of various peptides produced in the brain which make their way finally into the pituitary gland. The NSS includes vasopressin and oxytocin. These are produced in the hypothalamus and released from axonal endings close to the blood vessels, almost abutting them.

HISTOCHEMICAL METHODS FOR PITUITARY GLAND

Slidder's Orange Fuchsin Method (Slidders, 1961)

Fixation

Any general fixative

Reagents Required

Celestine blue B
Ferric alum
Glycerine
Orange G
Phosphotungstic acid

Preparation of Reagents

Solution 1 (Celestine blue solution)

Celestine blue B	500 mg
Ferric alum	5.0 g
Glycerine	14.0 ml
Distilled water	100.0 ml

First dissolve alum in distilled water, add celestine blue and boil, cool, filter and then add glycerine.

Solution 2

95 per cent alcohol	100.0 ml
Orange G	0.5–0.7 g
Phosphotungstic acid (PTA)	2.0 g

First dissolve PTA in alcohol and saturate solution with orange G.

Procedure

1. Deparaffinize and hydrate slides to water.
2. Transfer to solution 1.

3. Rinse in distilled water.

4. Stain with Mayer's haemalum for 5 minutes.

5. Wash.

6. Differentiate in acid alcohol.

7. Stain with solution 2.

8. Rinse in distilled water.

9. Transfer to 0.5 per cent acid fuchsin for 5 minutes.
   (0.5 g/100 ml 0.5 per cent acetic acid).

10. Rinse in distilled water.

11. Transfer to 1 per cent PTA for 5 minutes (1 g/100 ml water).

12. Rinse in distilled water.

13. Stain in 1.5 per cent light green for 2 minutes (1.5 g/1.5 per cent acetic acid).

14. Rinse in distilled water.

15. Dehydrate, clear and mount.

Result

Acidophils	Orange-yellow
Basophils	Reddish purple
Chromophobe cells	Grey
Nuclei	Blue-black
Erythrocytes	Yellow

Aldehyde Fuchsin/PAS Method (Elftman, 1959a, 1959b)

Fixation

5 per cent mercuric chloride	100.0 ml
(5 g/100 ml water)	
Potassium chromium sulphate	5.0 g
Formalin	5.0 ml

Prepare fresh and fix overnight.

Reagents Required

Basic fuchsin
Paraldehyde
HCl
Orange G

Preparation of Reagents

Solution 1 (Aldehyde fuchsin) (*see* Chapter 9)
Solution 2 Schiff's reagent
Solution 3 (Orange G)

Orange G	3.0 g
Distilled water	100.0 ml (pH adjusted to 2.0 with few drops of acetic acid)

Procedure

1.   Deparaffinize and bring slides to 70 per cent alcohol.
2.   Transfer to solution 1 for 30 minutes.
3.   Rinse in 70 per cent alcohol.
4.   Oxidize in 1 per cent periodic acid for 15 minutes.
5.   Wash in distilled water.
6.   Place in solution 2 for 30 minutes.
7.   Rinse in tap water.
8.   Stain in solution 3 for 1 minute.
9.   Wash, dehydrate, clear and mount.

Result

Thyrotrophs (β cells)	Dark purple
Gonadotrophs	Red
Acidophils	Orange
Nuclei	Stained

Chrome–Haematoxylin/Phloxine (Pearse, 1960)

Fixation

Bouin's fixative, Susa fixative, Steeve fixative

Reagents Required

Bouin's solution
Chrome alum
Potassium permanganate
Sulphuric acid
Oxalic acid
Haematoxylin
HCl, Phloxine B
Phosphotungstic acid

Preparation of Reagents

Solution 1 (Bouin's solution)

Bouin's solution	100.0 ml
Chrome alum	3.0 g

Solution 2 ($KMnO_4$–Sulphuric acid)

2.5 per cent potassium permanganate	10 ml
5 per cent aqueous sulphuric acid	10 ml
Distilled water	80 ml

Solution 3 (Oxalic acid)

Oxalic acid	150 mg
Distilled water	100.0 ml

Solution 4 (Chromium haematoxylin)
a) Potassium dichromate   150 mg
   Distilled water   100.0 ml
   Conc. H_2SO_4 0.15 ml

b) Haematoxylin   500 mg
   Distilled water 50.0 ml

When dissolved, add 3 per cent potassium chromium sulphate (50.0 ml). Mix well and add the following.

5 per cent potassium dichromate   2.0 ml
$N/2\ H_2SO_4$ 20 ml
(2.5 ml/100 ml water)

Allow to ripen for 48 hours. Store in refrigerator.

Solution 5 (Acid alcohol)
Conc.HCl   1.0 ml
70 per cent alcohol 100.0 ml

Solution 6
Phloxine B   500 mg
Distilled water 100.0 ml

Solution 7 (Phosphotungstic acid (PTA)
PTA   5.0 g
Distilled water 100.0 ml

Procedure

1. Deparaffinize and hydrate slides to water.
2. Place in solution 1 at 37°C for 1–24 hours.
3. Wash in running water until sections are colourless.
4. Oxidize in solution 2 for 3 minutes.
5. Wash in distilled water.
6. Bleach in solution 3.
7. Wash in running water.
8. Transfer to solution 4 for 10 minutes.
9. Differentiate in solution 5.
10. Wash.
11. Place in solution 6 for 2 minutes.
12. Treat with solution 7 for 2 minutes.
13. Wash.
14. Dehydrate, clear and mount.

Results

Neurosecretory substance   Dark purple
Nuclei   Light purple

Cameron and Steele Method (1959)

Fixation

Any general fixative.

Reagents Required

Potassium permanganate
Conc. H_2SO_4
Aldehyde fuchsin (*see* Chapter 9)
Light green
Orange G
Chromotrope 2R
PTA
Glacial acetic acid
Sodium bisulphate

Preparation of Reagents

Solution 1 (Potassium permanganate)

Potassium permanganate	300 mg
Distilled water	100.0 ml
Conc. H_2SO_4	0.3 ml

Solution 2 (Sodium bisulphite)

Sodium bisulphite	2.5 g
Distilled water	100.0 ml

Solution 3 Aldehyde fuchsin (*see* Chapter 9)

Solution 4 (Halmi's mixture)

Light green	200 mg
Orange G	1.0 g
Chromatrope 2R	500 mg
PTA	500 mg
Glacial acetic acid	1.0 ml
Distilled water	100.0 ml

Procedure

1. Deparaffinize and hydrate slides to water.
2. Oxidize in solution 1 for 3 minutes.
3. Wash in distilled water.
4. Bleach in solution 2.
5. Wash in running water.
6. Transfer to 70 per cent alcohol and later to solution 3.
7. Rinse in 95 per cent alcohol.
8. Stain in Halmi's mixture (solution 4).
9. Wipe back of slides.
10. Differentiate in 95 per cent alcohol.

11. Rinse in 95 per cent alcohol.
12. Dehydrate, clear and mount.

Results

Beta cells	Dark purple
Delta cells	Green
Acidophil granules	Orange

Methyl Blue Eosin B Technique (Glenner and Lillie, 1957b)

Fixation

10 per cent neutral buffered formalin, Helly or Zenker

Reagents Required

Eosin B
Methyl blue
Sodium acetate
Glacial acetic acid
Methyl blue
Iodine
Sodium thiosulphate

Preparation of Reagents

Solution 1

Eosin	1 g
Distilled water	100.0 ml

Solution 2

Methyl blue	1.0 g
Distilled water	100.0 ml

Solution 3 (Acetate buffer 0.01M) (pH 4.5–4.6). (*see* Chapter 5).

Solution 4 (Eosin methyl blue)

Solution 1	8.0 ml
Solution 2	2.0 ml
Solution 3	30.0 ml

Prepare fresh every time.

Solution 5

Iodine	500 mg
70 per cent alcohol	100.0 ml

Solution 6 (Bleaching solution)

Sodium thiosulphate	5.0 g
Distilled water	100.0 ml

Procedure

1. Deparaffinize and hydrate slides.
2. Treat with solution 5 to remove mercury.

3. Wash in water and bleach in solution 6.
4. Transfer to solution 4 at 60°C for 1 hour.
5. Wash in running water.
6. Dehydrate, clear and mount.

Result

Acidophil granules	Dark red
Basophil granules	Dark blue
Chromophobes	Pink

Performic Acid/Alcian Blue/PAS/Orange G Method for Human Hypophyses (Adams, 1956)

Fixation

Formol mercury,
10 per cent neutral buffered formalin

Reagents Required

Formic acid
Hydrogen peroxide
Conc. H_2SO_4
Acetic acid
Disodium phosphate
Periodic acid
Schiff's reagent (*see* Chapter 9)
Orange G
Alcian blue
1N HCl

Preparation

Solution 1 (Oxidizing solution)

Formic acid	8.0 ml
Hydrogen peroxide	31.0 ml
Conc. H_2SO_4	0.22 ml

It has to be used immediately or else decomposition sets in. Keep at 25°C.

Solution 2 (Alcian blue solution)

Alcian blue	1.0 g
Distilled water	100.0 ml

pH should be 1.2. To get this, add 1N HCl.

Solution 3 (Periodic acid)

Periodic acid	500 mg
Distilled water	100.0 ml

Solution 4 (Schiff's reagent) (*see* Chapter 9)

Solution 5 (Orange G)

100 ml saturated solution of orange G

Instead of solution 1, peracetic acid is another alternative.

Glacial acetic acid	95.5 ml
Hydrogen peroxide (30 per cent)	259.0 ml
Conc. H_2SO_4	2.2 ml

Allow it to settle for 72 hours and then add 40 mg of disodium phosphate (Na_2HPO_4). Keep at 5°C.

Procedure

1. Deparaffinize and hydrate slides to water.
2. Pass the slides in solution 1 or alternative for 10 minutes.
3. Rinse in running water.
4. Rinse in distilled water giving 2 changes.
5. Transfer to solution 2 for 1 hour.
6. Rinse in distilled water.
7. Transfer to solution 3 for oxidation for 5 minutes.
8. Rinse in distilled water.
9. Transfer to solution 4 for 15 minutes at 25°C.
10. Rinse in distilled water.
11. Stain in orange G (solution 5) for 5 minutes.
12. Rinse in distilled water.
13. Dehydrate, clear and mount.

Result

Acidophils	Orange
Mucoid granules of the R type	Magenta
Mucoid granules of the S type	Purple-blue

Performic Acid/Alcian Blue for Neurosecretory Substance (NSS) (Lillie and Fullmer, 1976)

Fixation

Picroformalin or picroglutaraldehyde

Reagents Required

(As in previous technique)

Preparation of Reagents

Solution 1 (Performic acid)
As in previous technique

Solution 2 (Alcian blue)

Sulphuric acid (90 per cent)	5.4 ml
Distilled water	94.6 ml

pH should be 0.2. To this add 3.0 g of alcian blue 8GS. Heat gently, cool and filter.

Procedure

1. Deparaffinize and bring slides to water.
2. Oxidize as in previous procedure.
3. Wash in distilled water.
4. Transfer to solution 2 for 1 hour.
5. Wash gently in distilled water.
6. Dehydrate rapidly, clear and mount.

Result

NSS is stained bright blue.

Chrome Haematoxylin/Phloxine for Neurosecretory Substance (Bargmann, 1950)

Procedure is given in Chapter 20.

Glenner's Lillie's Method for Pituitary

Fixation

10 per cent neutral buffered formalin

Reagents Required

Eosin B
Aniline blue
Citric acid
Disodium phosphate

Preparation of Reagents

Solution 1 (Staining solution)

1 per cent eosin R	8.0 ml
(1.0 g/100 ml water)	
1 per cent aqueous Aniline blue	2.0 ml
(1 g/100 ml water)	
Citric acid 0.1 M	1.1 ml
Disodium phosphate 0.2 M	0.9 ml
Distilled water	28.0 ml

Solution 2 (Acetone/xylene solution)

Acetone	50.0 ml
Xylene	50.0 ml

Procedure

1. Deparaffinize and hydrate slides to water.
2. Transfer to solution 1 at 60°C for 1 hour.
3. Wash in water.

4. Dehydrate through 50 per cent, 80 per cent and 100 per cent acetone.
5. Clear in solution 2.
6. Mount in Canada balsam.

Result

β-cell granules	Blue to black
Acidophils	Dark red
Chromophobe granules	Pale pink or grey

Monroe–Frommer Method for Pituitary (Monroe and Frommer, 1966)

Fixation

Zenker's fluid

Reagents Required

Tannic acid
Basic fuchsin
Aniline
Phosphomolybdic acid
Alcian blue 8GX

Preparation

Solution 1 (10 per cent tannic acid solution)
Tannic acid   10.0 g
Distilled water   100.0 ml

Solution 2 (1 per cent basic fuchsin stock solution)
Basic fuchsin   1.0 g
100 per cent alcohol   20.0 ml
Distilled water   80.0 ml

Solution 3 (Basic fuchsin working solution)
Basic fuchsin stock   50.0 ml
Distilled water   50.0 ml

Solution 4 (1 per cent aniline solution)
Aniline   1.0 ml
100 per cent alcohol   90.0 ml
Distilled water   10.0 ml

Solution 5 (1 per cent phosphomolybdic acid solution)
Phosphomolybdic acid   1.0 g
Distilled water   100.0 ml

Solution 6 (1 per cent Alcian blue 8GX solution)
Alcian blue 8GX   1.0 g
Distilled water   100.0 ml
Filter before use.

Procedure

1. Dewax and hydrate slides to water.
2. Place in solution 1 for 10 minutes.
3. Wash in running water.
4. Place in solution 3 for 5 seconds.
5. Wash in tap water.
6. Differentiate in solution 4.
7. Transfer to solution 5 for 30 seconds.
8. Rinse in distilled water.
9. Transfer to solution 6 for 30 seconds.
10. Rinse in distilled water.
11. Dehydrate, clear and mount.

Result

Acidophils	Red
Basophils	Green
Delta	Purple
Collagen fibre	Bluish green
NS granules	Deep red

Wilson Ezrin Method for Pituitary (Wilson and Ezrin, 1954)

Fixation

10 per cent neutral buffered formalin

Reagents Required

Periodic acid
Basic fuchsin
Sodium metabisulphite
Hydrochloric acid
Orange G
Phosphotungstic acid
Methyl blue

Preparation

Solution 1 (1 per cent periodic acid solution)
    Periodic acid   1.0 g
    Distilled water   100.0 ml

Solution 2 (Schiff's reagent (*see* Chapter 9)

Solution 3 (Sulphurous acid solution)
    10 per cent sodium metabisulphite   60.0 ml
    1N Hydrochloric acid   50.0 ml
    Distilled water   100.0 ml

Solution 4 (1 per cent orange G solution)
    Orange G   1.0 g
    Distilled water   100.0 ml

Solution 5 (5 per cent phosphotungstic acid (*see* Chapter 19)

Solution 6 (1 per cent methyl blue solution)
    Methyl blue   1.0 g
    Distilled water   100.0 ml

Procedure

1. Deparaffinize and hydrate slides to water.
2. Transfer to solution 1 for 5 minutes.
3. Rinse in distilled water.
4. Transfer to solution 2 for 15 minutes.
5. Wash in running water.
6. Place in solution 3 giving 3 changes.
7. Wash in running water.
8. Transfer to solution 4 for 1 minute.
9. Wash in running water.
10. Transfer to solution 5 for 30 seconds.
11. Wash.
12. Place slides in solution 6 for 1 minute.
13. Wash.
14. Dehydrate, clear and mount.

Result

Beta-granules	Red
Gamma granules	Purple
Acidophils	Yellow

Herlant (1960) Pituitary Stain I

Fixation

Zenker formol or any other general fixative.

Reagents Required

Erythrosin
Aniline blue
Orange G
Glacial acetic acid
Acid alizarine blue
Aluminium sulphate

Preparation of Reagents

Solution 1 (Erythrosin)
    Erythrosin B   1.0 g
    Distilled water   100.0 ml

Solution 2 (Mallory 2)
  Aniline blue   500 mg
  Orange G   2.0 g
  Distilled water   100.0 ml
  Glacial acetic acid 1.0 ml

Solution 3 (Acid alizarine blue)
  Acid alizarine blue 500 mg
  Aluminium sulphate 10.0 g
  Distilled water   100.0 ml

Boil these ingredients in a flask for 5 minutes. Make it up to 100 ml with distilled water, filter and add thymol.

Solution 4 (Phosphotungstic acid)
  Phosphotungstic acid 5.0 g
  Distilled water   100.0 ml

Procedure

1. Dewax and hydrate slides to water.
2. Place in solution 1 for 5 minutes.
3. Rinse in distilled water.
4. Place in solution 2 for 10 minutes.
5. Rinse in distilled water.
6. Place in solution 3 for 10 minutes.
7. Rinse in distilled water.
8. Transfer to solution 4 for 10 minutes.
9. Rinse in distilled water.
10. Differentiate in 70 per cent alcohol.
11. Dehydrate, clear and mount.

Result

Somatotropes	Yellow
Gonadotrophs	Violet
Thyrotrophs	Dark blue
Prolactin cells	Red
Nucleus	Dark blue to violet

Herlant (1960) Pituitary Stain II

Fixation

Hollande Bouin

Reagents Required

Potassium permanganate
Conc. H_2SO_4
Alcian blue 8GX

Glacial acetic acid
Sodium metabisulphite
Periodic acid

Preparation of Reagents

Solution 1 (Permanganate solution)

2.5 per cent potassium permanganate	10.0 ml
5 per cent aq. sulphuric acid	10.0 ml
Distilled water	100.0 ml

Mix just before use.

Solution 2 (Alcian blue) (pH 3.0)

Alcian blue 8GX	1.0 g
Distilled water	100.0 ml
Glacial acetic acid	1.0 ml

Solution 3 (Alcian blue) (pH 0.2)

Alcian blue 8GX	1.0 g
10 per cent sulphuric acid	100.0 ml

Warm the solution until stain dissolves. Cool and filter.

Solution 4 (Sodium metabisulphite)

Sodium metabisulphite	5.0 g
Distilled water	100.0 ml

Solution 5 (Periodic acid)

Periodic acid	1.0 g
Distilled water	100.0 ml

Procedure

1. Dewax and hydrate slides to water.
2. Oxidize in solution 1 for 5 minutes.
3. Rinse in distilled water.
4. Bleach in solution 4 for 2 minutes.
5. Wash in running water.
6. Place in solution 2 or 3 for 30 minutes.
7. Wash in running water for 10 minutes.
8. Treat with solution 5 for 10 minutes.
9. Wash in running water.
10. Transfer to Schiff's reagent for 30 minutes.
11. Wash in solution 4 for 5 minutes.
12. Wash in running water.
13. Dehydrate, clear and mount.

Result

Sections stained with solution 2 (Alcian blue at pH 3.0)

Thyrotrophs	Blue
Gonadotrophs	Blue

With PAS, they stain feebly.

Sections stained with solution 3 (Alcian blue at pH 0.2)

Gonadotrophs	Light blue
Thyrotrophs	Deep blue

With PAS, gonadotrophs are stained violet.

Ewen (1962) Modification of Cameron and Steele (1959) Aldehyde Fuchsin Technique

Fixation

Bouin's with 0.5–1.0 per cent trichloroacetic acid instead of acetic acid, Helly.

Reagents Required

Basic fuchsin
Potassium permanganate
Sulphuric acid
Sodium bisulphite
Light green
Orange G
Chromotrope 2R
Glacial acetic acid

Preparation of Reagents

Solution 1 (Aldehyde fuchsin) (*see* Chapter 9)
Solution 2 (Potassium permanganate)

Potassium permanganate	300 mg
Distilled water	100.0 ml
Conc. sulphuric acid	0.3 ml

Solution 3 (Sodium bisulphite 2.5 per cent)

Sodium bisulphite	2.5 g
Distilled water	100.0 ml

Solution 4 (Halmi, 1950)

Light green	400 mg
Orange G	1.0g
Chromotrope 2R	500 mg
Glacial acetic acid	1.0 ml
Distilled water	100.0 ml

Procedure

1. Dewax and hydrate slides to water.
2. Oxidize in solution 2 for 5 minutes.
3. Rinse in distilled water.
4. Bleach in solution 3 for 3 minutes.
5. Wash in running water.
6. Dip in 70 per cent alcohol.
7. Transfer to solution 1 for 10 minutes.

8. Wipe off back of slide and rinse in 95 per cent alcohol.
9. Differentiate in acid alcohol and then in water.
10. Mordant in a mixture containing

Phosphotungstic acid	4.0 g
Phosphomolybdic acid	1.0 g
Distilled water	100.0 ml

11. Rinse in water.
12. Transfer to solution 4 (Halmi's mixture) for 1 hour.
13. Differentiate in acid alcohol.
14. Rinse in 95 per cent alcohol.
15. Dehydrate, clear and mount.

Result

Gonadotrophs	Dark purple
Thyrotrophs	Green
Acidophilic granules	Orange
Nucleoli	Bright red

Lead Haematoxylin Method (Solcia *et al.*, 1969)

Fixation

Any aldehyde fixative

Reagents Required

Lead nitrate
Ammonium acetate
Haematoxylin

Preparation of Reagents

Solution 1 (Lead solution)

5 per cent aq. lead nitrate	1 part
Saturated aq. ammonium acetate	1 part

Add 2 ml of 40 per cent formalin to every 100 ml of filtrate. Keeps for several weeks at room temperature.

Solution 2 (Working solution)

Haematoxylin	0.200 mg
95 per cent ethanol	1.5 ml
Solution 1	10.0 ml
Distilled water	10.0 ml

Keep it for 30 minutes Filter and make up to 75 ml with distilled water and use immediately.

Procedure

1. Dewax and hydrate slides to water.
2. Stain in solution 2 for 3 hours at 37°C.

3. Wash in running water for 10 minutes
4. Dehydrate, clear and mount.

Result

Pancreatic islet A and D cells
Thyroid cells ⎱ Dark blue
Pituitary MSH and ACTH cells

Chrome Haematoxylin–Bargmann Modification for Neurosecretory Substance (NSS) (Pearse, 1968)

Fixation

Bouin's fixative, Susa fixative or Steeve's fixative. Bouin's is preferable.

Reagents Required

Chrome alum
Potassium permanganate
Sulphuric acid
Oxalic acid
Hydrochloric acid
Phloxine B
Phosphotungstic acid

Preparation of Reagents

Solution 1 (Bouin's chrome alum)
    Bouin's solution   100.0 ml
    Chrome alum   4.0 g

Solution 2 (Potassium permanganate–sulphuric acid)
    2.5 per cent aq. potassium permanganate   1 part
    5 per cent aq. sulphuric acid   1 part
    Distilled water   8 parts

Solution 3 (Oxalic acid)
    Oxalic acid   1.0 g
    Distilled water   100.0 ml

Solution 4 (Chromium haematoxylin) (*see* Chapter 20)

Solution 5 (Acid alcohol)
    Conc. hydrochloric acid   1.0 ml
    70 per cent alcohol   100.0 ml

Solution 6 (Phloxine B)
    Phloxine B   500 mg
    Distilled water   100.0 ml

Solution 7 (Phosphotungstsic acid)
    Phosphotungstic acid   5.0 g
    Distilled water   100.0 ml

Procedure

1. Dewax and hydrate slides to water.
2. Mordant in solution 1 at 37°C for 24 hours.
3. Wash in running water until sections are colourless.
4. Oxidize in solution 2 for 3 minutes.
5. Wash in distilled water for 1 minute.
6. Bleach in solution 3 for 1 minute.
7. Wash in running water.
8. Stain in solution 4 for 10 minutes.
9. Differentiate in solution 5 for 30 seconds.
10. Wash in running water for 3 minutes.
11. Transfer to solution 6 for 2 minutes.
12. Treat with solution 7 for 2 minutes.
13. Wash in running water for 5 minutes.
14. Dehydrate, clear and mount.

Result

Neurosecretory substance	Deep purple
Nuclei	Light purple
Background	Pinkish red

Comment

Instead of chrome haematoxylin, McGuire and Opel (1964) used the following stain after potassium permanganate oxidation.

Resorcin fuchsin	1.0 g
70 per cent ethanol	98.0 ml
Conc. HCl	2.0 ml

It lasts for 20 days.

Peracetic Acid

Peracetic acid is used as an oxidizing reagent (Tan, 1973), the oxidation time being 10–15 minutes.

Glacial acetic acid	72.0 ml
Hydrogen peroxide	226.0 ml
Sulphuric acid	2.0 ml

Use it for 3 days.

Periodic Acid/Schiff–Orange G Method (after Hotchkes, 1948)

Fixation

Any general fixative

Preparation of Reagents

Solution 1 (Periodic acid)

Periodic acid	0.4 g
Ethanol	35.0 ml

Distilled water   10.0 ml
0.2 M sodium acetate   5.0 ml
    (27.2 g + 1000 ml water)

First dissolve periodic acid in ethanol and distilled water and then add 0.2 M sodium acetate. Store at 0°C to be used at room temperature. When solution turns brown, it has to be discarded.

Solution 2 (Reducing solution)
  Potassium iodide   1.0 g
  Sodium thiosulphate   1.0 g
  Ethanol   30.0 ml
  Distilled water   20.0 ml
  20 per cent aq. hydrochloric acid   20.0 ml

Dissolve potassium iodide and sodium thiosulphate in ethanol–distilled water mixture. Add hydrochloric acid and store at 0°C. Lasts for a fortnight.

Solution 3 (Orange G solution)
  Orange G   2.0 g
  5 per cent phosphotungstic acid   100.0 ml

Mix and allow to stand for 24 hours Filter before use.

Solution 4 (Schiff's reagent) (*see* Chapter 9)

Procedure

1. Dewax and bring sections to distilled water.
2. Wash in 70 per cent alcohol.
3. Transfer to solution 1 for 5 minutes.
4. Rinse in 70 per cent ethanol.
5. Transfer to solution 2 for 3–5 minutes.
6. Again rinse in 70 per cent alcohol.
7. Transfer to solution 4 for 6 minutes.
8. Wash in running water.
9. Stain nuclei with haematoxylin until nuclei are blue.
10. Stain in solution 3 for 10 seconds.
11. Differentiate in tap water for 30 seconds.
12. Dehydrate, clear and mount.

Result

Mild pituitary colloids	Magenta
Nuclei	Blue-black
Acidophil cells	Yellow
Chromophobes	Pale blue or grey

OFG Method for Cells of the Anterior Pituitary (Slidders, 1961)

Fixation

Helly's fluid, formaldehyde or Bouin's (selective) fixative. Sublimate is preferred.

Preparation of Reagents

Solution 1 (Orange G)

Orange G	500 mg
Phosphotungstic acid	2.0 g
Absolute alcohol	95.0 ml
Distilled water	5.0 ml

Solution 2 (Acid fuchsin)

Acid fuchsin	500 mg
Glacial acetic acid	0.5 ml
Distilled water	99.5 ml

Procedure

1. Dewax and bring sections to water.
2. Treat with celestine blue haematoxylin.
3. Wash.
4. Differentiate in acid alcohol.
5. Wash in running tap water.
6. Rinse in 95 per cent alcohol.
7. Treat with solution 1 for 2 minutes.
8. Wash in running water.
9. Transfer to solution 2 for 2–5 minutes.
10. Rinse in tap water.
11. Differentiate in tap water for 30 seconds.
12. Dehydrate, clear and mount.

Result

Acidophils	Orange-yellow
Basophils	Magenta
Chromophobes	Pale greyish green
Nuclei	Blue-black

Br AB-OFG Method for Cells of the Anterior Pituitary (Slidders, 1961)

Fixation

Any general fixative

Preparation of Reagents

Solution 1 (Bromine water)

10 per cent aq. hydrobromic acid	45.0 ml
2.5 per cent potassium permanganate	5.0 ml

Solution 2 (Alcian blue)

Alcian blue	100 mg

Conc. sulphuric acid	1.0 ml
Glacial acetic acid	9.0 ml
Distilled water	90.0 ml

First mix alcian blue in sulphuric acid, slowly add glacial acetic acid, stir well, and add distilled water. Filter.

Procedure

1. Dewax and bring sections to water.
2. Transfer to solution 1 for 5 minutes.
3. Wash in running tap water for 5 minutes.
4. Rinse in distilled water.
5. Transfer to solution 2 for 1 hour.
6. Wash in tap water.
7. Proceed as for OFG method (*see* page 728).

Result

Acidophils	Orange-yellow
Basophils (S)	Dark green-blue
Basophil cells (R)	Magenta-red
Chromophobe cells	Pale grey-green
Nuclei	Grey-blue

Carmoisine orange G–Wool Green Method for Differentiating Acidophil Cells (Brookes, 1968)

Preparation of Solutions

Solution 1

Carmosoine	1.0 g
1 per cent acetic acid	100.0 ml

Solution 2

Phosphotungstic acid	2.0 g
95 per cent ethanol	100.0 ml

Saturate orange G in this solution.

Solution 3

Wool green S	500 mg
0.5 per cent acetic acid	100.0 ml

Procedure

1. Dewax and bring sections to water.
2. Treat with 10 per cent aq. copper sulphate for 2 hours at 44°C.
3. Treat with solution 1 for 30 minutes.
4. Wash in distilled water.
5. Rinse in 95 per cent alcohol.
6. Transfer to solution 2 for 5–30 minutes.

7. Rinse in distilled water.
8. Bring back slides to solution 1 for 5 minutes.
9. Rinse in distilled water.
10. Counterstain with solution 3 for 10 minutes.
11. Rinse in distilled water and treat with 1 per cent acetic acid for 2 minutes.
12. Dehydrate, clear and mount.

Result

Somatotropes	Yellow
Red blood cells	Red
Basophils	Green

Masked Metachromasia Method (Solcia *et al.*, 1968)

Fixation

Formaldehyde, paraformaldehyde, glutaraldehyde, Bouin's or Helly's fixative.

Preparation of Solutions

Solution 1 (Azure A)
Azure A   5 mg
Distilled water   100.0 ml

Solution 2 (Toluidine blue solution)
Toluidine blue   10 mg
20 mM McIlvaine's buffer   100.0 ml

Solution 3 (Acid)
0.2 M hydrochloric acid

Procedure

1. Dewax and bring down sections to water.
2. Hydrolyse in solution 3 at 60°C for 3–4 hours (for sections fixed in formaldehyde, paraformaldehyde and Bouin's) or 12 hours at 60–65°C (for sections fixed in glutaraldehyde or Helly's).
3. Wash in distilled water.
4. Treat with either solution 1 or 2 for 6 hours.
5. Wash in distilled water.
6. Mount in glycerine jelly or blot, dry, dehydrate in isopropanol, clear and mount.

Result

Endocrine cell granules are stained purple-red.

Masson–Hamperl Argentaffin Method (Singh, 1964)

Fixation

Formaldehyde
Glutaraldehyde

Preparation of Reagents

Solution 1 (Silver nitrate)
    Silver nitrate   10.0 g
    Distilled water   100.0 ml

To 10.0 ml of this solution, add concentrated ammonia dropwise. A precipitate is formed which slowly dissolves by continuous addition of ammonia. To this, again add 10 per cent silver nitrate (solution 1) dropwise until opalescent solution is formed. For working, use 1 ml of this with 10 ml of distilled water. Prepare fresh before use.

Procedure

1.   Dewax and bring sections down to water.
2.   Transfer to pre-heated working solution at 60°C for 15–30 minutes. Examine sections until they are light brown.
3.   Wash in distilled water.
4.   Place in 1 per cent aq. sodium thiosulphate for 1 minute.
5.   Wash in running water.
6.   Counterstain with 0.5 per cent aq. neutral red.
7.   Wash in tap water.
8.   Dehydrate, clear and mount.

Result

Argentaffin granules   Black
Nuclei   Red

Churukian and Schenk's Method for Argyrophils (Churukian and Shenk, 1979)

Fixation

Formaldehyde

Preparation of Reagents

Solution 1 (Acidified water)
    Add 0.3 per cent aqueous citric acid to 500 ml of distilled water. The pH should be 4.2.

Solution 2 (Silver nitrate solution)
    Silver nitrate   500 mg
    Solution 1   200.0 ml

Solution 3 (Reducing solution)
    Sodium sulphate   5.0 g
    Hydroquinone   1.0 g
    Distilled water   100.0 ml

Procedure

1.   Dewax and bring slides to solution 1.
2.   Transfer sections to solution 2 (pre-heated at 58°C) for 2 hours.
3.   Rinse in distilled water.

4. Treat with pre-heated solution 3 at 58°C for 5 minutes.
5. Wash in distilled water.
6. Again place sections in solution 2 at 58°C for 10 minutes.
7. Wash.
8. Again place sections in solution 3 at 58°C for 5 minutes.
9. Rinse in distilled water.
10. Dehydrate, clear and mount.

Result

Argyrophil cell granules are stained black.

Alcoholic Silver Nitrate Method for Argyrophil Cells especially D Cells (α_1) cells of the Pancreas (Hellerstorm and Hellman, 1960)

Fixation

Bouin's fixative or formaldehyde followed by fixation in Bouin's for 2 hours at 37°C.

Preparation of Reagents

Solution 1

Silver nitrate	10.0 g
Distilled water	10.0 ml
95 per cent alcohol	90.0 ml
M nitric acid	0.1 ml

pH of the solution should be 5.0 to be adjusted with concentrated ammonium hydroxide.

Solution 2 (Developing solution)

Pyrogallic acid	5.0 g
95 per cent alcohol	95.0 ml
40 per cent formaldehyde	5.0 ml

Procedure

1. Bring sections after dewaxing to water.
2. Dehydrate in 95 per cent alcohol.
3. Treat with solution 1 at 37°C overnight, free from light.
4. Rinse in 95 per cent alcohol for 10 seconds.
5. Treat with solution 2 for 1 minute.
6. Rinse in 95 per cent alcohol giving 3 changes.
7. Rinse in absolute alcohol.
8. Mount in resinous mount.

Result

D cell granules (α_1) of pancreas	Black
A and B cells	Negative
Argyrophil cells	Positive

Table 21.1   Histochemical techniques applied for the demonstration of endocrine glands— pancreas

Technique	Fixative	Alpha cells	Beta cells	Delta cells	Argyrophilic alpha cells
Mallory Heidenhain's Azan	Bouin's or Helly	Bright red	Orange-brown	Dark blue	
Mallory Heidenhain's Azan after permanganate oxidation	Any general fixative other than alcohol	Bright red	Cannot be distinguished	Cannot be distinguished	
Phosphotungstic acid Haematoxylin (PTAH)	Bouin's or formalin	Deep blue			
Post-coupled *p*-dimethylamino benzaldehyde	Neutral buffered formalin	Intense blue			
Aldehyde fuchsin	Formalin, Bouin's, no dichromate fixative.		Deep purple		
Alcoholic silver nitrate	Bouin's, Romeis				Impregnated with silver
Maldonados method	Bouin's without acetic acid	Purple	Violet-blue	Light blue	

Table 21.2 Histochemical techniques applied for the demonstration of chromophobes— Pituitary

Technique	Fixative	Acidophils	Basophils	Thyrotrophs	Gonado trophs	Chromop hobes	Beta cells	Delta cells	Gamma granules	NSS
Slidder's Orange Fuchsin	Any general	Orange-yellow	Reddish purple			Grey				
Aldehyde Fuchsin/PAS	5% mercuric chloride – 100 ml + 5 g potassium chromium sulphate + formalin — 5 ml (Fix overnight)	Orange		Dark purple	Red					
Chrome-haematoxylin/ Phloxine	Bouin's, Susa, Steeve									Deep purple
Cameron and Steele method	Any general	Orange					Dark purple	Green		
Methylene blue Eosin method	10% neutral buffered formalin Helly or Zenker	Dark red	Dark blue			Pink				

(Contd.)

Table 21.2 (Continued)

Technique	Fixative	Acidophils	Basophils	Thyrotrophs	Gonado trophs	Chromopho bes	Beta cells	Delta cells	Gamma granules	NSS
Performic acid/Alcian blue/PAS/ Orange G	Formol mercury, 10% neutral buffered formalin	Orange								
Glenner's Lillie's method	10% neutral buffered formalin	Dark red	Blue to black			Pink or grey				
Monroe-Frommer method	Zenker	Red	Green					Purple		
Wilson Ezrin method	10% neutral buffered formalin	Yellow	Red						Purple	
Herlant's stain I	Zenker's formol			Dark blue	Violet					
Herlant's stain II	Hollande, Bouin's			Deep blue	Blue					
Even's modification of aldehyde fuchsin	Bouin's Helly	Orange		Green	Dark purple					
Performic acid/Alcian blue	Picro formalin									Bright blue

Table 21.3 Histochemical techniques applied for the demonstration of enzymes—Adrenals

Technique	Fixative	Chromaffin	Epinephrine	Norepine-phrine	Entero-chromaffin	Argyrophil cells	Argentaffin	Adrenal cortex
Sudan black B method	10% neutral buffered formalin							Greenish black
Perchloric acid/ naphthoquinone	10% neutral buffered formalin							Dark blue
Chromaffin reaction	Muller's, Orth's	Brown	Dark brown	Light brown to golden yellow				
Ferric ferricyanide reaction	Alcoholic formalin, Carnoy	Dark blue						
Enterochromaffin reaction	Muller's solution				Deep yellow			
Grimelius Argyrophil reaction	Bouin's formalin					Dark brown or black		
Methenamine Silver Argentaffin	Formalin, Bouin's						Black	
Alkaline diazonium method	Bouin's, formalin				Orange-red			
Clara's haematoxylin	Bouin's, 10% NBF				Blue to black			

HAEMATOLOGICAL STUDIES

A specialized connective tissue which produces blood cells and simultaneously removes worn-out blood cells is the haemopoietic tissue. Categorization of the haemopoietic tissue is as follows:

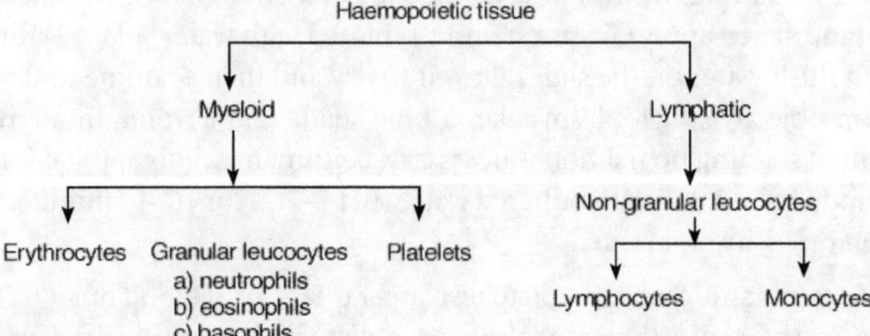

BLOOD

Blood is a form of fluid tissue and of connective tissue type and is always in circulation in the body. In higher animals, it is a viscous complex and in lower animals, it is thin and watery. It is a major transporting medium and is always in movement in life. It is heavier than water, red in colour except in lower animals like *Amphioxus* and *Leptocephalus*. It is saltish in taste and with a viscosity 5 times greater than water. On centrifugation it separates into two main components, a light yellowish supernatant plasma (55 per cent) and a reddish cellular portion, the corpuscles (45 per cent), at the bottom.

Plasma is straw-coloured (light yellowish), homogeneous and sticky. It has some coagulants such as fibrinogen and anticoagulants like heparin. The liquid part of the plasma, after removing the clot, is known as serum. The organic components of blood are proteins, amino acids, nitrogenous waste products such as ammonia, urea, uric acid, enzymes, hormones, antibodies, fatty acids, alcohols (cholesterol), vitamins, metabolic products (xanthine, hypoxanthine and creatine, etc.), neutral fats (phospholipids) and sugar (like glucose). The inorganic components are chlorides, bicarbonates, sulphates, phosphates of sodium, potassium, calcium and magnesium besides iron, iodine and phosphorus. Gaseous components of blood include oxygen, carbon dioxide, and traces of nitrogen.

The proteins which form 7 per cent of the plasma are albumins, globulins and fibrinogens (3 : 1 ratio). Functions of plasma proteins are closely associated with nutrition, transport and physico-chemical aspects. Corpuscles are of three types and float in plasma. They include red blood corpuscles or erythrocytes, white blood corpuscles or leucocytes and blood platelets or thrombocytes.

Blood Examination

Prepare a thin smear by finger prick. A small drop is spread on a clean slide and with a spreader, a uniform smear is made. Allow it to dry. Then stain with Giemsa or Leishman's stain. For staining with Giemsa, fix the slide in methanol for 5 minutes, then allow it to stand. Then apply Giemsa stain 1 : 1 diluted with water (pH 7.2) for 45 minutes. Then flush water on the slide, allow it to dry and then examine under the oil immersion objective. Cytoplasm takes a blue shade and chromatin a red shade. For Leishman staining no fixation is necessary. Leishman stain is applied for a minute or less and then diluted with buffered water (pH 7–7.2) for 10–15 minutes. The smear is then dried and examined.

Thick smear examination enables you to test larger quantity of blood but it has its disadvantage in the sense that morphology is distorted resulting in non-identification. A drop of blood is collected from a finger and put on a non-greasy glass slide and spread with the corner of another slide to form a uniformly thick smear. Smear is dried. Giemsa, Leishman, Wright's JSB stains yield good results.

Wright's Stain

Solution 1

Methylene blue	800 mg
Azure 1	500 mg
Disodium hydrogen phosphate	6.25 g
Distilled water	500 ml

Solution 2

Eosin water soluble	1.0 g
Na_2HPO_4	5.0 g
KH_2PO_4	6.25 g
Distilled water	500 ml

Grind the ingredients and dissolve. Allow the stain to ripen for 4 hour and filter.

EXAMINATION OF BLOOD SAMPLES

Parasites like *Trypanosoma* spp., *Plasmodium* spp., *Babesia* spp., *Leishmania donovani* and microfilariae are some of the parasites which are recovered from the blood.

Identification is based on the examination of permanent thick and thin blood films. Films can be prepared from fresh whole blood. Giemsa stain, Wright's stain or a combination of both and Delafield's haematoxylin stains are recommended.

Clear non-greasy slides should be used for the preparation of all blood films. Blood films must be prepared within one hour after the blood is drawn. Otherwise organism morphology may not be clear. Blood films should be stained as soon as possible. Delay of more than 3 days may result in failure to demonstrate staining characteristics of individual species.

The most common stains are of two types. In Wright's stain, the fixative is combined with the staining solution. As a result of this both fixation and staining occur at the same time. Thick films are lysed while processing.

Giemsa stain is the other one. Here the fixative and stain are separate. So the thin film is fixed with absolute methanol before staining. Thick films will be lysed during the staining process.

When slides are removed from either staining, they are air-dried in a vertical position. Then they can be examined under oil immersion by placing oil directly on the uncovered blood film. If permanent slides are required for record they should be mounted in permount and a cover glass is used.

Giemsa stain is commercially available as stock solution or in the form of powder. Stock solution is diluted with buffer 1 : 20 for thin film (staining time 20 minutes) or 1 : 50 for thick films (staining time 50 minutes). Phosphate buffer is used to dilute the solution. It should be neutral or slightly alkaline (pH 7.0–7.2).

Giemsa stain gives the following results.

- Erythrocytes are stained pale red.
- Nuclei of leucocytes are stained purple.
- Cytoplasm and eosinophil granules are stained deep pink or purple.

In malarial parasites

- Cytoplasm is stained blue.
- Nuclear material is stained red to purple red.

In thick blood films, blood cells concentrate in the centre of the film. Initially the slide should be examined under low magnification (10× objective) to detect microfilariae. *Babesia* spp., *Trypanosoma* spp. and malarial parasites are best detected under oil immersion (100× objective). Presence of brown pigment granules may indicate the presence of malarial parasites. RBC will concentrate at the periphery of the thick film and this gives a clue for malarial diagnosis.

In thin films, initial examination is under low magnification. Microfilariae occur in very small numbers in thin films. They are commonly found at the edge of the film or at the farther end of the film because they are carried to these sites while spreading whereas RBCs are drawn out into one single distinctive layer of cells.

These can be examined for malarial parasite, *Babesia* spp. and *Trypanosoma* spp. Examination of these parasites is under high magnification (100× objective).

HISTOCHEMICAL METHODS FOR EXAMINATION OF BLOOD

Wright's Stain Method

Fixation

Fix dry film in methyl alcohol

Reagents Required

Monobasic potassium phosphate
Dibasic sodium phosphate
Sodium thiosulphate

Preparation of Reagents

Solution 1

Monobasic potassium phosphate	6.63 g
Anhydrous dibasic sodium phosphate	2.56 g
Distilled water	1000.0 ml

Solution 2

Sodium thiosulphate	100–200 mg
Distilled water	1000.0 ml

It can be used after one hour.

Procedure

1. Flood slides with a few drops of Wright's stain for 1–2 minutes.
2. Flood with few drops of solution 1 or solution 2 for 1–2 minutes.
3. Dip in distilled water.
4. Blot and dry and put a cover glass.

Results

Erythrocytes	Pink
Nuclei	Blue to purple
Basophilic granules	Deep purple
Platelets	Violet to purple
Lymphocytes	Reddish
Eosinophilic granules	Red to red-orange

Wright's stain is used differently for cold-blooded and warm-blooded animals.

Wright's Stain for Cold-blooded Vertebrates (Heady and Rogers, 1962)

1. Flood slide with Wright's stain for 2 minutes.
2. Flood with equal volume of citrate buffer for 6–7 ppm.
3. Wash in distilled water, blot and mount.

Wright's Stain for Birds

1. Flood slides with Wright's stain (3 g of powder in 500 ml of methyl alcohol) for 8 minutes.
2. Simultaneously flood slides with formalin (concentrated formalin 0.25 ml + 500 ml water).
3. Flood slides with distilled water adjusted to pH 6.8 with 0.25 per cent sodium carbonate.
4. Blot and again rinse in a mixture of ether + methyl alcohol (1 : 10)
5. Blot, dry and mount.

Result

Erythrocytes	Pinkish yellow
Thrombocytes	Grey-blue
Lymphocytes	Blue
Monocytes	Light blue
Heterophils	Yellowish to brownish
Eosinophils	Yellowish to brownish
Basophils	Dark purple

Wright's Stain for Birds (Santamarina, 1964)

Preparation of Reagents

Solution 1 (Wright's stain)

Wright's stain powder	3.3 g
Absolute methyl alcohol	500.0 ml

Solution 2 (Formalin)

40 per cent formaldehyde	0.25 ml
Distilled water	500.0 ml

Final pH should be 6.8. This can be done with addition of 0.25 per cent sodium carbonate or 0.25 per cent HCl.

Procedure

1. Flood dry smears with solution 1 for 5–8 minutes.
2. Slowly add solution 2. A metallic sheen is observed.
3. Pour off fluid and dip in distilled water adjusted to pH 6.8.
4. Blot.
5. Differentiate in 1 : 1 ether/methyl alcohol for 6–10 minutes.
6. Clear and mount.

Result

Erythrocytes	Cells	Yellow to purplish
	Chromatin	Purple
Thrombocytes	Cytoplasm	Grey-blue
	Chromatin	Purple

Lymphocytes	Granules	Blue
	Chromatin	Purplish red
Monocytes	Granular cytoplasm	Light blue
	Chromatin	Purple
Heterophils	Cells (rods)	Yellow-brown
	Chromatin	Light purple
Eosinophils	Almost similar to heterophils	
Basophils	Dark	Purple granules

THICK SMEARS

A large drop of blood is taken from the finger with a needle prick. Make a circle of the blood smear with a finger. Dry the slide.

Field–Wartman Thick Film Method (Field, 1941; Wartman, 1943)

Reagents Required

Methylene blue
Azure A
Dibasic sodium phosphate
Monobasic potassium phosphate
Eosin Y

Preparation of Reagents

Solution 1

Methylene blue	800 mg
Azure A	500 mg
Dibasic sodium phosphate (anhydrous)	5.0 g
Monobasic potassium phosphate	6.25 g
Distilled water	500.0 ml

Solution 2

Eosin Y	1.0 g
Dibasic sodium phosphate (anhydrous)	5.0 g
Monobasic potassium phosphate	6.25 g
Distilled water	500.0 ml

Dissolve phosphate salts and later add other ingredients.

Procedure

1. Place slides in solution 1 for 2 seconds.
2. Dip in distilled water until excess does not come from slide.
3. Transfer to solution 2 for 1 second.
4. Rinse in distilled water.
5. Dry and mount.

Result

| Cytoplasm | Pale blue |
| Nuclei | Dark blue |

Eosinophilic granules	Red
Neutrophils	Purple
Basophils	Deep purple

Wilcox Thick Film Method

Preparation of Reagents

Solution 1 (Wright–Giemsa stock solution)

    Giemsa   2.0 g
    Glycerine   100.0 ml

Heat for 2 hours in a water bath and add aged Wright's staining solution (100.0 ml).
Allow to stand overnight and add 800 ml Wright's staining solution. Filter.

Solution 2

    Wright–Giemsa stock solution   1 part
    Neutral distilled water   9 parts

Solution 3 (Buffer pH 7.0)

    Dibasic sodium phosphate   9.5 g
    Distilled water   1000.0 ml

Solution 4

    Monobasic potassium phosphate   9.7 g
    Distilled water   1000.0 ml

Solution 5 (Working solution)

    Solution 3   61.1 ml
    Solution 4   38.9 ml
    Distilled water   900.0 ml

Procedure

1. Stain the smears for 10 minutes.
2. Flush the slides with solution 5.
3. Place slides in neutral distilled water for 1 minute.
4. Dry slides.

HISTOCHEMICAL METHODS FOR BLOOD TISSUE ELEMENTS AND INCLUSION BODIES

Giemsa–Walbach's Modification (Mallory, 1944)

Fixation

Any general fixative

Reagents Required

Glacial acetic acid
Giemsa powder
Methyl alcohol
White wood stain

Preparation of Reagents

Solution 1

    70 per cent alcohol   100.0 ml
    Glacial acetic acid   0.5–1.0 ml

Solution 2 (Giemsa stock solution)

    Giemsa powder   1.0 g
    Methyl alcohol   66.0 ml
    Glycerol   666.0 ml

Add Giemsa stain to glycerol and keep in an oven adjusted to 60°C for 2 hours. Later add methyl alcohol.

Solution 3 (Working solution)

    Stock solution (Giemsa)   2.5 ml
    Methyl alcohol   3.0 ml
    Distilled water   100.0 ml

Solution 4 (Rosin stock solution)

    White wood rosin   10.0 g
    Absolute alcohol   100.0 ml

Solution 5

    Rosin stock   5.0 ml
    95 per cent ethyl alcohol   40.0 ml

Procedure

1. Deparaffinize and hydrate slides to water.
2. Place in solution 1 for 5 minutes.
3. Wash in running water.
4. Rinse in distilled water.
5. Transfer to solution 3 for 1 hour.
6. Transfer to fresh Giemsa stain for overnight.
7. Differentiate in rosin alcohol (Solution 5).
8. Rinse in 90 per cent alcohol.
9. Dehydrate, clear and mount.
10. Rinse in distilled water.

Result

Nuclei   Reddish purple
Inclusion bodies   Blue or purplish blue

Giemsa stain (Cramer *et al.*, 1973)

Fixation

Any general fixative

Reagents Required

Azure II eosin
Azure II

Azure B eosin
Azure A eosin

Preparation of Reagents

Solution 1 (Azure–eosinate stock solution)

Azure II eosin	2.0 g
Azure II	1.0 g
Azure B eosin	1.0 g
Azure A eosin	500 mg

Dissolve these ingredients in a mixture of equal parts of glycerine and methanol (250 ml) and allow to stand at room temperature overnight, shake all night.

Working solution 2a (Staining time 2 hours)

Stock solution	5.0 ml
Distilled water	65.0 ml

Adjust with 1 per cent acetic acid to make final pH to be 4.8 to 5.2.

Working solution 2b (Overnight stain)

Stock solution	3 to 5 drops
Distilled water	65.0 ml

Adjust with 1 per cent acetic acid to make pH to be 6.5 to 6.8.

Procedure

1. Dewax and hydrate slides to water.
2. Place in working solution 2a (2 hours staining) in 2b (overnight).
3. Immerse in 1 per cent acetic acid and take out at once.
4. Blot and rinse in absolute alcohol (until bluish).
5. Clear and mount.

Results

Nuclear chromatin	Dark blue
Eosinophil granules	Red-orange
Mast cell granules	Dark purple
Erythrocytes	Pink to red
Connective tissue	Pink to light purple
Lymphocytes	Blue cytoplasm

Jenner Giemsa for Malaria (McChung, 1939)

Fixation

Any general fixative

Reagents Required

Jenner's stain
Methyl alcohol
Giemsa powder
Glycerol

Preparation of Reagents

Solution 1 (Jenner's stain)

Jenner's stain	200 mg
Methyl alcohol	100.0 ml

Solution 2

Giemsa powder	3.8 g
Methyl alcohol	75.0 ml
Glycerol	25.0 ml

Add Giemsa powder to alcohol and warm for 2 hours at 60°C, then add alcohol.

Solution 3 (Working solution)

Giemsa stock solution	10.0 ml
Distilled water	100.0 ml

Procedure

1. Deparaffinize and bring slides to 50 per cent alcohol.
2. Flood sections with solution 1 (equal amounts of Jenners and distilled water).
3. Flood sections with Giemsa stain (solution 3) for 20 minutes.
4. Rinse in distilled water and later in acetic acid.
5. Dehydrate.
6. Clear in acetone–xylene mixture.
7. Clear in xylene and mount.

Result

Chromatin	Red
Cytoplasm	Blue
Pigment	Yellow

Giemsa Staining for Cold-blooded Vertebrates (Pienaar, 1962)

1. Fix the dry smears in methyl alcohol.
2. Make a dry smear as already described and fix it in Leishman's or Jenner's or Wright's stock stain for 5 minutes.
3. Transfer to buffered distilled water (pH 6.2–6.8) for 5 minutes.
4. Transfer to dilute Giemsa stain (1 : 10 ratio with distilled water) for 20–30 minutes.
5. Again rinse in buffered distilled water.
6. Rinse in distilled water.
7. Blot, dry and mount.

Giemsa Staining for Warm-blooded Birds (Lucos and Jamroz, 1961)

1. Fix dry smears in Wright's stain for 4 minutes.
2. Transfer to distilled water for 5 minutes.
3. Again wash in running water giving 2 changes for 2 minutes.
4. Place in dilute Giemsa (15 drops of stain + 10 ml of distilled water) for 20 minutes.
5. Remaining procedure is the same as above.

Reticulocyte Staining (Brecher, 1949)

Reagents Required

New methylene blue N
Potassium oxalate

Preparation of Reagents

Solution 1

New methylene blue N	500 mg
Potassium oxalate	1.6 g
Distilled water	100.0 ml

Procedure

1. Mix equal amounts of stain and fresh or heparinized blood on a slide. Draw this mixture into a capillary pipette. Allow to stand for 10 minutes.
2. Expel mixture on a slide and mix.
3. Make thin smears and dry.

Results

Reticulocytes	Deep blue
Red cells	Greenish blue

Remarks

If slides fade out, fix in methanol and stain again with Wright's stain.

Thompson (1961) and Simpson *et al.* (1970) recommend Rhodanile blue for the demonstration of Heinz bodies. For birds, Lucas and Jamroz (1961) used the following method. A drop of blood is mixed with a drop of 1 per cent brilliant cresyl violet in 0.85% sodium chloride. After 2 minutes it is smeared, dried and stained with Wright's stain.

HAEMOGLOBIN STAINING

Buffalo Black for Haemoglobin (Puchtler *et al.*, 1964)

Fixation

Any general fixative

Reagents Required

Tannic acid
Phosphomolybdic acid
Methanol
Glacial acetic acid
Buffalo black NBR

Preparation of Reagents

Solution 1 (Tanninic acid)

Tannic acid	5.0 g
Distilled water	100 ml

Solution 2 (Phosphomolybdic acid)
    Phosphomolybdic acid   1.0 g
    Distilled water   100 ml

Solution 3 (Buffalo black)
    Saturate Buffalo black in methanol, glacial acetic acid (9 : 1) and let it stand for 48 hours. It lasts for several months.

Procedure

1. Deparaffinize and hydrate slides to water.
2. Place in solution 1 for 10 minutes.
3. Wash in distilled water (3 changes).
4. Transfer to solution 2 for 10 minutes.
5. Wash in distilled water (3 changes).
6. Stain in solution 3 for 5 minutes.
7. Differentiate in methanol–acetic acid (9 : 1) 2 changes.
8. Place in absolute ethanol, clear and mount.

Results

Haemoglobin   Dark blue or purplish black
Other tissues   Yellow

Canol reaction (Dunn, 1946)

Fixation

Buffered formalin (pH 7) for 48 hours.

Reagents Required

Cyanol
Zinc powder
Glacial acetic acid
Hydrogen peroxide (30%)

Preparation of Reagents

Solution 1 (Cyanol stock solution)
    Cyanol   1.0 g
    Distilled water   100 ml
    Zinc powder   1.0 g
    Glacial acetic acid   2 ml

Bring to boiling point in a short time. The solution gets decolorized. It lasts for several weeks.

Solution 2 (Working solution)
    Filter 10 ml of stock (solution 1) and then add
        Glacial acetic acid   2 ml
        Hydrogen peroxide (30%)   1.0 ml

Procedure

1. Deparaffinize and hydrate slides to water.
2. Place in solution 2 for 5 minutes.
3. Rinse in distilled water.
4. Counterstain in red nuclear stain.
5. Dehydrate clear and mount.

Results

Haemoglobin Dark blue to brownish grey

Foetal Haemoglobin (Modified from Kiossoglau *et al.*, 1963)

Fixation

Air-dry blood smears for 1 hour.
Fix in absolute methanol for 5 minutes.
Fix in 80% ethanol for 5 minutes.

Reagents required

McIlvaine buffer citric acid phosphate (pH 3.2 to 3.4)
Eosin y
Ponceau s
Acetic acid

Preparation of Reagents

Solution 1
McIlvaine buffer citric acid phosphate pH 3.2 to 3.4
Solution 2

Eosin y	500 mg
Distilled water	100 ml
or	
Ponceau	500 mg
1% aqueous acetic acid	100 ml

Procedure

1. Rinse fixed slides in tap water and then in distilled water.
2. Treat in solution 1 at 37°C for 5 minutes.
3. Rinse in tap water.
4. Place in solution 2 for 3 minutes.
5. Rinse in tap water for few seconds. Allow to stand in a vertical position.

Results

Cells containing foetal hemoglobin	Deep pink to red
Adult cells	Colourless or pale

PREPARATION OF PERMANENT WHOLE MOUNTS OF INVERTEBRATES OR THEIR PARTS

23

All invertebrates cannot be treated alike as far as fixatives and their application are concerned. However, certain generalized applications to the most common forms can be made. Fixatives cannot be applied directly to all forms since some groups, for example coelenterates, withdraw their tentacles, appendiculated forms contract their appendages and soft-bodied forms roll up into balls. To avoid this, certain anaesthetizing and narcotizing agents have to be administered before killing and fixing the specimens. As soon as narcotization is complete, one can proceed with fixation before the death of the organism.

SOME NARCOTIZING AND ANAESTHETIZING AGENTS

Magnesium chloride or magnesium sulphate These two are extensively used narcotizing agents, widely used in anaesthetizing sea anemones, corals, tunicates, annelids and nudibranchs. Magnesium sulphate is tied in a bag and suspended in water to touch its surface. Another way is to irrigate directly with 30 per cent magnesium sulphate. To ensure complete narcotization, lightly pricking with needle (without damaging the tissues) is necessary. If there is no response, then the organism is considered sufficiently narcotized.

Menthol In a bowl of water, sprinkle menthol on the surface and narcotize big animals by leaving overnight. Menthol (45.0 g) mixed with chlorol hydrate (55.0 g) works well. The powders are ground and sprinkled on the surface of water. It is good for sessile animals such as bryozoans, coelenterates like hydra and even some flukes. Larger animals require longer exposure. When contractions and movements cease, the organism can be fixed. For some molluscs, tunicates, hydrozoans, turbellarians and annelids, chlorol hydrate alone is sufficient.

Cocaine   A 1 per cent solution (1 g/100 ml water) is added to water so as to touch the surface of nudibranchs, rotifers and bryozoans. Eucaine hydrochloride also serves the purpose.

Chloretone   It is a slow narcotizer 0.33 to 1.0 per cent that could be used for narcotizing. It is especially useful for monogeneans.

Chloroform   For aquatic animals, both marine and fresh water, it can be sprinkled on the surface of water. Insects and arachnids are narcotized this way.

Ether and alcohol   Ether can be used as chloroform for insects whereas alcohol is added dropwise to water and the proportion should not exceed 10 per cent. Among terrestrial animals, it is good for earthworms and it is also recommended for freshwater forms.

Asphyxiation   Water is boiled to remove all oxygen. Cool water, and immerse animals in the non-oxygenated water. This is especially good for gastropods.

Cold treatment   The organisms are kept in a freezing chamber or in a mixture of ice and salt or in ice-cool water until relaxation occurs. Later specimens are transferred to lukewarm water before fixing.

Hanley's Solution (Gray, 1954)

Water	90.0 ml
Ethyl cellosolve-1	10.0 ml
Eucaine hydrochloride	300 mg

The solution is added to water containing animals. Recommended for rotifers and hydrozoans.

Propylene Phenoxetol (Owen, 1955; Owen and Steedman, 1956, 1958)

Take 5 ml of this solution and add to 50–20 ml of sea water where animals are kept. This solution is recommended for molluscs.

SOME METHODS OF HANDLING

Porifera

The following mixture can be used for small poriferans.

Osmium tetroxide	2.5 g
Mercuric chloride	9.0 g
Water	250.0 ml

Large sponges can either be fixed in Gilson's or Carnoy's fluid.

Decalcifying fluid for calcareous sponges

70–80 per cent alcohol + 3 per cent hydrochloric acid (for silicious sponges).

80 per cent alcohol + 5 per cent hydrochloric acid (for calcareous sponges).

After decalcification and desilicification, respectively transfer them to 80 per cent alcohol. A paraffin-coated container is best suited for the procedure.

Desilicification in a solution containing
> 80 per cent alcohol 100.0 ml
> 5 per cent hydrochloric acid added dropwise

This acid should be added to alcohol slowly in a wide-mouthed container which should be coated internally with paraffin wax. Animals can be later transferred to 80 per cent alcohol.

Coelenterates

Techniques differ from group to group.

Hydra Place hydra in a petri dish. When the animals are properly extended pour onto them a solution containing 95.0 ml and saturated mercuric chloride and 5.0 ml glacial acetic acid.

This solution should be poured starting from basal disc towards oral end. This will prevent the withdrawal of tentacles.

Sea anemones First anaesthetize animals with either menthol or 30 per cent magnesium chloride. These solutions should be added slowly for a period of one or two hours. When no contraction is evident, pipette out excess solution, leaving sufficient solution to cover and then add Susa or Bouin's or any other fixative having mercuric chloride or 10 per cent formalin. After having properly fixed, transfer them to fresh fixative. Animals harden later.

Scyphozoans To jelly fish in sea water add slowly 10 per cent formalin, with constant stirring. The concentration should be 10 ml to 100 ml water. After 3–4 hours, transfer specimens to fresh formalin.

Medusae Medusae should be anaesthetized with either cocaine or magnesium chloride. Continue adding the anaesthetizer until contractions cease. Transfer them to 10 per cent neutral buffered formalin.

Corals Corals have to be narcotized with magnesium sulphate and later fixed in a solution containing saturated mercuric chloride and 5 per cent acetic acid. This solution should be heated before fixing the animal.

Apart from all these methods, simple freezing may engender proper fixation.

Platyhelminthes

Planarians Animals have to be starved at least two days before fixation. Put planarian in a small quantity of water in a petri dish. When properly extended, add a drop or two of 2 per cent nitric acid directly. Then add Gilson's fluid or saturated mercuric chloride in saline. Transfer it to a fresh fixative. Dawar (1973) recommended the following relaxing and fixing agent.

Magnesium sulphate 2.50 g
Formaldehyde 4.5 ml
Conc. nitric acid 2.0 ml
Distilled water 200.0 ml

This has to be conducted at room temperature for 24 hours. From our experience, it has been observed that hot Susa is good for specimens which are to be sectioned.

Trematodes These animals can first be anaesthetized with menthol. First soak the filter paper in Gilson's fluid and place it on a glass plate. Take another soaked paper and place it on the worm. Apply proper pressure (taking caution not to destroy the tissue). The two glass plates could be tied with rubber bands at either end. Leave it in this position at least for 12 hours in fixative. Remove the glass plates and filter paper and transfer the worm to fresh fixative and leave overnight.

For small trematodes, put them in a small dish, and shake with small quantity of 0.5–1.0 salt solution. After 5 minutes, add saturated mercuric chloride containing 2 per cent acetic acid and again shake for several minutes. Change to fresh fixative for 6–12 hours. Remove the worm and place it on a slide and tie it with 2 rubber bands, or if it is too small with a coverslip. Pipette fixative between two slides and then leave the slides in a petri dish containing fresh fixative (Gower, 1939). For sectioning, stretch specimen on a dry filter paper and stretch it stroking gently with camel hair brush. The specimen now remains stuck to the filter paper. Now put appropriate fixative (Susa, Bouin's, FCA, etc.) dropwise. Let the specimen remain for one hour in the fixative.

Cestodes Put the worm first in a freezing chamber until relaxed (overnight). Then flatten the worm in between two slides. Tie thread around the slides to keep them together (Demke, 1952). If a museum specimen mount is required, bind the worm on a long glass plate and pour fixative on it. Some time after chilling they can be relaxed in 70 per cent alcohol or hot 5 per cent formalin. Small cestodes can be placed on a tall container with an applicator stick and fixative poured on it. When they hang straight, immerse them in fixative.

Nemertines

Nemertines are best fixed in a solution containing saturated mercuric chloride and acetic acid in the ratio 95 : 5.

Rotifers

These are narcotized with the following fluid containing

2 per cent benzamine hydrochloride 3 parts
Distilled water 6 parts
Cellosolve 1 part

After narcotization is complete, add 10 per cent neutral buffered formalin.

Nemathelminthes

Nematodes   Worms are initially placed in physiological saline. Solution is shaken well. To straighten nematodes, pour hot glycerine alcohol copiously on the specimen (80 per cent alcohol and 10 per cent glycerine). They are then transferred to glycerol (80 per cent alcohol and 10 per cent glycerol).

If small nematodes are to be treated, place them in a watch glass and apply gentle heat taking care not to destroy the tissue and then transfer them to a solution containing

95 per cent alcohol	12.20 ml
Glacial acetic acid	10.0 ml
Formalin	6.0 ml
Distilled water	40.0 ml

If a few drops of saturated aqueous picric acid are added, the worm takes up colour. If a whole mount has to be prepared, place the animal in glycerol–alcohol mixture. Allow the alcohol to evaporate. When alcohol is completely evaporated, only pure glycerine is left. Now the worm can be mounted in glycerine jelly. The same procedure can be followed for helminth ova.

When blood has to be tested for microfilaria, take blood smears at midnight as it is done for any other smear. Dry the slide and fix it any mercuric-chloride-containing fixative and stain with Delafield's haematoxylin. An alternate method to smearing is to dry the blood, dehaemoglobinize in 5 per cent acetic acid, air-dry, fix in methyl alcohol and stain with Giemsa.

Bryozoa

Bryozoans are to be anaesthetized first with menthol. Marine bryozoans can be fixed in a solution containing 7–10 ml of 10% chromic acid (10 g + 100 ml water, 10 ml of 10 per cent acetic acid to make 100 ml. Freshwater bryozoans can be fixed in 10 per cent formalin.

Annelida

When earthworms are to be sectioned, starve them for two days on moist filter paper to keep the intestine free of grit or any other hard material. Instead of starving them they can be fed cornmeal and agar (1 : 1) or chopped lettuce for a couple of days (Cocke, 1938) or with agar only (Becker and Raudabush, 1935). When no more grit is evident in faeces, place the worms in a petri dish, add water, straighten the worm and then siphon out 50 per cent alcohol on it, so that the solution is 10 per cent. After narcotizing, fix it either in Bouin's or Susa. Wash the worms thoroughly and transfer to tertiary butyl alcohol (24 hours) and then transfer to butyl alcohol saturated with paraffin (56–60°) for 24 hours and then pure paraffin for 24 hours and then embed.

Marine animals should be kept in clear sea water for 2 days and then anaesthetized with chloroform and fixed in either Bouin's or mercuric chloride fixative.

Arthropoda

Put cotton soaked either in ether or chloroform or potassium cyanide in a screw-capped bottle or a wide-mouthed bottle. Place the insect inside the bottle. As soon as the insect is anaesthetized, place it on a glass plate and spread the appendages with another plate and drain fixative in between the slides.

For whole mounts, rapidly penetrating fixative should be used. Picro-sulphuric, sublimate fixatives, mixtures containing nitric acid, alcohol or Bouin's could be used.

For whole mounts, clearing of exoskeleton is essential and difficult. Body contents have to be transparent or to be removed. Lactophenol mounting will serve the purpose. As the arthropod cuticle is tough and heavily pigmented, the pigment has to be bleached by hydrogen peroxide for 12 hours. Fleas and ticks have to be bleached with 10 per cent potassium hydroxide for 8–12 hours. This clears the body contents. Acid corrosives are preferred by some because it does not soften the integuments as much as alkaline corrosive.

Acid corrosives

Glacial acetic acid	1.0 ml
Chlorol hydrate	1.0 g
Water	1.0 ml

Vyas (1972) recommends a fluid containing

Glycerine	12.0 ml
Formaldehyde	2.0 ml
Distilled water	100.0 ml

Add few crystals of thymol.

This serves to preserve both exoskeleton and soft parts. When the specimens are large this fluid can be injected directly. Potassium hydroxide is not advisable for small delicate insects. Equal parts of chlorol hydrate and phenol are suggested to clear them. Leave them in this fluid for two weeks. If clearing is not satisfactory, transfer the solution to 40°C for two days. Later transfer them to absolute alcohol and mount in lactophenol.

Because of the chitinous cuticle, sectioning becomes difficult. Dioxane method or double embedding is preferred. Sometimes tissue blocks may be soaked overnight or for several days.

Modified Carnoy

Absolute isopropyl alcohol	6 parts
Chloroform	3 parts
Formic acid	1 part

Fixation, dehydration and infiltration are processed under reduced pressure.

Refer Beckel (1959) for sectioning methods. For better sections, see Barros-Pita (1971), Roden (1975), Nelson (1974) and Kimmel and Jee (1975).

Molluscs

Snails are put in boiled water and propylene phenoxetol is added. For bivalves the shells should be decalcified with 3–4 per cent nitric acid. If the soft parts are to be fixed, keep the valves apart and put the entire animal in the fixative. After proper fixation, dissect out the soft parts and fix again in a fresh fixative.

To decalcify bivalve shells, fix in 5 per cent formaldehyde overnight and decalcify with 2 per cent acetic acid. After decalcification, shell is put in 5 per cent formaldehyde for 5 hours and then washed in water overnight. The material is transferred to 70 per cent ethanol, dehydrated and cleared in creosote. Anderson (1971) recommends a decalcifier fixative (2 : 1) for good cellular details. All molluscs with calcareous shells can be treated in a similar manner.

Nudibranchs   Onto the nudibranchs kept in sea water, add 1 per cent neutral buffered formalin slowly dropwise every 15 minutes.

Freshwater molluscs   Slowly warm the water in which the animals are held. This will make them extend the foot. When there is no response for needle prick, fix them in Gilson's, Bouin's or Zenker.

Echinoderms

Narcotize echinoderms by sprinkling menthol on the surface of water. Magnesium chloride or magnesium sulphate can be added to water. Initially inject the fixative (mercuric chloride acetic acid) into the top of the ray. This will spread evenly into the tube feet. Then drop the animal in fixative (Moore, 1962).

STAINING INVERTEBRATES

Staining of sections depends on the choice of the stain and relative fixation. Delicate transparent forms like hydra, hydroids, daphnids, bryozoans, medusae, flukes, tapeworms, small annelids, tunicates and ammocetous larvae can be stained with alum carmine or Kornhauser haematein as an alternative.

Some trematodes and tapeworms have thick muscle layer apart from tegument and the body appears opaque and the internal anatomy remains obscure. In such cases, stain the worms, dehydrate, clear in good oil like cedar wood oil or terpenol, on a glass plate and put it in a dish and keep the dish under a binocular microscope. Alternate method is to stain and then wash in water. Then transfer to 0.5 per cent potassium permanganate. Then the worm turns to greenish brown colour. Remove it immediately to distilled water for 5 minutes, then transfer the worms to 2–3 per cent oxalic acid. It is bleached and the greenish brown sheen is lost. Wash in running water, dehydrate, clear and mount.

After destaining, the specimen can be turned blue with sodium carbonate solution.

Other references concerning invertebrate staining are: Beckel and Roudabush, 1945; Galigher and Kozloff, 1964; Gatenby and Beams, 1950; Gray, 1954; Mahoney, 1968 and Pantin, 1946.

PREPARATION OF CHICK EMBRYOS

Break the shell of the egg with the handle of a scissors at the air space. Cut the shell along the long axis avoiding damage by scissors. Keep the egg on the embryo cup and immerse in physiological saline at 37°C. Carefully remove half of the shell. The embryo can be seen floating on the yolk. With small scissors cut quickly around. Grip one edge with forceps and gently slip a watch glass under the chick embryo. Lift the watch glass with the embryo in a little saline. Any yolk particles coming along with the embryo can be pipetted out. Straighten out the embryo, ensuring that there are no folds. Gently pipette saline on the embryo; then draw out all saline, traces of yolk and vitelline membrane with a pipette.

Cut filter paper into small triangles and place them along the circular margin of the area all around, to stretch. With a pipette, apply Bouin's or Gilson's fixative. Leave overnight to harden the embryo. Remove embryo from the dish and transfer to fresh fixative and wash it. Alum carmine or Ehrlich haematoxylin are suitable stains. After Ehrlich haematoxylin staining, destain appropriately and blue with sodium bicarbonate or ammonia. Wash, dehydrate, clear and mount. While placing the coverslip, avoid undue pressure. This method could be applied to reptilian and mammalian embryos and amphibian larvae.

WHOLE MOUNTS

Whole mounts of small specimens are usually done in resins, toluene, glycerine jelly or gum arabic.

Whole Mount of Small Organisms

Fixation

1. Place specimen in 70 per cent alcohol.
2. Transfer to 95 per cent alcohol.
3. Transfer to bleaching fluid.

70 per cent alcohol	10.0 ml
Chlorox	4 drops
Water	500 ml
H_2O_2	50.0 ml

Fixation time is 24 hours.

Procedure

1. Wash with 70 per cent alcohol giving 3 changes.
2. Dehydrate in 95 per cent alcohol.
3. Clear in creosote and xylene mixture.
4. If suddenly put in mountant, the tissue may become brittle. To avoid this, impregnate gradually by adding a drop of mountant each day, and stir carefully.
5. Now mount the specimen. If the specimen is large, it should have support under the cover glass. Support may be in the form of cover glass bits or cutting glass rods into circles or squares. Place a drop of mountant on slide and place specimen in the centre of mount and put cover glass. If many specimens are to be put on the same slide, there is a chance of drifting. To avoid this, air-dry the mountant and then place the coverslip.

If celloidin mounting is required, dehydrate through absolute alcohol, alcohol ether and then into thin celloidin. Put specimens in petri dish with celloidin. After the solvent evaporates, squares of celloidin are cut, dehydrated, cleared and mounted. The advantage of this method is that no supports are required while mounting with a coverslip. PVA is also used as a mounting medium (Rubin, 1951). Courtright (1966) used polyester resins.

Glycerol Jelly Mounts

Most materials fixed in various fixatives or even frozen sections may be mounted directly from water to glycerol jelly. If there is danger of specimens collapsing, transfer the material from 70 per cent alcohol or water into a mixture containing 10–15 per cent glycerol in alcohol or water. Keep the dish containing glycerol alcohol open. Alcohol evaporates, and the material can then be mounted in glycerol jelly.

The staining fixatives used for Flukes, embryo and Hydra are the following.

Worms : Carnoy or Gilson's
Embryos : Zenker's or Bouin's
Hydra : Formol acetic or saturated mercuric chloride acetic (95/5)

STAINING OF WHOLE MOUNTS

A number of stains are available.

Grenacher Borax Carmine (Galigher, 1934)

Reagents Required

Carmine   3.0 g
Borax   4.0 g
Distilled water 100.0 ml

Boil until carmine dissolves. Then allow it to stand for some time and then add 100.0 ml of alcohol.

Procedure

1. Bring down material to 50 per cent alcohol.
2. Transfer to Borax carmine solution and keep for 4 hours overnight.
3. Add conc. hydrochloric acid dropwise to carmine (overnight) until precipitate has settled.
4. Add equal volume of 3 per cent HCl in 70 per cent alcohol and mix thoroughly. Draw precipitated carmine with pipette. Repeat it until most of the carmine is removed.
5. Add acid alcohol so that tissue is destained.
6. Then transfer specimens to 80 per cent alcohol for 1 hour.
7. Dehydrate.
8. Clear in a mixture of absolute alcohol and creosote and then into pure creosote xylene.
9. Mount.

Mayer's Carmalum (Cowdry, 1952)

Preparation of Reagents

Stock (Carmine stock)

Carmine	1.0 g
Ammonium alum	10.0 g
Distilled water	200 ml

When dissolved, filter and add 1.0 ml of formalin.

Working solution

Carmine stock	5.0 ml
Glacial acetic acid	0.4 ml
Distilled water	100.0 ml

Procedure

1. Stain for 2 days. No destaining is necessary.
2. Dehydration procedure is same as for previous technique.

Haematein (Kornhauser, 1930)

Preparation of Reagent

Stock

Haematein	500 mg
95 per cent alcohol	10.0 ml
Potassium aluminium sulphate	500.0 ml
(5 per cent)	

Put haematein in a mortar. Add alcohol and grind it and then add it to potassium aluminium sulphate. Haematein stain is recommended for flatworms. Hitherto alum cochineal was popular but alum carmine is better.

Procedure

1.   Stain for 10–15 hours.
2.   Place in 70 per cent alcohol.
3.   Destain with acid alcohol.
4.   Blue in alkaline alcohol or ammonia or sodium bicarbonate in 70 per cent alcohol.
5.   Dehydrate, clear and mount.

For small organisms alum haematoxylin is used. Celestine blue (Demke, 1952) and trichrome stain (Chubb, 1963) are also popular.

Cochineal Haematoxylin

Preparation of Reagent

Alum Cochineal	3 parts
Potash alum	30.0 g
Cochineal	30.0 g
Distilled water	100.0 ml
Delafield haematoxylin	1 part
Distilled water	25 parts

Let it stand for few hours. Filter and use.

Procedure

1.   Wash with 50 per cent alcohol after fixation.
2.   Wash in distilled water for 10 minutes.
3.   Keep in staining solution overnight.
4.   Place in 70 per cent alcohol.
5.   Differentiate in acid alcohol (1 part acid with 99 parts alcohol) until internal organization is clear.
6.   Transfer to 70 per cent alcohol.
7.   Blue in a solution of 70 per cent alcohol with few drops of lithium bicarbonate.
8.   Dehydrate.
9.   Clear in a mixture of absolute alcohol and cedar wood oil (1 : 1) for 1 hour.
10.   Clear in pure cedar wood oil.
11.   Mount in synthetic dye.

WHOLE MOUNTS OF PROTOZOA

Chen (1944) Cover Glass Method

Fixation

Most favourite fixative is Schaudin's at 50–56°C for 15 minutes. Other recommended fixatives are Bouin's, Champy's, Flemmings and Worcester's. For fixing *Stentor, Sporostomum* and *Vorticella*, consult Merton (1932) also.

Amoeba and *Paramecium* must be centrifuged. Pour off the supernatant and then add Schaudin's fixative or any other fixative. Amoeba settles down at the bottom of the culture dishes. Decant most of the culture medium and pour Schaudin's (50–60°C) over the organism. After it cools down, pour an equal volume of 85 per cent alcohol. Transfer the solution to a centrifuge tube for centrifugation. Pour off the fixative, wash several times with alcohol. After each wash, centrifuge.

Procedure

1. After washing in 70 per cent alcohol, wash with 80–85 per cent alcohol for 15 minutes.
2. Take a cover glass and smear albumen on it. Keep the albuminized side up on slide. Place few drops of alcohol containing organisms with a pipette in the centre of the cover glass. Alcohol will begin to evaporate, and specimens are in contact with albumen. When the centre of the cover glass is still moist, add few drops of 95 per cent alcohol on the specimen and slowly keep the cover glass in a petri dish containing 95 per cent alcohol.
3. Transfer the cover glass carefully to absolute alcohol and later flood it with 1 per cent celloidin. Excess celloidin is drawn off. Place it in 70–80 per cent alcohol till it is stained.
4. Staining with haematoxylin, Carmine or Feulgen stains depends on the study to be carried out.
5. Dehydrate, clear and put some mounting medium on the slide and place the cover glass with specimen on the medium.
6. Subbed cover glasses are good.

Smyth (1944) devised the following method. After fixation, the usual procedure of passing through graded series of alcohols up to absolute alcohol is followed. Then a drop of this is put on an albuminized slide. Place the slide in absolute alcohol. After water, flood with 1 per cent celloidin. Blot excess celloidin and air-dry. Later it can be stained with alum carmine, haematoxylin or Feulgen.

Agrell (1958) recommends the following method. Place minute embryos on albuminized slides. Allow them to dry, and during this process, they become flattened. Dip into absolute alcohol and then into fixative. Place them in 95 per cent alcohol vapour for 1 minute Then fix. This coagulates the embryos and attaches them.

Paramecium can be preserved intact with normal shape without contraction by adding copper sulphate or acetate.

Merton Method (Kirby, 1947)

1. Put a drop containing paramecia on an albuminized slide for 30 seconds.
2. Put an equal amount of 1 per cent copper sulphate for 7–8 minutes or 3 per cent copper acetate for 45 seconds.
3. Draw off part of flint and suspend slide over 2 per cent osmic acid for 46 seconds.
4. Now add saturated solution of mercuric chloride on the surface of the organism for 10 minutes.

5. Dip in a solution containing 70 per cent alcohol and a small amount of iodine for 10 minutes.
6. Wash with distilled water.
7. Stain.

Prescott and Carrier Method (1964)

1. Place a drop of few amoebae on subbed slides.
2. Now place a drop of fixative (70 per cent alcohol or acetic alcohol) onto a cover glass and place it on amoebae.
3. Freeze in liquid nitrogen for 15 seconds. Take off cover glass. If liquid nitrogen is not available, fix in 50 per cent aqueous nitric acid and freeze in dry ice.
4. Rinse the slide in 95 per cent alcohol, and air-dry.

STAINING OF PROTOZOA

A number of staining methods have been described.

1. To stain fibrillar elements, use iron haematoxylin and the fixative should contain chromium. Warm the stain at 50°C and destain with 10 per cent H_2O_2 (Kidder, 1933).
2. To stain flagella, Giemsa stain is used (Rothenbacher and Hitchcock, 1962) and also Loeffler's stain (Kirby, 1947).
3. In ciliates to stain cilia, basal granules and connecting filaments, Gelei osmium–toluidine blue method is adopted (Kirby, 1947; Pitelka, 1945).
4. For silver line system, the best techniques of Chatton and Lwoff (1930, 1935, 1936), Frankel and Heckman (1968), Gelei (1932, 35) and Klein (1926) are adopted. Also see Corliss (1953) for silver method. Protozoans are excellent subjects for vital staining and enzyme techniques.
5. Micro- and macronuclei can be differentiated by safranin–fast green method of Schiff *et al.* (1967).

Borror (1968) Nigrosin Method

Preparation of Reagents

Solution 1

Saturated aqueous mercuric chloride	10.0 ml
Glacial acetic acid	2.0 ml
Formaldehyde	2.0 ml
Tertiary butanol	10.0 ml

Solution 2

Formaldehyde	20.0 ml
Nigrosin (water solution)	4.0 g
Distilled water	100.0 ml

Solution 3 (Working solution)
   Solution 1   12 parts or 120 cc
   Solution 2   1 part or 10 cc

Procedure

1.   Place a cluster of organisms on a slide.
2.   Add a drop of solution 3 for 3–4 seconds.
3.   Wash.
4.   Dehydrate, clear and mount.

Result

Ciliary organelles are stained black.

Sectioning Protozoa

Stone and Cameron (1964) modification of Kimball and Perdue (1955) Agar method.

1.   Select a glass tubing which is sealed at one end. Then pour melted agar into it. Chill until agar hardens.
2.   Pipette out a thick concentration of organisms on top of the agar and then pipette out the fixative.
3.   Specimens settle on the agar. Then draw the fixative. Add some more agar to which eosin is added so that small organisms could be easily located. Chill the agar.
4.   With a pipette, siphon out water with force to loosen the block.
5.   Dehydrate, clear and embed the block of agar.
6.   When the block is sectioned, the cells could be located because of the eosin colour.

Dry Mounts

Radiolarians and foraminiferans require opaque type of slide mounting. Glue these dry specimens on a black background and cover it with a cover glass and some supporting ring around them. Gray (1964b) recommends the following method. Take two cardboard pieces of the size of a slide. Paint with black paint or cover with a strip of black paper. Make a 5/8 of an inch hole on one cardboard piece. Make a 7/8 of inch square on the other cardboard and stick the two pieces by dry mounting tissue. Now place a drop of gum in the centre and with a brush pick up a specimen and place it in the centre of a drop of gum. Blow on it because moisture is necessary to make it adhere. When the specimen has adhered, cover the hole. Depression slides can also be used.

Animal Parasites

Haematoxylin/eosin method is good for sections of tissue parasitized by protozoa or helminths. PAS method is excellent since protozoa and helminth worms are strongly PAS-positive due to stored glycogen. Methenamine silver method is very effective for flagellates. The scolices and hooks are best shown with haematoxylin background. Kinney *et al.* (1971) used acid-fast staining to stain hooklets.

SMEAR TECHNIQUE FOR INTESTINAL PROTOZOA (CONCENTRATED SMEARS)

Arensbarger and Markell (1960)

Procedure

1. Take 1 ml of faeces in a tube and add 15 times its volume of water. Mix well and strain through two layers of gauze in a funnel. Collect in a small centrifuge tube. Add 1 ml of ether, then shake the tube and fill with water.
2. Centrifuge (2500 rpm) and decant supernatant fluid.
3. Add 2–3 ml of normal saline and shake. Allow it to settle. Fill tube with normal saline up to 1 cm.
4. Decant the supernatant. Take a small trace of the original faecal matter with stick (applicator stick) and mix well with sediment at the bottom of tube.
5. Transfer the material with an applicator stick onto a clean slide. Make a smear. Immediately fix in Schaudinn's.
6. Stain with one of the following.

Goldman (1949) Smears

Fixation

Schaundinn's fixative for 15 minutes.

Reagents Required

Haematoxylin
Ferric ammonium sulphate
Glacial acetic acid
Concentrated sulphuric acid

Preparation

Stock solution 1 (Haematoxylin)

10 per cent haematoxylin in 95 per cent alcohol	1.0 ml
(10 g/ 100 ml 95 per cent alcohol)	
95 per cent ethyl alcohol	99.0 ml

Stock solution 2 (Alum solution)

Ferric ammonium sulphate	4.0 g
Glacial acetic acid	1.0 ml
Concentrated sulphuric acid	0.12 ml
Distilled water	100.0 ml

Working Solution

Equal parts of stock 1 and stock 2 solutions. A purple colour is seen, but within a short time it turns brown. Then filter. When the solution turns greenish black it should be discarded.

Procedure

1. Treat slides with 70 per cent alcohol and later with iodine alcohol.
2. Wash in 70 per cent alcohol till brown colour disappears.
3. Stain in haematoxylin for 3–5 minutes.
4. Wash in running water for 30 minutes.
5. Dehydrate, clear and mount.

Result

Protozoa take up a black nuclear stain.

Kohn Stain—Combination of Both Fixative and Stain (Faust *et al.*, 1970)

Reagents Required

Methanol
Glacial acetic acid
Phenol
Phosphotungstic acid
Chlorazol black E

Preparation of Reagent

Solution 1 (Basic solution)

90 per cent alcohol	170.0 ml
Methanol	160.0 ml
Glacial acetic acid	20.0 ml
Phenol	12.0 ml
1 per cent phosphotungstic acid	12.0 ml
Distilled water	618.0 ml

Take 5 g of chlorazol black E in a mortar and grind it. Add small quantity of solution 1. Add some more solution to the paste. Allow to settle and pour off supernatant into a container. Go on adding solution 1 gradually till the paste turns into solution. Add the remaining solution 1 and allow it to ripen for 4–6 weeks. Filter and store it. This solution is also available commercially.

Procedure

1. Treat sections with solution 1 for 2–3 hours. Time varies with dilution. (Stain positive basic solution in the ratio 1 : 1 which requires overnight; 2 : 1 ratio requires 4 hours; if the ratio is 1 : 2 it requires overnight; 1 : 3 also require overnight).
2. Dehydrate in 95 per cent alcohol for 10–15 seconds.
3. Dehydrate, clear and mount.

Result

Protozoa	Grey-green, grey or black
Cysts	Grey-green
Nuclei	Dark green

Kessel (1925) and Chen (1944a) Smear (modified)

Fixation

Schaudin's at 40°C for 15 minutes.

Reagents Required

Ferric ammonium sulphate
Haematoxylin
Lugol's iodine
Sodium thiosulphate
Lithium carbonate

Preparation

Solution 1 (Iron alum)
Ferric ammonium sulphate   4.0 g
Distilled water   100.0 ml

Haematoxylin stock

Solution 2
Haematoxylin   1.0 ml
Absolute ethyl alcohol   100.0 ml

Allow it to ripen for several months. There are methods to hasten the process. But it is always better to use only ripened stain for a length of time.

Solution 3 (Working solution)
Solution 2 (Haematoxylin stock)   0.5 ml
Distilled water   99.5 ml

Add few drops of saturated lithium carbonate.

Procedure

1. Place slides in 70 per cent alcohol (from fixatives) for 3 minutes.
2. Treat with a Lugol's iodine solution for 3 minutes.
3. Wash.
4. Decolorize in 5 per cent sodium thiosulphate for 2 minutes.
5. Wash in running water.
6. Mordant in solution 1 (iron alum) at 40°C for 15 minutes.
7. Wash.
8. Place in solution 3, haematoxylin working solution, at 40°C for 15 minutes.
9. Wash.
10. Destain in 2 per cent iron alum (2 g/100 ml). Cool.
11. Wash, dehydrate, clear and mount.

Results

Nuclei–Chromatoidal bodies are stained blue-black.

Lawless' Rapid Methods (1953)

Fixation

Schaudinn's

Reagents Required

Chromatrope-2 R
Light green
Fast green
Phosphotungstic acid
Glacial acetic acid

Preparation of Reagents

Solution 1 (Staining solution)

Chromatrope 2 R	600 mg
Light green yellowish	150 mg
Fast green	150 mg
Phosphotungstic acid	700 mg
Glacial acetic acid	1.0 ml
Distilled water	100.0 ml

Add glacial acetic acid to three dyes and later add phosphotungstic acid. Allow it to stand for 30 minutes before use.

Procedure

1. Fix a portion of stool (1 part stool + 3 parts fixative) for 15 minutes to about 1 hour. Decant excess fixatives, keep it in a vial and shake vigorously. Cover the vial with gauze. Now take out moist residue and spread it on a slide. Dry it in air and transfer the slide to iodine alcohol. These smears can be stored.
2. Decolorize in 70 per cent alcohol giving 2 changes, 1 minute in each.
3. Stain in solution 1 for 10 minutes.
4. Differentiate in acetic acid alcohol (1 drop acetic acid/10 ml alcohol) for 20 seconds.
5. Dehydrate, clear and mount.

Result

Background	Green
Cysts	Black-green
Engulfed red cells	Green, red or black
Helminth eggs	Red

Remarks

If the cysts appear refractory to stain, it means that fixation is incomplete. Bouin's fixative is preferred for karyosomes (Hajian, 1961) Phloxine–toluidine blue is used for Leishmania, intestinal protozoa and microfilariae (Tomlinson and Grocott, 1944), whereas PAS combined with haematoxylin and light green was recommended by Silva (1961) and carmine by Mariweather (1934).

MICROWAVE HISTOLOGY

24

Microwave ovens have become popular recently and are in use in a number of laboratories. Started in 1980, the use of microwave ovens it has now spread to many laboratories. The main advantage of microwave oven is that it hastens fixation, dehydration and staining of biological material. Some refinements have yet to be made in microwave technology. Microwave processing accelerates diffusion, can stabilize proteins and can speed up chemical linkage. Kok and Boon (1992) in their book *Microwave Cook book for microscopists* have discussed all aspects of microwave use in the histology laboratory.

Some important guidelines for using an oven include the following.

1. Laboratory ovens should not be used for heating food items.
2. Avoid metal containers.
3. Lids of containers should not be air-tight and they should be vented.
4. Inflammables like alcohols and clearing agents should not be heated.
5. Leakage should be monitored regularly.
6. Rectangular containers especially plastic and nylon containers are preferred. Glass containers are likely to break.
7. Oven should always be kept clean by wiping spills.
8. Containers with stains should be kept in plastic bags to prevent spilling.
9. People fitted with pacemakers should avoid using microwave ovens.

Recommended microwave components are:

1.   Microwave without grill
2.   Temperature control ($\pm$ 1°C)
3.   Oven with more than 500 watts
4.   Temperature probable
5.   Nitrogen or air bubble system for agitation and rotation of load
6.   Extraction system for venting fumes into fumehood

For slides of dry tissues, place them in a wooden slide box with lid on and place it in the oven for 2½ to 3 minutes (depends on the voltage). The temperature should be set at 60°C.

FIXATION OF TISSUE IN MICROWAVE PROCESSING

Microwave oven rapidly decreases the time of fixation. Specimens to be processed and sectioned can be microwaved for 5 to 15 minutes. This is followed by hydration procedure. This saves 2 to 8 hours of fixation time. Kok and Boon (1992) have described the method of preparing the brain for sectioning in a short time by using microwave oven. The oven temperature should be monitored at 50°C and 450 watts.

1.   Keep the brain in a plastic bag and sprinkle normal saline on it.
2.   Remove the brain from the plastic bag and cut into slices of 2×3 cm.
3.   Now keep the slices in a plate and sprinkle saline over them and again cover with plastic.
4.   Microwave the tissue for 15 minutes.
5.   Cut slices of brain of 2×2 cm size and prepare for paraffin embedding.
6.   Immerse these smaller sections in 10 per cent neutral buffered formalin or 3–5 hours or microwave them for 15 minutes.
7.   Proceed with regular processing.

Same method can be followed for previously perfused brain. A jar containing the saline can be used. Use 280 watts of power for 15 minutes.

Wash brain in 10 per cent sucrose in 0.1 M citrate buffer (pH 7.1) for 6 to 12 hours. Frozen sections can be cut.

TISSUE PROCESSING USING MICROWAVE

Before microwave processing, the tissue should be freed from formalin. Rinse the tissue in running water and immerse in 50 per cent alcohol for 1 hour. Plastic capsules must be used for processing. Vacuum paraffin impregnation is suggested for the final infiltration.

1.   Keep the tissue in absolute alcohol at 67°C for 1/2 an hour. With small batch of samples the time should be 5 to 15 minutes. For a large batch, the time should be 1 hour.

2. Again keep the tissue in isopropanol at 74°C for half an hour. With a small batch the time should be 3 to 15 minutes and with a large batch, 30 minutes.

3. After a 30–60-minute vacuum infiltration, embed as usual.

Haematoxylin/eosin staining may be hastened by microwave treatment.

Procedure

1. Hydrate slides to water.
2. Immerse slides in either Harris or Gill's haematoxylin and keep it in oven for 30 seconds.
3. Follow the routine procedure.
4. Rinse in water.
5. Immerse slides in microwave with 50 ml eosin for 20 seconds.
6. Dehydrate, clear and mount.

Result

Cytoplasm   Pink or shades of red
Nuclei   Dark blue

Grocott's Methenamine Silver Nitrate (Brinn Modification, 1983)

Fixation

10 per cent neutral buffered formalin

Reagents Required

Chromic acid
Sodium metabisulphite
Methenamine
Gold chloride
Sodium thiosulphate
Nuclear fast red
Light green

Procedure

1. Dewax and hydrate slides to water.
2. Transfer slides to a coplin jar containing 50 ml of 10 per cent chromic acid with a loose lid for 50 seconds.
3. Wash the slides.
4. Differentiate slides in 1 per cent sodium metabisulphite for 60 seconds.
5. Rinse in distilled water.
6. Transfer slides to a plastic Coplin jar containing 50 ml of working methenamine solution for 1 minute.
7. Remove slides and examine microscopically for intensity.
8. Rinse in distilled water.
9. Tone in gold chloride for 10 seconds.
10. Rinse in distilled water.

11. Transfer to sodium thiosulphate for 1 minute.
12. Wash in water.
13. Counterstain either in 0.2 per cent nuclear fast red or 1 per cent light green for 2 minutes.
14. Wash in water.
15. Dehydrate, clear and mount.

Result

Fungi and *Pneumocystis*	Black
Background	Pink and red

Perl's Iron Stain (1867) Modified by Kok and Boon (1992)

Fixation

Neutral buffered formalin

Reagents Required

Potassium ferrocyanide
Nuclear fast red

Preparation of Stains

Solution 1

Potassium ferrocyanide	250 mg
Distilled water	50.0 ml

Solution 2

Nuclear fast red	1.0 g
Distilled water	100.0 ml

Procedure

1. Dewax and hydrate slides to water.
2. Transfer slides to solution 1 (keep the slides in a loosely capped jar) for 45 seconds.
3. Remove from microwave.
4. Wash in water.
5. Counterstain with solution 2 for 2 minutes.
6. Wash.
7. Dehydrate, clear and mount.

Result

Iron is stained dark blue.

Luxol Fast Blue (Kennedy Modification)

Procedure

1. Dewax and hydrate slides to water.
2. Stain in Luxol fast blue for 1 minute.
3. Rinse in water.

4. Counterstain in 1 per cent neutral red for 1 minute.
5. Rinse in water.
6. Dehydrate, clear and mount.

Result

Myelin   Blue
Nuclei   Red

Rapid Papanicolaou Method

The usual routine method itself does not require much more time than the microwave method.

Procedure

1. Cover thin smear with fixative and microwave for 20 seconds (80 per cent power).
2. Rinse in distilled water.
3. Cover the smear with Gill's haematoxylin and microwave for 20 seconds.
4. Rinse in tap water.
5. Rinse in 50 per cent alcohol.
6. Cover smear with eosin Azure and microwave for 20 seconds at 80 per cent power level.
7. Dehydrate, clear and mount.

Result

Acidophilic cells   Red to orange
Nuclei   Blue

Many stains could be accelerated and they are numerous. Microwaves are being used in immunocytochemistry. Care should be taken not to overheat. The first persons to introduce microwave over the shorter incubation time were Leong and Milios (1986). For detailed procedures consult Kok and Boon (1992). Good information on routine staining of histological tissues (Brenn, 1983) acid and alcohol fast staining (Hafiz *et al.*, 1985) immunoperoxidase stains of lymphocyte antigens (Leong and Milios, 1986), immunoperoxidase (Login *et al.*, 1987) for antigen retrieval (Lan *et al.*, 1995) and for microwave equipment and safety concerns (Login and Dvorak, 1994) is available.

SUGGESTED READINGS

1. Brinn (1983) for routine staining of histological tissues.
2. Hafiz *et al.* (1985) for acid and alcohol fast staining.
3. Leong and Milios (1986) for immunoperoxidase staining of lymphocyte antigens.
4. Login *et al.* (1987) for immunoperoxidase.
5. Lan *et al.* (1995) for antigen retrieval.
6. Login and Dvorak (1994) for microwave equipment and safety concerns.

ULTRAHISTOCHEMISTRY

25

In recent times ultrahistochemistry has developed rapidly enabling to localize the enzymes and substances akin, histochemically, with electron microscope. In light microscopy, the end product could be easily localized because it is coloured. On the other hand in ultra histochemistry it is quite difficult to aim at the end product unless it is electron-dense. Added to these, fixation and fixatives of choice are very essential. Most of the fixatives chosen and used for ultrahistochemistry are deleterious to chemical substances to be demonstrated. But fortunately some of the end products resulting in light microscope histochemistry, in addition to being coloured and insoluble, are also electron-dense with little modification. Such methods could be applied to electron microscopy.

FIXATION

Fixatives used in light microscopy do not stand a chance in electron microscopy. Not that they damage the ultrastructural preservation, but although the technique can be easily demonstrated, the ultrastructural preservation is very poor, and hence the fixative is almost useless. For example formol calcium which takes the pride as a choice fixative in enzyme histochemistry at light microscopic level, fails miserably to preserve ultrastructural integrity of the tissue.

Most of the fixatives used in electron microscopy damage the tissue as well as chemical substances to be histochemically demonstrated. For example osmium tetroxide is used both as a fixative and stain but unfortunately it damages the tissue. So it is evident that the fixative of choice in electron microscopy should localize the substance to be demonstrated and also make the tissue ultrastructurally visible and protect the resulting product or reaction product starting from dehydration to electron beam.

To pinpoint such a qualified fixative is very difficult, so the histochemists selected combination fixatives. Very often, aldehyde fixatives such as glutaraldehyde and paraformaldehyde are used as first fixatives. After that, histochemical technique is performed and the tissue is later post-fixed in osmium tetroxide. Osmium tetroxide fixation renders the tissue more electron-dense.

Maximum information on a tissue could be obtained provided the sections are ultrathin. However thin the sections are for light microscopy, they are unusable for electron microscopy since they are too thin and are liable to be damaged by the electron beam. Resins are electron-beam-resistant and hence the tissue is embedded in resins. In other words, resin protects the tissue from electron beam.

As in light microscopy, fixation is very important as it decides the quality of the final picture. Choice of the fixative is very important as also the size of block. The suitable size of the block is 1 mm or even smaller.

For most of the morphological studies in electron microscopy, aldehyde fixatives like glutaraldehyde and paraformaldehyde are used. Osmium tetroxide is the next one. These fixatives act as cross-linkage agents by forming chemical bonds with cell substance. Glutaraldehyde is a very effective cross-linking agent for proteins whereas osmium tetroxide is effective for phospholipids. Glutaraldehyde also fixes cytoplasm.

To get best results, tissues must be first fixed (primary fixation) in glutaraldehyde at 4°C for 4 hours and this is followed by secondary fixation in osmium tetroxide. These fixatives are used in buffer solutions. After dehydration, the tissue is embedded in resins, which may be epoxy resins, polyester resins or methacrylate. Of these epoxy resins are frequently used in electron microscopy. This permits ultrathin sections of 500–600 Å thick which are cut with glass or diamond knife on an ultramicrotome.

Till now, the histochemical techniques described for light microscopy have to be altered for electron microscopy. Tissues have to be treated in a different way for the demonstration of enzymes and other chemical substances to be demonstrated.

Glutaraldehyde and osmium tetroxide which are the most common fixatives used in ultrastructural identification destroy the enzymes. Freezing techniques also destroy the morphological structure. Resin embedding for ultrathin sections is also not suitable for ultrahistochemistry, and the resin surrounding the sections does not permit the reagents to penetrate.

It is advisable to apply histochemical techniques to frozen sections either fixed or unfixed.

FROZEN SECTIONS

Aldehydes stabilize the protein components and osmium, the lipid components (for details refer Hayat, 1989). The quality of fixation depends upon tonicity, concentration of fixative, temperature and duration of fixation. To overcome all these, fixatives are made up in buffer solutions. The choice of the buffer is also very important. The choice buffers are phosphate and cacodylate buffers. Zeitherionic buffers using HEPES (*N*-hydroxyethyl piperazine-N´-2-ethane sulphonic acid), MOPS (3- [-morpholino] propane sulphonic acid) and PIPES (Piperazine-N₃-N´- bis 2-ethane sulphonic acid) are also preferred by many workers (Massie *et al.*, 1972; Johnson, 1985).

As far as the tonicity of the fixing solutions is concerned, it is a subject of controversy. It was felt that osmolarity of the final fixative can be varied over a wide range and the important factor was the tonicity of the buffer vehicle.

Concentration of the fixative on tissue varies but it is a fact that low concentrations and longer duration of fixation result in rupture of the cell, and diffusion. Aldehyde fixative concentration ranges between 2 and 4 per cent and osmium tetroxide between 1 and 2 per cent solution.

As far as temperature is concerned, fixation is mostly carried out at 4°C, but room temperature is also advised by some workers since penetration rate at 4°C is slow. Autolysis is kept at a low rate at room temperature, but 4°C is essential for electron cytochemical demonstration of enzymes. Fixation at temperature higher than room temperature increases the rate of autolysis, so it is not preferred. Fixatives and buffers are stored at 4°C. If fixation is to be carried out at room temperature, freshly prepared solutions are allowed to attain room temperature before they are put to use.

Duration of fixation should be very short and at the same time long enough to allow complete penetration of the fixative. Duration varies with the type of the tissue, the block size and buffer vehicle.

Long *et al.* (1985), and Login and Dvorak (1985) described microwave-assisted fixation. This type enables fixation time to be reduced to seconds. Marti *et al.* (1985) have reported that 25 m.mol of calcium and magnesium chloride in cacodylate buffer has improved microwave-assisted fixation. They also recommended secondary fixation to be followed 5–20 minutes after microwave-assisted fixation. (Also refer Hopwood and Milne (1991).)

ALDEHYDE FIXATIVES

Glutaraldehyde is a popular fixative among aldehyde fixatives and it is superior to others. Sabatini *et al.* (1963) introduced this fixative in a 4 per cent cacodylate buffer. The most commonly used concentration is 2.5–4 per cent.

Among buffers, cacodylate buffer is mostly preferred over phosphate buffer since it contains arsenic which prevents bacterial growth, but many investigators recommend phosphate buffer. It has its own demerits since it causes artifacts (Bullock, 1984).

2.5 Per cent Glutaraldehyde in 0.1 M Cacodylate Buffer Stock Solution

Sodium cacodylate   21.4 g
Distilled water   900.0 ml

To this, add 40 ml of 1.0 M hydrochloric acid. The final pH should be 7.4. Then add 0.1 M calcium chloride while agitating, and 3 g of sucrose and stir well. Make this solution to 1000 ml.

The fixative is prepared by adding 5 ml of 2.5 per cent glutaraldehyde to 50 ml of stock solution.

2.5. Per cent Glutaraldehyde in 0.12 M Millonig's Phosphate Buffer

Stock Solutions

Solution 1   2.26 per cent sodium orthophosphate ($NaH_2PO_4 \cdot 2H_2O$)
Solution 2   2.52 per cent sodium hydroxide
Solution 3   0.5 M calcium chloride
Solution 4   25 per cent glutaraldehyde

Procedure

First mix 41.5 ml of solution 1 with 8.5 ml of solution 2. pH should be 7.3–7.4. To this add 0.1 ml of solution 3 dropwise. While agitating, a precipitate is formed and it will disappear after shaking.

Now take 45.0 ml of this solution and add 5.0 ml of solution 4.

FORMALDEHYDE FIXATIVES

4 Per cent Paraformaldehyde in 0.1 M Cacodylate Buffer

Solution 1
    Paraformaldehyde   8.0 g
    Distilled water   100.0 ml

Heat this with continuous stirring in a fume cupboard. Now add 1.0 M sodium hydroxide dropwise with stirring until the solution is clear. 5–8 drops should be sufficient. Cool.

Solution 2
    0.2 M sodium cacodylate   50.0 ml
    0.2 M hydrochloric acid   27.0 ml
pH should be 7.4.

Now add 1.0 ml of 0.5 M calcium chloride dropwise while agitating.

Add 1.5 g sucrose.

The fixative is prepared by adding 25.0 ml of solution 1 with 25.0 ml of solution 2.

Millonig's Phosphate-buffered Formalin (Carson *et al.*, 1973)

Fix tissue for 2–16 hours.
    Sodium dihydrogen orthophosphate   1.86 g

    Tap water   90.0 ml
    Sodium hydroxide   0.42 g

Stir well and then add 10 ml of 40 per cent formaldehyde.

Paraformaldehyde–Glutaraldehyde Fixative (Karnousky, 1965)

Solution 1
    Paraformaldehyde   8.0 g
    Distilled water   100.0 ml

Heat to 60°C while stirring in a fume cupboard. To this, add 5–8 drops of 1.0 M sodium hydroxide dropwise until solution is clear, and cool.

Solution 2
    0.2 M sodium cacodylate   50.0 ml
    0.2 M HCl   2.7 ml

pH should be 7.4. To this add 5.0 g sucrose. Add 10 ml of 25 per cent glutaraldehyde to 25 ml of solution 1. Make up to 50 ml with solution 2. To this mixture, add 25 mg of calcium chloride.

OSMIUM FIXATIVES

1 Per cent Osmium Tetroxide in 0.12 M Phosphate Buffer (Millonig, 1962)

Solution 1
    Sodium dihydrogen orthophosphate   2.26 per cent
Solution 2
    Sodium hydroxide   2.52 per cent
Solution 3
    Glucose   5.4 per cent

Take 41.5 ml of solution 1 and add 8.5 ml of solution 2. pH should be 7.3–7.4. To this mixture, add 45 ml of solution 3. Add 0.5 g of osmium tetroxide.

SECONDARY FIXATION

Wash the aldehyde-fixed samples thoroughly giving 3 changes in the buffer. Later fix in 1 per cent aqueous osmium tetroxide.

Dehydration

Dehydration is done in graded series of solvents. Tissue is passed from distilled water rinse following secondary fixation into 40 per cent ethanol and later in increasing concentrations of ethanol to 100 per cent ethanol.

    40 per cent ethanol   5 minutes
    70 per cent ethanol   10 minutes
    90 per cent ethanol   10 minutes
    100 per cent ethanol   30 minutes

If ethanol is miscible with the embedding medium, the tissue is placed in that mixture directly from absolute ethanol. If ethanol is immiscible, then place the tissue in a solvent which is miscible with both ethanol and embedding medium. This is the transitional fluid. Xylene, toluene and 1 : 2 epoxypropane are some transitional fluids. For routine work, 1 : 2 epoxypropane can be used with two 5-minute changes.

EMBEDDING MEDIA

Plastic-based embedding media are used for transmission electron microscopy (also refer Nunn, 1970).

Newman *et al.* (1950) introduced butyl methacrylate as the first acrylic medium. More about this is discussed by German and Stevans (1996).

Epoxy resin embedding medium is commonly used in many laboratories. Epoxy-resin-embedded tissue is stable in the electron beam and are strong to be mounted on uncoated grids. This medium consists of epoxy resin, a hardener, plasticizer and an accelerator with different viscosities. Penetration rate is also different.

Araldite

This is used as an embedding medium in combination with a hardener dodecenyl succinic anhydride (DDSA), amine accelerator (tridimethylamino-methyl phenol DMO30) and a platicizer (dibutyl phthalate). Resin hardener reaction is accelerated by accelerator, and the cutting quality of the block is controlled by the plasticizer (Glauert *et al.*, 1956).

Araldite Embedding Mixture (Glauert and Glauert, 1958)

Araldite CY212	10.0 ml
Hardness DDSA	10.0 ml
Accelerator DMP30	0.5 ml
Dibutyl phthalate	1.0 ml

First mix the resin and hardener, stir well, then add dibutyl phthalate, and stir well until the plasticizer is dispersed. Finally add the accelerator and stir the mixture thoroughly.

Details of araldite embedding schedule and Epon embedding schedule can be obtained from Robinson and Gray (1996).

ULTRASTRUCTURAL DEMONSTRATION OF ENZYMES

Alkaline Phosphatase Lead Citrate Method (Mayahara *et al.*, 1967)

Fixation

Small blocks of tissue are fixed in 2 per cent glutaraldehyde in 0.1 M cacodylate buffer (pH 7.4) mixed with 8 per cent sucrose. Fixation time is 1 hour at 4°C. This is followed by washing the blocks with cacodylate.

Cut 40-µm thick cryostat sections or sections with a vibratome.

Reagents Required

Glutaraldehyde
0.1 M cacodylate buffer (pH 7.4)
Sucrose
Sodium β-glycerophosphate
Magnesium sulphate
Alkaline lead citrate solution
0.2 M tris buffer (pH 8.5)

Preparation of Reagents

Solution 1 (Incubating medium)

0.2 M tris buffer (pH 8.5)	1.4 ml
0.1 M sodium β-glycerophosphate	2.0 ml
0.15 M magnesium sulphate	2.6 ml
Alkaline lead citrate solution (pH 10.0)	4.0 ml
Sucrose	0.500 mg

Procedure

1. Transfer sections directly to solution 1 (incubating medium) for 15 minutes at room temperature.

2. Post-fix the sections in 1 per cent osmium tetroxide for 1 hour at 4°C.

3. Dip in tris buffer.

4. Dehydrate in alcohol.

5. Embed in Epon (Luft, 1961).

EPON Embedding (Luft, 1961)

Reagents Required

Epon 812 (Resin)
DDSA (Dodecenyl succinate anhydride) (hardener)
MNA (Methylnadic anhydride (hardener)
DMP 30 [(2,4,6-tri(dimethylaminomethyl phenol)] (accelerator)
Propylene oxide
Alcohols (graded series)

Preparation of Reagents

Solution 1 (Epon mixture)

Epon 812	62.0 ml
DDSA	100.0 ml

Solution 2 (Epon mixture 2)

Epon 812	100.0 ml
MNA	89.0 ml

Store these two solutions at 4°C. Mix these two solutions for final embedding medium. These solutions are fixed as follows:

Solution 1	6.0 ml
Solution 2	4.0 ml
DMP 30	0.15 ml

2 : 3 ratio of these solutions controls the hardening of the blocks. Blocks or sections are dehydrated prior to embedding.

Procedure

1. Place blocks in 70 per cent alcohol for 15 minutes.
2. Place blocks in fresh 70 per cent alcohol for 15 minutes.
3. Transfer blocks to 90 per cent alcohol for 15 minutes.
4. Place blocks in 90 per cent alcohol for 10 minutes.
5. Transfer blocks to 100 per cent alcohol for 15 minutes.
6. Place blocks in 100 per cent alcohol for 20 minutes.
7. Transfer to propylene oxide for 15 minutes giving 2 changes.
8. Transfer to a mixture of equal parts of propylene oxide and embedding mixture for 60 minutes at room temperature.
9. Place the block in a mixture of 1 part of propylene oxide and 2 parts of embedding medium for 1 hour at room temperature.
10. Transfer them to embedding medium for 2 hours at 37°C.
11. Embed in capsules and polymerize at 60°C for 18 hours.

Remarks

Propylene oxide is inflammable and toxic. Epon is hygroscopic, so care must be taken to prevent uptake of water.

Acid Phosphatase (Hanker *et al.*, 1971)

Fixation

Small pieces of tissue or blocks are fixed in a mixture of equal parts of 8 per cent formaldehyde, 2 per cent calcium chloride and 0.2 M acetate buffer (pH 5.6) at 4°C for 8–48 hours.

Place the blocks in a mixture containing 0.44 M sucrose and 1 per cent gum acacia for 8–48 hours at 4°C.

Reagents Required

Di-dicyclohexylammonium-2-naphthylthiophosphate (SSNTP)
Dimethyl formamide
Sodium acetate
Acetic acid
Sodium citrate
Copper sulphate
Potassium ferricyanide
Formaldehyde

Sucrose
Gum acacia
Thiocarbohydrazide
Osmium tetroxide

Preparation of Reagents

Solution 1 (Incubating medium)

DDNTP	5 mg
Dimethyl formamide	0.1 ml

Mix these two thoroughly and then add the following in the order given below:

0.008 per cent sodium acetate	7.9 ml
2.7 per cent acetic acid	0.25 ml
2.9 per cent sodium citrate	0.6 ml
0.75 percent copper sulphate	1.25 ml
0.17 percent potassium ferricyanide	1.25 ml

Copper sulphate and potassium ferricyanide should be added dropwise. Final pH should be 5.5.

Procedure

1. Remove the block from sucrose and place them in solution 1 for 10 minutes.

2. Dip and take out from distilled water giving 2 changes.

3. Place them in 0.5 per cent solution of 0.5 per cent thiocarbazide. This should be prepared fresh for 5 minutes.

4. Rinse in distilled water for 10 times.

5. Post-fix blocks in 2 per cent osmium tetroxide at 50°C for 15 minutes.

6. Dip in distilled water.

7. Dehydrate and embed in Epon as described above.

Aryl Sulphatase: Barium NCS Method (Hopsu-Havu *et al.*, 1967)

Fixation

Fix small blocks in 5 per cent glutaraldehyde in 0.1 M cacodylate buffer at a pH 7.4 at 4°C for 1 hour. After fixation, wash blocks of tissue in a mixture of cacodylate buffer (pH 7.4) and 7.5 per cent sucrose overnight.

Reagents Required

Glutaraldehyde
0.1 M cacodylate buffer (pH 7.4)
Sucrose
2-Hydroxy 5-nitrophenyl sulphate
0.1 M acetate buffer (pH 5.5)
Barium chloride
Acetic acid

Preparation of Reagents

Solution 1 (Incubating medium)

2-hydroxy 5-nitrophenyl sulphate	0.160 g
Distilled water	4.0 ml
0.1 M acetate buffer (pH 5.5)	12.0 ml
5 per cent barium chloride	4.0 ml

Final pH should be 5.5. Adjust with 0.2 M acetic acid.

Procedure

1. Transfer blocks to cacodylate buffer.
2. Transfer block to solution 1 for 1 hour at 37°C.
3. Rinse again in a mixture of cacodylate buffer (pH 7.4) and 7.4 per cent sucrose for 1 hour.
4. Post fix again in 2 per cent osmium tetroxide in cacodylate buffer for 1 hour at 4°C.
5. Dehydrate.
6. Embed using the steps given in the following method.

    Embedding method

    i.   Place tissue in 70% alcohol for 15 minutes.
    ii.   Transfer to fresh 70% alcohol for 15 minutes.
    iii. Place tissue in 90% alcohol for 15 minutes.
    iv.   Transfer to fresh 90% alcohol for 10 minutes.
    v.   Place tissue in 100% ethanol for 15 minutes.
    vi.   Transfer tissue to 100% ethanol for 20 minutes.
    vii. Place tissue in propylene oxide, giving 2 changes each of 15 minutes.
    viii. Replace with equal parts of propylene oxide and embedding mixture for 60 minutes at room temperature.
    ix.   Transfer tissue to 1 part propylene oxide and 2 parts embedding medium for 60 minutes at room temperature.
    x.   Replace with embedding medium only for 2 hours at 37°C.
    xi.   Embed in capsules and polymerize at 60°C for 18 hours.

Results

Enzyme sites are darkly stained.

26

CELLULAR ELEMENTS

ARGENTAFFIN AND CHROMAFFIN CELLS

Argentaffin and chromaffin cells usually occur in the pyloric glands of the stomach, and at the bases of intestinal glands. They were mistaken to be cells involved in carcinoid tumours (Masson, 1928) called as argentaffin tumours. The argentaffin cells react to histochemical tests (Lillie and Glenner, 1960). Argentaffin tissue differs from chromaffin tissue proper. Chromaffin is a dark granular material normally found in the adrenal medulla. Chromaffin reaction is positive only after treatment with potassium dichromate without prior treatment with formalin.

Gelatin (Benditt and Wong, 1957) and plasma (Vassor and Culling, 1962) models could be prepared by changing the concentration of 5-H (5-hydroxytryptamine) which after formalin fixation may be used. Argentaffin and diazo reactions are more specific. Diazo reaction in combination with Schmorl's ferric ferricyanide reaction is recommended.

HISTOCHEMICAL METHODS FOR CELLULAR ELEMENTS

Fixation

Formalin or formol saline is the best or Zenker formol. Alcohol dissolves argentaffin granules.

Silver Impregnation Method (Fontana technique)

Reagents Required

    Silver nitrate
    Ammonia

Gram's iodine
Sodium thiosulphate
Gold chloride
Safranin

Preparation of Reagents

Solution 1

10 per cent silver nitrate (10 g/100 ml water) 25 ml

To this, add 0.880 strong ammonia dropwise until the precipitate formed dissolves. Then add 25 ml distilled water. Leave it for 24 hours in a dark bottle and filter. After a fortnight, renew the solution.

Procedure

1. Bring frozen or paraffin sections to water.
2. Transfer to Gram's iodine for 5 minutes.
3. Place in 3 per cent sodium thiosulphate for 3 minutes.
4. Wash in distilled water.
5. Treat with solution 1 (silver solution) in a covered container for 18–48 hours.
6. Rinse in distilled water.
7. Tone in gold chloride (0.2 per cent) for 10 minutes.
8. Rinse in distilled water.
9. Fix in 3 per cent sodium thiosulphate for 2 minutes.
10. Wash.
11. Counterstain in 1 per cent safranin for 1 minute.
12. Wash in tap water.
13. Dehydrate, clear and mount.

Result

Argentaffin granules	Black
Melanin	Black
Other constituents	Different shades of red and pink

Diazo Reaction (Clayden, 1955)

Fixation

Formol saline

Reagents Required

Fast red salt B
Lithium carbonate
Haematoxylin

Preparation of Reagents

Solution 1 (Fast red)

Fast red salt B	1.0 g	
Distilled water	100.0 ml	

Solution 2 (Lithium carbonate)
   Lithium carbonate saturated

Solution 3
   To 5 ml of solution 1, add 2 ml of solution 2. Place in a cool place (4–5°C) for 10 minutes. This will be solution 3.

Procedure

1. Deparaffinize and hydrate slides to water.
2. Transfer and stain in solution 3 (staining) at 4°C for 2 minutes.
3. Rinse in distilled water.
4. Wash in running water.
5. Counterstain with haematoxylin.
6. Wash in tap water.
7. Dehydrate, clear and mount.

Results

Argentaffin cell granules	Orange-red
Nuclei	Black

Uric Acid Staining: Argentaffin Method (Gomori, 1952a)

Fixation

95 per cent alcohol

Reagents Required

Methenamine
Silver nitrate
Boric acid
Borax
Sodium thiosulphate

Preparation of Reagents

Solution 1 (Methenamine silver solution)
   5 per cent silver nitrate   5.0 ml
      (5 g/100 ml water)
   3 per cent methenamine   100.0 ml

When the two are added, a precipitate is formed which gradually dissolves. Keep in a dark cool place. It lasts for several months.

Solution 2 (Buffer pH 9.0)
   M/5 Boric acid (12.368 g/1000 ml water)   20.0 ml
   M/20 Borax (19.071 g/1000 ml water)   80.0 ml

Solution 3 (Working solution)
   Solution 1   30.0 ml
   Solution 2   8.0 ml

Solution 4

5.0 g of sodium thiosulphate in 100.0 ml distilled water.

Procedure

1. Deparaffinize and hydrate slides to water.
2. Transfer to solution 3 (working solution) (pre-warmed to 37°C) for 30 minutes.
3. Dip in distilled water.
4. Fix in solution 4 for 3 minutes.
5. Wash in running water for 5 minutes.
6. Counterstain, if desired.
7. Dehydrate, clear and mount.

Result

Uric acid crystals are stained black.

Enterochromaffin Cell Stain: Fontana Method (Culling, 1957)

Fixation

Zenker formol is preferable. Alcohol fixation is not advisable.

Reagents Required

Silver nitrate
Ammonia
Methenamine
Boric acid
Gold chloride
Safranin O

Preparation of Reagents

Solution 1 (Silver solution)

Silver nitrate   2.5 g
Distilled water   25.0 ml

Add slowly 28 per cent ammonia to 25.0 ml of 10 per cent silver nitrate (solution 1). A precipitate is formed which gradually dissolves. Add 25.0 ml of distilled water. Store in a dark bottle for 24 hours, filter. It will last for a fortnight.

Solution 2 (Methenamine silver solution)

3 per cent methenamine (3 g/100 ml water)   100.0 ml
5 per cent silver nitrate (5 g/100 ml water)   5.0 ml

First a white precipitate is formed which dissolves after shaking. It lasts for several months.

Solution 3

M/5 Boric acid (12.368 g/1000 ml water)   80.0 ml
M/5 Borax (19.07 g/1000 ml water)   20.0 ml

pH should be 7.8 to 8.0.

Solution 4 (Staining solution)
    Solution 2   30.0 ml
    Solution 3   8.0 ml

Solution 5
    1 per cent gold chloride (1 g + 100 ml water)   10.0 ml
    Distilled water   100.0 ml

Solution 6 (Safranin)
    Safranin   1.0 g
    Distilled water   100.0 ml

Add a few drops of glacial acetic acid.

Procedure

1. Deparaffinize and bring slides to water.
2. Place in Lugol's solution for 30 minutes to 1 hour.
3. Wash in running water.
4. Bleach in 5 per cent sodium thiosulphate for 3 minutes (5 g + 100 ml water).
5. Wash in running water for 5 minutes.
6. Transfer to solution 1 (silver nitrate).

   If desired, instead of solution 1, solution 4 can be used for 3–5 hours at 60°C and for 12–24 hours at 37°C.
7. Rinse in distilled water.
8. Tone in solution 5 (gold chloride) for 3 minutes.
9. Rinse in distilled water.
10. Fix in 5 per cent sodium thiosulphate for 3 minutes.
11. Wash in running water for 5 minutes.
12. Counterstain in solution 6 (safranin O) for 1 minute.
13. Rinse in 70 per cent alcohol.
14. Dehydrate, clear and mount.

Result

Argentaffin granules   Black
Melanin   Black

Diazo–Safranin Method (Lillie *et al.*, 1953)

Fixation

10 per cent formaldehyde buffered with 2 per cent calcium acetate. Fixation time is 3 days.

Reagents Required

Safranin O
Disodium phosphate
Sodium nitrate
Hydrochloric acid

Preparation of Reagents

Solution 1
(Stock solution (Diazotized safranin))

Safranin O	3.6 g
Distilled water	60.0 ml
NHCl	30.0 ml

This solution lasts for several weeks.

Solution 2 (N sodium nitrite)

Sodium nitrite	6.9 g
Distilled water	100.0 ml

Always keep in the refrigerator. It lasts for at least 3 months.

Solution 3 (Disodium phosphate)

Disodium phosphate	14.2 g
Distilled water	1000.0 ml

Solution 4 (Working solution)
Add 5 ml of cold solution 2 (sodium nitrate) to ice-cold solution 1. First the solution becomes deep blue. Leave the solution (safranin) at 5°C for 15 minutes for diazotization. Take 1.0 ml of this solution and add to it 40.0 ml of ice-cold disodium phosphate (solution 3) and it has to be used immediately (pH 7.7).

Procedure

1. Deparaffinize and hydrate slides to water.
2. Transfer slides to solution 4 (this has to be pre-chilled) for 5 minutes.
3. Destain in acid alcohol (1 ml of conc. HCl in 95 ml of 70 per cent alcohol) giving 3 changes for 15 seconds.
4. Wash excess acid with water.
5. Dehydrate, clear and mount.

Result

Enterochromaffin granules are stained black.

Ferric Ferricyanide Method (Modified Schmorl Technique) (Lasky and Greco, 1948; Lillie and Burtner, 1953)

Fixation

10 per cent formalin buffered with 2 per cent calcium acetate. Fix for 72 hours.

Reagents Required

Potassium ferricyanide
Ferric chloride
Safranin O

Preparation of Reagents

Solution 1 (Ferric ferricyanide)
1 per cent potassium ferricyanide (1 g/100 ml of water)   10.0 ml

| 1 per cent ferric chloride (1 g/100 ml of water) | 75.0 ml |
| Distilled water | 15.0 ml |

Solution 2

1.0 g of safranin in 100.0 ml water.

Procedure

1. Deparaffinize and hydrate slides to water.
2. Transfer to solution 1 (ferric ferricyanide) for 5 minutes.
3. Rinse in distilled water giving 3 changes.
4. Counterstain in solution 2 (safranin), and wash.
5. Dehydrate, clear and mount.

Results

Enterochromaffin granules are stained deep blue. (Suggested readings for enterochromaffin granules are Lillie (1955, 1956c, 1960, 1961), Lillie *et al.* (1953) and Lillie *et al.* (1961)

Azo Coupling Method (Gurr, 1958)

Fixation

Formalin or Bouin's fixative

Reagents Required

Garnet GBC salt

Borax

Preparation of Reagent

Solution 1

Garnet GBC salt	0.5 g
Distilled water	100.0 ml
Borax (Saturated solution)	2.5 ml

Procedure

1. Dewax and hydrate slides to water.
2. Treat with solution 1 for 30 to 60 seconds.
3. Wash in running water for 30 seconds.
4. Counterstain with Mayer's haematoxylin for 3 minutes.
5. Wash in running water for 5 minutes.
6. Dehydrate, clear and mount.

Results

| Argentaffin granules | Red |
| Nuclei | Blue |

Suggested readings for enterochromaffin granules are Lillie (1955, 1956c, 1960, 1961); Lillie *et al.* (1953); Lillie *et al.* (1961); Epple and Brunn (1986) and Schriebman (1986).

Table 26.1 Histochemical techniques applied for the demonstration of cellular elements

Technique	Fixative	Argentaffin granules	Enterochromaffin granules	Uric acid crystals
Silver impregnation method	Formalin or formol saline	Black		
Diazo reaction	Formol saline	Orange-red		
Argentaffin method	95 per cent alcohol			Black
Fontana method	Zenker formol	Black		
Diazo-safranin method	10 per cent formaldehyde buffer with 2 per cent calcium acetate		Black	
Azo coupling method	Formalin, Bouin's	Red		
Ferric ferricyanide method	10% formalin buffered with 2% calcium acetate		Deep blue	

CHROMOSOMES

Chromosomal study has grown widely in the last 15 years. The information available is so massive and diverse that it is not possible to cover here. Only the basic chromosome methods are given. Further details can be obtained from Halman (1989), Alberts *et al.* (1994) and Darnell *et al.* (1994).

It may be convenient and appropriate to begin with giant chromosomes. Giant chromosomes are found in somatic tissue of insects and reach the largest size in the salivary glands of the larvae. *Drosophila* larvae are frequently used to demonstrate these giant chromosomes especially for the cross-striations and the banding patterns. Barley (1964) found that larvae of black flies (*Simulium vittatum*) could also be used to demonstrate such banding. In 1964 Evans *et al.*, have described a method for squash technique. They chose mouse testis but the method is applicable to insect larvae also.

CHROMOSOME SQUASH METHODS

Procedure

1. Immerse testis of mouse in sodium citrate solution (2.2 per cent w/v). Swirl the testis to remove fat.
2. Take out and immerse in fresh sodium citrate solution. Tease the contents from the tubules and transfer the supernatant fluid to a centrifuge tube.
3. Centrifuge at 500 rpm for 5 minutes. Discard the supernatant and resuspend in 1 per cent sodium citrate.
4. Again centrifuge (500 rpm) for 5 minutes.
5. Discard the supernatant as much as possible and resuspend with 0.25 ml of fixative (3 : 1 methanol/glacial acetic acid). Add more fixative, then centrifuge for 5 minutes.
6. Discard supernatant, place one drop of suspension on a slide, allow it to dry and observe under phase-contrast microscope.

    Giemsa stain is used for staining.

Result

Chromosomes are stained blue.

For other methods of staining refer Chandley (1988).

Acetocarmine or Aceto-orcein

Boil 500 mg of carmine in 100 ml of 45 per cent acetic acid for 2–4 minutes. Cool and filter.

Preparation of Reagents

Working Solution

Stock solution	25.0 ml
45 per cent acetic acid	50.0 ml

Belling (1926) suggested addition of few drops of ferric hydrate in 50 per cent acetic acid.

Aceto-orcein Solution

Add 2 g of orcein to 45 ml of hot acetic acid. When cool, add 55 ml of distilled water. Lacour (1941) used 2 per cent orcein in 70 per cent acetic acid.

Alternate method is to mix equal volumes of acetic acid in 85 per cent lactic acid. Boil and add 1–2 g of 100.0 ml solution. Cool and filter (Yerganian, 1963).

Procedure

1. Usually large sluggish *Drosophila* larvae are preferred. The larva is put in physiological saline on a slide. Hold the posterior end with a forceps and pull out the mouthparts with a needle. Salivary glands are attached to the mouthparts.
2. Place the glands in Carnoy fixative or place them directly on a slide and add a small amount of stain (albuminized).
3. Place the coverslip on the slide and apply pressure.
4. Seal edges with paraffin wax.

Result

With acetocarmine, chromosomes are stained red.

With aceto-orcein chromosomes are stained purple.

Before making squashes, these glands are stored in Carnoy (Barley, 1964). Glands are treated with 1:1 hydrochloric acid–absolute alcohol mixture for 2–3 minutes. The glands are again put in Carnoy before staining.

To get good squashes, proper application of pressure is very important. The middle of the cover glass can be pressed with the thumb and the thumb is rolled towards the edges of the cover glass until no more fluid oozes out. Berrois (1994) described a method to obtain standardized results.

Permanent Mounts

Subbed slides are good for permanent mounts.

Smith Method (1947)

Procedure

1. Paraffin edge is dewaxed by xylene.
2. Slides are soaked in equal parts of acetic acid and 95 per cent alcohol.
3. Transfer to equal parts of 95 per cent alcohol and tertiary butyl alcohol for 2 minutes.
4. Place pure tertiary butyl alcohol giving 2 changes of 2 minutes each.
5. Blot and then add a thin resin mountant on slide and place the cover glass.

Nolte Method (1948)

Procedure

1. Place the slide in a dish so that the edge of the cover glass dips into 95 per cent alcohol, for 12 hours.
2. Immerse in 95 per cent alcohol for 1–2 hours.
3. Gently remove the cover glass.
4. Drain off excess alcohol, add a drop of mounting medium and replace with a clean cover glass.

Lacto-propionic Orcein Method (Dyer, 1963)

Preparation of Reagents

Stock Solution

Orcein	2.0 g
Lactic acid	50.0 ml
Propionic acid	50.0 ml

Working Solution

Stock solution	45.0 ml
Distilled water	55.0 ml

Procedure

1. Fix in the following fluid for 5 minutes.

Absolute ethyl alcohol	10 parts
Glacial acetic acid	2 parts
Chloroform	2 parts
Formalin	1 part

2. Macerate in 1N HCl at 60°C for 5 minutes.
3. Treat with working solution.
4. Squash.

Result

Chromosomes are stained deep reddish purple.

For further details on chromosomes, suggested references are Alberts *et al.* (1994), Darnell *et al.* (1994) and Thorgaard and Disney (1990).

Other Stains for Squashes and Spreads

1. Aceto basic fuchsin (Tanaka, 1961)
2. Aceto iron haematoxylin (Lowry, 1963; Wittman, 1962, 1963, 1965)
3. Acridine orange (Schiffer and Vaharu, 1962)

4. Carbol fuchsin (Carr and Walker, 1961)
5. Feulgen (Rafalko, 1946; Sachs, 1953)
6. Gomori and haematoxylin (Melander and Wingstrand, 1953)
7. Inverted Feulgen (Koulischer and Mulnard, 1962)
8. Lactic acetic orcein (Welshons *et al.*, 1962)
9. Lacto-propionic–Orcein (Dyer, 1963)
10. Iron haematoxylin (Griffin and McQuarrie, 1942; Chen, 1944)
11. Iron haematoxylin with acetocarmine (Austin, 1959)
12. Sudan black B (Cohen, 1949)
13. Sudan black with acetocarmine (Bradley, 1957)

Klingar and Hammond Method (1971)

Fixation

Blood smears are fixed in methyl alcohol for 1–2 minutes, oral smears in absolute alcohol ether (1 : 1) for 2–24 hours. Fix tissue blocks in Davidson's fixative.

Davidson's fixative

95 per cent alcohol	30.0 ml
Formaldehyde	20.0 ml
Glacial acetic acid	10.0 ml
Distilled water	30.0 ml

Reagents Required

Pinacyanol
Monosodium potassium phosphate
Dibasic sodium phosphate

Preparation of Reagents

Solution 1

Pinacyanol	250 mg
70 per cent alcohol	100.0 ml

Solution 2 (Wright buffer)

Monosodium potassium phosphate	6.63 g
Dibasic sodium phosphate	3.2 g
Distilled water	1000.0 ml

Procedure

1. Hydrate slides to water.
2. Keep slides in 5N HCl: smears for 2 minutes and sections for 3–6 minutes.
3. Wash in running water.
4. Transfer to solution 1 for 45 seconds.
5. Differentiate in solution 2 for 45 seconds.
6. Wash in running water for 5 seconds.

7.   Dehydrate in isopropanol giving 2 changes of 1 minute each.

8.   Clear and mount.

Result

Chromatin is stained blue.

Guard's Method for Sex Chromatin (Guard, 1959)

Fixation

95 per cent alcohol

Reagents Required

Biebrich scarlet
Phosphotungstic acid
Glacial acetic acid
Fast green
Phosphomolybdic acid
Haematoxylin

Preparation of Reagents

Solution 1 (Biebrich scarlet solution)

Biebrich scarlet	1.0 g
Phosphotungstic acid	300 mg
Glacial acetic acid	5.0 ml
50 per cent alcohol	100.0 ml

Solution 2 (Fast green solution)

Fast green	500 mg
Phosphomolybdic acid	300 mg
Phosphotungstic acid	300 mg
Glacial acetic acid	5.0 ml
50 per cent alcohol	100.0 ml

Solution 3 (Harris haematoxylin (*see* Chapter 7)

Harris haematoxylin	0.5 ml
50 per cent alcohol	100.0 ml

Procedure

1.   From 95 per cent alcohol fixative, transfer to 70 per cent alcohol.

2.   Transfer to solution 1 for 2 minutes.

3.   Rinse in 50 per cent alcohol.

4.   Differentiate in solution 2 for 4 hours.

5.   Differentiate in 50 per cent alcohol.

6.   Dehydrate, clear and mount.

Result

Sex chromatin	Red
Background	Green

Suggested readings are Barr *et al.* (1950), Klinger (1958), Moore *et al.* (1953), Moore and Bar (1955), Moore (1962), Theogaard and Derney (1990), Alberts *et al.* (1994) and Darnell *et al.* (1994).

SEX CHROMATIN STAINING

Sex differences in neural tissues of cat were revealed by Barr and Bertram (1949). Many investigators later evinced interest in the cells involving other parts of the body. Sex chromatin has been identified in more than 50 per cent of such cells in the female. In the male only 5 per cent of the cells show this. However, it is not homologous to sex chromatin of female (Hammerton, 1961). Female sex chromatin is in the form of a darkly stained mass near nuclear membrane.

There is some difference in chromatin structure between male and female lymphocytes and this was unmarked by Riis (1957). Similar differences were noticed by Moldovanu (1961) and Sommer and Picketi (1961) in lymphocytes and monocytes. Vaginal mucous membrane cells give better results than cells from other parts because the chromatin is slightly larger and cytological details are sharp.

Sex chromatin is composed largely of DNA and any specific test for DNA is suitable to demonstrate sex chromatin. Haematoxylin and eosin technique gives best results especially in tissue sections. [Details are given on *The Sex Chromatin*, Ed. K.L. Moore (1966).]

Good smear preparations are necessary. The scrapings of the tissue are smeared on clean slides and this is fixed for 15 minutes in a solution containing equal parts of ether and ethyl alcohol.

HISTOCHEMICAL METHODS FOR SEX CHROMATIN

Klinger and Ludwig Method (1957)

Fixation

Blood smears are fixed in methyl alcohol for 1–2 minutes, oral mucosal smears in absolute alcohol, ether (1 : 1) for 2–24 hours, tissue blocks in Davidson fixative for 24 hours.

Davidson Fixative

95 per cent ethyl alcohol	30.0 ml
Formalin	20.0 ml
Glacial acetic acid	10.0 ml
Distilled water	30.0 ml

Reagents Required

Thionin
Sodium acetate
Sodium barbiturate
HCl

Preparation of Reagents

Solution 1 (Buffered thionin)
Saturated solution of thionin in 50 per cent alcohol.

Solution 2

Sodium acetate	9.714 g
Sodium barbiturate	14.714 g
CO_2-free distilled water	500.0 ml

Solution 3

Hydrochloric acid (sp. gravity (1–19))	8.50 ml
Distilled water	991.50 ml

Solution 4 (Working solution) (pH 5.7)

Solution 1	40.0 ml
Solution 2	28.0 ml
Solution 3	32.0 ml

Procedure

1. Dewax and hydrate slides to water.
2. Place in 5N HCl for hydrolysis for 20 minutes at 20–25°C.
3. Rinse in distilled water.
4. Treat with solution 4 for 15–60 minutes.
5. Rinse in distilled water.
6. Rinse in 50 per cent, 70 per cent alcohols to drain off excess stain.
7. Dehydrate, clear and mount.

Result

Sex chromatin	Deep blue
Nucleus and chromatin	Lightly stained

Cytoplasm remains colourless.

Klinger and Hammond Method (1971)

Fixation

Blood smears to be fixed in methyl alcohol for 1–2 minutes.
Mucosal smears to be fixed in absolute alcohol–ether (1 : 1) for 2–24 hours.
Tissues are to be fixed in Davidson fixative (*see* Chapter 1)

Reagents Required

Pinacyanol
Monosodium potassium phosphate
Dibasic sodium phosphate

Preparation of Reagents

Solution 1 (Pinacyanol)

Pinacyanol	250 mg
70 per cent alcohol	100.0 ml

Solution 2

    Monobasic potassium phosphate 6.63 g
    Dibasic sodium phosphate   3.20 g
    Distilled water   100.0 ml

Procedure

1. Deparaffinize and hydrate slides to water.
2. Hydrolyse slides in 5N HCl for 6 minutes.
3. Wash in water.
4. Treat with solution 1 for 45 seconds.
5. Differentiate in solution 2 for 45 seconds.
6. Wash in running water.
7. Dehydrate in isopropanol giving 2 changes (2 minutes each).
8. Clear and mount.

Results

Chromatin is stained blue.

Comments

The principle is same in both Feulgen and Klinger and Ludwig methods.

Other references in this context are Guard (1959), Beckert and Garner (1966), Lennox (1956), Makowski *et al.* (1956), Vernino and Laskin (1960), Barr *et al.* (1950), Klinger (1958), Marberger *et al.* (1955), Marwah and Weinmann (1955), Moore and Barr (1955), Moore *et al.* (1953), Taylor (1963).

Cresyl Echt Violet Method (Moore, 1962)

Fixation

Ether ethyl alcohol (1:1)

Reagents Required

Cresyl echt violet

Preparation

Cresyl echt violet (1 g in 100 ml water)   1 per cent

Procedure

1. Immerse slides from fixatives into 70 per cent alcohol, 50 per cent and water.
2. Transfer them to 1 per cent cresyl echt violet for 5–8 minutes.
3. Differentiate in 95 per cent alcohol.
4. Differentiate in absolute alcohol.
5. Clear in xylol and mount in DPX.

Result

Sex chromatin is seen as a dark-stained dot near the nuclear membrane.

Comments

Smears have to be studied under oil immersion. At least 100 well-formed nuclei should be examined. If there is no sex chromatin, the smear must be from a male. If 40–60 per cent of nuclei in a smear contain the dark-stained images, then it may be interpreted as a smear from female.

Guard's Method for Sex Chromatin (1959)

Fixation

10 per cent formalin

Reagents Required

Biebrich scarlet
Phosphotungstic acid
Glacial acetic acid
Fast green FCF
Phosphomolybdic acid

Preparation of Reagents

1. Biebrich scarlet stain

Biebrich scarlet	1.0 g
PTA	300 mg
Glacial acetic acid	5.0 ml
50 per cent alcohol	100.0 ml

2. Fast green FCF

Fast green	500 mg
Phosphomolybdic acid	300 mg
Phosphotungstic acid	300 mg
Glacial acetic acid	5.0 ml
50 per cent alcohol	100.0 ml

Procedure

1. Rinse slides after fixation in 70 per cent alcohol for 2 minutes.
2. Stain in solution 1 (Biebrich scarlet) for 2 minutes.
3. Rinse in 50 per cent alcohol.
4. Immerse in solution 2 (Fast green) for 4 hours until cytoplasm and nuclei are green.
5. Rinse in 50 per cent alcohol.
6. Dehydrate, clear and mount.

Result

Sex chromatin	Bright red
Pycnotic nuclei	Bright red
Cytoplasm and nuclei	Green

Table 26.2   Histochemical techniques applied for the demonstration of chromosomes, sex chromatin

Technique	Fixative	Chromosomes	Sex chromatin
Acetocarmine	Physiological saline	Red	
Acetoorcein	Physiological saline	Purple	
Lactopropionic acid	10 parts absolute alcohol + 2 parts of glacial acetic acid + 2 parts chloroform + 1 part formalin	Deep reddish purple	
Klinger and Hammond method	Davidson fixative		Blue
Klingar and Ludwig method	Davidson fixative		Deep blue
Cresyl Echt method	Ether 1 part + ethyl alcohol 1 part		Dark-Istained dot
Guard's method	10 per cent formalin 95% alcohol		Bright red

CYTOPLASMIC CONSTITUENTS AND CELL PRODUCTS

DICTYOSOMES (GOLGI APPARATUS)

Golgi (1898) found the dictyosomes in nerve cells. A few specific methods are available for the demonstration of Golgi. Generally it is lost in routine fixation and staining. To get precise results, it is necessary to modify the fixing and staining time. Of the many methods, Da Fano Cajal method is specific. With osmium and silver method, Golgi appears in the form of a dark net which is diffuse. These reactions show the presence of lipids and phospholipids. Golgi also has a low concentration of acid phosphatase and alkaline phosphatase. The composition of Golgi varies from cell to cell and depends much upon the activity of the cell. Most of the fixatives suggested are those containing heavy metals but it has been shown that light metals or organic salts are also equally good.

HISTOCHEMICAL METHODS FOR GOLGI APPARATUS

Ludford's Osmium Tetroxide Method
(Lillie, 1954a; Cowdry, 1952)

Fixation

Osmic sublimate	18 hours
Osmic acid (1 g + 100 ml water)	50.0 ml
Mercuric chloride sublimate with 0.37 g of sodium chloride	50.0 ml

Reagents Required

Osmic acid
Mercuric chloride
Sodium chloride

Procedure

1. Wash tissue blocks in distilled water for 30 minutes.
2. Impregnate in 2 per cent osmic acid at 30°C for 3 days.
3. Impregnate in 2 per cent osmic acid at 35°C for 1 day.
4. Impregnate in 1 per cent osmic acid at 35°C for 1 day.
5. Impregnate in 0.5 per cent osmic acid at 35°C for 1 day.
6. Wash in distilled water for 24 hours.
7. Dehydrate, clear and embed.
8. Cut 6–7 μm thick sections and dry.
9. Deparaffinize, clear and mount.

Result

Golgi apparatus   Black
Yolk and fat   Black

Remarks

During impregnation, solutions should be kept in a dark chamber. Temperatures should be maintained if solutions turn dark. They should be discarded.

Nassonov–Kolatchew Method (Nassonov, 1923, 1924)

Fixation

3 per cent potassium permanganate (3 g + 100 ml water)   10.0 ml
1 per cent chromic acid (1 g + 100 ml water)   10.0 ml
2 per cent osmic acid (2 g + 100 ml water)   5.0 ml

Reagents Required

Potassium permanganate
Chromic acid
Osmic acid

Procedure

1. After fixation (24 hours), wash in distilled water.
2. Transfer to 2 per cent osmic acid (aqueous) for 8 hours at 40°C and for 3–5 days at 35°C.
3. Wash in running water overnight.
4. Dehydrate, rapidly clear and embed.
5. Take 2–4 μm thick sections.
6. Deparaffinize.
7. Stain with carmine.

Result

Golgi	Black
Background	Yellow
Mitochondria	Red

Dafano Cajal's Method (Culling, 1957; Cowdry, 1952)

Fixation

Cobalt nitrate	1.0 g
Distilled water	100.0 ml
Formalin	6–15 percent
Fixation time	3–18 hours

Reagents Required

Cobalt nitrate
Formalin
Hydroquinone
Sodium sulphate
Silver nitrate
Sodium thiosulphate
Ammonium sulphocyanide
Gold chloride

Preparation of Reagents

Solution 1 (Ramony's Cajal's developer)

Hydroquinone	2.0 g
Formalin	6.0 ml
Distilled water	100.0 ml
Sodium sulphate	150 g

Solution 2

Silver nitrate	1.5 g
Distilled water	100.0 ml

Solution 3 (Toning solution) (Optional)

a.	Sodium thiosulphate	30.0 g
	Ammonium sulphocyanide	15.0 ml
	Distilled water	100.0 ml
b.	Gold chloride	1.0 g
	Distilled water	100.0 ml

Solution 4 (Working solution)
Equal parts of solutions 3a and 3b.

Procedure

1. Fix as already described.
2. Rinse in distilled water.
3. Place in solution 2 (silver nitrate) for 36–48 hours.

  4. Rinse rapidly in distilled water.

  5. Reduce in solution 1 for 6–12 hours.

  6. Wash in running water.

  7. Dehydrate, clear and embed as usual for paraffin.

  8. Cut 3–6 µm thick sections.

  9. Deparaffinize and hydrate.

10. Place in solution 4 for 5–10 minutes.

11. Wash in running water.

12. Dehydrate, clear and mount.

Result

Golgi apparatus	Black
Cytoplasm	Grey
Mitochondria	Dark grey

Direct Silver Method (Elftman, 1952)

Procedure

1. Immerse small blocks of tissue in silver nitrate solution in formalin (2 g of silver nitrate in 100 ml of 15 per cent formalin) for 2 hours.

2. Rinse quickly in distilled water.

3. Develop for 2 hours in

Hydroquinone	2.0 g
15 per cent formalin (15 ml + 85 ml water)	100.0 ml

4. Place them again in 10 per cent formalin overnight.

5. Wash, dehydrate and embed.

6. Cut 6–7 µm thick sections, mount and dry.

7. Deparaffinize, clear and mount.

Result

Golgi apparatus is stained black.

Sudan Black B Technique (Baker, 1949)

Fixation

Neutral formol saline

Reagents Required

Potassium dichromate
Sodium chloride
Gelatin
Sodium parahydroxybenzoate
Potash alum
Sudan black B

Preparation of Reagents

Solution 1
10 per cent formalin in 0.7 per cent sodium chloride

Solution 2 (Formol dichromate)

Potassium dichromate	2.2 g
Sodium chloride	700 mg
Distilled water	95.0 ml
Formalin (neutral)	5.0 ml

Solution 3 (Embedding medium)

Gelatin	4.0 g
0.2 per cent sodium parahydroxybenzoate	100.0 ml

Solution 4 (Formalum)

Potash alum	4.0 g
Distilled water	8.0 ml
Formalin	20.0 ml

Solution 5 (Sudan black B)
Saturated solution. Boil 500 mg of Sudan black B in 100.0 ml of 70 per cent alcohol for 10 minutes.

Solution 6 (Mayer's carmalum)

Carmine	2.0 g
5 per cent ammonia alum	100.0 ml

Boil for an hour, make up to 100.0 ml and later add few crystals of thymol to prevent formation of moulds.

Procedure

1. Fix small pieces of tissue in solution 1 for 1 hour.
2. Place the tissue in solution 2 without washing for 5 hours.
3. Transfer to 5 per cent potassium dichromate and leave for 18–24 hours at room temperature.
4. Place the same at 60°C for 24 hours.
5. Wash in running water overnight for 6 hours.
6. Infiltrate in solution 3 for 18 hours at 37°C.
7. Embed in solution 3. Cool in a refrigerator and harden the block in solution 4 (formalum) for 17–24 hours.
8. Cut 8–10 μm thick sections with freezing microtome.
9. Transfer sections to 70 per cent alcohol for 1 minute.
10. Transfer sections to solution 5 (Sudan black B) for 4 minutes.
11. Transfer sections to 50 per cent alcohol.
12. Wash in water for 1 minute.
13. Counterstain in solution 6 for 2–3 minutes.
14. Rinse in distilled water.
15. Mount with glycerine jelly.

Result

Golgi apparatus	Dark blue
Cytoplasm	Pale blue
Nuclei	Shades of red
Neutral fats	Dark blue

MITOCHONDRIA

Mitochondria are tiny complex cell organelles and they are bound by a double membrane, matrix and dense granulation. Matrix is mostly composed of proteins and lipids, and membranes with insoluble enzymes, proteins and phospholipids. Mitochondria contain numerous enzymes like oxidoreductases, cytochrome oxidase, succinic dehydrogenase, glutamic dehydrogenase and adenosine triphosphate.

HISTOCHEMICAL METHODS FOR MITOCHONDRIA

Altman's Method

Fixation

Fix in Regaud's solution for 4 days with 4 changes in refrigerator. Mordant in 3 per cent potassium dichromate. Change every second day. Wash overnight, dehydrate and embed.

Reagents Required

Potassium dichromate
Oxalic acid
Methyl green
Aniline
Acid fuchsin

Preparation of Reagents

Solution 1 (Altman's solution)
To saturated filtered solution of aniline in distilled water, add 10 per cent acid fuchsin (10 g + 100 ml water). Allow it to stand for 24 hours. This solution lasts for one month.

Solution 2

Methyl green	2.0 g
Distilled water	100.0 ml

Procedure

1. Deparaffinize and hydrate slides to water.
2. Oxidize in 1 per cent potassium permanganate (1 g + 100 ml water) for 30 seconds.
3. Bleach in 5 per cent oxalic acid (5 g + 100 ml water) for 30 seconds.
4. Rinse in several changes of distilled water.

5. Flood the slide with solution 1 for 6 minutes.
6. Dry the slide and rinse in distilled water.
7. Flood the slide with solution 2 for 2 seconds.
8. Dehydrate, clear and mount.

Result

Mitochondria is stained bright red.

Osmic Method (Newcomer, 1940)

Fixation

Zenker's solution	46 hours fixation
Potassium dichromate	1.25 g
Ammonium dichromate	1.25 g
Copper sulphate	1.0 g
Distilled water	100.0 ml

Reagents Required

Potassium dichromate
Ammonium dichromate
Copper sulphate
Benzene
Potassium permanganate
Oxalic acid

Procedure

1. Wash blocks of tissue overnight.
2. Impregnate in 2 per cent osmic acid (2 g + 100 ml water) for 4–6 days.
3. Wash overnight.
4. Dehydrate, clear in benzene and embed.
5. Cut 5 µm thick sections.
6. Deparaffinize and hydrate slides to water.
7. Oxidize in 1 per cent potassium permanganate for 5 minutes.
8. Rinse in distilled water.
9. Bleach in 3 per cent oxalic acid.
10. Wash in running water for 15 minutes.
11. Dehydrate clear and mount.

Result

Mitochondria is stained black.

Short Acid Fuchsin Method (Novelli, 1962)

Fixation

Fix in 10 per cent formalin or 1 per cent osmic acid for 24 hours at room temperature. Embed in paraffin and take 4 µm thick sections.

Reagents Required

Osmic acid
Acid fuchsin
Methyl blue
NHCl

Preparation of Reagents

Solution 1

Acid fuchsin	200 mg
Methyl blue	100 mg
NHCl	100.0 ml

Procedure

1. Deparaffinize and hydrate slides to water.
2. Stain in solution 1 for 5 minutes.
3. Rinse gently in distilled water.
4. Dehydrate quickly in absolute alcohol giving 2 changes.
5. Clear and mount.

Result

Mitochondria	Purple-red
Chromatin and collagen	Blue
Nucleoli	Brilliant red

Acid Haematein Method (Hori and Chang, 1963)

Reagents Required

Acetone
Uranyl nitrate
Chlorol hydrate
Borax
Potassium ferricyanide
Acid haematein
Potassium dichromate

Preparation of Reagents

Solution 1 (Acetone)

Acetone	50.0 ml
Uranyl nitrate	20 mg
Chlorol hydrate	20 mg

Solution 2 (Acid haematein)

Haematoxylin	50 mg
0.01 per cent sodium iodate	50 ml

Heat the solution, cool and add 1.0 ml of glacial acetic acid.

Solution 3 (Borax ferricyanide solution)

Borax	250 mg
Potassium ferricyanide	250 mg
Distilled water	100.0 ml

Keep in the refrigerator.

Procedure

1. Freeze-dry sections.
2. Transfer the sections into screw-capped bottles of acetone mixture (solution 1). This solution should be pre-chilled and left in ice overnight.
3. Rinse briefly in acetone at room temperature and mount on cover glass.
4. Transfer to 5 per cent potassium dichromate at 60°C for 5–6 hours.
5. Wash in distilled water.
6. Stain in solution 2 (Acid haematein) for 30 minutes.
7. Wash in distilled water.
8. Differentiate in solution 3 for 18 hours at room temperature.
9. Dehydrate, clear and mount.

Result

Mitochondria is stained blue-black.
Lipids may also stain.

Champy Kull's Method

Fixation

Fix in Zenker's formalin fluid for 24 hours. Post-treat with 3 per cent potassium dichromate for 4 days or in Champy's fluid for 24 hours, followed by post-treatment with 3 per cent potassium dichromate.

Champy's Fluid

3 per cent potassium dichromate	7.0 ml
1 per cent chromic acid	7.0 ml
2 per cent osmium tetroxide	4.0 ml

Reagents Required

Potassium dichromate
Acid fuchsin
Aniline blue
Toluidine blue
Aurantia

Preparation of Reagents

Solution 1

Saturated acid fuchsin in aniline blue

Solution 2

Toluidine blue	500 mg
Distilled water	100.0 ml

Solution 3 (Aurantia fluid)
    Aurantia   500 mg
    70 per cent alcohol   100.0 ml

Procedure

1. Hydrate slides to water.
2. Wash.
3. Flood slides with solution 1.
4. Rinse in distilled water.
5. Flood slides with solution 2 for 3 minutes.
6. Rinse in distilled water.
7. Flood slides with solution 3 for 2 minutes.
8. Flood slides with absolute alcohol.
9. Clear and mount.

Results

Mitochondria	Red
Nuclei	Blue
Cytoplasm	Yellow

Altman's Acid Fuchsin–Picric Acid Technique

Reagents Required

Acid fuchsin
Aniline water
Picric acid

Preparation of Reagents

Solution 1
As prepared in the previous case (Saturated acid fuchsin in aniline water)

Solution 2
Picric acid saturated in absolute alcohol   20–80 ml
30 per cent alcohol

Procedure

1. Hydrate slides to water.
2. Flood slides with solution 1.
3. Rinse in distilled water.
4. Differentiate in solution 2 until mitochondria are red.
5. Rinse in 90 per cent and 100 per cent alcohol.
6. Clear in xylol and mount.

Result

Mitochondria	Bright red
Nuclei	Yellow
Cytoplasm	Yellow

Cain's Method (Cain, 1948)

Fixation

Fix in Regaud's solution for 4 days, in fresh solution each day. After 4 days it is followed by post-chromation in 5 per cent $K_2Cr_2O_7$, for 8 days, fresh solution each day.

Reagents Required

Aniline
Acid fuchsin
Sodium carbonate
Concentrated hydrochloric acid
Methyl blue

Preparation of Reagents

Solution 1 (Aniline acid fuchsin)
Aniline	1.0 ml
Distilled water	20.0 ml
Acid fuchsin	4.0 g

Add aniline to distilled water and shake thoroughly. Let it stand for 24 hours. Filter and then add acid fuchsin and shake. This should be used within 24 hours.

Solution 2 (0.5 per cent sodium carbonate solution)
Sodium carbonate	100 mg
Distilled water	100.0 ml

Solution 3 (1 per cent hydrochloric acid)
Concentrated hydrochloric acid	1.0 ml
Distilled water	99.0 ml

Solution 4 (0.5 per cent methyl blue solution)
Methyl blue	500 mg
Distilled water	100.0 ml

Procedure

1. Dewax and hydrate slides to water.
2. Place slides in solution 1 for 15 minutes.
3. Differentiate in solution 2.
4. Rinse in solution 3.
5. Rinse in distilled water.
6. Counterstain in solution 4 for few seconds.
7. Differentiate in solution 3.
8. Wash in distilled water.
9. Dehydrate, clear and mount.

Result

Mitochondria	Bright red
Nuclei	Blue to green

Table 26.3   Histochemical techniques applied for the demonstration of cell constituents

Technique	Fixative	Golgi	Mitochondria	Nucleus	Cytoplasm	Yolk and Fat
Ludford's Osmium tetroxide method	Osmic sublimate, osmic acid	Black				Black
Nassonov-Kolatchew method	Osmic potassium permanganate chromic acid	Black	Red			
Dafano Cajal's method	Cobalt formalin	Black	Dark-Grey		Grey	
Direct silver method	Formalin	Black				
Sudan black B technique	Neutral formalin saline chloride	Dark blue		Shades of red	Pale blue	Dark blue
Altman's method	Regauds		Bright red			
Osmic method	Zenker's		Black			
Short acid fuchsin method	Formalin, osmic acid		Purple-red			
Acid haematein method	Freeze-dried		Blue-black			
Champy Kull's method	Zenker's formol		Red	Blue	Yellow	
Altman's acid Fuchsin/picric acid method	Regaud's		Bright red	Yellow	Yellow	
Cain's method	Regaud's		Bright red			

SUPRAVITAL STAINING

Vital stains are dyes which penetrate living cells and specifically stain certain organelles. One such is neutral red which stains certain granulations, cytoplasmic products, digestive vacuoles, secretory granules and so on. These vacuoles contain phospholipoprotein complexes, alkaline phosphatase, lipase, acid phosphatase (Koenig, 1963; Ogawa *et al.*, 1961). For vital staining of Golgi and mitochondria simultaneously, neutral red is used in conjunction with Janus green B. Janus green B stains mitochondria and is influenced by enzymatic activities of the cell.

HISTOCHEMICAL METHODS USING VITAL STAINS

Neutral red–Janus Green B Method

Stock solution can be stored in glass-stoppered bottles. Dilute solutions are made in small quantities and the mixed solution should be used immediately because it is not stable. Slides should be cleaned with dichromate solutions, washed in running water and then distilled water, and stored in 95 per cent alcohol.

Preparation of Reagents

Solution 1 (Neutral red)
Neutral red   500 mg
Absolute alcohol   100.0 ml

Solution 2 (Janus green B)
Janus green B   500 mg
Neutral absolute ethyl alcohol   100.0 ml

Procedure

1. Dilute neutral red solution in the ratio 1 : 10 with neutral absolute ethyl alcohol.
2. Mix 0.4 ml of solution 2 with 3 ml of dilute neutral red solution.
3. Flood slides with mixed dye uniformly so that it is thinly distributed. Dry the slides, free from dust.

Or

Keep the slides in a Coplin jar containing the mixture. Before keeping the slide gently warm it and plunge into the Coplin jar.

Result

Mitochondria   Green
Neutral red bodies   Red

EXFOLIATIVE CYTOLOGY

Periodically, cells from mucous membranes or renal tubules are cast off as a normal behaviour (desquamation or exfoliation). From the superficial epithelium, they slough off. Exfoliative cytology deals with such a phenomenon. However, in the deceased state, cells are shed as flakes or scales and methods for early diagnosis are necessary or sometimes urgent. Difficulties arise because of poor smear preparation or poor fixation. Smears should be immediately fixed when still moist, otherwise exfoliated cells degenerate rapidly. Fixation in a fluid containing equal parts of 95 per cent alcohol and ether for 15 minutes is good. Slides can remain in this solution for a week.

To study body fluids, take equal volumes of body fluids and 95 per cent alcohol and centrifuge (2000 rpm) for 3 minutes. Pour off supernatant, take a drop of the sediment and place it on an albuminized slide to make uniformly thin smears. Fix in ether alcohol for one hour. Air-drying should be avoided at any stage. Five per cent polyethylene glycol is sometimes added to the fixative slides (Ehrenrich and Kerpe, 1959).

Sample Collection and Treatment

Early morning urine sample is required for the study. More than 50 ml of the sample is collected and the sample should be examined immediately avoiding delay. Some workers add equal volume of 95 per cent alcohol to the fresh sample.

One of the following three techniques can be adopted.

1. *Millipore filter method (Solomon et al., 1950)*   Urine sample is filtered through the cellulose membrane (5 mm pore size) using a negative pressure of 25 mm Hg. These filters withhold the cells and the cells are mounted on glass slides and stained using Papanicolaou technique.

2. *Deden's method (Deden, 1954)*   Collect large volumes of urine (500 ml). Let it stand till sediment settles down at the bottom, if not sediment it in a separating funnel. Now take the sediment centrifuge, make smears of the sediment on several albuminized slides and stain using Papanicolaou technique.

3. *Sagi and Mackenzie method (1958)*   Collect 200 ml of urine and free it from urates, carbonates, phosphates and so on. Now distribute it into several tubes, centrifuge the tubes and pour off the supernatants. Now combine all sediments into one tube. Add little amount of distilled water and again centrifuge. Discard the supernatant. Now add a drop of a mixture of equal parts of saliva and glycerol to the sediment and stir. Take a few drops and place them on slides and spread evenly. Fix the smear in acetone and stain using Papanicolaou technique.

Other samples like sputum, pleural, gastric, peritoneal or spinal fluids containing cells may be treated by Dedens or Millipore methods, can be fixed in Bouin's fluid for 8–12 hours and later filtered through a very coarse filter paper which almost serves as a container. The paper funnel having the material is subjected to the usual procedures of dehydration, clearing and embedding in paraffin wax. When it is to be embedded, the folds of filter paper are opened and the material is scraped into a mould for blocking. Sections are cut and stained with Papanicolaou stain or haematoxylin eosin.

HISTOCHEMICAL TECHNIQUES FOR EXFOLIATIVE CYTOLOGY

Papanicolaou Method (1942, 1947, 1954, 1957)

Fixation

Ether/alcohol mixture

Reagents Required

Harris haematoxylin (without acetic acid)
Orange G
Phosphotungstic acid
Eosin Y
Bismark brown
Lithium carbonate
Light green
Eosin Azure

Preparation of Reagents

Solution 1 (Haematoxylin)
Harris haematoxylin (without acetic acid) (*see* Chapter 7)

Solution 2 (Orange G 6)
0.5 per cent orange G 6 in 95 per cent alcohol
(500 mg/100 ml alcohol)   100.0 ml
Phosphotungstic acid   15 mg

Solution 3 (Eosin Azure)
0.5 per cent light green yellowish in 95 per cent alcohol 45.0 ml
0.5 per cent Bismark brown in 95 per cent alcohol   10.0 ml
0.5 per cent Eosin Y in 95 per cent alcohol   45.0 ml
Phosphotungstic acid   200 mg
Saturated lithium carbonate   1 drop

Procedure

1. After fixation, rinse smears in distilled water.
2. Transfer to solution 1 (Harris haematoxylin) for 4 minutes.
3. Wash in tap water.
4. Differentiate in acid alcohol.
5. Blue in tap water or 1.5 per cent sodium carbonate.
6. Rinse in distilled water.
7. Treat with 95 per cent alcohol.
8. Stain with solution 2 (Orange G 6) for 2 minutes.
9. Rinse in 95 per cent alcohol giving 3 changes.
10. Stain with solution 3 (EA 36) for 1 minute.
11. Rinse in 95 per cent alcohol giving 3 changes.
12. Dehydrate, clear and mount.

Result

Nucleoli	Blue
Acidophilic cells	Red to orange
Basophilic cells	Green to blue
Blood vessels	Orange-red
Cells or fragments of tissue drenched by blood	Orange to orange-green

Fluorescent Acridine Orange Technique
(Bertalanffy and Bickis, 1956; Von Bertalanffy, Masin *et al.*, 1958)

Fixation

Smears are fixed in ether/alcohol for 30 minutes. Formalin is not preferred.

Reagents Required

Phosphate buffer (pH 6.0) (*see* Chapter 5)
Acridine orange
Calcium chloride
Acetic acid

Preparation of Reagents

Solution 1
    M/15 phosphate buffer (pH 6.0) (*see* Chapter 5)

Solution 2
    Acridine orange–100 mg in 100 ml phosphate buffer pH 6.0

Solution 3 (Differentiator)
    M/10 calcium chloride, differentiator (1.109 g in 100 ml distilled water)

Procedure

1. Hydrate slides to water.
2. Transfer to 1 per cent acetic acid for 6 seconds.
3. Rinse in distilled water.
4. Transfer to solution 2 (0.1 per cent acridine orange) for 3 minutes.
5. Wash in solution 1 (M/15 phosphate buffer pH 6.0) for 1 minute.
6. Differentiate in solution 3 (M/10 calcium chloride) for 30 seconds.
7. Mount in a drop of phosphate buffer (pH 6.0) and observe under fluorescent microscope.

Result

DNA fluoresces green.
RNA fluoresces red.

Rapid Acridine Orange Fluorescent Method

Riva and Turner (1962) introduced this method. Only 10 seconds are needed. Results are comparable to Bertalanffy's method. Some staining time is extended and differentiation time is decreased.

Fixation

Unfixed smears

Reagents Required

Acridine orange
Acetic acid
Merthiolate

Preparation of Reagents

Solution 1 (Staining solution)
    Acridine orange   25 mg
    2 per cent acetic acid   100.0 ml
Merthiolate is added to prevent mould formation.

Procedure

1.   Treat with solution 1 (acridine orange) for 5 seconds (agitate).
2.   Differentiate in 2 per cent ethyl alcohol in physiological saline for 2 seconds.
3.   Rinse and then mount in physiological saline.

Result

Same as for Bertalanffy's method.

Dart and Turner Method (1959)

Fixation

Alcohol-ether

Reagents Required

Sodium phosphate
Citric acid
Acridine orange
Tween 80

Preparation of Reagents

Solution 1 (McIlvaine's buffer)
    Sodium phosphate   10.081 g
    Citric acid   13.554 g
    Distilled water   1000.0 ml
Keep in refrigerator. It lasts for a week.

Solution 2 (Buffered acridine orange) (Stock solution)
    Acridine orange   100 mg
    Distilled water   100.0 ml
Add 2 ml of Tween 80 to 1000 ml of stock.

Solution 3
    Acridine orange   1 part
    McIlvaine's buffer 9 parts

Procedure

1. Hydrate slides to water.
2. Rinse in 1 per cent acetic acid 3 times, each of 1 minute duration (1 ml/99 ml water).
3. Rinse in distilled water.
4. Transfer to solution 1 (McIlvaine's buffer) for 3 minutes.
5. Treat with solution 3 (acridine orange) for 3 minutes.
6. Differentiate in McIlvaine's buffer for 4 minutes.
7. Blot excess material.
8. Mount with buffer.

27

It is always advisable to fix the tissue within an hour or two after death of the animal because autolysis sets in soon after death. Some suitable general fixatives are:

- Mercuric chloride–Formaldehyde
- 5 per cent formaldehyde in 0.9 per cent sodium chloride
- Zenker's fluid
- Helly's fluid

Suitable stains are:

Ehrlich haematoxylin/eosin

Heidenhain's iron haematoxylin/eosin

Masson's trichrome

Van Gieson

Ponceau-S

To get accurate cytological details, tissues should be fixed immediately after death while the body is still warm. Sometimes, injection of the fixative avoids autolysis.

ALIMENTARY CANAL

Good fixatives for the mucosa of the alimentary canal are

- Susa
- 5 per cent formaldehyde in 0.9 per cent sodium chloride
- Mercuric chloride–formaldehyde
- Zenker's fluid

For the intestine, the fixative should be injected with a syringe. Stomach, oesophagus and pharynx are cut open and put in fixatives. For cytological work, Flemming's fluid or Helly's fixative are useful. Dehydration occurs as in normal procedure. But clearing is done with benzene or cedarwood oil instead of xylene. Paraffin embedding is quite good but vacuum embedding is preferred. Standard histological techniques can be applied. Oxyntic cells of gastric mucosa are well shown with Mann's methylene blue/eosin or Masson's acid fuchsin aniline blue method.

Masson's Stain Method

Solution 1 (Acid fuchsin stain)
    Acid fuchsin   1.0 g
    Glacial acetic acid   1.0 ml
    Distilled water   100.0 ml

Solution 2 (Phosphomolybdic acid)
    Phosphomolybdic acid   1.0 g
    Distilled water   100.0 ml

Solution 3 (Aniline blue solution)

Boil 100 ml water and add 2–3 g of aniline blue and when saturated, add 2.5 ml of glacial acetic acid, cool and filter.

Procedure

1. Hydrate slides to water.
2. Stain nuclei with iron haematoxylin.
3. Transfer to solution 1 (Masson's acid fuchsin) for 5 minutes.
4. Rinse in distilled water.
5. Transfer to solution 2 for 5 minutes.
6. Flood slide with 5 to 6 drops of solution 3 (aniline blue solution) for 30 seconds.
7. Rinse in distilled water.
8. Transfer to 1 per cent aqueous acetic acid for 5 minutes.
9. Dehydrate.
10. Clear in xylene or toluene.
11. Mount in Canada balsam.

Result

Nuclei   Black
Cytoplasm   Red
Collagen fibres   blue

Comments

Fat absorption could be studied in material fixed in Flemings or stained with Sudan black. Mitochondria could be demonstrated by Heidenhain's iron haematoxylin or Metzner's method (fixation Helly). For Golgi, Cajal's or Da Fano are good.

CARTILAGE

Best fixatives for cartilage are Susa and 5 per cent formaldehyde.

Embedding is carried out in paraffin or celloidin.

Suitable stains are

- Ehrlich's haematoxylin/eosin
- Harris haematoxylin/eosin
- Nicolles carbol Thionin/safranin.

EYE

Eye is a composite organ containing many tissue layers, so the sooner the eye is fixed after death, the better. The entire eye is fixed in formol saline or Zenker's fluid for 24 hours. This is wrapped in a thin gutta percha membrane and frozen. When it is completely frozen, it is bisected with a sharp knife. As usual dehydrate and embed in celloidin. Take 15–25 μ sections. For staining, standard methods are applied for connective tissue, myelin and neuroglia. Retina could be stained with Heidenhain's iron haematoxylin (*see* Chapter 7), fat-soluble stains (*see* Chapter 5) and PAS technique (*see* Chapter 9).

EAR

External ear is elastic. Internal ear or cochlea requires special treatment. Fix the material within an hour after death. Inter-vascular injection is preferred. Fix the material in 5 per cent aqueous solution of trichloroacetic acid for 3 days. This fixative also acts as a decalcifier. Fixed material can be directly transferred to absolute alcohol to which little iodine has been added. Embed in paraffin. Zenker's fluid also serves as a fixative (one or two days). Post-treatment in iodine alcohol, usual dehydration, clearing and embedding are all routine procedures.

Staining Methods

- Ehrlich's haematoxylin/eosin
- Weigert's iron haematoxylin/eosin

PANCREAS

General microanatomical fixatives like Susa or Zenker's fluid could be used. Formalin or Zenker–formalin will give good results. As far as pancreas is concerned, autolysis sets in rapidly, so it is better to fix the tissue as soon as possible.

For routine examination, haematoxylin and eosin staining is quite adequate but for alpha (α), beta (β) and γ cells, islets of Langerhans and zymogen cells lining the alveoli, special methods are required. β cell granules are important in the formation of insulin and zymogen granules for pancreatic enzymes.

Mitochondria are demonstrated by Chamoy's technique. β cells stain pale orange-brown with trichrome stain, are negative to PAS and take purple colour with aldehyde fuchsin. Cell granules take a red stain with trichrome stain. D cells stain with aniline blue and light green. Zymogen granules are acidophilic and PAS-positive.

PITUITARY GLAND

Pituitary gland is situated at the base of the brain and is separated into two main parts a) pars anterior and b) pars nervosa.

1. Pars nervosa is composed of nervous tissue and can be demonstrated by the usual method.
2. Pars anterior comprises two types of cells.
    i. Chromophobe cells which do not have any stainable cells and form at least 50 per cent of the cells.
    ii. Chromophil cells, which in turn are divided into two types based on their staining affinities.
        a) α (alpha) or acidophilic cells which constitute at least 40 per cent and β (beta) cells or basophil cells which form the remaining part of cells.
        b) α cells are PAS-negative and take a red shade when stained with any trichrome stain. β cells are PAS-positive and take a red shade.

Slidder's Orange-Fuchsin Method

Reagents Required

Celestine blue
Ferric ammonium sulphate
Glycerine
Orange G
Phosphotungstic acid

Preparations of Reagent

Solution 1 (Celestine blue solution)

Celestine blue B	500 mg
Ferric ammonium sulphate	5.0 g
Glycerine	14.0 ml
Distilled water	100.0 ml

Dissolve iron alum in water. Add the celestine blue, for 5 minutes, cool, filter and then add glycerine.

Solution 2 (Orange G solution)

95 per cent alcohol	100 ml
Orange G	500–700 mg
Phosphotungstic acid	2.0 g

Dissolve phosphotungstic acid in alcohol and saturate solution with orange G.

Procedure

1. Hydrate slides to water.
2. Stain in solution 1 (celestine blue) for 5 minutes.
3. Rinse in water.

4. Transfer to Mayer's haemalum for 5 minutes.
5. Wash, and differentiate in acid alcohol.
6. Wash in distilled water.
7. Rinse in 95 per cent alcohol and stain in solution 2 for 2 minutes.
8. Stain in 0.5 per cent acid fuchsin (in 0.5 per cent acetic acid) for 5 minutes.
9. Rinse in distilled water.
10. Treat with 1 per cent phosphotungstic acid for 5 minutes.
11. Rinse in distilled water.
12. Transfer to 1.5 per cent light green for 2 minutes.
13. Rinse in distilled water.
14. Dehydrate, clear and mount.

Result

Nuclei	Blue-black
Acidophils	Orange yellow
Basophils	Reddish purple
Chromophobe cells	Grey
Erythrocytes	Yellow
Connective tissue	Green

SUPRARENAL GLANDS

Suprarenal glands are situated on the upper part of each kidney. These glands are vulnerable to autolysis and as such they must be fixed within one or two hours after death.

The medulla contains chromaffin tissue which possesses the property of turning dark brown after prolonged treatment with chromic acid. This is due to the reduction of chromic salts by the adrenaline. Fixation of adrenals is in a fixative with non-acid dichromate-containing fluid. Suitable general fixatives for routine anatomical details are Zenker, Helly's, Susa, formalin, etc. Staining can be done with haematoxylin/eosin or Weigert's and Van Gieson stains used for connective tissue (*see* Chapter 19) may also give good results.

Chromaffin

Fresh material is fixed in Regaud's (*see* Chapter 1) post chromation. Frozen sections of such tissue show the typical brown colour of the cells of medulla. Either frozen or paraffin sections can be stained with azure-eosin stain producing a characteristic yellowish green colour. A greenish colour is obtained by the suprarenal tissue (medullary) with Vulpian reaction.

Chromaffin tissue gives a greenish blue colour with Schmorl's ferric chloride–ferric cyanide test (*see* Chapter 26) and greyish red colour with PAS.

Adrenaline

This tissue is demonstrated by osmium tetroxide method. Thin slices of suprarenal is fixed in vapour of 2 per cent osmium tetroxide for 2–3 hours at 37°C. The tissue is rapidly dehydrated and embedded in paraffin wax. In this tissue, globules of fat which are blackened by osmium tetroxide are removed from sections with turpentine. The black granules which remain after turpentine treatment are adrenaline.

TEETH

General technique is as for bone (*see* Chapter 4). Ground bone sections are similarly prepared. Cutting of thin slices before grinding with a saw may be difficult. So a metal wheel the edges of which are impregnated with diamond dust should be used. Very thin sections cut may be directly mounted in H.S.R. or DPX.

Fixation may be in any general fixative, formalin being the best.

Staining methods are the same as that for bone (*see* Chapter 4)

LUNGS

Lung substance requires special treatment because the air inside has to be removed, before it can be satisfactorily embedded. Animals are killed by a blow on the head or by anaesthesia. The best way to kill the animal is to open a large blood vessel which should be rapidly performed.

Trachea should be exposed without opening the thoracic cavity. Insert a cannula and inject the fixative with a syringe. Lungs, heart and trachea are dissected and placed in fixative.

Fixatives to be used are Susa, mercuric chloride–formaldehyde, 5 per cent formaldehyde in 0.9 per cent sodium chloride, Bouin's fluid or Zenker's fluid.

Staining is by routine methods like Heidenhain's iron haematoxylin/eosin, Van Gieson and Orcein.

For alveolar epithelium, thick sections (35–40 μ) are required. Heidenhain's iron haematoxylin and eosin are used.

SKIN

Skin has two layers, the outer epidermis and inner dermis.

- Epidermis in turn has 2 layers—the malpighian layer which has connection with dermis and the outer horny layer consisting of dead cells.
- Dermis consists of connective tissue and contains the hair follicle, sweat glands and sebaceous glands.

Skin exhibits certain difficulties in processing and cutting sections. Routine fixatives like formalin, Zenker or Bouin's fluid can be used. Pieces of skin are first stretched on a core board otherwise they will curl. Immerse the cork into a fixative.

Skin becomes hardened during processing so it is better to go in for celloidin embedding. Since skin is a dense tissue, it should be left in the embedding medium for longer periods than normal. Haematoxylin and eosin are sufficient for normal histological study of the structures. Alcian blue and Alcian blue/PAS techniques are recommended for connective tissue/polysaccharides. Eleidin can be demonstrated by picronigrosin technique.

Picronigrosin technique for Eleidin

Procedure

1. Cut frozen sections.
2. Wash in water.
3. Transfer to saturated aqueous picric acid for 5 minutes.
4. Dip in distilled water.
5. Transfer to stain in aqueous nigrin for 1 minute.
6. Wash in distilled water.
7. Rinse in 96 per cent alcohol.
8. Clear in origanum.
9. Mount in Canada balsam.

Result

Eleidin	Black
Other elements	Yellow

BONE MARROW AND BLOOD-FORMING ORGANS

Bone marrow can be studied in the form of

1. Films, which are made and stained by the usual methods for blood films.
2. Impression smears, by pressing the unfixed material on a slide and fixing in Schaudin's fluid and staining in the same way as for blood films.
3. Needle biopsis—fixed in Susa or by Cappell, Hutchinson and Harvey-Smith method.
4. After fixation in Helly's fluid and washing to remove dichromate, bone marrow may have small spicules of bone which may be decalcified in Gooding, and Stewart's fluid.

Gooding and Stewart's Fluid

Formic acid	5–25.0 ml
Formalin	5.0 ml
Distilled water	100.0 ml

Cappell, Hutchison and Harvey-Smith Method

Procedure

1. Fix pieces of bone marrow in fresh prepared solution

   Zenker's fluid   45.0 ml
   Formalin   5.0 ml
   Formol saline   50.0 ml

2. Fix for 20 minutes.
3. Add 400 ml of distilled water to the container and allow the pieces to settle.
4. Remove the supernatant fluid and wash the pieces in distilled water giving 2 changes.
5. Remove water and add 100 ml of 70 per cent alcohol and leave the pieces for 24 hours.
6. Then dehydrate and clear in toluene.
7. Keep the pieces in a flat-bottomed tube and pour molten wax. Leave it for 2 hours with 2 changes. Cool the wax, free the block by breaking the dish and cut sections.

Leishman's Stain

Preparation of Reagent

Leishman's stain (Powder)   15 mg
Methyl alcohol   100.0 ml

Grind the powder in a mortar and pour methyl alcohol, then pour the alcohol in a bottle, now add more alcohol into the mortar and repeat the process.

Procedure

1. Hydrate slides to water.
2. Incubate phosphate buffer in pH 6.8 at 56°C for 30 minutes.
3. Transfer to Leishman's stain which is diluted $1:3$ with pH 6.8 buffer for 30 to 60 minutes.
4. Dip in pH buffer.
5. Blot the section.
6. Flood with xylol
7. Mount in DPX.

Results

Nuclei	Bluish red
Acidophil granules	Pink to red
Basophil granules	Blue
Red blood cells	Salmon pink

Maximow's Stain

Reagents Required

Eosin
Azure

Preparation of Reagents

Stock solution A   1: 1000 eosin
Stock solution B   1: 1000 azure II

Working Solution

Add 10 ml solution A to 100 ml of 6.8 pH phosphate buffer and then add 10 ml of solution B.

Procedure

1. Dehydrate slides to water.
2. Stain in Ehrlich's haematoxylin for 10 minutes
3. Incubate in pH 6.8 buffer.
4. Transfer to working solution.
5. Differentiate, clear and mount as in previous case.

Results

Same as for Leishman's stain.

May–Grunwald–Giemsa Technique

Procedure

1. Hydrate slides to water.
2. Transfer to pH 6.8 buffer for 30 minutes at 37°C.
3. Transfer to May–Grunwald–Giemsa stain for 15 minutes.
4. Rinse in buffer solution.
5. Transfer to Giemsa stain with dilution of 1:10 with pH 6.8 for 15 minutes.
6. Differentiate in buffer.
7. Differentiate in a mixture of glycerin–ether.
8. Again differentiate in buffer.
9. Dehydrate, clear in xylol and mount in DPX.

Results

Same as for Leishman's stain.

BONE

Bone should be decalcified and then embedded in celloidin. Paraffin embedding can also be done.

Schmorl's Picro-thionin Method

Fixation

Muller's fixative (see Chapter 1)
or formol saline

Procedure

1. Wash sections in water.
2. Transfer to saturated aqueous solution of thionin which contains 0.880 ammonia for 100 ml of stain.
3. Rinse in water.
4. Transfer to saturated aqueous phosphotungstic or phosphomolybdic acid for 30 seconds.
5. Wash.
6. Differentiate in 70 per cent alcohol.
7. Dehydrate, clear and mount.

Result

Ground substance	Yellow to brown
Lacunae and canaliculi	Dark brown
Cells	Red
Ground substance of cartilage	Purple

Schmorl's Thionin-Phosphotungstic Acid Method

Fixation

Fix with Muller's fluid or formol saline. Avoid mercury-containing fixatives.

Procedure

1. Bring sections to water.
2. Transfer to alkaline thionine for 30 minutes.
3. Rinse in water.
4. Place in saturated aqueous PTA or PMA for few seconds.
5. Wash in water.
6. Treat with 10 per cent ammonia in distilled water.
7. Differentiate in 90 per cent alcohol.
8. Dehydrate, clear and mount.

Results

Ground substance	Pale green
Lacunae	Blue or blue-black

OVARY

General fixatives like Susa, Zenker's and Bouin's fluid. For cytological observations Helly's or Flemming's without acetic acid is preferable. Staining is by Heidenhain's iron haematoxylin and Da Fano technique for Golgi elements. As far as ovary is concerned, paraffin embedding is preferred over celloidin embedding. Dehydration is as usual but clearing is in cedar wood oil or benzene. Good results are obtained with Masson's trichrome or Heidenhain's Azan technique.

Phloxine–Methylene Blue Method

Fixation

Any general fixative

Reagents Required

Phloxine
Glacial acetic acid
Methyl blue
Azure B
Borax

Preparation of Reagents

Solution 1 (Phloxine solution)

Phloxine	500 mg
Distilled water	100.0 ml
Glacial acetic acid	0.2 ml

Should be filtered before use.

Solution 2 (Methylene blue azure)

Methylene blue	250 mg
Azure B	250 mg
Borax	250 mg
Distilled water	100 ml

Procedure

1. Hydrate slides to water.
2. Place in solution 1 for 2 minutes.
3. Rinse in distilled water for 1 minute.
4. Transfer to solution 2 for 1 minute.
5. If required, destain in 0.2 per cent acetic acid.
6. Dehydrate, clear and mount.

Result

Nuclei	Blue
Plasma cell and cytoplasm	Blue
Other elements	Rose or red

SPLEEN AND LYMPH GLANDS

Smears may be made from spleen and lymph gland and stained with Leishman's or Giemsa's stain for sectional purposes. Zenker's or Helly's or formol saline may be used. For paraffin embedded material standard histological stains such as Heidenhain's Azan or Mallory can be used. Trabiculae and capsules are well demonstrated by Van Gieson or Mallory stain.

EMBRYOS

Fixation is in Susa, 4 per cent formaldehyde, and Zenker's, etc. Embryos are likely to shrink. To avoid this, dehydration should be carried out very rapidly. Dioxane gives good results. Xylene is to be avoided as clearing agent. Another important point to be noticed is to not transfer the material directly from xylene to paraffin. Embryos should be kept in a mixture of clearing agent and paraffin.

Hill's Method for Embryos

1. Place the dehydrated material in cedar wood oil overnight and then transfer to benzene for few hours.
2. Transfer to benzene–paraffin mixture overnight.
3. Embed in paraffin.

HAIR

For whole mounts, place hairs in a small tube filled with equal volumes of ether and absolute alcohol. Shake well to remove sebaceous matter and debris, clear in benzene, partially dry on filter paper and mount in balsam.

For sections treat as above with absolute alcohol and ether, place in 6 volumes of 2 per cent alcohol and 4 volumes of 5 per cent ammonia.

Soak in 10 per cent potassium hydroxide at 50°C for one to 2 minutes. Wash for few minutes in 5 per cent sulphuric acid and absolute alcohol.

Clear in benzene for 10–15 minutes

Dry with filter paper.

Impregnate with paraffin wax for 15 minutes and block.

Cut section 8–10μ in thickness.

LIVER

Fix in Susa, formol saline, Zenker's fluid. For staining, Heidenhain's iron haematoxylin can be used and Atlamann-Metzner method after Helly's fixation. For bile canaliculi, fix in formol saline and stain with Heidenhain's iron haematoxylin.

Liver areas attached to alimentary tract should be avoided. Immediately after death. In these regions, hydrogen sulphide is released from the intestine and haemoglobin breaks down to form iron sulphide.

KIDNEY

For cytological work on renal epithelium, kidney should be absolutely fresh.

Fixation

In Susa, formol saline, Zenker's fluid.

Staining

Masson's method is excellent after fixation in Susa. Altamann Metzner technique is good after Helly's fixation for mitochondria and for Golgi elements fixed by Da Fano method.

THYROID

Thyroid has the secretory epithelial cells and extracellular amorphous colloid. This colloid is a stored secretion and is made up of mainly proteins and carbohydrates. Some proteins are of basic type. Sakaguchi's reaction for arginine, ferric ferricyanide reaction for sulphydryl groups. It is PAS-positive, and may be containing mucopolysaccharides.

Fixation in AFA:

Acetic acid	1 part
Formalin	5 parts
Absolute alcohol	15 parts
Fix for 24 hours.	

INVERTEBRATE STAINING METHODS

28

Trematode Whole Mounts (Gower, 1939)

Fixation

FAA 70 per cent

Reagents Required

Carmine
Acetic acid
Potassium chlorate
Conc. hydrochloric acid

Preparation of Reagents

Solution 1 (Acidified carmine)

Carmine	10.0 g
45 per cent acetic acid	100.0 ml

Boil, cool and filter. This residue is acidified carmine.

Solution 2 (Working solution)

Solution 1	1.0 g
Alum	10.0 g
Distilled water	200.0 ml
Potassium chlorate crystals	100.0 g
Conc. HCl	0.1 ml
70 per cent alcohol	100.0 ml

Add HCl to potassium chlorate in a closed dish.

Procedure

1. Hydrate specimens to water.
2. Place the specimen in a dish containing solution 2 for 36 hours. Time schedule depends on the size of the specimen.
3. Wash in water.
4. Destain in acid alcohol.
5. Dehydrate and clear in cedar wood oil.
6. Mount.

Result

All principal organs take a deep rose shade. Parenchyma, cytoplasm and muscle remain unstained.

Demonstration of Nervous System in Whole Mounts of Trematodes and Cestodes (Kishore *et al.*, 1988a, 1988b, 1988c, 1988d, 1990)

Fixation

10 per cent neutral formalin. Specimens are flattened between 2 slides and fixed for 6 hours.

Reagents Required

5-Bromoindoxyl acetate
Tris buffer
Potassium ferrocyanide
Potassium ferricyanide
Calcium chloride

Preparation of Reagents

Solution 1 (Incubating medium)

1–2 per cent of 5-bromoindoxyl acetate	0.1 ml
0.1 M tris buffer (pH 6.8)	2.0 ml
0.05 M potassium ferrocyanide	1.0 ml
0.05 M potassium ferricyanide	1.0 ml
0.1 M calcium chloride	1.0 ml
Distilled water	5.0 ml

Mix bromoindoxyl in 0.1 ml alcohol and then add the following in the order given.

Procedure

1. After fixing for 6 hours, bring the specimen to water.
2. Place in a dish containing solution 1 (incubating medium) for 12 hours at room temperature.
3. Wash in distilled water.
4. Dehydrate through 70 per cent, 90 per cent, 95 per cent and absolute alcohol.
5. Clear in creosote.
6. Mount in DPX.

Result

Nervous system is stained deep indigo blue.

Egg Shell in Trematodes and Cestodes (Smyth, 1951)

Fixation

0.5 per cent formol saline

Reagents Required

Acidified carmine (*see* Chapter 26)
Alum
Malachite green
Orange G

Preparation of Reagents

Solution 1

Acidified carmine	1.0 g
Alum	10.0 g
Distilled water	200.0 ml

Dissolve by heating. Cool and filter.

Solution 2

Malachite green	500 mg
Distilled water	100.0 ml

Solution 3

Orange G	1.0 g
Absolute alcohol	99.0 ml

Procedure

1. Hydrate slides to water.
2. Place them in solution 1 for 2 hours.
3. Rinse in water.
4. Transfer to solution 2 for 2 minutes.
5. Wash and dehydrate.
6. Counterstain in solution 3 for 1 seconds.
7. Rinse in absolute alcohol.
8. Clear in xylene and mount.

Result

Egg shell	Green or greenish blue
Nuclei	Red
Cytoplasm	Pink

Staining Spines of Trematodes (Echinostomes) by Azure-1 Schiff reaction (Kasten, 1957) Modified by Hanumantha Rao and Murthy (1972)

Fixation

Unfixed

Reagents Required

Azure-1
Schiff's reagent (*see* Chapter 9)

Preparation of Reagents

Solution 1 (Azure/schiff)

Dissolve 1.0 g of Azure-1 in boiling distilled water. Remove from the flame and allow it to cool and then filter. Add 2.0 g of potassium metabisulphite and 20 ml of normal hydrochloric acid. Keep solution in dark for 24 hours.

Procedure

1. Place the cercaria on a clean slide.
2. Add few drops of Azure-1 Schiff reagent (final pH 4–4.3) for 10 minutes.
3. Rinse in water.
4. Rinse in absolute alcohol.
5. Clear in xylene and mount.

Result

Echinostome spines are stained blue.

Gomori's Aldehyde Fuchsin for Neurosecretory Cells in the Trematodes (Shyamasundari and Hanumantha Rao, 1975a; Shyamasundari, 1985)

Fixation

10 per cent formalin or Susa.

Reagents Required

Basic fuchsin
Conc. hydrochloric acid
Peraldehyde
Potassium permanganate
Oxalic acid

Preparation of Reagents

Solution 1 (Aldehyde fuchsin)

Basic fuchsin	500 mg
Boiling water	100.0 ml
Conc. HCl	1.0 ml
Paraldehyde	1.0 ml

Add basic fuchsin to 100.0 ml of boiling water, boil, cool and filter. Add 1.0 ml of conc. HCl and 1.0 ml of paraldehyde. Leave the bottle stoppered in a dark chamber till red colour of the fuchsin disappears and a violet colour is retained.

Solution 2

Potassium permanganate	300 mg
Distilled water	100.0 ml
Conc. H_2SO_4	0.3 ml

Solution 3

Oxalic acid	1.0 g
Distilled water	100.0 ml

Procedure

1. Deparaffinize and hydrate slides to water.
2. Oxidize in solution 2 for 1–2 minutes.
3. Wash.
4. Bleach in solution 3 till tobacco colour of permanganate is lost.
5. Wash in running water.
6. Stain in solution 1 for 10–30 minutes.
7. Differentiate in 95 per cent alcohol.
8. Dehydrate, clear and mount.

Result

Neurosecretory cells with neurosecretory substance	Purple
Mucosubstance	Purple

Demonstration of Phospholipids in Mehlis' Gland of Trematodes (Fasciola hepatica) by Applying Acid Haematein Method (Hanumantha Rao, 1959) (Plate 14, Figure 1)

Fixation

Fix the tissue in formol calcium for 24 hours followed by post-chromation with 2 per cent potassium dichromate for 24 hours at room temperature and 24 hours at 60°C.

Reagents Required

Haematein
Sodium iodate
Potassium ferricyanide
Sodium tetraformate

Preparation of Reagents

Solution 1 (Acid haematein solution)

Haematein	550 mg
1 per cent sodium iodate	1.0 ml
Distilled water	49.0 ml

Solution 2 (Differentiator)

Potassium ferricyanide	250 mg
Sodium tetraformate	250 mg
Distilled water	100.0 ml

Procedure

1. Deparaffinize and hydrate slides to water.
2. Stain in solution 1 (acid haematein) at 60°C for 5 hours.
3. Rinse in distilled water.
4. Transfer to solution 2 for differentiation for 10 hours at 37°C.

5.   Wash.

6.   Mount in glycerine jelly.

Result

Phospholipids are stained dark blue.
Also refer Hanumantha Rao (1960, 1963).

Application of Copper Phthalocyanin to *Fasciola* Mehlis' Gland for the Demonstration of Phospholipids (Hanumantha Rao, 1959) (Plate 15, Figure 2)

Fixation

The tissue is fixed in formol calcium and postchromated as in previous procedure.

Reagents Required

Luxol fast blue G
Lithium carbonate
Neutral red

Preparation

Solution 1 (Copper phthalocyanin)
   Luxol fast blue G   10 mg
   95 per cent alcohol   100.0 ml

Solution 2
   Lithium carbonate   50 mg
   Distilled water   100.0 ml

Solution 3
   Neutral red   1.0 g
   Distilled water   100.0 ml

Procedure

1.   Fix material (control) in weak Bouin's. After usual procedure, cut 8-µm sections and keep them in pyridine for 12 hours. Then wash.

2.   Bring both sections, i.e., formol-calcium-fixed and control to absolute alcohol.

3.   Place both sections in solution 1 for 6–18 hours at 60°C.

4.   Rinse in 70 per cent alcohol and wash in water.

5.   Differentiate in solution 2 for 30 minutes.

6.   Rinse in water.

7.   Counterstain in solution 3 for 10 minutes.

8.   Rinse in water.

9.   Dehydrate, clear and mount in Canada Balsam.

Result

Phospholipids stain blue.
Control sections show no phospholipid activity and remain unstained.

Catechol Technique for Vitellaria in Trematodes and Cestodes (Johri and Smyth, 1956)

Fixation

Tissue is fixed in 70 per cent alcohol or 10 per cent formalin. 70 per cent alcohol is preferable.

Reagents Required

Catechol

Preparation of Reagents

Solution 1

| Catechol | 100 mg |
| Distilled water | 100.0 ml |

This must be freshly prepared before use.

Procedure

1. Deparaffinize and bring slides to water.
2. Transfer to solution 1 for 60–90 minutes at 37°C or for 4–5 hours at room temperature.
3. Wash in distilled water for 15 minutes.
4. Dehydrate, clear and mount.

Result

Vitellaria is stained reddish brown.

Diazo Technique using Fast Red Salt B for Vitellaria in Trematodes and Cestodes (Johri and Smyth, 1956)

Fixation

Tissue is fixed in 70 per cent alcohol for 4 days.

Reagents Required

Fast Red salt B

Preparation of Reagent

Solution 1

| Fast Red Salt B | 100 mg |
| Distilled water | 100.0 ml |

This solution should be freshly prepared and filtered before use.

Procedure

1. Deparaffinize and hydrate slides to water.
2. Place slides in solution 1 for 10–40 minutes.
3. Wash in distilled water for 10–15 minutes.
4. Dehydrate, clear and mount.

Result

Vitellaria and vitelline ducts are stained orange to red.

Malachite Green Method for Egg Shell in Trematodes and Cestodes (Smyth, 1951)

Fixation

Tissue is fixed in 0.5 to 10 per cent formalin.

Reagents Required

Carmine
Alum
Malachite green
Orange G

Preparation of Reagents

Solution 1

Acidified carmine	1.0 g (*see* Chapter 26)
Alum	10.0 g
Distilled water	200.0 ml

Solution 2

Malachite green	500 mg
Distilled water	100.0 ml

Solution 3

Orange G	1.0 g
Absolute alcohol	99.0 ml

Procedure

1. Deparaffinize and hydrate slides to water.
2. Stain in solution 1 for 2 hours.
3. Rinse in water.
4. Stain in solution 2 for 2 minutes.
5. Dehydrate in 70 per cent and 90 per cent alcohol.
6. Differentiate in absolute alcohol.
7. Counterstain in solution 3 for 1 second.
8. Rinse in absolute alcohol, clear and mount.

Result

Egg shell material is stained green or greenish blue.

Demonstration of Disulphides in the Neurosecretory Cells of the Trematodes by Applying Performic Acid/Alcian Blue Technique (Shyamasundari and Hanumantha Rao, 1975a)

Fixation

Tissue is fixed in formalin, Carnoy's fluid, etc.

Reagents Required

Formic acid
Hydrogen peroxide
Sulphuric acid
Alcian blue

Preparation of Reagents

Solution 1 (Oxidizing solution)

98 per cent performic acid	40.0 ml
100 per cent hydrogen peroxide	4.0 ml
Sulphuric acid	0.5 ml

Solution 2 (Staining solution)

Alcian blue	1.0 g
98 per cent sulphuric acid	2.7 ml
Distilled water	47.2 ml

Procedure

1. Deparaffinize and hydrate slides to water.
2. Treat with solution 1 for 5 minutes.
3. Wash in tap water.
4. Dry sections.
5. Rinse in tap water.
6. Transfer to solution 2 (alcian blue) for 1 hour.
7. Wash in running water.
8. Counterstain if desired.
9. Wash.
10. Dehydrate, clear and mount.

Result

Sites with disulphide are stained dark blue.

Demonstration of Calcium by Alizarin Red S Method in Crustacean Intermoult Cuticle (Rama Devi *et al.*, 1991)

Fixation

Any general fixative is used, preferably neutral fixatives. Avoid acid fixatives.

Reagents Required

Sodium alizarin sulphonate
Hydrochloric acid

Preparation of Reagents

Solution 1 (Alizarin Red S)

Sodium alizarin sulphonate	500 mg
Distilled water	45.0 ml

Mix the solution thoroughly and add 5 ml of ammonia (1 part of 28 per cent ammonia and 99 parts of distilled water). Stir well while adding and the final pH should be 6.3–6.5. Adjust it with buffers.

Solution 2 (Differentiator)
Hydrochloric acid   0.1 ml
95 per cent alcohol   99.9 ml

Procedure

1.   Dewax and hydrate slides to water.
2.   Treat with solution 1 for 2 minutes.
3.   Wash in distilled water for 10 seconds.
4.   Rinse in solution 2 for 10 seconds.
5.   Rinse in 95 per cent and 100 per cent alcohols.
6.   Clear in xylene and mount in cedar wood oil.

Result

Calcium sites are stained orange-red.

Demonstration of Lipase with Gomori's Method in Crustacean Cuticle (Erri Babu *et al.*, 1985)

Fixation

Fresh frozen

Reagents Required

Tween 60 or 80
Tris buffer
Calcium chloride
Lead nitrate
Ammonium sulphide
Haematoxylin

Preparation of Reagents

Solution 1
2 per cent tween 60 or 80   5.0 ml
0.2 M tris buffer (pH 7.2)   20.0 ml
4 per cent calcium chloride   5.0 ml
Distilled water   20.0 ml

Solution 2
Lead nitrate   2.0 g
Distilled water   100.0 ml

Solution 3
Ammonium sulphide   1.0 g
Distilled water   100.0 ml

Procedure

1. Unfixed sections or chilled acetone-fixed sections (fixation time 24 hours).
2. Dehydrate in 2 changes of acetone (room temperature).
3. Clear in benzene giving 2 changes of 45 minutes each (maximum 2 hours).
4. Embed in paraffin.
5. Cut 5-μm sections and float on warm water at 35°C.
6. Bring them to wash.
7. Incubate in solution 1 at 37°C for 6–24 hours.
8. Rinse in distilled water.
9. Transfer to solution 2 for 10–15 minutes.
10. Treat with solution 3 for 1 minute.
11. Stain nuclei with haematoxylin.
12. Mount in glycerine jelly, or dehydrate in dichloroethylene and mount in Gurr's medium.

Results

Sites of lipase activity   Brown
Nuclei   Blue

Demonstration of Keratin with Rhodamine B in the Spermatophore Wall Capsule of Crustaceans (Hanumantha Rao *et al.*, 1989)

Fixation

Any general fixative may be used.

Reagents Required

Toluidine blue
Rhodamine B
McIlvaine's buffer
Na_2HPO_4 and citric acid

Preparation of Reagents

Solution 1
    Toluidine blue   100 mg
    Distilled water   100.0 ml

Solution 2 McIlvaine buffer (pH 3.6)
    0.2 M Na_2HPO_4 64.4 m
    0.1 M Citric acid   135.6 m

Solution 3
    Solution 2   100.0 ml
    Rhodamine B   100 mg

Procedure

1. Deparaffinize and hydrate slides to water.
2. Immerse slides in solution 1 for 10 minutes.
3. Wash in distilled water.
4. Transfer to solution 3 for 10 minutes.
5. Wash rapidly.
6. Dehydrate very rapidly, clear and mount.

Result

Sites of keratin activity are stained rose-red.

Demonstration of Elastin in the Stomodaeum of the Amphipod Crustacean with Weigert's Resorcin Fuchsin (Shyamasundari and Hanumantha Rao, 1973)

Fixative

Any general fixative may be used.

Reagents Required

Basic fuchsin
Resorcin
Ferric chloride
Hydrochloric acid

Preparation of Reagents

Solution 1

Basic fuchsin	2.0 g
Resorcin	4.0 g
Distilled water	200.0 ml
29 per cent ferric chloride (29 g/100 ml water)	25.0 ml
95 per cent alcohol	200.0 ml
Hydrochloric acid	4.0 ml

Boil water and add fuchsin and resorcin. While boiling, add 25.0 ml ferric chloride. Boil for 5 minutes, cool and filter. Discard the filtrate and dry the precipitate in the filter paper. After thorough drying bring the powder to another dish and add 200.0 ml of 95 per cent alcohol. Heat, take out the filter paper. When the precipitate is dissolved, cool, filter and add 95 per cent alcohol to make it to 200.0 ml. Add 4.0 ml of HCl.

Procedure

1. Dewax and bring slides to water.
2. Stain in solution 1 for 20–60 minutes.
3. Differentiate in 95 per cent alcohol.
4. Wash in tap water.
5. If desired, counterstain in haematoxylin.
6. Dehydrate and mount.

Result

Elastin lining is stained dark blue or black.

Also refer *Trinadh Babu et al.* (1989a).

Simultaneous Demonstration of Neurosecretory and Mucous Substances in the Amphipod (crustaceans) Tissue Section (Shyamasundari and Hanumantha Rao, 1975b)

Fixation

Methanol formaldehyde acetic acid, Bouin's or Susa

Regents Required

Basic fuchsin
Paraldehyde
Hydrochloric acid
Potassium permanganate
Oxalic acid
Mercurochrome

Preparation of Reagents

Solution 1
Aldehyde function (*see* Chapter 9)

Solution 2
1 g of potassium permanganate in 100 ml distilled water

Solution 3
1 g of oxalic acid in 100 ml of distilled water

Solution 4
Mercurochrome   500 mg
Distilled water   100.0 ml

Procedure

1. Dewax 2 sets of slides and hydrate to water.
2. Oxidize in solution 2 for 2 minutes.
3. Wash.
4. Bleach in solution 3 for 3 minutes.
5. Wash in running water.
6. Stain in solution 1 for 10–30 minutes.
7. Differentiate in 95 per cent alcohol.
8. Dehydrate one set, clear and mount.
9. Counterstain another set with solution 4 for 10 minutes. Wash, dehydrate and mount.

Result

Those stained in solution 1:
        Neurosecretory substance   Purple
        Mucosubstances   Purple
Those counter stained with solution 4:
        Neurosecretory substance   Brick red
        Mucosubstances   Purple

Demonstration of Disulphides in the Neurosecretory Cells of Crustaceans with Performic Acid/Alcian Blue Technique (Shyamasundari, 1977)

Fixation

Any general fixative, formol

Reagents Required

Performic acid
Sulphuric acid
Alcian blue

Preparation of Reagents

Solution 1
        Alcian blue   3.0 g
        2N sulphuric acid   100.0 ml (pH 0.2–0.3)

Solution 2
        98 per cent formic acid   40.0 ml
        30 per cent H_2O 4.0 ml
        Conc. H_2SO_4 0.5 ml
Allow the mixture to stand for 1 hour.

Procedure

1.   Dewax and hydrate slides to water.
2.   Blot.
3.   Immerse sections in solution 2 for 5 minutes.
4.   Wash gently in tap water.
5.   Rinse in 70 per cent alcohol and blot.
6.   Transfer to solution 1 at room temperature for 1 hour.
7.   Wash in distilled water.
8.   Counterstain if desired.
9.   Dehydrate, clear and mount.

Result

Sites containing cystine are stained deep blue.
Also refer *Erri Babu et al.* (1979, 1980a, 1980b; *Jalaja Kumari et al.*, 1980; *Trinadha Babu et al.*, 1989a; *Lalitha et al.*, 1993).

Application of Alcian Blue/Safranin for the Demonstration of Strongly Acidic Mucosubstance in the Crustacean Tegumental Glands (Shyamasundari, 1979)

Fixation

Cetylpyridinium chloride in 10 per cent formalin.

Reagents Required

Alcian blue
Safranin

Preparation of Reagents

Solution 1
    Alcian blue   500 mg
    3 per cent acetic acid   100.0 ml
Solution 2
    Safranin   250 mg
    0.125N HCl   100.0 ml

Procedure

1.   Deparaffinize and hydrate slides to water.
2.   Stain in solution 1 for 30 minutes.
3.   Wash in distilled water.
4.   Transfer to solution 2 for 30 seconds.
5.   Dehydrate rapidly, clear and mount.

Result

Most strongly acidic substances are stained red.

Hale's Colloidal Iron Method for Acid Mucosubstances in Crustacean Tissue (Shyamasundari, 1979)

Fixation

Cetylpyridinium chloride in 10 per cent formalin or 3 per cent calcium acetate in 10 per cent formalin.

Reagents Required

Ferric chloride
Hydrochloric acid
Potassium ferrocyanide
Acetic acid

Preparation of Reagents

Solution 1 (Stock colloidal iron solution)
    29 per cent ferric chloride (29 g/100 ml water)   2.2 ml
    Distilled water   125.0 ml

Boil distilled water. When it is boiling add ferric chloride and stir. The solution is dark red. At that point, remove it from heat and cool.

Solution 2
    Glacial acetic acid   5.0 ml
    Solution 1   20.0 ml
    Distilled water   15.0 ml

Solution 3 (Acid ferrocyanide mixture)
    Potassium ferrocyanide   2.0 g
    Conc. hydrochloric acid   2.0 ml
    Distilled water   98.0 ml
First add potassium ferrocyanide to water. After it dissolves, add hydrochloric acid.

Procedure

1. Deparaffinize and hydrate slides to water.
2. Rapidly rinse in 12 per cent acetic acid.
3. Transfer to solution 2 for 1 hour.
4. Again rinse in 12 per cent acetic acid.
5. Transfer to solution 3.
6. Wash in distilled water.
7. Dehydrate, clear and mount.

Result

Acid mucosubstances   Bright blue

[Also refer for crustacean tissue: *Shyamasundari* and *Hanumantha Rao* (1975a, 1975b): *Shyamasundari* (1977); *Trinadha Babu et al.* (1989a, 1989b, 1989c, 1989d, 1991, 1993); *Lalitha et al.* (1993a, 1993b, 1996); *Kameswaramma et al.* (1987, 1990); *Rama Devi et al.* (1987, 1991).]

Demonstration of Sulphated and Carboxylated Mucosubstances in the Oesophageal Glands of Crustaceans Alcian Blue/Alcian Yellow Technique (Shyamasundari, 1979)

Fixation

Susa or cetylpyridinium chloride

Reagents Required

Alcian blue 8GX
Alcian yellow
Hydrochloric acid
Neutral red

Preparation of Reagents

Solution 1
    Alcian blue 8GX   1.0 g
    N/5 Hydrochloric acid   100.0 ml

Solution 2

    Alcian yellow   1.0 g

    3 per cent acetic acid   100.0 ml

Solution 3

    0.5 per cent neutral red

Procedure

1. Deparaffinize and take down slides to water.
2. Rinse in N/5 hydrochloric acid.
3. Treat with solution 1 for 5–15 minutes.
4. Rinse in N/5 hydrochloric acid.
5. Transfer to solution 2 for 5–10 minutes.
6. Wash in water.
7. Counterstain with solution 3 for a few seconds.
8. Wash.
9. Dehydrate, clear and mount.

Result

Sites of sulphated mucins	Blue
Sites of carboxylated mucins	Yellow
Mixtures	Green

Demonstration of Elastin Type Protein in the Seminal Receptacle Wall by the Application of Spirit Blue Technique (Raghu, 1993)

Fixation

Helly's fluid

Reagents Required

Potassium permanganate
Sodium metabisulphite
Concentrated hydrochloric acid
Spirit blue
Aniline oil
Phosphotungstic acid
Picro fuchsin
Acid fuchsin

Preparation of Reagents

Solution 1 (Potassium permanganate)

    0.5 per cent potassium permanganate (500 mg in 100 ml water)   100.0 ml

    Concentrated sulphuric acid   3 drops

Solution 2 (Sodium metabisulphite)

    1 per cent sodium metabisulphite   100.0 ml

    Concentrated hydrochloric acid   1 drop

Solution 3 (Spirit blue)
    Spirit blue (aniline blue alcoholic-soluble)   750 mg
    70 per cent ethyl alchol   100.0 ml
    Aniline oil   2 drops

Solution 4 (Phosphotungstic acid)
    Phosphotungstic acid   5.0 g
    Distilled water   100.0 ml

Solution 5 (Picric acid–acid fuchsin)
    Saturated alcoholic picric acid   100.0 ml
    1 per cent acid fuchsin (1 g/100 ml water)   5.0 ml

Procedure

1. Deparaffinize and hydrate slides to water.
2. Oxidize in solution 1 for 1 minute.
3. Wash.
4. Bleach in solution 2 for 30 seconds.
5. Transfer to solution 3 for 2 minutes.
6. Wash in distilled water.
7. Dip in 70 per cent alcohol.
8. Treat with solution 4 for 5 minutes.
9. Wash in tap water.
10. Place in solution 5 for 5 seconds
11. Wash, dehydrate, clear and mount.

Result

Elastin   Bright blue
Collagen   Red

Confirmation of Sialomucins by the Application of Neuroaminidase Digestion Technique on the Tegumental Glands of Lobsters (Crustaceans) (Shyamasundari and Hanumantha Rao, 1978)

Fixation

Any general fixative may be used.

Reagents Required

Neuraminidase
Acetate buffer
Calcium chloride
Alcian blue

Preparation of Reagents

Solution 1 (Sialidase)
    Neuraminidase   1.0 ml

Acetate buffer (pH 5.5)   4.0 ml
Calcium chloride   50 mg

Solution 2 [Alcian blue solution (pH 2.5)] (*see* Chapter 9)

Procedure

1. Take 2 test and 2 control sections, dewax and bring them down to water.
2. Place 1 test and 1 control section in solution 1 (neuraminidase solution) for 18 hours at 37°C.
3. Incubate the other two sections in buffer solution at 37°C for 18 hours.
4. Wash in tap water.
5. Treat all sections with solution 2 (alcian blue) for 5 minutes.
6. Wash in water.
7. Counterstain if desired in Mayer's carmalum.
8. Wash in water.
9. Dehydrate, clear and mount.

Result

Sites of sialomucins unstained.
Other acid mucins   Blue
Nuclei   Red

Molluscan Tissues: Oyster Tissue to Demonstrate Collagen Fibres (Pauley, 1967)

Fixation

Zenker's solution

Reagents Required

Ammonium hydroxide
Acid fuchsin
Aniline blue
Orange G
Phosphotungstic acid

Preparation of Reagents

Solution 1
    Ammonium hydroxide   3 drops
    Water   1000 ml
Solution 2
    Acid fuchsin   500 mg
    Distilled water   100 ml
Solution 3
    Aniline blue   500 mg
    Orange G   2.0 g
    Phosphotungstic acid   1.0 g
    Distilled water   100.0 ml

Procedure

1. Deparaffinize slides and hydrate to water.
2. After removing mercury, wash in water.
3. Clear in 5 per cent sodium thiosulphate.
4. Wash in running water.
5. Transfer to Harris haematoxylin (*see* Chapter 7) for 20 minutes.
6. Rinse in distilled water.
7. Rinse in 1 per cent acid alcohol.
8. Transfer to solution 1 until sections turn deep blue.
9. Rinse in running water.
10. Transfer to solution 2 for 5 minutes.
11. Rinse in distilled water.
12. Transfer to solution 3 for 15 minutes.
13. Rinse in tap water.
14. Pass through 95 per cent alcohol and 100 per cent alcohol.
15. Clear in xylene and mount.

Results

Collagen	Blue
Cartilage	Blue
Epithelium	Orange-red
Nuclei	Reddish blue

Alcian Blue (pH 2.5) to Demonstrate Acid Mucopolysaccharides in Pelecypod Foot (Anisa Banu *et al.*, 1979a)

Fixation

Fix in 1 per cent cetylpyridinium chloride in 10 per cent formalin.

Reagents Required

Alcian blue 8GX
3 per cent acetic acid

Preparation of Reagents

Solution 1

Alcian blue 8 GX	1.0 g (pH 2.5)
3 per cent acetic acid	100.0 ml

Procedure

1. Deparaffinize and hydrate slides to water.
2. Stain in solution 1 (alcian blue) for 30 minutes (requires more time with ageing).
3. Wash in running water.
4. Dehydrate, clear and mount.

Result

Acidic mucopolysaccharides are stained deep blue.

Alkaline Phosphatase Activity in the Gland Cells of the Foot of Pelecypod: Calcium Cobalt Method (Anisa Banu et al., 1979b)
(Plate 17, Figure 1)

Fixation

Cold acetone, ether, paraffin or cold formalin frozen

Reagents Required

Sodium β-glycerophosphate
Diethyl sodium diethylbarbiturate
Calcium chloride
Magnesium sulphate
Cobalt nitrate
Yellow ammonium sulphide

Preparation of Reagents

Solution 1

Cobalt nitrate	2.0 g
Distilled water	100 ml

Solution 2 (Incubating medium)

3 per cent sodium β-glycerophosphate (3 g/100 ml water)	10.0 ml
2 per cent sodium diethyl barbiturate (2 g/100 ml water)	10.0 ml
Distilled water	5.0 ml
2 per cent calcium chloride (2 g/100 ml water)	20.0 ml
5 per cent magnesium sulphate (5 g/100 ml water)	1.0 ml

Procedure

1. Fix the material in cold acetone (4°C) for 24 hours.
2. Transfer the blocks to absolute alcohol (change every half an hour).
3. Transfer to ethanol–ether and to 1 per cent celloidin.
4. Clear in benzene.
5. Embed in paraffin wax avoiding high temperatures.
6. Dry slides at 37°C and store at 4°C.
7. Remove wax with light petroleum.
8. Pass to water via absolute acetone.
9. Incubate in solution 2 for 16 hours at 37°C.
10. Rinse in running water.
11. Place in solution 1 for 5 minutes.
12. Rinse in distilled water.
13. Treat with dilute solution of yellow ammonium sulphide for 2 minutes.
14. Wash in water.

15. Counterstain in 1 per cent eosin for 5 minutes.

16. Dehydrate, clear and mount.

Result

Sites of alkaline phosphatase activity are stained black.

Modified Lead Nitrate Method for Acid Phosphatase Activity in the Enzyme Glands of Pelecypods (Anisa Banu *et al.*, 1979b)

Fixation

Either cold acetone paraffin or cold formol calcium frozen section

Reagents Required

Sodium β-glycerophosphate
Acetate buffer
Lead acetate
Magnesium chloride
Ammoniacal silver nitrate
Sodium thiosulphate

Preparation of the Reagents

Solution 1 (Sodium veronol)
    Sodium β-glycerophosphate   2.0 g
    Distilled water   100.0 ml

Solution 2
    0.1 M acetate buffer (*see* Chapter 5)

Solution 3
    Lead acetate   2.0 g
    Distilled water   100.0 ml

Solution 4
    Magnesium chloride 1.5 g
    Distilled water   100.0 ml

Solution 5 (Incubating medium)
    Solution 1   2 vol
    Solution 2   1 vol
    Solution 3   1 vol
    Solution 4   0.3 vol

Procedure

1. After fixation in cold acetone, wash block in distilled water.

2. Mount blocks on cryostat tissue holder and cut 8–10 μm sections.

3. Mount sections on slides smeared with a mixture containing equal volume of gelatin (1 per cent) and 2 per cent formaldehyde. Allow sections to dry for 1 hour at 37°C.

4. Incubate slides in solution 5 for 2 hours.
5. Develop in ammoniacal silver nitrate solution for 30 minutes (add 28 per cent ammonia to 5 per cent silver nitrate).
6. Rinse in 5 per cent sodium thiosulphate for 5 minutes.
7. Dehydrate, clear and mount in a synthetic medium or mount directly in glycerine jelly.

Result

Sites of acid phosphatase activity are stained black.

Detection of Neutral Fats in the White Gland in the Pelecypod Foot by Applying Oil Red O Method (Anisa Banu *et al.*, 1980a)

Fixative

Formol calcium

Reagents Required

Oil red O
Isopropanol

Preparation of Reagents

Solution 1 (Stock solution)

Oil red O	500 mg
98 per cent isopropanol	100.0 ml

Solution 2 (Working solution)
Stock solution   6.0 ml
Distilled water   4.0 ml
Allow to stand for 24 hours and then filter.

Procedure

1. Cut frozen sections proceeding as in the previous case.
2. Rinse slides in water.
3. Rinse in 60 per cent isopropanol.
4. Transfer to solution 2 for 10–15 minutes.
5. Differentiate in 60 per cent isopropanol.
6. Wash in water.
7. Counterstain in Mayer's haemalum.
8. Wash in distilled water.
9. Mount in glycerine jelly.

Result

Site of neutral lipid activity	Red
Nuclei	Blue

Demonstration of Calcium by Applying Alizarin Red S to Calcium-cum-excretory Cells of Gastropod Digestive Gland (Umadevi *et al.*, 1981)

Fixation

Neutral formalin

Reagents Required

Sodium alizarin sulphate
Hydrochoric acid

Preparation of Reagents

Solution 1
    Sodium alizarin sulphate   500 mg
    Distilled water   45.0 ml

To this solution, add 28 per cent ammonia and stir well (1 part ammonia + 99 parts water). Final pH should be 6.3–6.5.

Solution 2 (Differentiator)
    Hydrochloric acid   0.1 ml
    Distilled water   99.9 ml

Procedure

1.   Hydrate slides after dewaxing to water.
2.   Transfer to solution 1 (Alizarin red S) for 2 minutes.
3.   Wash in distilled water.
4.   Differentiate in solution 2 (acid alcohol) for 1 second.
5.   Dehydrate, clear and mount in cedar wood oil.

Result

Sites of calcium deposits are stained orange-red.

Perl's Prussian Blue Technique to Demonstrate Iron in the Calcium-cum-excretory Cells of Digestive Gland of Snails (Umadevi *et al.*, 1981)

Fixation

Neutral formalin

Reagents Required

Potassium ferrocyanide
Hydrochloric acid
Neutral red

Preparation of Reagents

Solution 1
    Potassium ferrocyanide   2.0 g

Distilled water   100.0 ml

This solution should be prepared afresh.

Solution 2

     2 per cent hydrochloric acid   2.0 ml
     Distilled water   49.0 ml

Solution 3 (Working solution)
     Solution 1   25.0 ml
     Solution 2   25.0 ml

Procedure

1. Dewax and hydrate slides to water.
2. Transfer to freshly prepared solution 3 for 30 minutes.
3. Wash in water.
4. Counterstain if desired in neutral red.
5. Wash rapidly.
6. Dehydrate, clear and mount.

Result

Sites with ferric alum   Blue
Nuclei   Red

Application of Azure at pH 3.0 and pH 4.0 to Demonstrate Sialic Acid and Hyaluronic Acid in the Salivary Gland of the Gastropod (Rajalakshmi Bhanu *et al.*, 1981a)

Fixation

Cetyl pyridinium chloride in 10 percent formalin

Reagents Required

Azure A
Citric acid
Sodium hydrogen phosphate

Preparation of Reagents

Solution 1 (1 per cent Azure A solution)
     Azure A   1.0 g
     Distilled water   5000.0 ml

Solution 2 (10 per cent citric acid)
     M/10 citric acid   (*see* Chapter 5)

Solution 3
     M/5 Na_2HPO_4 (*see* Chapter 5)

Solution 4 [Staining solution Azure (pH 3.0)]
     Solution 1   48.0 ml
     Solution 2   1.65 ml
     Solution 3   0.35 ml

Solution 5
    Solution 1   48.0 ml
    Solution 2   1.25 ml
    Solution 3   0.75 ml

Procedure

1. Deparaffinize and hydrate slides to water (2 sets).
2. Stain one set in solution 4 for 30 minutes and the other in solution 5 for 30 minutes.
3. Dehydrate in graded series of alcohol.
4. Place in a mixture of xylene and alcohol.
5. Clear in xylene.
6. Mount in caprate or permount.

Result

Sites with sialomucin activity are metachromatic at pH 3.0.
Sites with hyalomucins are metachromatic at pH 4.0

Cresyl Fast Violet Stain for Oyster Tissue

Fixation

Ice-cold formaldehyde   150.0 ml
1.3 per cent calcium chloride   850.0 ml

Reagents Required

Cresyl-fast violet
Glacial acetic acid

Preparation of Reagents

Solution 1
    Cresyl-fast violet   5.0 g
    Distilled water   500.0 ml
    Glacial acetic acid (pH 3.7)   0.5 ml

Procedure

1. Deparaffinize and hydrate slides to water.
2. Stain in solution 1 for 20 seconds.
3. Wash in running water.
4. Dehydrate and clear.
5. Mount in technicon medium.

Result

Fungi   Blue
Nuclei   Blue
Bacteria   dark blue
Cartilage   Pink

For molluscan tissue also refer the following articles:

1. Anisa Banu *et al.* (1979a, 1979b, 1980a, 1980b, 1981, 1982).
2. Rajalakshmi Bhanu *et al.* (1980, 1981a, 1981b, 1982, 1983, 1984).
3. Umadevi *et al.* (1981, 1982, 1983, 1984a, 1984b, 1984c, 1987).
4. Lalitha *et al.* (1993).
5. Rupavathi *et al.* (1983, 1984).
6. Shyamasundari and Hanumantha Rao (1988).

Insect Chromosomes (Crosier, 1968)

Fixation

10% neutral buffered formalin

Reagents Required

Sodium chloride
Calcium chloride
Potassium chloride
Sodium bicarbonate
Orcein
Lactic acid
Acetic acid
Sodium acetate

Preparation of Reagents

Solution 1

Sodium chloride	14.0 g
Calcium chloride	0.4 g
Potassium chloride	200 mg
Sodium bicarbonate	0.2 ml
Water	1000.0 ml

Solution 2

Colcemid–Ringer, i.e., 0.05 per cent colcemid (ciba) in Solution 1.

Solution 3

Orcein	1.0 g
Lactic acid (85 per cent)	28.0 ml
Glacial acetic acid	22.0 ml

Procedure

1. Keep sections in solution 2 at 25°C for 5 hours.
2. Pass the slides through 1 per cent sodium nitrate for 20 minutes.
3. Keep sections in acetic methanol (1 : 3) for 30 minutes.
4. Transfer to a drop of 60 per cent acetic acid on a clean warmed slide and macerate if necessary.
5. Transfer to acetic methanol.

6. Place the slide in acetic ethanol (1 : 3) for 4 hours.
7. Rinse in 70 per cent ethanol.
8. Transfer to solution 3, apply coverslip and place in heat for 12 hours at 50°C.
9. Dehydrate, clear and mount.

Wismar's Quadrachrome Stain for Chitin (Wismar, 1966)

Fixation

Fix in the following solution
Formalin   20.0 ml
$HgCl_2$ 4 g
Acetic acid   5.0 ml
Distilled water 80.0 ml

Reagents Required

Potassium iodide
Iodine
Alcian blue
Acetic acid
Ferric chloride
Woodstain scarlet
Acid fuchsin
Saffron
Sodium thiosulphate

Preparation of Reagents

Solution 1 (Lugol's iodine)
Potassium iodide 6.0 g
Iodine   4.0 ml
Distilled water   100.0 ml

Solution 2
Alcian blue 8GX   1.0 g
1 per cent acetic acid 100.0 ml

Solution 3 (Verhoeff's haematoxylin)
5 per cent haematoxylin (5 g/100 ml absolute alcohol) 50.0 ml
10 per cent ferric chloride   20.0 ml
Solution 1   20.0 ml

Solution 4 (Woodstain Scarlet acid fuchsin)
Wood stain scarlet   100 mg
0.5 per cent acetic acid 100.0 ml

Solution 5
Acid fuchsin   100 mg
0.1 per cent acetic acid 100.0 ml

Solution 6
Solution 4 2 parts
Solution 5 1 part

Solution 7

Extraction of safranin   100.0 ml

Extract 6.0 g of safranin in 100 ml of 100 per cent alcohol at 58–60°C for 48 hours. Filter, decant and store.

Procedure

1. Deparaffinize and bring slides to 80 per cent alcohol.
2. Transfer to solution 1 for 5 minutes.
3. Decolorize in 5 per cent sodium thiosulphate for 5 minutes.
4. Wash in running water.
5. Transfer to solution 2 for 30 minutes.
6. Wash in running water for 1 minute.
7. Transfer to solution 3 for 4–6 hours.
8. Differentiate in 95 per cent ethanol for 3 minutes.
9. Transfer to solution 6 for 3 minutes.
10. Transfer to 1 per cent acetic acid solution for 1 minute.
11. Differentiate in 5 per cent phosphotungstic acid.
12. Keep in 1 per cent acetic acid for 1 minute.
13. Differentiate in 10 per cent iron chloride.
14. Dehydrate in 2 changes of ethanol.
15. Stain in solution 7 for 5 minutes.
16. Rinse in 2 changes of absolute alcohol.
17. Clear in xylene, and mount.

Result

Chitin   Red
Cytoplasm   Red
Nucleic acid Purple to black

Aceto-orcein for Insect Chromosomes (Lacour, 1941)

Fixation

Fresh material or acetic alcohol

Reagents Required

Orcein
Acetic acid

Preparation of Reagents

Solution 1

Orcein   1.0 g
Acetic acid   45.0 ml

Boil 45 ml of acetic acid and add 1 g of orcein. Cool and add 55.0 ml of distilled water.

Procedure

1. Dissect out tissues in staining solution.
2. The strength of stain solution should be increased (2 g orcein + 70 per cent acetic acid).
3. Stain for 10 minutes.
4. Apply coverslip.

Result

Chromosomes are stained purple.

Aldehyde Fuchsin for Neurosecretory Products in Insects (Ewen, 1962)

Fixation

Any general fixation (Bouin's or Helly)

Reagents Required

Basic fuchsin
Concentrated HCl
Paraldehyde

Preparation of Reagents

Solution 1 (Aldehyde fuchsin) (*see* Chapter 9)

Solution 2

Potassium permanganate	150 mg
Conc. H_2SO_4	0.1 ml
Distilled water	50.0 ml

Solution 3

Sodium bisulphite	2.5 g
Distilled water	100.0 ml

Solution 4 (Acid alcohol)

Ethanol	100.0 ml
Conc. HCl	0.5 ml

Solution 5

Phosphotungstic acid	4.0 g
Phosphomolybdic acid	1.0 g
Distilled water	100.0 ml

Solution 6

Light green	0.4 g
Orange G	1.0 g
Chromotrope 2R	500 mg
Glacial acetic acid	1.0 ml
Distilled water	100.0 ml

Procedure

1. Deparaffinize and hydrate slides to water.
2. Place slides in solution 2 for 1 minute.
3. Rinse in distilled water.
4. Bleach in solution 3.
5. Differentiate in 50 and 70 per cent alcohol.
6. Stain in solution 1 for 10 minutes.
7. Differentiate in 95 per cent ethanol.
8. Differentiate in solution 4 for 30 seconds.
9. Differentiate in 70 per cent alcohol.
10. Mordant in solution 5 for 10 minutes.
11. Rinse in distilled water.
12. Counterstain in solution 6 for 1 hour.
13. Differentiate in 0.2 per cent acetic acid in 95 per cent alcohol.
14. Dehydrate, clear and mount.

Result

Cytoplasm	Light green
Neurosecretory substance	Dark purple

Rapid Azan Method for Crustaceans (Hubschman, 1962)

Fixation

Bouin's picroformalin

Reagents Required

Azocarmine
Glacial acetic acid
Phosphotungstic acid
Aniline blue
Orange G

Preparation of Reagents

Solution 1

Azocarmine	1.0 g
Distilled water	100.0 ml

Dissolve azocarmine in distilled water and boil for 5 minutes. Cool and add 2 ml of glacial acetic acid and filter.

Solution 2

Phosphotungstic acid	10.0 g
Aniline blue	1.2 g
Orange G	4.4 g
Distilled water	1000.0 ml

Procedure

1.   Deparaffinize and hydrate slides to water.
2.   Keep in solution 1 for 15–30 minutes.
3.   Rinse in distilled water.
4.   Keep in aniline oil (1 ml in 100 ml alcohol) for 30 minutes.
5.   Wash in water.
6.   Transfer to solution 2.
7.   Wash.
8.   Dehydrate, clear and mount.

Result

Epicuticle	Red
Endocuticle	Blue
Epidermal cells	Yellowish pink
Nuclei	Orange
Hepatopancreas	Pale orange
Oocytes	Grey with red nucleoli
Sperms	Orange

MAST CELLS

Mast cells are in the form of granules found in connective tissue around blood vessels. Large number of them occur scattered in many tissues particularly in skin. These stain metachromatically with toluidine blue, azure A, Bismark brown and nuonen. Alcian blue methods also enable to distinguish them. Choice methods of detection of mast cells are the chloroacetate esterase method using either fast blue RR or pararosaniline.

HISTOCHEMICAL METHODS FOR MAST CELLS

Alcian Blue Safranin (Csaba, 1969)

Fixation

Any general fixative may be used.

Preparation of Reagents

Solution 1 (Alcian blue)

Alcian blue	900 mg
Safranin	45 mg
Ferric ammonium sulphate	1.2 g
Acetate buffer (pH 1.62)	250.0 ml

Procedure

1. Dewax and bring sections to water.
2. Treat with solution 1 for 15 minutes.
3. Rinse in water.
4. Dehydrate, clear and mount.

Result

Mast cells containing biogenic amines	Blue
Mast cells containing heparin	Red

Chloroacetate Esterase (Fast blue RR) Method (Burstone, 1957)

Fixation

The tissue is fixed in formalin

Reagents Required

Incubating medium

Naphthol AS-D acetate (dissolved in 0.5 ml dimethylformamide)	5 mg
Distilled water	25.0 ml
0.2 M tris buffer (pH 7.1)	25.0 ml
Fast blue RR salt	30 mg

Mix them in order, shake well and filter. Final pH should be below 7.1.

Procedure

1. Dewax and bring sections to water.
2. Incubate in the medium for 5 minutes at room temperature (5 minutes to several hours).
3. Rinse in water.
4. Counterstain in Mayer's haemalum for 10–15 minutes.
5. Wash in water and mount in glycerine jelly.

Result

Mast cells take an intense blue shade.

Chloroacetate Esterase (Pararosaniline) Method (Moloney *et al.*, 1960)

Fixation

Formalin

Preparation of Reagents

Solution 1 (Substrate solution)

Naphthol AS-D chloroacetate	10 mg
Dimethyl formamide	1.0 ml

Solution 2 (Buffer)

0.1 M Michaelis buffer (pH 6.8–7.6) 30.0 ml (*see* Chapter 5)

Solution 3 (Pararosaniline–HCl stock)

Pararosaniline hydrochloride	2.0 g
2N HCl	50.0 ml

Heat gently, cool and filter.

Solution 4 (Sodium nitrite)

Sodium nitrite	400 mg
Distilled water	10.0 ml

Solution 5 (Hexazotized pararosaniline solution)

Solution 3	0.4 ml

Solution 4 (freshly prepared)   0.4 ml

Mix well. Use within half a minute.

Incubating Medium

To 0.8 ml of solution 5, add 30 ml of 0.1 M solution 2 and the pH should be 6.3 which can be adjusted with HCl. Then add solution 1. The resulting solution is pink. Filter.

Procedure

1. Dewax and bring sections to water.
2. Incubate in incubating medium for 30 minutes to 1 hour.
3. Rinse in water.
4. Counterstain nuclei in Mayer's haematoxylin for 5 minutes.
5. Blue in Scot's tap water.
6. Rinse in water, dehydrate, clear and mount in DPX.

Result

Mast cells and neutrophils	Red
Myeloid cell	Pinkish red
Nuclei are	Blue-black

Luna's Method for Mast Cells

Fixation

The tissue is fixed in 10 per cent neutral buffered formalin.

Reagents Required

Basic fuchsin
Paraldehyde
Hydrochloric acid
Haematoxylin
Methyl orange

Preparation of Reagents

Solution 1 (Aldehyde fuchsin solution) (*see* Chapter 9)

Solution 2 (Weigert's haematoxylin) (*see* Chapter 7)

Solution 3 (Methyl orange solution)
Methyl orange   250 mg
95 per cent alcohol   100.0 ml

Procedure

1. Dewax and bring slides to 70 per cent alcohol.
2. Place in solution 1 for 30 minutes.
3. Rinse in 70 per cent alcohol.
4. Transfer to solution 2 for 1 minutes.

5.   Wash in running water.
6.   Rinse in 95 per cent alcohol.
7.   Counterstain in solution 3 for 5 minutes.
8.   Dehydrate, clear and mount.

Result

Mast cells	Purple
Elastin fibres	Purple
Background	Yellow

Unna's Method for Mast Cells (Mallory, 1961)

Fixation

Formol alcohol or 10 per cent neutral buffered formalin.

Reagents Required

Methylene blue
Potassium carbonate
Ether
Glycerine
Calcium chloride

Preparation of Reagents

Solution 1 (Polychrome–methylene blue)

Methylene blue	1.0 g
95 per cent alcohol	20.0 ml
Potassium carbonate	1.0 g
Distilled water	100.0 ml

Solution 2 (Glycerine–ether solution)

Glycerine	50.0 ml
Calcium chloride	10.0 g

Heat the solution until the salt dissolves.

Procedure

1.   Dewax and hydrate slides to distilled water.
2.   Place in solution 1 for 10 minutes.
3.   Wash in distilled water.
4.   Transfer to solution 2 which has been diluted 5 times with distilled water for 30 seconds.
5.   Rinse in distilled water.
6.   Dehydrate, clear and mount.

Result

Mast cell granules	Red
Others	Greenish blue

Thionin Method (Cook, 1961)

Fixation

Any general fixative may be used.

Reagents Required

Thionin
Acetate buffer

Preparation of Reagents

Solution 1

Thionin	500 mg
0.01 M acetate buffer	100.0 ml

Procedure

1. Deparaffinize and bring down slides to water.
2. Place in solution 1 for 30 minutes.
3. Rinse in water.
4. Dehydrate, clear and mount.

Result

Mast cells	Red-purple.
Nuclei	Faint blue-violet.

Quick Toluidine Blue Method

Fixation

Any general fixative may be used.

Reagents Required

Toluidine blue O

Preparation of Reagents

Solution 1

Toluidine blue O	0.2 g
60 per cent alcohol	100.0 ml

Procedure

1. Dewax and bring down slides to 60 per cent alcohol.
2. Place in solution 1 for 2 minutes.
3. Rinse rapidly in water.
4. Dehydrate in acetone giving 2 changes of 3 minutes each.
5. Clear in xylene and mount.

Result

Mast cells	Deep reddish purple
Background	Faint blue

Toluidine Blue (Conroy and Toledo, 1976)

Fixation

10 per cent neutral buffered formalin (10 ml formalin and 10 ml of 95 per cent alcohol).

Reagents Required

Toluidine blue O

Preparation of Reagent

Solution 1
  Toluidine blue O 100 mg
  Distilled water 100.0 ml
Prepare it in McIlvaine buffer. pH should be 6.8–7.2.

Procedure

1. Dewax and bring slides to water.
2. Place in Harris or Mayer's haematoxylin for 2 minutes.
3. Wash and blue it.
4. Place in eosin and rapidly rinse in 95 per cent alcohol.
5. Rinse in water.
6. Place in solution 1 for 10 seconds.
7. Rinse in distilled water.
8. Rinse in 95 per cent alcohol.
9. Dehydrate, clear and mount.

Result

Mast cells	Blue-purple
Collagen	Pink
Cytoplasm	Pink
Nuclei	Blue

Bismark Brown (Spatz, 1960)

Fixation

Fix the tissue in 10 per cent formalin.

Reagents Required

Bismark brown
HCl

Preparation of Reagents

Solution 1 (Bismark Brown)
  Bismark Brown   500 mg
  Absolute ethyl alcohol 80.0 ml
  1 per cent HCl   20.0 ml

Procedure

1. Deparaffinize and transfer to absolute alcohol.
2. Dip in 95 per cent, 90 per cent and 70 per cent alcohol.
3. Transfer to solution 1 for 30–90 minutes.
4. Dip in 70 per cent alcohol giving 3 changes of 2 seconds each.
5. Transfer to 95 per cent and 100 per cent alcohol.
6. Clear and mount.

Result

Mast cell granules are stained yellow-brown.

Neutral Red

Fixation

Fix the tissue in 10 per cent formalin.

Reagents Required

Neutral red
Ethyl alcohol
Haematoxylin
Alum

Preparation of Reagents

Solution 1
    Neutral red   500 mg
    50 per cent ethyl alcohol   100.0 ml
Solution 2
    Alum haematoxylin (*see* Chapter 7)

Procedure

1. Deparaffinize and bring slides to 70 per cent alcohol.
2. Transfer to solution 2 (alum haematoxylin) for 3 minutes.
3. Wash.
4. Stain in solution 1 (neutral red) for 10 minutes.
5. Dehydrate, in *n*-butyl alcohol.
6. Clear and mount.

Result

Mast cell granules	Red
Cartilage matrix	Red
Nuclei	Blue

Chrysoidin (Harada, 1957)

Fixation

Any general fixative may be used.

Reagents Required

Chrysoidin

Preparation of Reagents

Solution 1

Chrysoidin	500 mg
Distilled water	100.0 ml

Procedure

1. Deparaffinize and bring slides to 80 per cent alcohol.
2. Transfer to solution 1 (Chrysoidin) for 10 minutes.
3. Rinse in distilled water.
4. Dehydrate, clear and mount.

Result

Mast cell granules are stained deep brown to black.

Comments

There are other methods: Combination of trichrome with Giemsa (Toren, 1963), acridine orange, neutral red and basic fuchsin staining (Barlow, 1957). Methylene blue and basic fuchsin (Menzies, 1962).

Table 26.4   Histochemical techniques applied for the demonstration of mast cells

Technique	Fixative	Result
Alcian blue/Safranin	Any general fixative	Biogenic amines stained blue – Mast cells containing heparin stained red
Chlorolacetate esterase (Fast blue RR method)	Formalin	Intense blue
Chlorolacetate esterase (pararosaniline)	Formalin	Red
Luna's method	10 per cent neutral buffered formalin	Purple
Unna's method	10 per cent neutral buffered formalin	Red
Thionine method	Any general fixative	Red-purple
Quick method	Any general fixative	Deep reddish purple
Toluidine blue method	10 per cent neutral buffered formalin	Blue-purple
Bismark brown method	10 per cent formalin	Yellow-brown
Neutral red method	10 per cent formalin	Red
Chrysoidin method	Any general fixative	Deep brown to black

IMMUNOCYTOCHEMISTRY

30

The horizons of histochemistry are ever-expanding. The limits of biochemistry are broken down to attain more and more precise localization of chemical substances enmeshed in the tissues. Immunological aspects are pressed into science for the resolution of some chemical aspects of the tissues. Here and elsewhere, the central aim is to go into the details of immunology per se and explanation in detail would be a load on this book, which is already bulky. However certain basics will be mentioned.

Immunocytochemistry has advanced since 1940 when it was incepted. It is a main tool for diagnostic cellular pathology. Immunocytochemistry gained importance as a lever for identifying cellular or tissue constituents (antigen) with the help of antigen–antibody interaction. Antibody binding can be identified either by direct labelling of the antibody or by the use of secondary labelling methods.

Antigenic proteins carbohydrates or lipid molecules have many antibody-binding sites which are highly specific regions in the sense they have a number of amino acid units or monosaccharide units. These units are called epitopes.

Antibodies are serum proteins called immunoglobulins which are formed in the humoral immune system by plasma cells. In higher vertebrates the common antibodies are IgA, IgD, IgE, IgG and IgM. IgG is made up of 2 pairs of light and heavy polypeptide chains. These polypeptide chains are linked by disulphide bonds resulting in "Y" shaped structure. The amino acid sequence in the end regions of the arm varies considerably and these are called variable domains. This variability allows antibody to bind superficially to the antigen against which it was raised. Antigens already exist but the antibody has to be raised so to speak. Roughly speaking, antigen is described as a foreign body, the composition of which varies from context to context. Again, an antibody is an agent produced as a response to an antigen. In this scheme, there is a specific antibody for a corresponding specific antigen. When an antigen "x" enters the living animal, it is promptly carried by blood or lymph to lymph nodes. In the lymph nodes, lymphocytes (derived from bone marrow) react to "x" antigen molecules by becoming plasma cells whose main function is to synthesize specific "x" antibodies Antibodies are capable of combining with antigens (which are responsible for producing a reaction to produce the antibodies). From the site of production, the antibodies are carried to every nook and corner of the animal in preparation for neutralizing the antigen activity. When a medium containing an antigen is mixed with a solution containing a specific antibody, a precipitate results. This precipitate is an antigen–antibody complex.

ANTIGEN–ANTIBODY BINDING

Capra and Edmundson (1977) described the cavity formed by the amino acid side chains of the variable domains of an antibody. This cavity is chemically complimentary to a single type of antigen epitope. The antigen–antibody specificity is analogous to the lock (antibody)-and-key antigen arrangement. Van der Waals forces, hydrogen bonds and electrostatic forces make the antigen–antibody link stronger.

When antigens are experimentally introduced into an animal, the production of antibodies is provoked. Ordinarily, introduction of antigen as derived from infectious diseases is called inoculation. When antibody titre reaches a certain level of concentration, the animal is said to be immunized for that particular antibody. This is the subject of immunity and immunization.

There are two classes of antibodies, the monoclonal and polyclonal antibodies. Monoclonal antibodies are produced by clones of plasma (lymphoid cells), so the antibodies are immunologically similar and are capable of reacting with a specific epitope on the antigen against which they are raised. On the other hand, polyclonal antibodies are produced by different immunologically competent lymphoid cells. Consequently these antibodies are immunocytochemically dissimilar and bind at different structural parts of epitopes.

Usually rabbits, sheep and goats are the favourite experimental animals for raising these antibodies. Once the immune response has been achieved (mice) β-type lymphocytes are taken from the spleen or lymph nodes and fused with myeloma cells of mice to produce a hybrid cell. These cells can be cultured or can be transplanted into the peritoneal cavity of genetically similar mice. This results in the sequestration of unlimited quantities of identical antibodies.

Lectins are of non-immuno origin and are sugar-binding glycoproteins or proteins. They are derived from both plants and animals. Like antibodies, these can also be used as labels.

Cellular epitopes are highly selective and very specific. An appropriate antibody and equally agreeable label have revolutionized our efforts to pinpoint chemical substances in histopathological studies and in the applied arena of diagnosis in pathology. The great proliferation of publications in the last four or five decades is a witness to the ever-widening horizons and potentialities of immunocytochemistry.

For the past four decades there has been a steady refinement or evolution of newer methods. Labelling system or production of antibodies or even their involvement of synthetic peptides or certain innovations like trypsin treatment or paraffin sections to unmask certain antigens otherwise shut in the process of fixation and embedding (Huang *et al.*, 1976) have all been geared up to the best advantage by reducing technical difficulties.

LABELS

The most agreeable and relatively easily available and usable tags in immunocytochemistry are enzymes. Incubation with a chromogen with a standard histochemical method produces a reaction which is satisfactorily discernible with light microscope is the result. The number of enzymes and chromogens available give the investigator a wide choice for choosing the colours for the reaction and product. In earlier days, limited number of enzymes were being pressed into the service of immunocytochemistry. Modern methods and choice have brought many kinds of enzymes into the fold. Immunocytochemistry with horseradish peroxidase still enjoys popularity. This choice of enzymes goes very well in combination with the chromogen 3,3′-diaminobenzidine tetrahydrochloride (DAB). The end product is dark brown in colour (Graham and Karnovsky, 1966).

DAB is reported to be a carcinogen but the risk is not much (Weisberger *et al.*, 1978). Peroxidase 3-amino 9-ethyl carbazole (carcinogen) (Graham *et al.*, 1965; Kaplow, 1975), red alpha-naphthol pyronin (Taylor and Burns, 1974), red purple 4-chloro 1-naphthol (Nakane 1968)- blue. Hanker-Yates agent (Hanker *et al.*, 1977) dark blue are other chromogens.

A drawback seems to be the solubility of the reaction product in alcohol and xylene. Glycerine jelly or similar aqueous media have to be preferred.

Another alternative for horseradish peroxidase is the calf intestinal alkaline phosphatase. This has been in vogue since 1984 when Cordell *et al.* developed alkaline phosphatase (APAAP). Fast red TR along with AS-MX phosphate sodium salt produces a bright red product.

This is soluble in alcohol. New fuchsin yields a red insoluble product (Malik and Damon, 1982). Glucose oxidase is also used which gives a navy blue reaction end product (Suffin *et al.*, 1979). Bacterial derived beta-D-galactosidase has also been used as a tracer and this can be developed by using the indigogenic method which reaction results in the formation of turquoise blue end product (Bondi *et al.*, 1982).

Colloidal Metal Labels

This is a popular metal tracer. When used independently it has pink colour under transmitted light but in practice the size of the particle should be 20 mm for light microscopy (Ellis *et al.,* 1988).

To amplify the visibility of gold conjugate, 5-mm gold conjugates can be used (Holgate *et al.,* 1983). For polarized incident light microscopy both gold- and silver-enhanced gold conjugates are required (Ellis *et al.,* 1988). Colloidal gold is widely used for electron microscopy. Others like ferritin can also be used.

Radiolabels

In immunocytochemical studies, radioisotopes as tracers are required. Autoradiographic facilities should be available. Hunt *et al.* (1986) have discussed radioisotopes as tracers.

Jasani *et al.* (1981–1983) have indicated the use of haptens as bridging techniques–haptens like dinitrophenol and orsnalic acid are used. Linkage of hapten to the primary antibody to form a complex is the key (antihapten–antibody or hapten-labelled enzyme or hapten-labelled PAP complex). This is as sensitive as avidin–biotin technique.

In histological techniques, tissues are immersed in different fixatives to stop putrefaction and autolysis. The material is routinely processed for section cutting. In these procedures, antigenic isotopes are masked or destroyed. To offset this destruction, some of the labile antigenic sites frozen sections (which are fixed in cold acetone) can be employed. Masked antigens in routinely processed tissue can be revealed by using one of the following techniques.

- Proteolytic enzyme digestion
- Microwave antigen recovery
- Microwave and trypsin antigen retrieval
- Pressure cooker antigen retrieval

Sections are rendered wax-free and then passed into alcohols. It they are peroxidase-labelled, they are treated with 0.5 methanolic hydrogen peroxide (10–15 minutes) to remove indigenous peroxidase activity. Subsequently washing in running water and later in distilled water should be followed.

1. *Proteolytic enzyme digestion*   Huang *et al.* (1976), Curran Gregory (1977) and Mepham *et al.* (1979) described the procedures of pre-treating the routinely processed formalin-fixed sections with proteolytic enzymes. This enables the unmasking of some antigenic determinants. Routinely used enzymes are protease and trypsin.

   This enzyme digestion breaks down formalin cross-linking and the antigenic sites of some antibodies are recovered. Pure trypsin has very little digestion effect on formalin-fixed paraffin sections. Actually trypsin with impurities like chymotrypsin is active.

2. *Microwave antigen recovery*   This new technique was described by Shi *et al.* (1991) which was later modified by Cattorett *et al.* (1993). Many antigens may be recovered by microwave heating which were previously considered irretrievable or lost during routine histological processing. In microwave heating formalin-fixed (dewaxed) paraffin sections are boiled in 0.01 M citrate buffers (pH 6). Antibodies like Ki67 and MI BI (proliferation markers) can be used in frozen sections.

   Boiling is carried out in a coplin jar with enough fluid. Strong adhesives like vectabond or aminopropyltriethoxysilane (APES) have to be used to overcome damage due to severe heating.

3. *Pressure cooker antigen retrieval*   Norton (1993) and Norton *et al.* (1994) could obtain good results by replacing microwave a time consuming procedure with pressure cooker. But Miller *et al.* (1995) observed that microwave over heating time was unable to recover certain antigens and pressure cooker method was less time consuming.

PRACTICAL ASPECTS OF IMMUNOCYTOCHEMICAL STAINING

Broad dilution series is used when applying an untested antibody to a tissue section containing relevant antigen. Most of the commercially supplied primary antibodies and labelling systems are provided with a recommended dilution range. Primary antibody dilution has to be adjusted.

To avoid the formation of antigen–antibody complexes which will unnecessarily create problems, it is necessary to remove the unfound antibody before incubation in the next layer. This can be done by washing the section between antibody incubation in Tris buffer saline.

Incubation Methods

There is a chance of incubation of antiserum. To avoid this, incubation is carried out in moist atmosphere. For this, slides have to be placed on perspex strips which are placed in a trough containing water. Cross contamination can be avoided by leaving a small gap between adjacent slides.

Fixation and Paraffin Wax Block Immunocytochemistry

Prolonged fixation reduces immunoreactivity. Acidic fixatives like Bouin's fluid after 24 hours fixation destroys immunoreactivity of some diagnostic antibodies such as UCHLL (C45RO). Similar picture is seen with mercury-containing fixatives (Bilbe *et al.*, 1989: Gillibrand, 1991 Unpublished data). Even formalin preparation has similar effects. Other antibodies such as L26 (CD20) are less affected. However microwave antigen retrieval according to Singh *et al.* (1993) enables the retrieval of antigens after formalin fixation up to 2 months duration.

PREPARING PARAFFIN SECTIONS FOR IMMUNOSTAINING

1. 3–5-µ thick sections are cut and placed on clean glass slides. Adhesives like APES or vectabond are very successful and can be easily used.

2. If adhesives are not used, heating at 56°C overnight is necessary for avoiding damages especially while using microwave antigen recovery techniques. Adhesive coated sections require 1–2 hours heating at 56°C prior to staining. Heating on hot plate is not advisable since it is detrimental to antigens.

3. Dewax sections in xylene and bring to absolute alcohol.

4. Block endogenous peroxide activity by incubating in 0.5 per cent hydrogen peroxide in methanol for 10 minutes.

5. Dehydrate, wash well in running water and transfer to TBS.

ANTIGEN RETRIEVAL TECHNIQUES

Proteolytic Enzyme Method

The digestion media have to be prepared afresh before commencing the procedure. Proteolytic enzyme digestion is limited to sections fixed in formalin and paraffin embedded. This limitation will enable the antigens that are blocked by cross linking with formalin to be exposed for binding to the relevant antibody.

Digestion timings (optional) vary with the size of the specimen, fixative duration, temperature of fixative and rate of penetration of fixative. With such variability, there is no fixed digestion time, if so it does not achieve optional staining. Generally in the initial stages, uniform time established on controls is used.

Trypsin Method

This method involves slides being pre-heated at 37°C in distilled water prior to treatment with trypsin which is freshly prepared by employing 0.1 per cent calcium chloride in distilled water adjusted to pH 7.8 using 0.1 M NaOH and used at 37°C. The digestion time is predetermined. After digestion, the slides are first transferred to cold running water.

Protease Method

Similar to trypsin digestion, sections are pre-heated at 37°C but the pre-heating time is less in this case on routine paraffin wax tissue because the digestion time of protease is rapid than the rate of trypsin. 0.1 or 0.05 per cent solution of protease in distilled water adjusted to pH 7.8 (with 1 M NaOH) is suggested.

Pepsin Method

This solution is as follows. 0.4 per cent solution of pepsin in 0.01M HCl (pH 2.0). Digestion is carried at 37°C for 15–20 minutes, some antigens like basement membrane and proteins are stained in a better way.

Immunocytochemical Staining Techniques

At every step, washing in necessary. Sections have to be washed to avoid contamination of one reagent with another. Wash with tris buffer saline (TBS), i.e., flooding the slide and then drain excess buffer and blot-dry.

Preparation of TBS (0.00 M)

Distilled water	10 litres
Sodium chloride	80.0 g
Tris (hydroxymethylamine)	6.05 g
MHCl	44.0 ml

Adjust pH to 7.6 with 1 M HCl or 0.02 M tris solution.

Direct Immunopeptidase for Either the Traditional Conjugated or Enhanced Polymer One-step (EPOS) Technique

1. Keep section in TBS.
2. Drain off excess fluid and blot.
3. a) Incubate the slides in diluted peroxidase-linked primary antibody for 1–15 hours. This is achieved at 4°C or ambient temperature or

    b) Incubate in EPOX peroxidase pre-diluted primary antibody for 1–2 hours at ambient temperature.
4. Wash in TBS.
5. Transfer to freshly prepared DAB solution.
6. Again rinse in TBS and wash in water.
7. Counterstain in haematoxylin. Dehydrate, clear and mount.

Pluzek *et al.* (1993) introduced this method by using a polymer "backbone" to attach a large number of primary antibody molecular and peroxidase enzymes directly to the primary antibody. As in traditional direct technique there is similar sensitivity to the avidin–biotin technology but with few steps. Under the commercial name of enhanced polymer one-step staining (EPOS).

The advantages of this technique are that it is rapid especially for frozen section immunocytochemistry and sensitive enough to demonstrate small amounts of antigens. Much time is saved as compared to the three stage method. The reduction in steps also improves reproducibility.

Indirect technique for monoclonal antibody

1. Keep sections in TBS.
2. Drain off excess solution and blot-dry.
3. Apply optimally diluted primary monoclonal antibody for 30–60 minutes.
4. Wash slides in TBS.

5. Cover with an optimally diluted peroxidase-conjugated rabbit anti-mouse containing 1/25 dilution of normal human serum.
6. Wash slides in TBS.
7. Bring to incubation medium (freshly prepared DAB) for 10 minutes.
8. Rinse in TBS.
9. Wash in running water.
10. Counterstain in haematoxylin, dehydrate, clear and mount.

This is a less expensive and rapid method than the usual direct method. Horseradish peroxidase labelling is a common method with suitable chromogen substrate. In this method, first the optimally diluted primary antibody is applied which is followed by a labelled antibody which is directed against the first antibody.

Some modification has to be made for polyclonal primary antibody. Sections have to be incubated in 1/10 dilution. Normal swine serum should be drained. No washing is suggested. Another modification is, instead of rabbit anti-mouse antibody, peroxidase-conjugated swine anti-rabbit serum is used.

Peroxidase–Antiperoxidase (PAP) Technique

1. Keep sections in TBS.
2. As usual, drain off excess solution and blot-dry.
3. Incubate in optimally diluted rabbit anti-mouse antibody for 30 minutes.
4. Gently wash in TBS.
5. Transfer the optimally diluted rabbit anti-mouse antibody to incubation medium for 30 minutes.
6. Wash in TBS.
7. Transfer to incubation medium which is optimally diluted mouse peroxidase for 30 minutes.
8. Gently wash in TBS.
9. Transfer to freshly prepared DAB incubating medium.
10. Repeat step 8 and then wash in running water.
11. Counterstain with haematoxylin.
12. Dehydrate, clear and mount.

Mason *et al.* (1969) and Sternberger (1969) introduced immunoenzyme bridge method using enzyme-specific antibody. This method was replaced by another sophisticated method by Sternberger *et al.* (1970). In this method, a soluble peroxidase–antiperoxidase complex (PAP) was used. These complexes are formed from three peroxidase molecules and two antiperoxidase antibodies and they are used as a third layer in the staining method. A binding takes place between unconjugated primary antibody (e.g. rabbit anti-human IgG) and an antibody that is usually a swine anti-rabbit applied in excess. One of its two identical binding sites binds to the primary antibody and the other to the rabbit–PAP complex.

Some modifications are required for rabbit primary antibody. They are as follows:

a) After draining of excess TBS, incubate with 1/10 dilution of normal swine serum for 10 minutes. Drain off. No washing is suggested before applying primary antibody.

b) Instead of incubating in the optimally diluted rabbit anti-mouse antibody, use swine anti-rabbit antibody.

c) Instead of DAB incubating medium, use rabbit peroxidase antiperoxidase.

Labelled Avidin/Avidin–Biotin Complex Technique for Monoclonal Antibodies

1. Keep sections in TBS.
2. Drain excess solution and blot-dry.
3. Transfer to incubating medium which is optimally diluted primary antibody for 30–40 minutes.
4. Wash slides in TBS.
5. Transfer to incubating medium which is optimally diluted.
6. Wash gently in TB.
7. Transfer to optimally prepared labelled avidin or avidin /biotin complex for 30 minutes.
8. Wash gently in TBS.
9. Transfer to DAB substrate solution.
10. Wash in running water.
11. Counterstain in haematoxylin.
12. Dehydrate, clear and amount.

This labelled avidin/avidin biotin technique depends on the affinity of the glycoprotein avidin for biotin a vitamin with low molecular weight. It is the affinity of glycoprotein with vitamin. Avidin (M.W 67 kD) is present in the white of egg. This has a tertiary structure with four subunits. The four subunits possess four biotin-binding hydrophobic pockets. The oligosaccharide residues present in egg white, avidin and its charged properties are reported as giving it some affinity for some tissue components and it results in non-specific binding.

Liver tissue contains large amounts of biotin and this leads to problems. Streptavidin (M.W 60 kDa) can be extracted from the bacterial culture broth (*Streptomyces avidini*). Streptavidin is devoid of oligosaccharide residues and has neutral isoelectric point. In a way it is advantageous over the chicken egg variant. Commercially produced egg and bacterial variants show negligible difference in sensitivity and background (Happerfield *et al.*, 1993). Biotin-rich tissues such as liver, commercially available biotin blocking systems may be required when using the avidin/biotin technique to block unwanted biotin staining.

Biotin (Vitamin H) has the capacity to get conjugated to the antibodies and enzyme markers to the extent of 150 biotin molecules getting attached to one antibody molecule with the help of spacer arms. By spacing the biotins, the glycoprotein – avidin will bind and maximize its strong affinity for biotin. Peroxidase and alkaline phosphatase are the variants of the avidin–biotin system. These are either found directly to the avidin or streptavidin (Guesden *et al.*, 1979) or the enzymes are biotinylated and 75% of avidin-binding sites are occupied by the avidin or streptavidin–biotin complex (Hsu *et al.*, 1981). Commercially they are supplied as two separate reagents biotinylated label and avidin or streptavidin. They are added together, 30 minutes before use.

Since a large number of biotins get attached to a single antibody, a number of labelled avidin or streptavidin get bound to it. Due to this, sensitivity improves. This is superior to the previous enzyme technique and allows a higher dilution of the primary antibody.

Indirect Technique for Polyclonal Primary Antibody

1. Keep sections in TBS.
2. Drain off excess TBS, blot, dry and incubate in normal saline serum, diluted 1/10 with TBS for 10 minutes.
3. Transfer to incubating medium, i.e., optimally diluted rabbit primary antibody for 30–60 minutes.
4. Wash in TBS.
5. Transfer to incubating medium, i.e., optimally diluted alkaline phosphatase-conjugated secondary antibody for 30–40 minutes.
6. Wash gently in TBS.
7. Transfer to substrate solution of choice, for example fast red solution.
8. Wash in running water.
9. Counterstain and mount.

Remarks

The reaction end product may be enhanced by incorporating the following step after step 5:

5a) Gently wash off alkaline phosphatase conjugate tip off buffer.

5b) Apply alkaline phosphate-conjugated antibody directed against the species of the secondary antibody.

Alkaline Phosphatase-Antialkaline Phosphatase (APAAP) for Monoclonal Antibodies

1. Keep section in TBS.
2. Drain off normal swine serum and blot-dry.
3. Transfer to incubating medium, primary antibody at optional dilution, for 30–40 minutes.

4. Wash in TBS.
5. Transfer to the incubating medium, optimally diluted unconjugated rabbit anti-mouse bridge antibody for 30 minutes.
6. Incubate in alkaline phosphatase–anti alkaline phosphatase complex at the optimal dilution of 30 minutes.
7. Repeat step 5.
8. Incubate in substrate medium of choice, for example fast red salt solution.
9. Wash in running tap water.

Immunogold Techniques

TBS containing bovine serum albumin (BSA-TBS)

Tris (hydroxymethyl methylamine)	12.14 g
Sodium chloride	45.00 g
BSA	5.00 g
Sodium azide	6.5 g
Distilled water	5 litres

Adjust final pH to 8.2 with 1 M HCl

Indirect Immunogold Technique for Monoclonal Antibodies

1. Bring sections to distilled water.
2. Treat with Lugol's iodine for 5 minutes. Clear with 2.5 per cent sodium thiosulphate and wash in running water.
3. Keep sections in BSA-TBS (0.1 per cent bovine serum albumin in TBS at pH 8.2). Drain and blot-dry.
4. Transfer to incubating medium 1/20 normal goat serum (NGS) in BSA-TBS for 10 minutes.
5. Drain, blot and dry.
6. Transfer to primary antibody and optimally dilute in BAS-TBS for 30–60 minutes.
7. Gently wash in TBS.
8. Transfer to incubating medium, gold conjugated secondary antibody at optimal dilution in BSA-TBS, for 60 minutes.
9. Wash gently in TBS.
10. Wash in 0.1 M phosphatase-buffered saline (PBS) (pH 7.6) for 3–2 minutes changes.
11. Post-fix with 2 per cent glutaraldehyde in PBS for 10–15 minutes.
12. Wash in several changes of distilled water and enhance staining with silver.
13. Counterstain, dehydrate, clear and mount.

Silver Enhancement

The intensity of gold label may be enhanced by incubation of the sections in a silver solution.

Preparation of Reagents

1. Solution 1 (2 M citrate buffer stock)
    Trinatrium citrate   23.5 g
    Citric acid   25.5 g
    Distilled water   100.0 ml

2. Solution 2 (Silver solution)
    Silver lactate   110 mg
    Distilled water   15.0 ml

3. Solution 3 (Hydroquinone solution)
    Hydroquinone   950.0 mg
    Distilled water   15.0 ml

4. Solution 4 (Silver enhancement solution) (50 per cent gum acacia solution)
    Solution 2   15.0 ml
    Solution 3   15.0 ml
    Solution 1   10.0 ml
    Distilled water   60.0ml
    Solution 4   7.0 ml

Procedure

1. Rinse sections in solution 1 for 2 minutes.
2. Transfer to silver enhancement solution (freshly prepared and protected from light) for 3 minutes.
3. Fix in 1:10 solution of fixing solution for 2 minutes.
4. Wash in running water.

Immunogold Silver Staining Technique (IGSS)

Faulk and Taylor (1971) first introduced the use of colloidal gold as a label for immunocytochemistry. It finds a place in both direct and indirect methods and is widely used in ultrastructural immunolocalization. But at light microscopic immunocytochemical level it did gain importance (Holgate *et al.*, 1983). In this technique, by the addition of metallic silver layers the gold particles are enhanced. In this technique, silver lactate is used as the ion supplier, and any hydroquinone as the reducing agent in a "protective" colloid of gum arabic at pH 3.5 to produce slow-forming metallic silver with a tolerance for natural light. The staining intensity may be improved by pre-treatment with Lugol's iodine and sodium after dewaxing and rehydration. This technique supersedes PAP technique but there is a drawback with the formation of silver deposits in the background especially in the hands of a beginner. It causes great confusion. De May *et al.* (1986) suggested a modification of the old method.

Visualization (substrate solutions)

For visualization of peroxidase enzyme DAB (3.3′-diaminobenzidine tetrahydrochloride) and AEC (3-amino 9-ethyl) are used. These substrates are carcinogenic and care is taken while using them. Usually a fume cupboard is suggested.

DAB Method (Graham and Karnovsky, 1960)

Tris buffer (0.2 M tris containing 24.228 g/l)	12.0 ml
0.1 M HCl	19.0 ml
Distilled water	10.0 ml
DAB solution	
DAB	5 mg
Tris HCl buffer (pH 7.6)	10.0 ml
H_2O_2 (fresh)	0.1 ml

This solution is prepared just before use. Sections are kept in the solution and incubated at room temperature. A dark brown product develops within 10 minutes. This product is resistant to alcohols and xylene.

Modified DAB Method (with chemical enhancement)

Incubation medium

1.	DAB	150 mg
	Tris HCl	10.0 ml
2.	Imidazole	200 mg
	Tris HCl	10.0 ml

Add solutions 1 and 2 to 280 ml tris HCl. Incubate sections in this solution for 30 seconds. Add 8 drops of 30 per cent H_2O_2 and again incubate for 10 minutes. Imidazole is added to enhance the speed of the reaction and intensity of staining. Further enhancement is obtained by incubation of the sections in 0.5 per cent copper sulphate in 0.9 per cent sodium chloride. Cobalt chloride is added to DAB substrate for enhancement.

3-amino 9-ethyl Carbazole (Kaplow, 1975)

1)	3-amino 9-ethylcarbazole	10 mg
	Dimethyl sulphoxide	6.0 ml
	0.02 M acetate buffer (pH 5.0–5.2)	50.0 ml

2) Add 0.4 ml 0.35 H_2O_2 just before use. Rinse the sections in 0.02 M acetate buffer (pH 5.0), filter the substrate solution on the sections and incubate for 5–10 minutes at room temperature. The end product is red and it is not resistant to alcohol and xylene and so mounting medium should be aqueous.

Other Substrate Solutions for Peroxidase

These include 4-chloro-naphthol (Nakane, 1963), Hanker Yates reagent (Hanker *et al.*, 1977) and naphthol pyronin (Taylor and Burns, 1974).

VISUALIZATION SUBSTRATES FOR ALKALINE PHOSPHATASE METHODS

This is based on the coupling of substituted naphthol to a suitable azo dye. The most commonly used dyes are Fast red TR and hexazotized new fuchsin.

Fast red TR (Good for histological preparation)

Naphthol AS-MX phosphate-free acid	4 mg
N,N-dimethyl formamide	0.2 ml
M Tris HCl buffer (pH 8.2)	9.8 ml
Levamisole	2.4 g
Fast red TR salt	10 mg

First dissolve the naphthol AS-MX phosphate in *N,N*-dimethyl formamide and then add tris buffer. This is followed by addition of levamisole and Fast red TR. Now filter the solution onto the sections. Incubate for 10–20 minutes. The end product is bright, red in colour. It is soluble in alcohol, so mount in an aqueous medium. Instead of Fast red TR, Fast blue BB (4 mg) can be added and the end product is blue.

Alternatively, Fast red substrate solution is recommended for cytological preparations.

Solution Suggested by Ponder and Wilkinson (1981)

Naphthol AS-BI phosphoric and sodium salt	5 mg
N,N-dimethyl formamide	0.2 µl
Veronol acetate buffer (pH 9.2)	9.8 µl
Levamisole	2.5 µl
Fast red TR salt	5.0 µl

Dissolve AS-BI phosphate in *N,N*-dimethyl formamide in a glass vial. This is followed by the addition of buffer and levamisole, and Fast red TR is added at the end just before use.

Hexazotized New Fuchsin (Damon and Malik, 1982)

Naphthol AS-BI phosphate	5 mg
N,N-dimethyl formamide	60 ml
M-tris HCl buffer (pH 8.7)	10 ml
M levamisole	10 ml
Freshly prepared sodium nitrite	50 ml
Per cent new fuchsin in 2M HCl	20 ml

First add new fuchsin to sodium nitrite, mix well and then add tris buffer and levamisole. Just before use, add naphthol AS-BI phosphate dissolved in the *N,N*-dimethyl formamide. Now filter the solution on to the sections. Incubate for 20 minutes. The end product is red which is resistant to dehydration, xylene and mounting. This may not be true always, so it is better to use a aqueous medium. Modification of this method is prescribed by Stein *et al*. (1985).

Modified New Fuchsin Method

Solution 1

M2-amino 2-methyl 1,3-propanediol	18.0 ml
M-tris HCl buffer (pH 9.7)	50.0 ml
NaCl	600 mg
Levamisole	28 mg

Solution 2

Naphthol AS-BI phosphate	35 mg
N,N-dimethyl formamide	0.42 ml

Solution 3

New fuchsin (5 g 100 ml 2M HCl)	0.14 ml
Sodium nitrite (40 mg in 1 ml distilled water)	0.35 ml

Mix new fuchsin in freshly prepared sodium nitrite and incubate the mixture for 60 seconds at room temperature. Agitate continuously. First mix solution 1 and 2 and then add solution 3. Final pH should be 8.7 and adjust with HCl. Filter on to the sections and incubate for 20 minutes.

Nitroblue Tetrazolium Method for Alkaline Phosphatase (McGadey, 1970)

Solution (Buffer solution)

0.2M tris HCl (pH 9.5) containing 100 M $MgCl_2$.

Solution a

5 mg 5 bromo 4-chloro3-indoxyl phosphate is dissolved in 0.1 ml dimethyl formamide (DMF) and then in 1.0 ml of the above buffer.

Solution b

5 mg nitro blue tetrazolium is dissolved in 0.1 ml DMF. Solution a and are added with continuous stirring to 30 ml of the buffer and filtered. Incubate immediately for 20 minutes to 12 hours. A blue-black product is noticed at the site of alkaline phosphatase activity. This is soluble in alcohol and xylene. So an aqueous mounting medium is suggested.

Suggested Readings

Bullock and Petrusz (1982, 1983, 1985); Polak and Van Noorden (1983, 1986), Childs (1986), Incstar Corp (1989), Naish (1989), Elias (1990), Jasani and Schmid (1993) and Kirkham and Hall (1995).

REFERENCES

Adams, C.W.M., (1956)., *J. Histochem. Cytochem.*, 4, 23.

___________ (1957)., *J. Clin. Path.*, 10, 56.

___________ (1959)., *J. Path. Bact.*, 77, 648.

___________ (1960)., *J. Path. Bact.*, 80, 442.

___________ (1961)., *Nature, London*, 192, 331.

___________ (1965)., *Neurohistochemistry*, Ed. Adams, C.W.M., London, Elsevier.

Adams, C.W.M., Abdulla, Y.H., Bayliss, O.B., and Weller, R.O., (1966)., *J. Histochem. Cytochem.*, 14, 385.

Adams, C.W.M., and Bayliss, O.B., (1963)., *J. Path. Bact.*, 85, 113.

Adams, C.W.M., and Bayliss, O.B., (1974)., "Lipid Histochemistry," In: Glick, D. and Rosenbaum, R.M. (eds.)., *Techniques of Biochemical and Biophysical Morphology*, New York – Wiley Interscience, vol.2, 99–156.

Adams, C.W.M., Baylis, O.B. and Ibrahim, M.Z.H., (1963)., *J. Histochem. Cytochem.*, 11, 560.

Adams, C.W.M., and Sloper, J.C., (1955)., *Lancet.*, 1, 651.

Adams, C.W.M., and Sloper, J.C., (1956)., *J. Endocr.*, 13, 321.

Adamstone, F.B., and Taylor, A.B., (1948)., *Stain Tech.*, 23, 109.

Agrell, I.P.S., (1958)., *Stain Tech.*, 33, 265.

Alan Stevans., and Chalk, B.T., (1996)., "Pigments and Minerals" in *Theory and Practice of Histological Technique*, Eds. Bancroft, J.D. and Stevans, A., Churchill Livingstone.

Alberts, B.D., Bray, J., Lewis, M., Raff, K., Roberts, K., and Watson, J., (1994)., *The Molecular Biology of the Cell*, 3rd edn. New York, Garland Publishing Inc.

Alfert, M., and Geschwind, I.I., (1953)., *Proc. Natl. Acad. Sci.*, 39, 991.

Allison, R.T., (1973)., *Med. Lab. Tech.*, 30, 27.

Alpert, M., Jacobowitz, D., and Marks, B.H., (1960)., *J. Histochem. Cytochem.*, 8, 153.

Altman, R., (1890)., *Die, Elementarorganismen und ihre Bezichungen zur Zellen*, Liepzig.

Anderson, R.A., (1971)., *Stain Tech.*, 46, 267.

Andre, G., Wenger, J.B., Reballoso, D., Arrington, J., and Mehm, W., (1994)., *J. Histotechnol.*, 2, 137.

Anisa Banu., Shyamasundari, K., and Hanumantha Rao, K., (1979a)., *Mikroskopie.*, 35, 151.

___________ (1979b)., *Folia. Histoschem. Cytochem.*, 17, 395.

___________ (1980a)., *Histochem. J.*, 12, 553.

___________ (1980b)., *Proc. Ind. Acad. Sci.*, 89, 449.

___________ (1981)., *Proc. Natl. Acad. Sci.*, 51, 7.

___________ (1982)., *Life. Sci. Adv.*, 1, 1.

Anthony, E.E., Jr., (1931)., *Science.*, 73, 319.

Arensberger, K.E., and Markell, E.K., (1960)., *Amer. J. Clin. Path.*, 34, 50.

Aschoff, A., Fritz, N., and Hert, M., (1982)., In: *Axonal transport in physiology and pathology*, Weiss, D.G. and Gorio, A. (eds.), Berlin, Springer, p. 177.

Aubert, E., (1950)., *J. Publ. Hlth.*, 41, 31.

Austin, A.P., (1959)., *Stain Tech.*, 34, 69.

Ayoub, P., and Shiklar, G., (1963)., *J. Oral. Surg.*, 16, 580.

Bachler, B.S. and Alexander, W.F., (1952)., *Stain Tech.*, 27, 147.

Bahr, G.F., (1954)., *Exp. Cell. Res.*, 7, 457.

Bahr, G.F., Bloom, G., and Friberg, U., (1957)., *Exp. Cell. Res.*, 12, 343.

Baker, J.R., (1944)., *Quart. J. Micr. Sci.*, 85, 1.

__________ (1946)., *Quart. J. Micr. Sci.*, 87, 441.

__________ (1947)., *Quart. J. Micr. Sci.*, 88, 463.

__________ (1949)., *Quart. J. Micr. Sci.*, 90, 293.

__________ (1956)., *Quart. J. Micr. Sci.*, 97, 161.

__________ (1958)., *Principles of Biological Microtechnique*, London: Methuen.

__________ (1966)., *Principles of Biological Microtechnique*, London: Methuen.

Balough, K. Jr., (1965)., *J. Histochem. Cytochem.*, 13, 303.

Bancroft, J.D., (1963)., *Stain Tech.*, 38, 336.

__________ (1967)., *J. Med. Lab. Tech.*, 24, 309.

__________ (1975)., *Histochemical Techniques*, London and Boston, Butterworths.

Bancroft, J.D., and Cook, H.C., (1994)., *Manual of Histological Techniques and their Diagnostic Application*, Edinburgh: Churchill Livingstone.

Bancroft, J.D., and Cook, H.C., (1996)., "Carbohydrate" In: *Theory and Practice of Histological Techniques*, Bancroft, J.D. and Stevens, A. (eds.), Edinburgh: Churchill Livingstone.

Bancroft, J.D., and Hand, N.M., (1987)., *Enzyme Histochemistry*, Royal Microscopical Society Handbook 14 Oxford: Oxford Science.

Bancroft, J.D., and Palmer, J., (1996)., "Frozen and related sections." In: *Theory and Practice of Histological Techniques*, Bancroft, J.D., and Stevans, A. (eds.), Edinburgh: Churchill Livingstone.

Banny, T.M., and Clark, G., (1949)., *Stain Tech.*, 24, 223.

Bargman, W., (1950)., *Die elektive Microscopies*, 5, 289.

Barka, T., (1960)., *Nature. Lond.*, 187, 248.

Barka, T., and Anderson, P., (1963)., *Histochemistry Theory and Practice and Bibliography*, New York: Harper and Row Publishers Inc. Hoeber Med. Div.

Barley, D.A., (1964)., *Turtox News*, 44, 298.

Barlow, R.M., (1957)., *J. Path. Bact.*, 73, 272.

Barr, M.L., and Bettram, E.G., (1949)., *Nature Lond.*, 163, 676.

Barr, M.L., Bettram, F., and Lindsay, A.R., (1950)., *Anat. Rec.*, 107, 283.

Barrett, A.M., (1944)., *J. Path. Bact.*, 56, 135.

Barrnett, R.J., and Roth, W.D., (1958)., *J. Histochem. Cytochem.*, 6, 406.

Barrnett, R.J. and Seligman, A.M., (1951)., *Science New York*, 114, 579.

__________ (1952)., *J. Natn. Cancer, Inst.*, 13, 215.

__________ (1953/1954)., *J. Natn. Cancer. Inst.*, 14, 769.

__________ (1958)., *J. Biophys. Biochem. Cytol.*, 4, 169.

Barros-Pita, J.C., (1971)., *Stain Tech.*, 46, 171.

Bartholomew, J.W., (1962)., *Stain Tech.*, 37, 139.

Bartholomew, J.W., and Finkelstein, H., (1959)., *Stain Tech.*, 34, 147.

Bartholomew, J.W., Lechtman, M.D., and Finkelstein, J., (1965)., *J. Bact.*, 90, 1146.

Bartholomew, J.W., and Mittwer, T., (1950)., *Stain. Tech.*, 25, 103.

__________ (1951)., *Stain. Tech.*, 26, 231.

Bauer, H., (1932)., *Z. Zellforsch.*, 15, 225.

__________ (1933)., *Z. Mikros. Anat. Forsch.*, 33, 143.

Baylis, W.M., (1906)., *Biochem. J.*, 1, 175.

Bayliss, O.B., (1981)., *Acta. Histochem. Suppl. Band.*, XXIV., 5, 247.

Bayliss, O.B., and Adams, C.W.M., (1972)., *Histochem. J.*, 4, 505.

Beamer, P.R. and Firminger, H.L., (1955)., *Lab. Invest.*, 4, 9.

Beckel, W.E., (1959)., *Nature.*, 184, 1584.

Becker, E.R., and Roudabush, R.L., (1935)., *Brief Directions in Histological Technique.* Ames, Iowa: Collegiate Press Inc.

Becker, S.W., Draver, L.L., and Thatcher, H., (1935)., *Arch. Dermat. Syph.*, 31, 190.

Beckert, W.H., and Garner, J.G., (1966)., *Stain Tech.*, 41, 141.

Beek, R.M., (1955)., *El. Aliso.*, 3, 131.

Belanger, L.F., and Bois, P., (1964)., *Anat. Rec.*, 148, 573.

Belling, J., (1926)., *Biol. Bull. Woods Hole*, 50, 160.

Belt, W.D., and Hayes, E.R., (1956)., *Stain Tech.*, 31, 117.

Beltrami, C.A., Fabris, G., Marzola, A., Nency, L., Lanza, G.B., and Vezzandeni, P., (1975)., *Histochem. J.*, 7, 95.

Bendit, E.P., and Wong, R.L., (1957)., *J. Exp. Med.*, 105, 509.

Benhold, E., (1922)., *Minch. Med. Wschr.*, 2, 1537.

Bennett, H.S., and Watts, R.M., (1958)., *General Cytochemical Methods*, Ed. Danielli, J.F., New York, Academic Press.

Bensley, R.R., and Gersh, I., (1933)., *Anat. Rec.*, 57, 217.

Bergeron, J.A., and Singer, M., (1958)., *J. Biophysic. Biochem. Cytol.*, 4, 433.

Berrois, M., (1994)., *Biotechnic. Histochem.*, 2, 78.

Bert, F., (1966)., *Z. Wiss. Mikros.*, 23, 319.

Bertalanffy, L.von., and Bickis, I., (1956)., *J. Histochem. Cytochem.*, 4, 481.

Bertalanffy, F.D., von and Nagy, I., (1962)., *Med. Radiograph. Photograph.*, 38, 82.

Berube, G.R., Powers, M.M., and Clark, G., (1965)., *Stain Tech.*, 40, 53.

Berube, G.R., Powers, M.M., Kerkeny, J., and Clark, G., (1966)., *Stain Tech.*, 41, 73.

Best, I. (1906). *Z. Wiers. Mikrosk.*, 23, 319.

Bitensky, L., (1963)., *Quar. J. Micr. Sci.*, 104, 193.

Black, M.M., and Ansley, H.R., (1964)., *Science*, 43, 693.

Black, M.M., Ansley, H.R., and Mandla, R., (1964)., *Arch. Path.*, 78, 350.

Bloch, B., (1917)., *Arch. Dermat. Syph.*, 124, 129.

Block, M., Victoria, S., and Brown, J., (1953). *J. Lab. Clin. Med.*, 42, 145.

Bloor, W.R., (1925–1926)., *Chem. Rev.*, 2, 243.

Bodian, M., and Lake, B.D., (1963)., *British J. Surgery*, 50, 702.

Boelsma-Von Homte, E., (1965)., *Histochemie van Fosfolipiden in Verbrand met Atherosklerose van de aorta*. Thesis, Leiden.

Bondi, A., Chieregatti, G., Eusebi, V., Fulcheri, E., and Bussolati, G., (1982)., *Histochemistry*, 76, 153.

Borror, A.C., (1968)., *Stain Tech.*, 43, 293.

Bosch, R., (1966)., *Stain Tech.*, 41, 250.

Bourgeois, C., and Hubbard, B., (1965)., *J. Histochem. Cytochem.*, 13, 571.

Bowes, J.A. and Cator, C.W., (1966)., *J. Roy. Micr. Soc.*, 85, 193.

Boyd, G.A., (1955)., *Autoradiography in Biology and Medicine*. New York, Academic Press.

Brachet, J., (1940)., *c.r. Soc. Biol. Paris*, 133, 88.

——————— (1953)., *Quart. J. Micr. Sci.*, 94, 1.

Bradley, M.V., (1957)., *Stain Tech.*, 32, 85.

Brecher, G., (1949)., *Amer. J. Clin. Path.*, 19, 895.

Bridges, C.H., and Luna, L., (1957)., *Lab. Invest.*, 6, 357.

Briggs, N.L., (1958)., *Stain Tech.*, 33, 299.

Brinn, N., (1983)., *J. Histotech.*, 6, 125.

Brookes, L.D., (1968)., *Stain Tech.*, 43, 41.

Brown, J.H. and Brenn, L., (1931)., *Bull Johns Hopkins Hospital*, 48, 69.

Bruemmer, N., Carver, M.J., and Thomas, L.E., (1957)., *J. Histochem. Cytochem.*, 5, 140.

Bullock, G.R., (1984)., *J. Micros.*, 133, 1.

Bullock, G.R., and Petrusz, P. (eds.) (1982)., *Techniques in Immunocytochemistry*, Vol.1, New York, Academic Press.

__________ (1983)., *Techniques in Immunocytochemistry*, Vol.2, New York, Academic Press.

__________ (1985)., *Techniques in Immunocytochemistry*, Vol.3, New York, Academic Press.

Burrows, W., (1954)., *Text Book of Microbiology*. Philadelphia, W.B. Saunders Co.

Burstone, M.S., (1955)., *J. Histochem. Cytochem.*, 3, 32.

__________ (1957a)., *Amer. J. Clin. Path.*, 28, 429.

__________ (1957b)., *J. Nat. Cancer. Inst.*, 18, 167.

__________ (1957c)., *Arch. Path.*, 63, 164.

__________ (1958)., *J. Nat. Cancer. Inst.*, 6, 322.

__________ (1959)., *J. Histochem. Cytochem.*, 7, 112.

__________ (1960)., *J. Nat. Cancer. Inst.*, 24, 1199.

__________ (1962)., *Enzyme Histochemistry and its Application in the Study of Neoplasms.*. New York: Academic Press.

Burstone, M.S., and Folk, J.E., (1956)., *J. Histochem. Cytochem.*, 4, 217.

Burtner, H.J., and Lillie, R.D., (1949)., *Stain Tech.*, 24, 225.

Bussolati, G., and Bassa, T., (1974)., *Stain Tech.*, 49, 313.

Cain, A.J., (1947)., *J. Anat.*, 88, 467.

__________ (1948)., *Quart. J. Micr. Sci.*, 89, 229.

__________ (1950)., *Biol. Rev.*, 25, 73.

Cameron, M.L., and Steele, J.E., (1959)., *Stain Tech.*, 34, 265.

Capra, J.D., and Edmundson, A.B., (1977)., *Scientific American.*, 236, 50.

Carazzi, D., (1911)., *Z. Wis. Mickos. Microsk, Tech.*, 28, 273.

Carleton, H.M., and Leach, E.H., (1947)., *Histological Technique*. New York: Oxford University Press.

Carlo, R., (1964)., *J. Histochem. Cytochem.*, 12, 44.

Carr, K.B., Rambo, O.N., and Feichtmeir, T.V., (1961)., *J. Histochem. Cytochem.*, 9, 415.

Carr, D.H., and Walker, J.E., (1961)., *Stain Tech.*, 36, 233.

Carson, F., (1990)., *Histotechnology: A Self Instructional Text*. Chicago: ASC Press.

Carson, F.L., Martin, J.H., and Lynn, J.A., (1973)., *J. Clin. Path.*, 59, 365.

Cason, J.E., (1950)., *Stain Tech.*, 25, 225.

Cassel, W.A., and Hutchinson, W.G., (1955)., *Stain Tech.*, 30, 105.

Castaneda, M.R., (1939)., *Amer. J. Path.*, 15, 467.

Castel, P., (1936)., *Bull. Histol. Tech. Micr.*, 13, 106.

Cattoretti, G., Pileri, S., and Parravicini, C., (1993)., *J. Path.*, 171, 83.

Cerbulis, J., and Zittle, C.A., (1961)., *Analyt. Chem.*, 33, 1131.

Chambers, R.W., (1968)., *Arch. Path.*, 85, 18.

Chandley, A.C., (1988)., *Trends in Genetics*, 4, 79.

Chang, J.P., and Hori, S.H., (1961)., *J. Histochem. Cytochem.*, 9, 292.

Chapman, D.M., (1975)., *Stain Tech.*, 50, 25.

__________ (1977)., *Can. J. Med. Tech.*, 39, 65.

Chatton, E., and Lwoff, A., (1930)., *Compt. Rendez. Soc. Biol.*, 104, 834.

__________ (1935)., *Arch. De Zool. Exptl. Gen.*, 77, 1.

__________ (1936)., *Bull. Soc. Frenq. Micros.*, 5, 25.

Chayen, J. and Bitensky, L. (1991)., *Practical Histochemistry*, 2nd edn., Chichester: Wiley.

Chen, T., (1944)., *Stain Tech.*, 19, 83.

Childs, G.V. Ed. (1986)., *Immunocytochemical Technology*, New York: Alan R. Liss.

Chiquoine, A.D., (1954)., *J. Comp. Neurol.*, 100, 415.

__________ (1955)., *J. Histochem. Cytochem.*, 3, 171.

Christeller, E., (1927)., *Verh. Deutsch. Path. Ges.*, 22, 173.

Chu, C.H.U., Fogelson, M.A., and Swinyyard, C.A., (1953)., *J. Histochem. Cytochem.*, 1, 39.

Chubb, J.C., (1963)., *Sain Tech.*, 37, 179.

Churikian, C.J., (1993a)., *Histo-Logic.*, 1, 13.

__________ (1993b)., *Manual of Special Stains Laboratory of the University of Rochester Medical Centre,* Rochester. New York.

Churukian, C.J., and Schenk, E.A., (1976)., *Stain Tech.*, 51, 213.

__________ (1979)., *J. Histotech.*, 1, 102.

Ciacio, C., (1906)., *C.R. Soc. Biol.*, 60, 70.

Clara, M., (1935)., *Z. Zellforsch.*, 22, 318.

Clark, G., (1971)., Unpublished Material.

Clark, G., and Allard, L., (1983)., *History of Staining*, Baltimore: Williams and Wilkins.

Clark, G., Reed, C.S., and Brown, F.M., (1973)., *Stain Tech.*, 48, 189.

Clark, G., and Sperry, M., (1945)., *Stain Tech.*, 20, 23.

Clark, S.K., and Ward, J.W., (1934)., *Stain Tech.*, 9, 53.

Clark, W.M., and Lubs, A., (1917)., *J. Bact.*, 2, 1.

Clarke, G., (1953)., *Amer. J. Clin. Path.*, 24, 113.

Clayden, E.C., (1952)., *J. Med. Lab. Tech.*, 10, 103.

__________ (1955)., *Practical Cutting and Staining*, 3rd edn. London, J & A Churchill Ltd.

Clegg, J.A. (1965). *Ann. NY Acad. Sci.*, 18, 969.

Cocke, E.C., (1938)., *Science*, 87, 443.

Cohen, A.S., Calkins, E., and Levine, C., (1959)., *Amer. J. Path.*, 35, 971.

Cohen, I., (1949)., *Stain Tech.*, 24, 177.

Cole, E.C., (1943)., *Stain Tech.*, 18, 125.

Conn, H.J., (1946)., *Cuba Symposia*, 7.

__________ (1969)., *Biological Stains*, 8th edn. (ed.) Lillie, R.D., Baltimore, Williams & Wilkins.

Conroy, J.D., and Teledo, A.B., (1976)., *Vet. Path.*, 13, 78.

Cook, H.C., (1961)., *J. Med. Lab. Tech.*, 18, 3.

__________ (1972)., *Human Tissue Mucins* (Laboratory Aid Series), London, Butterworths.

__________ (1974)., *Manual of Histological Demonstration Techniques*, London, Butterworths.

Cook, S.F., and Erzacolin, H.E., (1962)., *J. Histochem. Cytochem.*, 10, 560.

Cordell, J.L., Falini, B., Erber, W.Wn., Abdulaziz, Z., McDonald, S., Purford, K.A.F., Ghosh, A.K., Stein, H. and Mason, D.Y. (1984)., *J. Histochem. Cytochem.*, 32, 219.

Corliss, J.O., (1953)., *Stain Tech.*, 28, 97.

Cort, W.W. (1912). *Trans. Amer. Micros. Soc., 31, 151.*

Courtwright, R.C., (1966), *Trans. Amer. Micros. Soc.*, 85, 319.

Cowdry, E.V., (1952)., *Laboratory Technique in Biology and Medicine*, 3rd edn., Baltimore, Williams & Wilkins & Co.

Crabb, E.D., (1949)., *Stain Tech.*, 24, 87.

Cramer, A.D., Rogers, E.R., Parker, J.W., and Lukes, R.J., (1973)., *Amer. J. Clin. Path.*, 60, 148.

Crevier, M., and Belanger, L.R., (1955)., *Science*, 122, 556.

Crosier, R.H., (1968)., *Stain Tech.*, 43, 171.

Crout, J.R., and Jennings, R.B., (1957)., *J. Histochem. Cytochem.*, 5, 170.

Crow, J., Gibbs, D.A., Cozens, W., Spellacy, E., and Watts, R.W.E., (1983)., *J. Clin. Path.*, 36, 415.

Csaba, G., (1969)., *Acta. Biol. Acad. Sci. Hungaricae.*, 20, 205.

Culling, C.F.A., (1949)., *Bull. Inst. Med. Lab. Tech.*, 14, 93.

__________ (1957)., *Handbook of Histopathological Technique*, London, Butterworth.

Culling, C.F.A., Reid, P.A.B., and Dunn, W.L., (1975)., *J. Histochem. Cytochem.*, 24, 1225.

Curran, R.C., (1961)., *Biochem. Soc. Symp.*, 20, 24.

Curran, R.C., and Collins, D., (1957)., *J. Path. Bact.*, 74, 207.

Curran, R.C., and Gregory,J., (1977)., *Experientia*, 33, 1400.

Curran, R.C., and Kennedy, J., (1955)., *J. Path. Bact.*, 70, 449.

Dahl, L.K., (1952)., *Proc. Soc. Exp. Biol., N.Y.*, 80, 474.

Dane, E.T., and Herman, D.L., (1963)., *Stain Tech.*, 38, 97.

Danielli, J.F., (1947)., *Sympos. Soc. Exp. Biol.*, 1, 101.

—————— (1950)., *Cold. Spr. Harb. Symp. Quant. Biol.*, 14, 32.

Daoust, R., (1964)., *J. Histochem. Cytochem.*, 12, 640.

Darnell, J., Lodish, H., and Baltimore, D., (1994)., *Molecular Cell Biology*, 3rd edn. New York: W.H. Freeman and Co.

Dart, L.H., and Turner, T.R., (1959)., *Lab. Invest.*, 8, 1513.

Davenport, H.A., (1960)., *Histological Histochemical Technique*, Philadelphia, Saunders.

Davenport, H.A., Windle, W.F., and Buch, R.H., (1934)., *Stain Tech.*, 9, 5.

Davis, B.J., and Ornstein, L., (1959)., *J. Histochem. Cytochem.*, 7, 297.

Dawar, B.L., (1973)., *Stain Tech.*, 48, 93.

Dayar, M.R., (1947)., *J. Bact.*, 43, 498.

Deden, C., (1954)., *Acta. Radiol. Suppl.*, 115, 1.

DeGalantha, E., (1935)., *Amer. J. Clin. Path.*, 5, 165.

DeMay, J., Hacker, G.W., DeWaele, M., and Springall, D.R., (1986)., "Gold Probes in Light Microscope". In: *Immunocytochemistry Modern Methods and Applications*, (eds.) Polak, J.H., and Van Noorden, S., Bristol: Wright, 71–89.

Demke, D.D., (1951)., *Stain Tech.*, 27, 135.

Dempster, W.T., (1960)., *J. Amer. Anat.*, 107, 59.

Denz, F.A., (1949)., *Quart. J. Micr. Sci.*, 90, 317.

DeRobertis, E., and Grasso, R., (1946)., *Endocrinology*, 38, 137.

Desbordes, J., Fournier, E., and Guyotjeannin, C., (1952)., *Ann. Inst. Pasteur (Paris)*, 83, 268.

Dickman, S.R., and Crockett, A.L., (1956)., *J. Biol. Chem.*, 220, 951.

Dienes, L., (1967)., *J. Bact.*, 93, 689.

Dienst, R.B., and Sanderson, E.S., (1936)., *Amer. J. Public Health*, 26, 910.

Dietch, A.D., (1961)., *J. Histochem. Cytochem.*, 9, 477.

Dixon, K.C., (1959)., *Amer. J. Path.*, 35, 199.

—————— (1962)., *Quart. J. Exp. Physiol.*, 47, 1.

Dixon, M., and Webb, E.C., (1958)., *Enzymes*, Longmans: London.

Dorling, J., (1980)., *J. Clin. Path.*, 33, 897.

Dotti, L.B., Paparo, G.P., and Clarke, E., (1951)., *Amer. J. Clin. Path.*, 21, 475.

Duguid, J.P., (1951)., *J. Path. Bact.*, 63, 673.

Dunn, R.C., (1946)., *Arch. Path.*, 41, 676.

Dunn, R.C., and Thompson, E.C., (1945)., *Arch. Path.*, 39, 49.

—————— (1946)., *Stain Tech.*, 21, 65.

Dyer, A.F., (1963)., *Stain Tech.*, 38, 85.

Eager, R.P., (1970)., *Brain Res.*, 22, 137.

Eastwood, H., and Cole, K.R., (1971)., *Stain Tech.*, 46, 208.

Ehrenrich, T., and Kerpe, S., (1959)., *J. Amer. Med. Assoc.*, 170, 94.

Ehrlische, P., (1888)., *Z. Wis. Meckro. Mikrop. Tech.*, 3, 150.

Einarson, L., (1932)., *Amer. J. Path.*, 8, 295.

—————— (1951)., *Acta. Path. Micro. Scandinaveca*, 28, 82.

Elftman, H., (1952)., *Stain Tech.*, 27, 47.

—————— (1954)., *J. Histochem. Cytochem.*, 2, 1.

—————— (1957)., *Stain Tech.*, 32, 29.

—————— (1959a)., *J. Histochem Cytochem.*, 7, 93.

__________ (1959b)., *J. Histochem Cytochem.*, 7, 98.

Elias, J.M., (1969)., *Stain Tech.*, 44, 201.

__________ (1990)., *Immunohistopathology. A Practical Approach to Diagnosis*, Chicago, ASCP Press.

Ellis, I.O., Bell, J., and Bancroft, J.D., (1988)., *J. Histochem. Cytochem.*, 36, 121.

Ellender, M., and Lojda, Z., (1973)., *Histochemie.*, 36, 149.

Epple, A., (1967)., *Stain Tech.*, 42, 53.

Epple, A., and Brinn, J.E., (1986)., *Pancreatic Islets. Vertebrate Endocrinology; Fundamentals and Biomedical Implications*, Pang, P.K.T. and Schreibman, M.P., (eds.), Academic Press: New York.

Eranko, O., and Palkama, A., (1961)., *J. Histochem. Cytochem.*, 9, 585.

Ericsson, J.L.E., (1965)., *Lab. Invest.*, 14, 1.

Erribabu, D., Shyamasundari, K., and Hanumantha Rao, K., (1979)., *Z. Zellforsch.*, 93, 1085.

__________ (1980a)., *Ind. J. Exp. Biol.*, 18, 263.

__________ (1980b)., *Folia. Histochem. Cytochem.*, 18, 42.

__________ (1985)., *J. Exp. Mar. Biol. Ecol.*, 88, 120.

Evans, E.P., Breckon, G., and Ford, C.E., (1964)., *Cytogenetics*, 3, 289.

Everett, M.M., and Miller, W.A., (19974)., *Histochem. J.*, 6, 25.

Ewen, A.B., (1962)., *Trans. Amer. Micr. Soc.*, 81, 94.

Faulk, W., and Taylor, G., (1971)., *Immunochemistry*, 8, 1081.

Faulkner, R.R., and Lilie, R.D., (1945)., *Stain Tech.*, 20, 81.

Faust, E.C., Russell, P.F., and Jung, R.C., (1970)., *Clinical Pathology*, 4th edn. Philadelphia: Lea and Febiger.

Feder, N., and Sidman, R.L., (1950)., *J. Histochem. Cytochem.*, 6, 401.

Feigl, F., (1954)., "Spot Tests", Vol.1, *Inorganic Applications*. Elsevier, Amsterdam.

Fernando, J.C., (1961)., *J. Insect. Sci. Tech.*, 7, 1.

Feulgen, R., (1914)., *Zetschrift. Physiol. Cheme.*, 92, 154.

Feulgen, R., and Rossenbeck, H., (1924)., *Z. Physiol. Chem.*, 135, 203.

Feulgen, R., and Vorr, K., (1924)., *Pflugers. Arch. Ges. Physiol.*, 206, 389.

Field, J.W., (1941)., *Trans. Roy. Soc. Trop. Med. Hyg.*, 35, 35.

Filipe, M.I., and Lake, B.D., Eds. (1983)., *Histochemistry in Pathology*, Edinburgh: Churchill, Livingstone.

Fischler, C., (1904)., *Zbl, allg. Path. Anat.*, 15, 913.

Fishman, W.H., and Baker, J.R., (1956)., *J. Histochem. Cytochem.*, 4, 570.

Fite, G.I., (1938)., *Amer. J. Path.*, 14, 491.

__________ (1940)., *J. Lab. Clin. Med.*, 25, 743.

Flitney, F.W. (1966)., *J. Roy. Micr. Soc.*, 85, 353.

Folin, O., and Ciocalteu, V., (1927)., *J. Biol. Chem.*, 73, 627.

Folin, O., and Marenzi, A.D., (1929)., *J. Biol. Chem.*, 83, 89.

Fontana, A., (1926)., *Dermatol.*, 2, 46, 291.

Francis, C.M., (1953)., *Nature, London*, 171, 701.

Frankel, H.H., and Peters, R.L., (1964)., *Amer. J. Clin. Path.*, 42, 324.

Frankel, J., and Heckmann, K., 1968)., *Trans. Amer. Micr. Soc.*, 87, 317.

French, D., and Edsall, J.T., (1945)., *Adv. Protein. Chem.*, 2, 277.

Fullmer, H.M., and Lillie, R.D., (1956)., *Stain Tech.*, 31, 27.

Gabe, M., (1976)., *Histological Techniques*, Paris, Masson/New York, Springer-Verlag.

Galigher, A.E., (1934)., *The Essentials of Practical Microtechnique*, Privately published.

Galigher, A.E., and Kozloff, E.N., (1964)., *Essentials of Practical Microtechnique*, Philadelphia: Lea and Febiger.

Garvey, W., Fathi, A., Bigelow, F., Jimenez, C., and Carpenter, B. (1987)., *J. Histotech.*, 10, 245.

__________ (1990)., *J. Histotech.*, 13, 279.

__________ (1991)., *J. Histotech.*, 14, 39.

Gatenby, J.B., Beams, H.W., (1950)., *The Microtomists*, Vade-Mecum, London, J & A Churchill.

Gatter, K.C., Falini, B., and Mason, D.Y., (1984)., "The use of monoclonal antibodies in histopathological diagnosis"., In: *Recent Advances in Histopathology*, 12, Antony, P.P. and MacSween, R.N.M. (eds.), Edinburgh, Churchill, Livingstone, pp.35–67.

Gelei, J.V., (1932)., *Arch fur. Protistenkunde*, 77, 219.

__________ (1935)., *Arch fur. Protistenkunde*, 84, 446.

George, J.C., Ambadkar, P.M., (1963)., *J. Histochem. Cytochem.*, 11,420.

George, J.C., and Iype, P.T., (1960)., *Stain Tech.*, 35, 151.

Gerebtzoff, M.A., (1959)., *Cholinesterases*, Oxford, Perman.

Germain, J., and Stevans, A., (1996)., "Resin embedding media" In: *Theory and Practice of Histological Technique*, Eds. Bancroft, J.D. and Stevans, A., Churchill, Livingstone.

Gersch, I., (1932)., *Anat. Rec.*, 53, 309.

Gershstein, L.M., (1958)., In *Histochemical Methods in Normal and Pathological Morphology*, eds. Purtugalov, V.V., and Strukob, A.L., Moskva.

Geyer, G., (1962)., *Acta. Histochem.*, 14, 307.

__________ (1965)., *Acta. Histochem.*, 20, 130.

Gill, G.W., Frost, J.K., and Miller, K.A., (1974)., *Acta. Cytologia.*, 18, 300.

Glauert, A.M., and Glauert, R.H., (1958)., *J. Biophys. Biochem. Cytol.*, 4, 191.

Glauert, A.M., Rogers, G.E., and Glauert, R.H., (1956)., *Nature (London)*, 178, 803.

Glenner, G.G., (1957)., *Amer. J. Clin. Path.*, 27, 1.

Glenner, G.G., Bartner, H.J. and Brown, G.W., (1957). *J. Histochem. Cytochem.*, 5, 59.

Glenner, G.G., Keiser, H.R., Bladen, H.A., Cuatrecasas, P., Eanes, E.D., Ram, J.S., Hanfer, J.N. and Delellis, R.A., (1968)., *J. Histochem. Cytochem.*, 16, 633.

Glenner, G.G., and Lilie, R.D., (1957a)., *J. Histochem. Cytochem.*, 5, 279.

__________ (1957b)., *Stain Tech.*, 32, 187.

__________ (1959)., *J. Histochem. Cytochem.*, 5, 279.

Glick, D., (1949)., *Techniques of Histo- and Cytochemistry*, New York, Interscience Publishers.

Goad, A., and Sylven, (1969)., *Brit. J. Cancer.*, 23, 52.

Goldman, M., (1949)., *Stain Tech.*, 24, 57.

Goldschmidt, R., (1909). *Zool. Anz.*, 34, 481.

Goldstein, D.J. and Horobin, R.W., (1974)., *Histochem. J.*, 6, 157.

Gomori, G., (1936)., *Amer. J. Path.*, 12, 655.

__________ (1939)., *Anat. Rec.*, 74, 439.

__________ (1941a)., *Arch. Path.*, 32, 169.

__________ (1941b)., *Amer. J. Path.*, 17, 395.

__________ (1941c)., *J. Cell. Comp. Physiol.*, 17, 71.

__________ (1945)., *Proc. Soc. Exp. Biol, N.Y.*, 58, 362.

__________ (1946)., *Tech. Bull. Reg. Med. Tech.*, 7, 115.

__________ (1948)., *Proc. Soc. Exp. Biol. N.Y.*, 68, 354.

__________ (1950a)., *Amer. J. Clin. Path.*, 20, 665.

__________ (1950b)., *Amer. J. Clin. Path.*, 20, 662.

__________ (1950c)., *Amer. J. Clin. Path.*, 20, 665.

__________ (1951)., *J. Lab. Clin. Med.*, 37, 526.

__________ (1951a)., *Methods in Medical Research*, 4, 1, Chicago.

__________ (1952a)., *Microscopic Histochemistry*, Chicago, University of Chicago Press.

__________ (1952b)., *Int. Rev. Cytol.*, 1, 323.

__________ (1956)., *J. Histochem. Cytochem.*, 4, 453.

Gordon, H., and Sweets, H.H.Jr., (1936)., *Amer. J. Path.*, 12, 545.

Gowali, F.M., (1995)., *Lab. Med.*, 26, 476.

Gower, W.C., (1939)., *Stain Tech.*, 14, 31.

Gradwohl, R.B.H., (1963)., *Clinical Laboratory Methods and Diagnosis.* II. St. Louis, C.V. Mosby Graupner, H. and Weissberger, A.

Graff, S., (1916)., *Zeutbl. Path.*, 27, 318.

Graham, R.C.Jr. and Kornovsky, M.J., (1966)., *J. Histochem. Cytochem.*, 14, 291.

Graham, R.C.Jr., Ladholm, U., and Karnovsky, M.J., (1965)., *J. Histochem. Cytochem.*, 13, 150.

Gray, P. (1952)., *Handbook of Basic Microtechnique*, New York, The Blakinston Co.

__________ (1954)., *The Microtomists Formulary and Guide*, New York, The Blakiston Co.

__________ (1964b)., *Handbook of Basic Microtechnique*, 3rd edn. New York, Blakiston Co.

Gray, P.H.H., (1926)., *J. Bact.*, 12, 273.

Gray and Wess (1952)

Green, F., (1990)., *Sigma-Aldrich handbook of Stains*, St. Louis, Sigma.

Greenblatt, R.A., and Manautou, J.M., (1957)., *Amer. J. Obste. Gynaec.*, 74, 629.

Greep, R.O., Fisher, C.J., and Morse, A., (1948)., *J. Amer. Dent. Assoc.*, 36, 427.

Gridley, M.F., (1951)., *Amer. J. Clin. Path.*, 21, 897.

__________ (1953)., *Amer. J. Clin. Path.*, 23, 303.

__________ (1957)., *Manual of Histologic and Special Staining Techniques*, Armed Forces Institute of Pathology, Washington, D.C.

Griffin, L.E., and McQuarrie, A.M., (1942)., *Stain Tech.*, 17, 41.

Grimelius, L., (1968)., *Acta. Soc. Med. Ups.*, 73, 243.

Groat, R.A., (1949)., *Stain Tch.*, 24, 157.

Grocott, R.G., (1955)., *Amer. J. Clin. Path.*, 25, 975.

Grogg, E., and Pearse, A.G.E., (1952)., *Nature, London*, 170, 578.

Gross, M., (1952)., *J. Clin. Path.*, 22, 1034.

Grundland, I., Bulliard, H., and Maillet, M., (1949)., *C.R. Soc. Biol. Paris*, 143, 771.

Guard, H.R., (1959)., *Amer. J. Clin. Path.*, 32, 145.

Guesdon, J.L., Terynick, T., and Avrameas, S., (1979)., *J. Hisochem. Cytochem.*, 27, 1131.

Guillery, R.W., Shira, B., and Webster, K.G., (1961)., *Stain Tech.*, 36, 9.

Gurr, E., (1956)., *A Practical Manual of Medical and Biological Staining Techniques.* New York, Interscience.

__________ (1958)., *Methods of Analytical Histology and Histochemistsry*, London, Leonard Hill.

__________ (1969)., *Encyclopaedia of Microscopic Stains.* London: Edward Gurr, Baltimore Williams and Wilkins.

Haddock, N.K., (1948)., *Research,* 1, 685.

Hafiz, S., Spencer, R.C., Lee, M., Gooch, H., and Duerden, B., (1985)., *J. Clin. Path.*, 38, 1073.

Hajian, A., (1961)., *Tech. Bull. Reg. Med. Tech.*, 31, 92.

Hale, G.W., (1946)., *Nature, London.*, 157, 802.

Hale, L.J., (1958)., *Biological Laboratory Data Methuen's Monographs.* London, Methuen and Co., New York, John Wiley & Sons Inc.

Hall, M.J., (1960)., *Amer. J. Clin. Path.*, 34, 313.

Halmi, N.S., (1950)., *Endocrinology*, 47, 289.

Halnan, C., (1989)., *Cytogenetics of Animals.* Wallington, United Kingdom: CAB International.

Hammerton, J.L., (1961)., *Sex Chromatin and Human Chromosome. Intern. Reg. Cytol.*, 12, Eds. Bourne, G.H. and Danielle, J.F., Academic Press, New York.

Hammett, F.S. and Chapman, S.S., (1938)., *J. Lab. Clin. Med.*, 24, 293.

Hammond, G.F., and Beckman, B., (1978)., *Stain Tech.*, 53, 13.

Hance, R.T., and Green, F.J., (1961)., *Stain Tech.*, 36, 253.

Hancox, N.M., (1957)., *Expt. Cell. Res.*, 13, 263.

Hanker, J.S., Anderson, W.E., and Bloom, F.E., (1971)., *Science, New York.*, 175, 991.

Hanker, J.S., Yates, P.E., Metz, C.B., and Rustini, A., (1977)., *J. Histochem.*, 9, 789.

Hanumantha Rao, K., (1959a)., *J. Parasit.*, 45, 347.

——————— (1959b)., *Experientia*, 25, 464.

——————— (1960)., *Parasitology*, 50, 349.

——————— (1963)., *Parasitology*, 52, 1.

Hanumantha Rao, K., and Murthy, A.S., (1972)., *Stain Tech.*,17, 163.

Hanumantha Rao, K. and Shyamasundari, K., (2003)., *Adhesives from Marine Mussels: Possible Biotechnological Applications*, pp. 349–382, Eds. Fingerman, M. and Nagabushanam, R., Oxford University Press and IBH Publications Ltd., New Delhi.

Hanumantha Rao, K., Trinadha Babu, B., and Shyamasundari, K., (1989)., *Curr. Sci.*, 58, 213.

Happerfield, L.C., Bobrow, L.G., Bains, R.M., and Miller, K.D., (1993)., *J. Histochem. Cytochem.*, 14, 291.

Harada, K., (1957)., *Stain Tech.*, 32, 183.

——————— (1973)., *Stain Tech.*, 48, 269.

Harris, H.F., (1900)., *J. Appl. Micr. Lab. Meth*, 3, 777.

Harris, M.B.K., (1930)., *Science*, 72, 275.

Hartz, P.H., (1947)., *Amer. J. Clin. Path.*, 17, 750.

Hayat, M.A., (1989)., *Principles and Techniques of Electron Microscopy. Biological Application*, 2nd edn, MacMillan, London.

Heady, J., and Rogers, T.G., (1962)., *Iowa. Acad. Sci. Proc.*, 69, 587.

Heath, I.D., (1961), *Nature, London*, 191, 1370.

Heidenhain, M., (1896)., *Z. Wis. Mikros. Micros. Tech.*, 13, 186.

Hellerstorm, C., and Hellman, B., (1960)., *Acta. Endocrinol.*, 35, 518.

Hellman, B., Hellerstorm, C., (1960)., *Z. Zellforsch.*, 42, 278.

Herlant, M., (1960)., *Bull. Micr. Applique*, 10, 37.

Hess, M., and Hollander, F., (1947)., *J. Lab. Clin. Med.*, 32, 905.

Hess, R., Scarpelli, D.G., and Pearse, A.G.E., (1958)., *Nature, London*, 181, 1531.

Hicks, J.D. and Mutthaei, E., (1958)., *J. Path. Bact.*, 75, 473.

Highman, B., (1946)., *Arch. Path.*, 41, 559.

Hillarp, N.A., and Hokfelt, B., (1955)., *J. Histochem. Cytochem.*, 3, 1.

Hillarp, N.A., Hokfelt, B., and Nilson, B., (1954)., *Acta Anat.*, 21, 155.

Hilleman, H.H., and Lee, C.H., (1953)., *Stain Tech.*, 28, 285.

Hirano, A., and Zimmerman, H.M., (1962)., *Arch. Neurol.*, 6, 114.

Holczinger, L., (1959)., *Acta. Histochem.*, 8, 167.

Holde, P., and Isler, H., (1958)., *J. Histochem. Cytochem.*, 6, 265.

Holgate, C., Jackson, P., Cowen, P., and Bird, C., (1983)., *J. Histochem. Cytochem.*, 36, 742.

Hollander, D.H., (1963)., *J. Histochem. Cytochem.*, 11, 118.

Holmes, W., (1943)., *Anat. Rec.*, 86, 157.

Holt, S.J., (1954)., *Proc. Roy. Soc. Series*, B142, 160.

——————— (1959)., *Exp. Cell. Res. Suppl.*, 7, 1.

Holt, S.J., and Hicks, R.M., (1961)., *J. Biophys. Biochem. Cytol.*, 11, 31.

Holt, S.J., Hobbiger, E.E., and Pawan, G.L.S., (1960)., *J. Biophy Biochem. Cytol.*, 7, 383.

Holt, S.J., and Withers, R.F.J., (1952)., *Nature, London*, 170, 1012.

Holzer, W., (1921)., *Neural, Psychiat*, 69, 354.

Hopsu-Havu, V.K., Arstila, A., Helminan, H.J., Kalimo, H.O., nd Glenner, G.G., (1967)., *Histochemie*, 8, 54.

Hopwood, D., (1967)., *J. Anat.*, 101, 83.

——————— (1968)., *J. Anat.*, 103, 581.

——————— (1969)., *Histochem. J.*, 1, 323.

Hopwood, D., and Milne, G., (1991)., In: Harris, J.R. (ed.) *Electron Microscopy, A Practical Approach*. Oxford, IRL Press.

Hori, S.H., (1963)., *Stain Tech.*, 38, 221.

Hori, S.H., and Chang, J.P., (1963)., *J. Histochem. Cytochem.*, 11, 115.

Horobin, R.W., (1982)., *Histochemistry: An explanatory outline of histochemistry and Biophysical staining*. Stuttgart, Fischer &London, Butterworths.

__________ (1988)., *Outstanding Histochemistry Selection, Evaluation and Design of Biological Stains*. Horwood, Chichester.

__________ (1990)., "An overview of the theory of staining", In: *The Theory and Practice of Histological Techniques*, Ed. Bancroft, J.D.

__________ (1996)., "The theory of staining and its practical implication", In: *Theory and Practice of histological techniques*, Eds. Bancroft, J.D. and Stevans, A., 4th edn., Churchill Livingstone.

Horobin, R.W., Flemming, L., and Kevill-Davies, I.M., (1974)., *Stain Tech.*, 49, 207.

Horton, W.A., (1984).,, *Ckollagen. Related Res.*, 4, 231.

Horton, W.A., Dwyer, C., Goering, R. and Dean, D.C. (1983)., *J. Histochem. Cytochem.*, 31, 417.

Hotchkiss, R.D., (1948)., *Arch. Biochem.*, 16, 131.

Hsu, S.M., Raine, L., and Fanger, H., (1981)., *J. Histochem. Cytochem.*, 29, 577.

Huang, S., Minassian, H., and Moore, J.D., (1976)., *Lab. Invest.*, 35, 383.

Hubschman, J.H., (1962)., *Stain Tech.*, 37, 379.

Hueck, W., (1912)., *Beitr. Path. Anat.*, 54, 68.

Hughesdon, P.E., (1949)., *J. Roy. Micro. Soc.*, 69, 1.

Hukill, P., and Putt, A., (1962)., *J. Histochem. Cytochem.*, 11, 490.

Humphrey, A.A., (1935)., *Arch. Path.*, 20, 256.

Hunt, S.P., Allanson, J., and Mantyh, P.W., (1986)., "Radioimmunochemistry", In: *Immunochemistry modern methods and applications*, 2nd edn. Polak, J.M., and Van Noorden, S. (eds.), Bristol, Wright, pp. 99–114.

Hupt, A.W., (1930)., *Stain Tech.*, 5, 97.

Hutchinson, H.E., (1953)., *Blood*, 8, 236.

Hyashi, M., Nakajama, J., and Fishman, W.H., (1964)., *J. Histochem. Cytochem.*, 12, 293.

Incstar Corp (1989)., *Instructions to Immunocytochemistry*, Stillwater, M.N., Incstar Corp.

Irwin, D.A., (1955)., *Arch. Indust. Health.*, 12, 218.

Jagatic, J., and Weiskopf, R., (1966)., *Arch. Path.*, 82, 430.

Jalaja Kumari, C., Hanumantha Rao, K., and Shyamasundari, K., (1980)., *J. Zootomy*, 21, 67.

Jasani, B., and Schmid, K.W., (1993)., *Immunocytochemistry in Diagnostic Pathology*. Edinburgh: Churchill: Livingston.

Jasani, B., Thomas, N.D., Newman, G.R., and Williams, E.D., (1983)., *Immunological Communications*, 12, 51.

Jasani, B., Wyunford-Thomas, D., and Williams, E.D., (1981)., *J. Clin. Path.*, 34, 1000.

Jeanloz, R.W., (1960)., *Arthritis Rhem.*, 3, 223.

Johnson, T.J.A., (1985)., *J. Electron. Micro. Tech.*, 2, 129.

Johri, G.N., and Smyth, J.D., (1956)., *Parasitology*, 46, 107.

Jorpes, J.E. and Gardell, S., (1948)., *J. Biol. Chem.*, 176, 267.

Kallman, J., (1971)., *Stain Tech.*, 46, 210.

Kameswaramma, A.L., Shyamasundari, K., and Hanumantha Rao, K., (1987)., *Natn. Symp. Physiol. Crust.*, 105, 110.

__________ (1990)., *Rev. Idro. Biol.*, 3, 781.

Kaplow, L.S., (1963)., *Amer. J. Clin. Path.*, 37, 439.

__________ (1975)., *Amer. J. Clin. Path.*, 63, 451.

Kaplow, L.S., and Burstone, M.S., (1964)., *J. Hisochem. Cytochem.*, 12, 805.

Karnovsky, M.J., (1965)., *J. Cell. Biol.*, 27, 137A.

Kasten, F.H., (1957)., *Stain Tech.*, 33, 39.

Keilig, L., 1964)., *Vinchows. Arch. Path. Anat. Physiol.*, 312, 405.

Kelly, J.W., Morgan, P.N., and Saini, N., (1962)., *Arch. Path.*, 73, 70.

Kenney, M., Dyckman, J., and Aronson, S.M., (1970)., *Stain Tech.*, 46, 160.

Kerr, D.A., (1938)., *Amer. J. Clin. Path. Suppl.*, 2, 63.

Kessel, J.F., (1925)., *China. Med. J.*, 1, 57.

Kidder, G.W., (1933)., *Arch. Fur. Protisten.*, 79, 1.

Kiernan, J.A., (1990)., *Histological and Histochemical Methods.* 2nd edn. Oxford: Pergamon Press.

_________ (1996)., *Biotech. Histochem.*, 71, 304.

Kimbal, R.F., and Perdue, S.W., (1955)., *Exp. Cell. Res.*, 27, 405.

Kimmel, D., and Jee, W.S.S., (1975)., *Stain Tech.*, 50, 83.

Kiossoglou, K.A., Wolman, I.J., and Garrison, M.J., (1963)., *Blood.*, 21, 553.

Kirkham, N., and Hall, P. (eds.)., (1995)., *Progress in Pathology*, Edinburgh: Churchill-Livingston.

Kirby, H., (1947)., *Methods in the study of Protozoa.* University of California Syllabus Series. Berkely, University of California.

Kishore, B., Shyamasundari, K., and Hanumantha Rao, K., (1988a)., *Folia. Morphol.*, XXVI, 59.

_________ (1988b)., *Folia. Morphol.*, XXXVI, 107.

_________ (1988c)., *Rev. Iber. Parasitol.*, 46, 145.

_________ (1988d)., *Rev. Parasitol.*, VXLIX, 123.

_________ (1990)., *Ind. J. Parasitol.*, 13, 91.

Klein, B.M., (1926)., *Zool. Anz.*, 67, 160.

Klieneberger-Nobel, E., (1962)., *Pleuropneumonia like Organisms (PPLO) Mycloplasmataceae.* New York, Academic Press.

Klinger, H.P., (1958)., *Exp. Cell. Res.*, 14, 207.

Klinger, H.P., and Hammond, D.O., (1971)., *Stain Tech.*, 46, 43.

Klinger, H.P., and Ludwig, K.S., (1957)., *Stain Tech.*, 32, 235.

Klüver, H., Barrera, E., (1953)., *J. Neuropath. Exp. Neurol.*, 12, 400.

Koelle, G.B., and Friedenwald, J.S., (1949)., *Proc. Soc. Exp. Biol. N.Y.*, 70, 617.

Koenig, H., (1963)., *J. Histochem. Cytochem.*, 11, 120.

Kok, L.P., and Boon, M.E., (1992)., *Microwave Cookbook for Microscopists. Art and Science of Visualization.* Leiden, Netherlands: Coulomb Press.

Koneff, A., (1938)., *Stain Tech.*, 13, 49.

Koplow, L.S., (1975)., *Amer. J. Clin. Path.*, 63, 451.

Kornhauser, S.I., (1930)., *Stain Tech.*, 5, 13.

_________ (1945)., *Stain Tech.*, 20, 33.

Koski, J.P., and Reyes, P.F., (1986)., *J. Histotech.*, 9, 265.

Koulischer, L., and Mulnard, J., (1962)., *Lancet*, I, 917.

Krajian, A.A., (1939)., *Amer. J. Symphilol.*, 23, 617.

Krajian, A.A., and Gradwohl, R.B.H., (1952)., *Histopathological Technique.* St. Louis, C.V. Mosby Co.

Kramer, H., and Windrum, G.M., (1954)., *J. Histochem. Cytochem.*, 2, 196.

_________ (1955)., *J. Histochem. Cytochem.*, 3, 227.

Krichesky, B., (1931)., *Stain Tech.*, 6, 97.

Kristensen, H.K., (1948)., *Stain Tech.*, 23, 151.

Kruth, H.S., Vaughan, M., (1980)., *J. Lipid. Res.*, 21, 123.

Krutsay, M., (1962)., *Stain Tech.*, 35, 283.

Kultschizky, N., (1897)., *Arch. Mikrosk. Anat. Forsch.*, 2, 163.

Kurnick, N.B., (1955)., *Intn. Rev. Cytol.*, 4, 221.

Kutlik, I.E., (1957)., *Acta. Histochem.*, 4, 141.

Lacour, L., (1941)., *Stain Tech.*, 16, 169.

Laidlaw, G.F., (1929)., *Amer. J. Clin. Path.*, 5, 239.

Laidlaw, G.F., and Blackberg, S.N., (1932)., *Amer. J. Path.*, 8, 491.

Lake, P.S., (1970)., *Hisochemical J.*, 2, 441.

__________ (1974)., *Proc. Roy. Micr. Soc.*, 9, 113.

Lalitha, M., Shyamasundari, K., and Hanumantha Rao, K., (1993a)., *Rev. Idrobiol.*, 32, 23.

__________ (1993b)., *Rev. Idrobiol.*, 32, 35.

__________ (1996)., *Adv. Biol.*, 15, 189.

Lalitha, M., Uma Devi, C., Hanumantha Rao, K., and Shyamasundari, K., 1993)., *Rev. Idrobiol.*, 32, 1.

Lamb, D., (1969)., *Intracellular Development and Secretion in the Normal and Morbid Bronchial Tree*. Ph.D. Thesis, University of London.

Lan, H.Y., Mu, W., Nokolic-Patterson, D.J., and Atkins, R.C., (1995)., *J. Histochem. Cytochem.*, 43, 97.

Landing, B.H., Uzman, L.L., and Whipple, A., (1952)., *Lab. Invest.*, 1, 456.

Langman, J.M., (1995)., *J. Histotech.*, 2, 131.

Lapham, L.W., Johnstone, M.A., and Brundjar, K.H., (1964)., *J. Neuropath. Exp. Neurol.*, 23, 156.

Laskey, A.M., (1950)., *Stain Tech.*, 27, 233.

Lasky, A., and Greco, J., (1948)., *Arch. Path.*, 46, 83.

Lawless, D.K., (1953)., *Amer. J. Trop. Med.*, 2, 1137.

Lawrie, G.W. and Leblond, c.P., (1982a)., *J. Histochem. Cytochem.*, 30, 973.

__________ (1982b)., *J. Histochem. Cytochem.*, 30, 983.

Lawrie, G.W., Leblond, C.P., Martin, G.R. and Silver, M.H., (1982)., *J. Histochem. Cytochem.*, 30, 991.

Leatham, A. and Atkins, N., (1983).,. *J. Clin. Path.*, 36, 747.

Leatham, A., Dokal, I., and Atkins, N., (1983)., *Diagnotic. Histopathol.*, 6, 171.

Leaver, R.W., Evans, B.J., and Corrin, B., (1977)., *J. Clin. Path.*, 30, 290.

Lechtman, M.D., Bartholomew, J.W., Phillips, A., and Russo, M., (1965)., *J. Bact.*, 89, 848.

Leib, E., (1947). *J. Clin. Pathol.*, 17, 413.

Leibnitz, H.L., (1964)., *Acta Histochem.*, 17, 73.

Leifson, E., (1960)., *Atlas of Bacterial Flagellation*. New York: Academic Press.

Lemberg, M.R., (1959)., *Physiol. Rev.*, 49, 48.

Lendrum, A.C., (1947)., *J. Path. Bact.*, 59, 399.

__________ (1951)., *Recent Advances in Clinical Pathology*, 2nd edn., London: Churchill, 535.

Lendrum, A.C., Fraser, D.S., Slidders, W., and Henderson, R., (1962)., *J. Clin. Path.*, 15, 401.

Lendrum, A.C., and McFarlane, D., (1940)., *J. Path. Bact.*, 50, 38.

Lendrum, A.C., Slidders, W., and Fraser, D.S., (1969)., *Ned. Tijdschr. Geneesk.*, 113, 374.

Lennox, B., (1956)., *Stain Tech.*, 31, 167.

Leong, A.S.Y., Daymon, E., and Milos, J., (1985)., *J. Path.*, 146, 313.

Leong, A.S.Y., and Milos, J., (1986)., *J. Path.*, 148, 183.

Leppi, T.J., and Spicer, S.S., (1967)., *Anat. Rec.*, 159, 179.

Leske, R., and Mayersbach, H., (1969)., *J. Histochem. Cytochem.*, 17, 527.

Levene, C., Feng, P., (1964)., *Stain Tech.*, 39, 39.

Lewis, P.R., (1962)., *Histochem.*, 2, 423.

Lewis, P.R., and Grillo, T.A.I., (1950)., *Histochemie.*, 1, 391.

Lhotka, J.F., and Van Ferriera, A., (1950)., *Stain Tech.*, 25, 27.

L' Hoste, R., and Torres, M.A., (1995)., *Lab. Med.*, 26, 210.

Lieb, E., (1948)., *Arch. Path.*, 45, 559.

Lillie, R.D., (1940)., *Stain Tech.*, 15, 17.

__________ (1947)., *Bull. Int. Med. Mus.*, 27, 33.

__________ (1949)., *Bull. Intern. Ass. Med. Mus.*, 29, 153.

__________ (1951)., *Amer. J. Clin. Path.*, 21, 484.

__________ (1952a)., Josiah Macy Foundation. New York 3rd Conf. Connective Tissues, p. 11.

__________ (1952b)., *Arch. Path.*, 54, 220.

__________ (1954a)., *Histologic Technique.* 2nd edn. New York and Maidenhead, McGraw Hill.

__________ (1954b)., *Histopathological Technique and Practical Histochemistry*, 3rd edn., New York, The Blakiston Co.

__________ (1955)., *J. Histochem. Cytochem.*, 3, 453.

__________ (1956a)., *Stain Tech.*, 31, 151.

__________ (1956b)., *J. Histochem. Cytochem.*, 4, 377.

__________ (1956c)., *J. Histochem. Cytochem.*, 4, 118.

__________ (1957a)., *J. Histochem. Cytochem.*, 5, 279.

__________ (1957b)., *J. Histochem. Cytochem.*, 5, 188.

__________ (1957c)., *Arch. Path.*, 64, 100.

__________ (1960)., *J. Histochem. Cytochem.*, 9, 44.

__________ (1961)., *J. Histochem. Cytochem.*, 9, 184.

__________ (1962)., *J. Histochem. Cytochem.*, 10, 763.

__________ (1965)., *Histopathologic Techniques and Practical Histochemistry*, New York: McGraw Hill.

__________ (1977)., *Conns Biological Stains.* 9th edn. St. Louis, Sigma.

Lillie, R.D., and Ashburn, L.L., (1943)., *Arch. Path.*, 36, 432.

Lillie, R.D., and Burtner, H.J., (1953)., *J. Histochem. Cytochem.*, 1, 87.

Lillie, R.D., and Fullmer, H.M., (1976)., *Histopathologic Technique and Practical Histochemistry*, 4th edn. New York: McGraw Hill.

Lillie, R.D., and Glenner, G.G., (1957)., *J. Histochem. Cytochem.*, 5, 311.

__________ (1960)., *Amer. J. Path.*, 36, 623.

Lillie, R.D., Glenner, P.R.Jr., and Welsh, R.A., (1961)., *Stain Tech.*, 36, 361.

Lillie, R.D., Greco, J., and Hensen, J.P., (1953)., *J. Histochem. Cytochem.*, 1, 154.

Lillie, R.D., Greco, J., Hensen, J.P., and Cason, J.C., (1961)., *J. Histochem. Cytochem.*, 9, 11.

Lillie, R.D., and Henderson, R., (1968)., *Stain Tech.*, 43, 121.

Lillie, R.D., Lasky, A. Greco, J., Burtner, H.J., and Jones, P., (1951)., *Amer. J. Clin. Path.*, 21, 711.

Lieb, E., (1947)., *Amer. J. Clin. Path.*, 17, 413.

Lison, L., (1936)., *Histochemic Animale, Paris, Gauthier-Villard.*

Lison, L., and Dagnelie, J., (1935)., *Bull. Histol. Appl. Physiol. Path.*, 12, 85.

Ljungberg, O., (1970)., *Acta. Parasitol. Microbiol. Scandi.*, 78, 618.

Llewellyn, B.D., (1970)., *J. Med. Lab. Tech.*, 27, 308.

Lloyd, B., Brenn, N., and Burger, P.C., (1985)., *J. Histotech.*, 8, 155.

Login, G.R., and Dvorak, A.M., (1985)., *Amer. J. Path.*, 120, 230.

__________ (1994)., *Microwave Tool Book.* Boston: Beth Isreal Hospital, Department of Pathology.

Login, G.R., Schnitt, S.J., and Dvorak, A.M., (1987)., *Lab. Invest.*, 57, 585.

Lorch, I.J., (1946)., *Nature Lond.*, 158, 269.

Lowry, R.J., (1963)., *Stain Tech.*, 38, 149.

Loyez, M., (1910)., *Compt. Rendz. Sci. Soc. Biol.*, 69, 511.

Lucas, A.M., and Jamroz, C., (1961)., *Atlas of Avian Hematology. Agriculture Monograph* 25. U.S. Department of Agriculture, Washington, D.C., 271 pp.

Luft, J.H., (1961)., *J. Biophys. Biochem. Cytol.*, 9, 409.

Luna, L.G., (1960)., *Manual of Histologic Staining Methods of the Armed Forces.* Institute of Pathology. 2nd edn., New York: McGraw Hill.

__________ (1968)., *Manual of Histologic Staining Methods of the Armed Forces.* Institute of Pathology. 3rd edn., New York: McGraw Hill.

Lynch, M.J., and Inwood, M.J.H., (1963)., *Stain Tech.*, 38, 259.

Lyon, H., (1991)., *Theory and Strategy of Histochemistry. A guide to the selection and understanding of techniques.* Berlin: Springer-Verlag.

MacCallum, W.G., (1919)., *JAMA.*, 72, 193.

MacChiavello, A., (1937)., *Rev. Chil. Hig.*, 1, 101.

Madoff, S., (1960)., *Isolation and Identification of PPLO. Ann. N.Y. Acad. Sci*, 79, 383.

Maengwyn-Davies, G.D., Friedenwald, J.S., and White, R.J., (1952)., *J. Cell. Comp. Physiol.*, 39, 395.

Mager, M., McNary, W.F.Jr., and Lionetti, F., (1953)., *J. Histochem. Cytochem.*, 1, 493.

Mahon, G.S., (1937)., *Neurol. Psychiatry.*, 38, 103.

Mahoney, R., (1968)., *Laboratory Techniques in Zoology*, London, Butterworth.

Majno, G., and Rouiller, C., (1951)., *Virchows. Arch.*, 321, 1.

Makowski, E.L., Prem, K.A., Kaiser, I.H., (1956)., *Science*, 123, 542.

Malik, N.J., and Daymon, M.E., (1982)., *J. Clin. Path.*, 35, 1092.

Mallory, F.B., (1897)., *J. Exp. Med.*, 2, 529.

__________ (1900)., *J. Exp. Med.*, 5, 15.

__________ (1938)., *Pathological Techniques.* W.B. Saunders, Philadelphia, Hafiner, New York.

__________ (1944)., *Pathological Techniques.* Philadelphia, W.B. Saunders.

__________ (1961)., *Pathological Technique.* New York, Hafner Publishing Co., 169 AFIP.

Mallory, F.B., and Wright, J.H., (1924)., *Pathological Technique*, 8th edn. Philadelphia, W.B. Saunders Co.

Mann, G., (1894)., *Z. Wiss. Mikrosk.*, 11, 479.

Marberger, E.B., Rita, A., and Nelson, W.O., (1955)., *Soc. Exp. Biol. Med. Proc.*, 89, 488.

Margolis, G., and Pickett, J.P., (1956)., *Lab. Invest.*, 5, 459.

Marsland, T.A., Glees, P., and Erikson, L.B., (1954)., *J. Neuropath. Exp. Neurol.*, 13, 587.

Marti, R., Wild, P. *et al.*, (1985)., *J. Histochem. Cytochem.*, 35, 1415.

Marti, W.J., and Johnson, B.H., (1951)., *Amer. J. Clin. Path.*, 21, 793.

Marwah, A.S., and Weinmann, J.P., (1955)., *J. Periodontology*, 26, 11.

Mason, J.E., Pfifer, R.F., Spicer, S.S., Swallow, R.A., and Drukin, R.B., (1969)., *J. Histochem. Cytochem.*, 17, 563.

Massie, H.R., Samis, H.V., and Baird, M.B., (1972)., *In Vitro*, 7, 191.

Massignani, A., and Malferrari, R., (1961)., *Stain. Tech.*, 36, 5.

Masson, P., (1928)., *Amer. J. Path.*, 4, 181.

Mayahara, H., Hiramo, H., Saito, T., and Ogawaka, K., (1967)., *Histochemie.*, 11, 88.

Mayer, P., (1896)., *Uber. Schlumfarbung mitzool. Stel. Neapel.*, 12, 303.

Maxwell, M.H., (1978)., *J. Med. Lab. Sci.*, 35, 401.

McClung, C.E., (1939)., *Handbook of Microscopical Technique.* New York, B. Hoeber & Co.

__________ (1950)., *Handbook of Microscopical Technique*, 3rd edn. Ruth McClung Jones (ed.), Paul B. Hoeber, New York.

McCormick, W.F., and Coleman, S.A., (1962)., *Tech. Bull. Reg. Med. Tech.*, 32, 113.

McGadey,J., (1970)., *Histochemie*, 23, 180.

McGee-Russell, S.M., (1958)., *Nature Lond.*, 175, 301.

McGuire, S.R., and Opel, H., (1969)., *Stain Tech.*, 44, 235.

McManus, J.F.A., (1946)., *Nature Lond.*, 158, 202.

McManus, J.F.A., (1948)., *Stain Tech.*, 23, 99.

McManus, J.F.A., and Cason, J.E., (1950)., *J. Exp. Med.*, 91, 651.

McManus, J.F.A., and Mowry, R.W., (1964)., In: *Staining Methods: Histologic and Histochemical.* London, Harper and Row.

McMullen, L., Walker, M.M., Bain, L.A., Karim, Q.N., and Baron, J.H., (1987)., *J. Clin. Path.*, 464.

McNamara, W.L., Murphy, B., and Gore, W.A., (1940)., *J. Lab.Clin. Med.*, 25, 874.

Melander, Y., and Wingstrand, K.G., 1953)., *Stain Tech.*, 28, 217.

Menschik, Z., (1953)., *Stain Tech.*, 28, 13.

Menton, W.L., Junge, J., and Green, M.H., (1944)., *J. Biol. Chem.*, 153, 471.

Menzies, D.W., (1961)., *Stain Tech.*, 36, 285.

——————— (1962)., *Stain Tech.*, 37, 43.

——————— (1963)., *Stain Tech.*, 38, 157.

Mepham, B.L., Frater, W., and Mitchell, B.S., (1979)., *Histochem. J.*, 11, 345.

Meriweather, L.S., (1934)., *Proc. Staff. Meetings Mayo Clinic Mayo Foundation*, 9, 95.

Merryman, H.T., (1960)., *Ann. New York Acad. Sci*, 85, 630.

Merton, A., (1932)., *Archiv. Für. Protistinkunde.*, 77, 449.

Metcalf, R.I., and Paton, R.L., (1944)., *Stain Tech.*, 19, 11.

Mettler, F.A., (1932)., *Stain Tech.*, 7, 95.

Mettler, F.A., and Hanada, R.E., (1942)., *Stain Tech.*, 17, 111.

Meyer, K., (1953)., In: *Some Conjugated Proteins*, Symposium, Rutger's University, New York.

——————— (1966)., Introduction Mucopolysaccharides. *Fedn. Proc.*, 25, 1032.

Meyer, K., Linker, A., Davidson, E.A., and Weissman, B., (1953)., *J. Biol. Chem.*, 205, 611.

Meyer, K., and Palmer, J., (1934)., *J. Biol. Chem.*, 107, 629.

Meyer, K., and Rapport, M., (1951)., *Science N.Y*, 113, 596.

Meyer, L., (1864)., *Ann. Chim.*, 132, 156.

Meyer, P., (1896)., *Uber. Schlumfarbung. Mit. Zool. Stel. Neapet.*, 12, 303.

Michaelis, L., (1930)., *J. Biol. Chem.*, 81, 34.

Miller, K., Auld, T., Jessup, E., Rhodes, A., and Ashton-Key, M., (1995)., *Adv. Anat. Path.*, 2, 60.

Milligan, M., (1946)., *Amer. J. Clin. Path.*, 10, 184.

Millonig, G., (1962)., Fifth International Congress in *Electron Microscopy*. New York: Academic Press, 2, 8.

Mitchell, B.S., (1975)., *Med. Lab. Tech.*, 32, 331.

Mittwer, T., Bartholomew, T.W., and Kallman, B.J., (1950)., *Stain Tech.*, 25, 169.

Mohos, S.C., and Skoza, L., (1970)., *Exp. mol. Path.*, 12, 316.

Moldovanu, G., (1961)., *Revue Francaise. D' Etudes Cliniques et Biologiques.*, 6, 165.

Moller, O., (1951)., *Acta path. Micro. Scand.*, 28, 127.

Moloney, W.C., McPherson, K., and Fliegelman, L., (1960)., *J. Histochem. Cytochem.*, 8, 200.

Monk, C.R., (1938)., *Science*, 88, 174.

Monroe, C.W., and Frommer, J., (1966)., *Stain Tech.*, 41, 248.

Monroe, C.W., and Spector, B., (1963)., *Stain Tech.*, 38, 187.

Moog, F., (1943)., *J. Cell. Comp. Physiol.*, 22, 233.

Moore, A.R., (1962)., *World Bulletin 1, Wards Natural Science Establishment.*

Moore, G.W., (1951)., *J. Inst. Med. Lab. Tech.*, 9, 105.

Moore, K.L., (1962)., *Acta. Cytol.*, 6, 1.

——————— (1966)., Ed. Of *Sex Chromatin*, Philadelphia and London, Saunders.

Moore, K.L., and Barr, M.L., (1955)., *The Lancet*, 269, 57.

Moore, K.L., Graham, M.A., and Barr, M.L., (1953)., *Surg. Gynec. Obst.*, 96, 641.

Mori, M., Takada, T., and Okamoto, J., (1962)., *Histochemie.*, 2, 427.

Morris, R.E., and Benton, R.S., (1956)., *Amer. J. Clin. Path.*, 26, 882.

Mota, I., Ferri, A.G., and Yoneda, S., (1956)., *Quat. J. Micr. Sci.*, 97, 251.

Mowry, R.W., (1956)., *J. Histochem. Cytochem.*, 4, 407.

——————— (1958)., *Lab. Invest.*, 7, 566.

_________ (1959)., *J. Histochem. Cytochem.*, 7, 288.

_________ (1963)., *Ann. N.Y. Acad. Sci.*, 106, 402.

Mukherji, M., Deb, C., and Sen, P.B., (1960)., *J. Histochem. Cytochem.*, 8, 189.

Muller, G., (1955)., *Acta. Histochem.*, 2, 68.

Muller, L.K., and Jacks, T.H., (1975)., *J. Histochem. Cytochem.*, 23, 107.

Murdock, S.E., (1945)., *J. Tech. Methods*, 25, 71.

Murdock, T.F., and Fratkin, T.D., (1988)., *Lab. Med.*, 19, 109.

Nachlas, M.M., Crawford, D.T., Goldstein, T.P., and Seligman, A.M., (1958)., *J. Histochem. Cytochem.*, 6, 445.

Nachlas, M.M., Crawford, D.T., and Seligman, A.M., (1957)., *Proc. Roy. Micr. Soc.*, 5, 264.

Nachlas, M.M., and Seligman, A.M., (1949)., *J. Nat. Cancer. Inst.*, 9, 415.

Nachlas, M.M., Prinn, W., and Seligman, A.M., (1956)., *J. Biophys. Biochem. Cytol.*, 2, 487.

Nachlas, M.M., Tsou, K.C., Desouza, E., Cheng, C.S., and Seligman, A.M., (1957)., *J. Histochem. Cytochem.*, 3, 420.

Nachlas, M.M., Walker, D.G., and Seligman, A.M., (1958)., *J. Biophys. Biochem. Cytol.*, 4, 29.

Naish, S.J. Ed. (1989)., *Handbook of Immunocytochemical Staining Methods.* Carpentaria, C.A.: Dak Corporation.

Nakane, P.K., (1968)., *J. Histochem. Cytochem.*, 16, 557.

Naoumenko, J., and Feigin, I., (1974)., *Stain Tech.*, 49, 153.

Nassar, T.K., and Shanklin, W.M., (1951)., *Stain Tech.*, 26, 13.

Nassar, T.K., and Shanklin, W.M., (1961)., *Arch. Path.*, 71, 611.

Nassonov, D.N., (1923)., *Archiv fur Mikros. Anat. Ent. Mech..*, 97, 136.

_________ (1924)., *Archiv fur Mikros. Anat. Ent. Mech..*, 100, 433.

Nauta, W.J.H., and Gygax, P.A., (1951)., *Stain Tech.*, 26, 5.

Neelsen, F., (1883)., *Centralbl. Med. Wiss.*, 21, 497.

Neleon, E.V., (1974)., *Stain Tech.*, 49, 117.

Newcomer, E.A., (1940)., *Stain Tech.*, 15, 89.

_________ (1953)., *Science,* 118, 161.

Newman, S.B., Borysko, E., and Swerdlow, M., (1950)., *J. Applied Physiol.*, 12, 67.

Nolte, D.J., (1948)., *Stain Tech.*, 23, 21.

Nonidez, J.F., (1939)., *Amer. J. Anat.*, 65, 361.

Norton, A.J., Jordan, S., and Yoemans, P., (1994)., *J. Path.*, 173, 371.

Norton, W.T., Korey, S.R., and Brotz, M., (1962)., *J. Histochem. Cytochem.*, 10, 83.

Novelli, A., (1962)., *J. Histochem. Cytochem.*, 10, 102.

Novoteny, G.E.K., and Novoteny, E., (1974)., *Stain Tech.*, 49, 273.

_________ (1977)., *Stain Tech.*, 52, 97.

Nunn, R.E., (1970)., *Electron Microscopy: Preparation of Biological Specimens.* London, Butterworths.

Ogawa, K., Mizuno, N., and Okamoto, M., (1961)., *J. Histochem. Cytochem.*, 9, 625.

Okamoto, K., and Utamura, M., (1938)., *Trans. Soc. Path. Japan*, 20, 573.

Okun, M.R., (1967)., *J. Invest. Derm.*, 48, 424.

Okun, M.R., Edelstein, L.OR.N., Hamada, G., and Donnelan, B., (1970)., *J. Invest. Derm.*, 55, 1.

Oster, K.A., and Schlossman, N.C., (1942)., *J. Cell. Comp. Physiol.*, 20, 373.

Owen, G., (1955)., *Nature Lond.*, 175, 434.

Owen, G., and Steedman, H.F., (1956)., *Quart. J. Micr. Sci.*, 97, 319.

_________ (1958)., *Proc. Malac. Soc. Lond.*, 33, 101.

Padawar, J., (1959)., *J. Histochem. Cytochem.*, 7, 352.

Paddy, J.F., (1970)., *Metachromasy of dyes in solution.* NATO Advanced Study Institute. St. Margherita (Report) Ed. Balasz, E.A., London, Academic Press.

Page, K., (1965)., *J. Med. Lab. Tech.*, 22, 224.

Page, W.G., and Green, R.G., (1942)., *Cornell. Vet.*, 32, 265.

Palmer, M.W., and McDonald, L.W., (1963)., *Amer. J. Clin. Path.*, 40, 633.

Palmgren, A., (1948)., *Acta. Zool.*, 29, 377.

Pantin, C.F.A., (1946)., *Notes on Microscopical Techniques for Zoologists.* Cambridge University Press.

Papanicoloau, G.N., (1942)., *Science*, 173, 821.

___________ (1947)., *J. Natl. Cancer. Inst.*, 7, 357.

___________ (1954)., *Atlas of Exfoliative Cytology.* Cambridge Harvard University Press.

___________ (1957)., *C.A. Bull. Cancer Progr.*, 7, 125.

Pappenheim, A., (1899)., *Virchows. Arch. Path. Anat. Physiol.*, 157, 19.

Patrick, R., (1936)., *Science*, 83, 85.

Patten, S.F., and Brown, K.A., (1958)., *Lab. Invest.*, 7, 209.

Pauley, G.B., (1967)., *J. Invest. Path.*, 9, 268.

Pearse, A.G.E., (1951)., *J. Clin. Path.*, 4, 1.

___________ (1953a)., *Histsochemistry Theoretical and Applied.* London: J & A Churchill.

___________ (1953b)., *J. Path. Bact.*, 66, 331.

___________ (1957)., *J. Histochem. Cytochem.*, 5, 420.

___________ (1960)., *Histochemistsry: Theoretical and Applied.*, Boston, Little Brown & Co.

___________ (1963)., *J. Sci. Instsuments.*, 40, 176.

___________ (1968)., *Histochemistry: Theoretical and Applied.* 3rd edn. Vol. 1, London: J & A Churchill Ltd.,

___________ (1972)., *Histochemistsry: Theoretical and Applied.*, J.A. Churchill.

___________ (1980)., *Histochemistry: Theoretical and Applied,* 4th edn. Edinburgh, Churchill – Livingstone, Vol. 1.

___________ (1985)., *Histochemistsry: Theoreticl and Applied.* Secaucus, N.J: Churchill-Livingston.

Pearse, A.G.E., and Stoward, P.J. (eds.)., (1991)., *Histochemistry: Theoretical and Applied.*, Vol. 3, 4th edn. Edinburgh, Chuchill Livingstone.

Pearson, B., (1958)., *J. Histochem. Cytochem.*, 6, 112.

Pearson, B., Paul, W., and Andrews, M., (1963)., *Lab. Invest.*, 12, 712.

Peltier, L.F., (1954)., *J. Lab. Clin. Med.*, 43, 321.

Perls, M., (1867)., *Virchows. Arch. Path. Anat.*, 39, 42.

Peterson, M., and Leblond, C.P., (1964)., *J. Cell. Biol.*, 21, 143.

Pickett, J.P., Bishop, C.M., Chick, E.W., and Baker, R.D., (1960)., *Amer. J. Clin. Path.*, 34, 197.

Pickett, J.P., and Klavins, J.V., (1961)., *Stain Tech.*, 36, 371.

Pienaar, Uys de Villiers (1962)., *Haematology of African Reptiles.* Johnnesberg: Witwatersrand University press.

Pinkus, H., (1944)., *Arch. Dermat & Syph.*, 49, 355.

Pitelka, D.R., (1945)., *J. Morphol.*, 76, 179.

Pizzalato, P., (1964)., *J. Histochem. Cytochem.*, 12, 333.

Pluzek, K.J., Sweeney, E., Miller, K., and Isaacson, P.G., (1993)., *J. Path.* 169 (Suppl), Abstract 220.

Polak, J.M., and Van Noorden, S. (Eds.) (1983)., *Immunocytochemistsry, Practical Applications in Pathology and Biology,* Bristol: Wright.

___________ (1986)., *Immunocytochemistsry, practical applications in pathology and Biology.* 2nd Edn., Bristol, Wright.

Poley, R.W., and Forbes, C.D., (1964)., *Arch. Path.*, 77, 325.

Ponder, B.a., and Wilkinson, M.M., (1981)., *J. Histsochem. Cytochem.*, 29, 981.

Powell, E.W., and Brown, G., (1975)., *Mikroscopie*, 31, 77.

Powers, M., and Clark, G., (1955)., *Stain Tech.*, 30, 83.

___________ (1963)., *Stain Tech.*, 38, 289.

Powers, M., Clark, G., Darrow, M., and Emmel, V.M., (1960)., *Stain Tech.*, 35, 19.

Preman, J., (1954)., *Stain Tech.*, 29, 105.

Prescott, D.M., and Carrier, R.F., (1964)., *Experimental Procedures and Cultural methods for Euplotes curystomus and Amoeba proteus – Methods in Cell Physiology*, 3rd edn. Ed. Prescott, D.M., New York: Academic Press.

Presnell, J.K., and Schreibman, M.P., (1997)., *Humason's Animal Tissue Techniques*, 5th edn. John Hopkins University Press, Baltimore & London.

Proescher, F., (1934)., *Stain Tech.*, 9, 33.

Puchtler, H., Meloan, S.N., and Terry, M.S., (1969)., *J. Histochem. Cytochem.*, 17, 110.

Puchtler, H., Rosenthal, S.I., and Sweat, F., (1964)., *Arch. Path.*, 78, 76.

Puchtler, H., and Sweat, F., (1962)., *Arch. Path.*, 73, 245.

_____ (1963)., *Arch. Path.*, 75, 588.

_____ (1964)., *Stain Tech.*, 39, 163.

_____ (1965)., *J. Histochem. Cytochem.*, 13, 693.

Puchtler, H., Sweat, F., and Levine, M., (1962)., *J. Histochem. Cytochem.*, 10, 355.

_____ (1963)., *Amer. J. Clin. Path.*, 40, 334.

Puchtler, H., Waldrop, F.S., Meloan, S.N., Terry, M.S., and Conner, H.M., (1970)., *Histochemie.*, 21, 97.

Pugh, D., and Walker, P.G., (1961)., *J. Histochem. Cytochem.*, 9, 105.

Putt., F.A., (1948)., *Arch. Path.*, 45, 72.

_____ (1951)., *Amer. J. Clin. Path.*, 21, 92.

_____ (1971)., *Yale J. Biol. Med.*, 43, 279.

Pydakula, H.A., Herman, E., (1955)., *J. Histochem. Cytochem.*, 3, 170.

Quintarelli, G., Gifonelli, J.A., and Zito, R., (1971)., *J. Histochem. Cytochem.*, 19, 648.

Quintarelli, G., Scott, J.E., and Dellovo, M.C., (1964a)., *Histochemie.*, 4, 86.

_____ (1964b)., *Histochemie.*, 4, 99.

Rafalko, S., (1946)., *Stain Tech.*, 21, 91.

Raghu, J., (1993)., *Some aspects of Biochemistsry and Histochemistry of **Cardisoma carnifex** (Herbst) (Brachyura: Gecarcinidae)*. Dissertation submitted for the award of Ph.D. degree, Andhra University.

Rajyalakshmi Bhanu, R.C., Shyamasundari, K., and Hanumantha Rao, K., (1980)., *Proc. Nat. Acad. Sci.*, 50, 38.

_____ (1981a)., *Monitore. Zool. Ital.*, 15, 239.

_____ (1981b)., *Acta. Histochem. Cytochem.*, 14, 516.

_____ (1982)., *Proc. Ind. Acad. Sci*, 91, 407.

_____ (1983)., *Z. Mikros. Anat. Forsc.*, 97, 535.

_____ (1984)., *Adv. Bioscien.*, 3, 59.

Ralph, P.H., (1941)., *Stain Tech.*, 16, 105.

Rama Devi, K.R.L.S., Shyamasundari, K., and Hanumantha Rao, K., (1987)., *Natn. Symp. Physiol. Crust.*, 99, 104.

_____ (1991)., *Arch. Ital. Anat. E,mbryol.*, 96, 121.

Randolph, L.F., (1935)., *Stain Tech.*, 10, 95.

Ravetto, C., (1964)., *J. Histochem. Cytochem.*, 12, 44.

_____ (1968)., *J. Histochem. Cytochem.*, 16, 663.

Rawlin, T.E., and Takahashi., (1947)., *Stain. Tech.*, 22, 99.

Reid, P.E., Livingstone, D.J., and Dunn, W.L., (1972)., *Stain Tech.*, 47, 101.

Rhinehart, J.F., (1952–53)., *J. natn. Cancer. Inst.*, 13, 212.

Richards, O.W., (1941)., *Amer. J. Clin. Path.*, 5, 1.

Richards, O.W., Kline, E.K., and Leach, R.E., (1941)., *Amer. Rev. Tuberculosis.*, 44, 255.

Riis, P., (1957)., *Acta. Haematol..*, 18, 168.

Ritter, H.B., and Oleson, J.J., (1950)., *Amer. J. Path.*, 26, 639.

Ritter, M.A., (1986)., "Raising and testing monoclonal antibodies for immunocytochemistsry", In: *Immunocytochemistry Modern Methods, Applications*, Polak, J.M. and Van Noorden, S. (eds.), Bristol, Wright.

Riva, H.L., and Turner, T.R., (1962)., *J. Obstet. Gynec.*, 20, 45.

Robinow, C.F., (1942)., *Proc. Roy. Soc. Lond.*, 130, 299.

__________ (1944)., *J. Hyg.*, 43, 413.

Robinson, G., and Gray, T., (1996)., "Electron Microscopy 2 Practical Procedures", In: *Theory and Practice of Histological Techniques*, Eds. Brancroft, J.D., and Stevans, A., Churchill and Livingstone.

Roden, D.B., (1975)., *Stain Tech.*, 50, 207.

Rodriquez, J., Deinhardt, F., (1960)., *Virology.*, 12, 316.

Rojkind, M., Cordero-Hernandez, J., and Ponce, P., (1984)., *Myelofibrosis. Biol. Connective tissue*, New York, AR Liss 103.

Romeis, B., (1948)., *Mikroskop. Tech. Leibniz Verlog, Munchen.*

Roque, A.L., Jafarey, N.A., and Coutler, P., (1965)., *Exp. Mol. Path.*, 4, 266.

Rost, F.W.D., Ewen, S.W.B., (1971)., *Histochem. J.*, 3, 207.

Rothenbacher, H.J., and Hitchcock, D.J., (1962)., *Stain Tech.*, 37, 111.

Rubin, R., (1951)., *Stain Tech.*, 26, 257.

Rupavathi, P., Hanumantha Rao, K., and Shyamasundari, K., (1983)., *Proc. Ind. Acad. Sci.*, 93, 35.

__________ (1984)., *Folia. Morphol.*, XXXII, 185.

Russell, R., (1973)., *Lab. Med.*, 4, 40.

Ruttenberg, A.M., Cohen, R.B., and Seligman, A.M., (1952)., *Science N.Y.*, 116, 539.

Ruttenberg, A.M., Ruttenberg, S.H., Monis, B., Teague, R., and Seligman, A.M., (1958)., *J. Histochem. Cytochem.*, 6, 122.

Ruttenberg, A.M., and Seligman, A.M., (1955)., *J. Histochem. Cytochem.*, 3, 455.

Ryu, E., (1963)., *Jpn. J. Microb.*, 7, 81.

Sabatini, D.D., Bensch, K., and Barrnett, R.J., (1963)., *J. Cell. Biol.*, 17, 19.

Sabatini, D.D., Miller, F., and Barrnett, R.J., (1964)., *J. Histochem. Cytochem.*, 12, 57.

Sachs, L., (1953)., *Stain Tech.*, 28, 169.

Sagi and Mackenzie (1958)., *Cancer Cytol.*, 1, 24.

Sakaguchi, S., (1925)., *J. Biochem. Tokyo*, 5, 25.

Salthouse, T.N., (1958)., *Canad Entamol.*, 90, 555.

Sanger, F., (1945)., *Biochem. J.*, 39, 507.

__________ (1950)., *Cold. Spr. Harb. Syp. Quant. Biol.*, 12, 142.

Santamarina, E., (1964)., *Stain Tech.*, 39, 267.

Saunders, A.M., (1964)., *J. Histochem. Cytochem.*, 12, 164.

Scarpeli, D.G., Hess, R., and Pearse, A.G.E., (1958)., *J. Biophys. Biochem. Cytol.*, 4, 147.

Schaeffer, A.B., and Fulton, M., (1933)., *Science*, 77, 194.

Schajowicz, F., and Cabrine, R.L., (1955)., *J. Histochem. Cytochem.*, 3, 122.

__________ (1956)., *Stain Tech.*, 31, 129.

__________ (1958)., *Stain Tech.*, 34, 59.

Schiff, R., Quinn, L.Y., and Bryan, J.H.D., (1967)., *Stain Tech.*, 42, 75.

Schiffer, L.M., and Vaharu, T., (1962)., *Tech. Bull. Reg. Med. Tech.*, 32, 91.

Schleifstein, J., (1937)., *Amer. J. Pub. Health.*, 27, 1283.

Schmidt, R.W., (1956)., *Lab. Invest.*, 5, 306.

Schmitz-Moorman, P., (1969)., *Histochemie.*, 20, 78.

Schmorl, G., (1928)., *Die pathologisch-histologischen Untersuchungsmethoden.* Ed.15 F.C.W. Vogel, Leipzig.

Schneider, J.D., (1963)., *Bull. Med. Tech.*, 33, 195.

Schreibman, M.P., (1986)., *The Pituitory gland. Vertebrate Endocrinology. Fundamental and Biomedical Implications.* Ed. Pang, P.K.T. and Schreibman, M.P. New York, Academic Press, 1, 11–56.

Schultz, A., (1924–25)., *Zbl. Allg. Path. Anat.*, 35, 314.

Scott, J.E., (1976)., *Histochemistry*, 48, 1084.

Scott, J.E., and Dorling, J., (1965)., *Histochemie.*, 5, 221.

Seligman, A.M., Chauncey, H.H., and Nachlas, M.M., (1951)., *Stain Tech.*, 26, 19.

Seligman, A.M., Karnovsky, M.J., Wasserkrug, H.L., and Hanker, J.S., (1968)., *J. Cell. Biol.*, 38, 1.

Seligman, A.M., Tsou, K.C., Ruttenberg, S.H., and Cohen, R.B., (1954)., *J. Histochem. Cytochem.*, 2, 209.

Shapiro, S., and Sohns, L., (1994)., *J. Histotech.*, 17, 127.

Shi, S.R., Key, M.E., and Kalra, K.L., (1991)., *J. Histochem. Cytochem.*, 39, 741.

Shimizu, N., and Kumamoto, (1952)., *Stain Tech.*, 27, 97.

Shyamasundari, K., (1977)., *Int. J. Cell. Molec. Biol.*, 22, 79.

———————— (1979)., *Z. Zellforsch.*, 93, 417.

———————— (1985)., *Riv. Parasitol.*, XLVI, 331.

Shyamasundari, K., and Hanumantha Rao, K., (1973)., *Curr. Sci.*, 42(4), 134.

———————— (1975a)., *Z. Parasitenk.*, 47, 103.

———————— (1975b)., *Acta. Histochem.*, 54, 272.

———————— (1978)., *Folia Histochem. Cytochem.*, 16, 247.

———————— (1988)., *Marine Biodeterioration.*, 45.

Shum, M.W., and Hon, J.K.Y., (1969)., *J. Med. Lab. Tech.*, 26, 38.

Silva, E.E.M., (1961)., *Stain Tech.*, 36, 196.

Simmons, J.S., and Gentkow, C.J., (1944)., *Laboratory Methods of the United States Army*, 5th edn., Philadelphia, Penn. Lea and Fehiger, 572.

Simpson, C.F., Carlisle, J.W., and Mallard, L., (1970)., *Stain Tech.*, 45, 221.

Simpson, W.L., (1941)., *Anat. Rec.*, 80, 329.

Singh, I., (1964)., *Anat. Anz.*, 115.

Singh, N., Wortherspoon, A.C., Miller, K.D., and Isaacson, P.g., (1993)., *Ann. N.Y. Acad. Sci.*, 254, 559.

Slidders, W., (1961)., *J. Path. Bact.*, 82, 532.

———————— (1969)., *J. Micr.*, 90, 61.

Slidders, W., Fraser, D.S., Smith, R., and Lendrum, A.C., (1958)., *J. Path. Bact.*, 75, 466.

Sloper, J.C., and Adams, C.W.M., (1956)., *J. Path. Bact.*, 72, 587.

Smith, E.J., Puchtler, H., and Sweat, F., (1966)., *Lab. Invest.*, 15, 1141.

Smith, L., (1947)., *Stain Tech.*, 22, 17.

Smyth, J.D., (1944)., *Science*, 100, 62.

———————— (1951)., *Stain Tech.*, 26, 255.

Snook, T., (1944)., *Anat. Rec.*, 89, 413.

Solcia, E., Capella, C., Vassallo, G., (1969)., *Histochemie.*, 20, 116.

Solcia, E., Vassalo, G., and Capella, C., (1968)., *Stain Tech.*, 43, 257.

Solomon, C., Amelos, R.D., Hyman, R.H., Chaisen, R.J., and Europa, D.L., (1958)., *J. Urol.*, 80, 374.

Sommer, J.R., and Pickett, J.P., (1961)., *Arch. Path.*, 71, 669.

Sorensen, and Walbum (1920)

Sorvari, T.E., and Lauren, P.A., (1973)., *Histochem. J.*, 5, 405.

Sorvari, T.E., and Sorvari, R.M., (1969)., *J. Histochem. Cytochem.*, 17, 291.

Southgate, H.W., (1927)., *J. Path. Bact.*, 30, 729.

Spatz, M., (1960)., *Amer. J. Clin. Path.*, 34, 285.

Spicer, S.S., (1960)., *J. Histochem. Cytochem.*, 8, 18.

———— (1961a)., *Amer. J. Clin. Path.*, 36, 393.

———— (1961b)., *Stain Tech.*, 36, 337.

———— (1963)., *Ann. N.Y. Acad. Sci, 103, 322.*

———— (1965)., *J. Histochem. Cytochem.*, 13, 211.

Spicer, S.S., Horn, R.G., and Leppi, T.J., (1967)., "In the connective tisssue" on *Int. Acad. Path. Monograph,* Ed. Wagner, B.W., Baltimore, Williams and Wilkins.

Spicr, S.S., Leppi, T.J., and Stoward, P.J., (1965)., *J. Histochem. Cytochem.*, 13, 599.

Spicer, S.S., and Lillie, R.D., (1959)., *J. Histochem. Cytochem.*, 7, 123.

———— (1961)., *Stain Tech.*, 36, 365.

Spicer, S.S., and Meyer, E., (1960)., *Amer. J. Clin. Path.*, 33, 453.

Spicer, S.S., Neubecker, R.D., Warren, L., and Henson, J.G., (1962)., *J. Natn. Cancer. Inst.*, 29, 963.

Spicer, S.S., and Warren, L., (1960)., *J. Histochem. Cytochem.*, 8, 135.

Staple, P.H., (1967)., *J. Histochem. Cytochem.*, 15, 552.

Staples, T.C., (1991)., *National Society Histotechnology Workshop.*

Staples, T.C., and Clark, L.P., (1990)., *J. Histotech.*, 13, 137.

Staples, T.C., and Grizzle, W.E., (1986a)., *Stain Tech.*, 62, 41.

———— (1986b)., *Lab. Med.*, 17, 532.

Steedman, M., (1947)., *Quart. J. micr. Sci.*, 88, 123.

Steedman, H.F., (1950)., *Quart. J. micr. Sci.*, 91, 477.

———— (1960)., *Section Cutting in Microscopy.* Oxford Blackwell.

Stein, H., Gatter, K., Asbahr, H., and Mason, D.Y., (1985)., *Lab. Invest.*, 52, 676.

Sternberger, L.A., (1969)., *Mikroscopie.*, 25, 346.

Sternberger, L.A., Hardy, P.H., Cuculis, J.J., and Meyer, H.G., (1970)., *J. Histsochem. Cytochem.*, 27, 1424.

Sternberger, M., Cohen-Forterre, L., and Peyroux, J., (1985)., *Deabete et Metabolisme (Paris),* 11, 27.

Stone, G.E., and Cameron, I.L., (1964)., "Methods for using Tetrahymena in studies of the normal cell cycle", *Methods in Cell Physiology,* 1st edn. Prescott, D.M., New York, Academic Press.

Stovall, W.D., and Black, C.E., (1940)., *Amer. J. Clin. Path.*, 10, 1.

Stoward, P.J., (1963)., D.Phil. Thesis, University of Oxford.

(1967)., *J. Roy. Micr. Soc, 87, 215.*

Stoward, P.J., and Burns, J., 1967)., *Histochemie.*, 10, 230.

Straus, W., (1964)., *J. Histochem. Cytochem.*, 12, 462.

Suffin, S.C., Muck, K.B., Yong, J.C., Lewin, K., and Porter, D.D., (1979)., *Amer. J. Clin. Path.*, 102, 113.

Summer, B.E.N., (1988)., *Basic Histochemistry.* Chichester: Wiley.

Swank, R.L., and Davenport, H.A., (1934a)., *Stain Tech.*, 9, 11.

———— (1934b)., *Stain Tech.*, 9, 129.

———— (1935a)., *Stain Tech.*, 10, 45.

———— (1935b)., *Stain Tech.*, 10, 87.

Sweat, F., Meloan, S.N., and Puchtler, H., (1968)., *Stain Tech.*, 38, 179.

Sweat, F., and Puchtler, H., (1965)., *Arch. Path.*, 80, 613.

Taft, E.B., (1951)., *Expt. Cell. Res.*, 2, 322.

Takada, K., Tani, T., and Mori, M., (1960)., *Jap. J. Clin. Path.*, 8, 237.

Tan, K.H., (1973)., *Stain Tech.*, 48, 146.

Tanaka, R., (1961)., *Stain Tech.*, 36, 325.

Tandler, C.J., (1974)., *Stain Tech.*, 49, 147.

Taylor, A.I., (1963)., *Lancet.*, 1, 912.

Taylor, C.R., (1993)., *Immunomicroscopy A diagnostic Tool for the Surgical Pathologist.* Philadelphia, Saunders.

Taylor, R.D., (1966)., *Amer. J. Clin. Path.*, 46, 472.

Taylor, C.R., and Burns, J., (1974)., *J. Clin. Path.*, 27, 14.

Tekeuchi, T., and Kuriaki, H., (1955)., *J. Histochem. Cytochem.*, 3, 153.

Terner, G.Y., and Hayes, E.R., (1961)., *Stain Tech.*, 36, 265.

Thiele, H.J., (1967)., *Acta. Histochem.*, 26, 373.

Thomas, J.A., (1941)., *Comp. Rend. Sci. Soc. Biol.*, 135, 935.

Thompson, E.C., (1961)., *Stain Tech.*, 36, 38.

Thompson, S.W., (1966)., *Selected Histochemical and Histopathological Methods.* Springfield, Thomas.

Thorgaard, G.H., and Disney, J.E., (1990)., "Chromosome preparation and analysis", In: *Method for Fish Biology*, Eds. Schreck, C.B. and Moyle, P.B., Bethseda, M.D., American Fisheries Society, 171–190.

Tilden, I.L., and Tanaka, M., (1945)., *Amer. J. Clin. Path.*, 9, 95.

Timm, F., (1958)., *Dat. Zeischriff. Ges. Gruchtliche. Medizun.*, 46, 706.

Tock, E.P.C., and Pearse, A.G.E., (1965)., *J. Roy. Micros. Soc.*, 84, 519.

Tokuda, Y., Nakamura, T., Satonaka, K., Maeda, S., Doi, K., Baba, S., and Sugiyama, T., (990)., *J. Clin. Path.*, 43, 748.

Tomlinson, W.J., and Grocott, R.G., (1944)., *Amer. J. Clin. Path.*, 14, 316.

Toren, D.A., (1963)., *Stain Tech.*, 38, 249.

Traunt, J.P., Brett, W.A., and Thomas, W., (1962)., *Henry Ford. Hosp. Bull.*, 10, 287.

Trevan, D.J., and Sharrock, A., (1951)., *J. Path. Bact.*, 63, 326.

Trinadha Babu, B., Shyamasundari, K., and Hanumantha Rao, K., (1987)., *Nat. Symp. Physiol. Crust.*, 66.

__________ (1989a)., *Curr. Sci.*, 58, 213.

__________ (1989b)., *Folia. Morphol.*, 37, 274.

__________ (1989c)., *Ind. J. Inv. Zool. Aq. Biol.*, 1, 3.

__________ (1989d)., *Folia. Morphol.*, 37, 79.

__________ (1989e)., *Proc. Natn. Sci. Acad.*, 855, 15.

__________ (1989f)., *Folia. Morphol.*, 1, 4.

__________ (1989g)., *Curr. Sci.*, 58, 213.

__________ (1989h)., *Folia. Morphol.*, 37, 79.

__________ (1991)., *Uttar Pradesh J. Zool.*, 11, 136.

__________ (1993)., *Uttar Pradesh J. Zool.*, 11, 136.

Turchini, J.P., and Mallet, P., 1965)., *J. Histochem. Cytochem.*, 13, 405.

Uma Devi, C., Hanumantha Rao, K., and Shyamasundari, K., (1981)., *Proc. Ind. Acad. Sci.*, 90, 307.

__________ (1982)., *Ind. J. Zootomy*, 23, 135.

__________ (1983)., *Ind. Natn. Acad. Sci.*, 49, 20.

__________ (1984a)., *Ind. Natn. Sci. Acad.*, 3, 213.

__________ (1984b)., *Folia Morphol.*, XXXII, 153.

__________ (1984c)., *Proc. Natn. Acad. Sci.*, 54, 213.

__________ (1984c)., *Proc. Natn. Acad. Sci.*, 47, 323.

Unna, P.G., (1902)., *Mh. Prakt. Derm.*, 35, 76.

Vacca, J., (1985)., *Lab. Man. Histochem.*, Hagerstown, MD: Lippincott-Raven.

Vacek, Z., and Plackova, A., (1959)., *Stain Tech.*, 34, 1.

Vallance-Owen, J., (1948)., *J. Path. Bact.*, 60, 325.

Van Diujn, P., (1953)., *J. Histochem. Cytochem.*, 1, 143.

Van Gieson, I., (1889)., *New York Med. J.*, 50, 57.

Van Kosa, J., 1901)., *Beitre. Path. Anat.*, 29, 163.

Vassar, P.S., and Culling, C.F.A., (1959)., *Amer. Arch. Path.*, 68, 487.

__________ (1962)., *Can. Med. Ass. J.*, 87, 299.

Vedal, O.R., Aya, T., and Sandberg, A.A., (1971)., *Stain Tech.*, 46, 89.

Verhoeff, F.H., 1908)., *J. Amer. Med. Assoc.*, 58, 876.

Vernino, D.M., and Laskin, D.M., 1960)., *Science*, 132, 675.

Villyaneuva, A.R., (1979)., *J. Histotech.*, 2, 24.

Von Bertalanffy, L., Masin, M., and Masin, F., (1958)., *Cancer*, 11, 873.

Voss, H., (1940)., *Z. mikr. Anat. Forsch.*, 49, 51.

Vyas, A.B., 1972)., *Turtox News.*, 49, 25.

Wachstein, M., and Meisel, E., (1956)., *J. Histsochem. Cytochem.*, 4, 592.

__________ (1957)., *Amer. J. Clin. Path.*, 27, 13.

__________ (1960)., *J. Histochem. Cytochem.*, 8, 387.

__________ (1964)., *J. Histochem. Cytochem.*, 12, 538.

Wachstein, M., and Zak, F.G., (1946)., *Amer. J. path.*, 22, 603.

Wachtler, K., and Pearse, A.G.E., (1966)., *Z. Zellforsch. Mikros. Anat.*, 69, 326.

Wade, H.W., (1952)., *Amer. J. Path.*, 28, 157.

__________ (1957)., *Stain Tech.*, 32, 287.

Walpole, G.S., (1914)., *J. Chem. Soc.*, 105, 2501.

Walter, E.K., Smith, J.L., Israel, G.W., and Gager, W.E., (1969)., *Br. J. Vener. Dis.*, 45, 6.

Warburton, M.J., Mitchell, D., Ormerod, E.J., and Rudland, P. (1982)., *J. Histochem. Cytochem.*, 30, 667.

Waranke, R.A., Gattes, K.C., and Mason, D.Y., 1983)., *Rec. Adv. Clin. Immunol.*, 3, 163.

Warren, L., (1959)., *J. Biol. Chem.*, 234, 1951.

Warren, T.N., and McManus, J.F.A., (1951)., *Exp. Cell. Res.*, 2, 703.

Wartenberg, H., (1956–57)., *Acta. Histochem.*, 3, 145.

Wartman, W.B., (1943)., *Army. Med. Bull.*, 68, 173.

Watson, J.D., and Crick, F.H.C., (1953)., *Nature Lond.*, 171, 737.

Weaver, H.L., (1955)., *Stain Tech.*, 30, 63.

Weigert, K., (1904)., *Z. Weissen. Mikro. Mikrop. Tech.*, 21, 1.

Weil, A., and Davenport, H.A., (1930)., *Z. Ges. Neurol. Psychiatr.*, 154, 228.

__________ (1935)., *Arch. Neurol. Psychiatry.*, 30, 75.

Weiner, N., (1960)., *Arch. Biochem. Biophys.*, 91, 182.

Weisberger, E.K., Russfield, A.B., Homberger, F., Weisberger, J.H., Boger, E., Van Dongen, C.G., and Chu, K.C., (1978)., *J. Env. Path. Toxicol.*, 2, 325.

Weiss, L.P., Tsou, K.C., and Seligman, A.M., (1954)., *J. Histochem. Cytochem.*, 2, 29.

Welshons, W.J., Gibson, B.H., and Scandyln, B.J., (1962)., *Stain Tech.*, 37, 1.

West, K.P., Pattis, H.a., Fletcher, A., and Walker, F. (1982)., *J. Clin. Path.*, 35, 239.

White, P.B., 1947)., *J. Path. Bact.*, 59, 334.

Wilder, H.C., (1935)., *Amer. J. Path.*, 11, 817.

Wilson, W.D., and Ezrin, E., (1954)., *Amer. J. Path.*, 30, 891.

Williams, G., and Jackson, D.S., (1956)., *Stain Tech.*, 31, 189.

Windaus, T., (1910)., *Z. Physics. Chem.*, 65, 110.

Wittman, W., (1962)., *Stain Tech.*, 37, 27.

__________ (1963)., *Stain Tech.*, 38, 217.

Wismar, B.I., (1966)., *Stain Tech.*, 41, 309.

Wolbach, S.B., Todd, J.L., and Palfrey, F.W., (1922)., *The Etiology and Pathology of Typhus*, Cambridge, Mass, Harward University Press.

Wolman, M., (1970)., *Pathology Annual.*, 5, 381.
Woohsman, H., and Hartrodt., (1965)., *Histochemies.*, 4, 366.
Wynnchuk, M., (1993)., *Histotech.*, 2, 143.
Yamada, K., (1962)., *Acta Histochem.*, 15, 50.
Yasue, T., (1969)., *Acta Histochem. Cytochem.*, 2, 83.
Yasuma, A., and Itchikawa, T., (1953)., *J. Lab. Clin. Med.*, 41, 296.
Yetwin, I.J., (1944)., *J. Parasitol.*, 30, 201.
Yoshiki, S., (1962)., *Bull. Tokyo den Coll.*, 3, 14.
Yergaman, G., (1957)., *Amer. J. Hum. Genetics*, 9, 42.
Zerdt, C.H., (1988)., *Parasitology Today*, 4, 15.
Ziehl, F., (1882)., *Disch. Med. Wochenschr.*, 8, 451.
Zlotnik, I., (1953)., *Nature London*, 172, 962.
Zorzoli, A., (1948)., *Anat. Rec.*, 102, 445.
Zugibe, F.T., (1970)., *Diagnostic Histochemistsry*, St. Louis, C.V. Mosby.
__________ (1970)., *Histochem. J.*, 2, 191.
Zwemer, K., (1933)., *Anat. Rec.*, 57, 41.

AUTHOR INDEX

SUBJECT INDEX